The Infrared and Submillimetre Sky after COBE

NATO ASI Series

Advanced Science Institutes Series

A Series presenting the results of activities sponsored by the NATO Science Committee, which aims at the dissemination of advanced scientific and technological knowledge, with a view to strengthening links between scientific communities.

The Series is published by an international board of publishers in conjunction with the NATO Scientific Affairs Division

A Life Sciences	Plenum Publishing Corporation
B Physics	London and New York
C Mathematical	Kluwer Academic Publishers
and Physical Sciences	Dordrecht, Boston and London
D Behavioural and Social Sciences	
E Applied Sciences	
F Computer and Systems Sciences	Springer-Verlag
G Ecological Sciences	Berlin, Heidelberg, New York, London,
H Cell Biology	Paris and Tokyo
I Global Environmental Change	

NATO-PCO-DATA BASE

The electronic index to the NATO ASI Series provides full bibliographical references (with keywords and/or abstracts) to more than 30000 contributions from international scientists published in all sections of the NATO ASI Series.
Access to the NATO-PCO-DATA BASE is possible in two ways:

– via online FILE 128 (NATO-PCO-DATA BASE) hosted by ESRIN,
Via Galileo Galilei, I-00044 Frascati, Italy.

– via CD-ROM "NATO-PCO-DATA BASE" with user-friendly retrieval software in English, French and German (© WTV GmbH and DATAWARE Technologies Inc. 1989).

The CD-ROM can be ordered through any member of the Board of Publishers or through NATO-PCO, Overijse, Belgium.

Series C: Mathematical and Physical Sciences - Vol. 359

The Infrared and Submillimetre Sky after COBE

edited by

M. Signore

and

C. Dupraz

Ecole normale supérieure, Paris, France and
Observatoire de Meudon, Meudon, France

Springer-Science+Business Media, B.V.

Proceedings of the NATO Advanced Study Institute on
The Infrared and Submillimetre Sky after COBE
Les Houches, France
March 20–30, 1991

ISBN 978-94-010-5080-7   ISBN 978-94-011-2448-5 (eBook)
DOI 10.1007/978-94-011-2448-5

TABLE OF CONTENTS

IV. ANISOTROPIES OF THE CMB

V. INSTRUMENTATION

VI. CONCLUSION

AUTHOR INDEX

SUBJECT INDEX

PREFACE

This volume consists of invited lectures and seminars presented at the NATO Advanced Study Institute *"The Infrared and Submillimetre Sky after COBE"*, which was held at the Centre de Physique Théorique of Les Houches (France) in March 1991.

The school has been planned by a Scientific Organizing Committee. It was organized with the aim of providing students and young researchers with an up-to-date account of the Cosmic Microwave Background, the Cosmic Infrared Background (if any), and the infrared emission of the Galaxy, after the early results from COBE (Cosmic Background Explorer). It was attended by about sixty researchers from many countries.

The lectures and seminars represent a complete coverage of our present knowledge and understanding of : the Early Universe, Large-Scale Structure, Dust in Galaxies, Infrared to Submillimetre Backgrounds, CMB Anisotropies, complementary observations and instrumentation problems, etc. Most of these lectures are reproduced in this volume. Unfortunately, a few lecturers have chosen not to submit their manuscript.

I would like to express my gratitude to the Scientific Affairs Division of NATO (North Atlantic Treaty Organization) and to the Theoretical Physics and Astrophysics Sections of the CNRS (Centre National de la Recherche Scientifique) for their generous support. Further help was obtained from the DRET (Direction des Recherches, Etudes et Techniques), the CNES (Centre National d'Etudes Spatiales) and the IN2P3 (Institut National de Physique Nucléaire et de Physique des Particules), which I hereby gratefully acknowledge.

Special thanks are due to Brigitte Rousset and Danièle Choupin for taking care of everyone so well at the Centre de Physique Théorique in the beautiful setting of Les Houches.

Finally, many thanks to all the invited speakers whose papers appear below.

Monique Signore
Director of the School

Scientific Organizing Committee :
    M.G. HAUSER, USA ; F. MELCHIORRI, Italy ;
    D. SCIAMA, Italy ; M. SIGNORE, France.

Co-Editors :
    M. SIGNORE and C. DUPRAZ, France.

LIST OF PARTICIPANTS

Xavier BARCONS : *Universidad de Cantabrià, Santander, SPAIN*
James G. BARTLETT : *University of California at Berkeley, Berkeley, USA*
Alain BLANCHARD : *Observatoire de Paris-Meudon, Meudon, FRANCE*
Samuele BOTTANI : *Università "La Sapienza", Roma, ITALY*
François BOUCHET : *Institut d'Astrophysique de Paris, Paris, FRANCE*
François BOULANGER : *Ecole Normale Supérieure, Paris, FRANCE*
Alberto CAPPI : *Osservatorio Astronomico di Bologna, Bologna, ITALY*
Bernard CARR : *Queen Mary College, London, UNITED KINGDOM*
Laura CAYON-TRUEBA : *Universidad de Cantabrià, Santander, SPAIN*
Rolf CHINI : *Max-Planck Institut für Radioastronomie, Bonn, GERMANY*
Fabrice DEBBASCH : *Ecole Normale Supérieure, Paris, FRANCE*
Paolo DE BERNARDIS : *Università "La Sapienza", Roma, ITALY*
Antonella DE LUCA : *Università "La Sapienza", Roma, ITALY*
Marco DE PETRIS : *Università "La Sapienza", Roma, ITALY*
Christophe DUPRAZ : *Ecole Normale Supérieure, Paris, FRANCE*
Pierre ENCRENAZ : *Ecole Normale Supérieure, Paris, FRANCE*
Michele EPIFANI : *Università "La Sapienza", Roma, ITALY*
Maria-Cristina FALVELLA : *Università "La Sapienza", Roma, ITALY*
Massimo GERVASI : *Università "La Sapienza", Roma, ITALY*
Krzysztof M. GORSKI : *Princeton University, Princeton, USA*
Michel GUELIN : *Institut de Radioastronomie Millimétrique, Grenoble, FRANCE*
Stephen HANCOCK : *Cavendish Laboratory, Cambridge, UNITED KINGDOM*
Robin HARMON : *California Institute of Technology, Pasadena, USA*
Michael A. JANSSEN : *Jet Propulsion Laboratory, Pasadena, USA*
Roman JUSZKIEWICZ : *N. Copernicus Astronomical Center, Warszawa, POLAND*
Jean-Michel LAMARRE : *Institut d'Astrophysique Spatiale, Orsay, FRANCE*
Ariane LANCON : *Institut d'Astrophysique de Paris, Paris, FRANCE*
Anthony LASENBY : *Cavendish Laboratory, Cambridge, UNITED KINGDOM*
Charles LINEWEAVER : *Lawrence Berkeley Laboratory, Berkeley, USA*
Vladimir LUKASH : *Space Research Institute, Moscow, USSR*
Bruno MAFFEI : *Institut d'Astrophysique Spatiale, Verrières-le-Buisson, FRANCE*
Roberto MAOLI : *Università "La Sapienza", Roma, ITALY*
Enrique MARTINEZ-GONZALEZ : *Universidad de Cantabrià, Santander, SPAIN*
Silvia MASI : *Università "La Sapienza", Roma, ITALY*
Sophie MAUROGORDATO : *Observatoire de Paris-Meudon, Meudon, FRANCE*
Francesco MELCHIORRI : *Università "La Sapienza", Roma, ITALY*
Tahar MELLITI : *Institut d'Astrophysique de Paris, Paris, FRANCE*

Bianca OLIVO-MELCHIORRI : *Università "La Sapienza", Roma, ITALY*
Jamila OUKBIR : *Observatoire de Paris-Meudon, Meudon, FRANCE*
Georg RAFFELT : *MPI für Physik und Astrophysik, München, GERMANY*
William T. REACH : *University of California at Berkeley, Berkeley, USA*
Maria SAKELLARIADOU : *Université Libre de Bruxelles, Bruxelles, BELGIUM*
Pierre SALATI : *LAPP, Annecy-le-Vieux, FRANCE*
Norma SANCHEZ : *Observatoire de Paris-Meudon, Meudon, FRANCE*
Christoph SCHMID : *ETH–Hönggenberg, Zürich, SWITZERLAND*
Patrick SEGUIN : *Ecole Normale Supérieure, Paris, FRANCE*
Monique SIGNORE : *Ecole Normale Supérieure, Paris, FRANCE*
Joseph SILK : *University of California at Berkeley, Berkeley, USA*
Haydeh SIROUSSE-ZIA : *Institut Henri-Poincaré, Paris, FRANCE*
George SMOOT : *Lawrence Berkeley Laboratory, Berkeley, USA*
Francesco SYLOS-LABINI : *Università "La Sapienza", Roma, ITALY*
Luis Francisco TENORIO : *Lawrence Berkeley Laboratory, Berkeley, USA*
Luigi TOFFOLATTI : *Universidad de Cantabrià, Santander, SPAIN*
Marie-Agnès TREYER : *University of California at Berkeley, Berkeley, USA*
Michael TURNER : *Fermi Laboratory, Batavia, USA*
Gabriele VENEZIANO : *CERN, Genève, SWITZERLAND*
Boqi WANG : *University of California at Berkeley, Berkeley, USA*
David E. WILLMES : *University of Florida, Gainsville, USA*
Edward L. WRIGHT : *University of California at Los Angeles, Los Angeles, USA*

LIST OF LECTURERS

Xavier BARCONS
   Departamento da Fisica Moderna, Universidad de Cantabrià
   Avenida de Los Castros S/N, E–39005 Santander, SPAIN
Alain BLANCHARD
   DAEC, Observatoire de Paris-Meudon
   F–92195 Meudon principal Cedex, FRANCE
François BOUCHET
   Institut d'Astrophysique de Paris
   98bis boulevard Arago, F–75014 Paris, FRANCE
François BOULANGER
   Laboratoire de Radioastronomie Millimétrique, Ecole Normale Supérieure
   24 rue Lhomond, F–75231 Paris Cedex 05, FRANCE
Bernard CARR
   School of Mathematical Sciences, Queen Mary College
   Mile End Road, London E1 4NS, UNITED KINGDOM
Rolf CHINI
   Max-Planck Institut für Radioastronomie
   Auf dem Hügel 69, D–5300 Bonn 1, GERMANY
Paolo DE BERNARDIS
   Dipartimento di Fisica, Università "La Sapienza"
   Piazzale Aldo Moro 2, I–00185 Roma, ITALY
Pierre ENCRENAZ
   Laboratoire de Radioastronomie Millimétrique, Ecole Normale Supérieure
   24 rue Lhomond, F–75231 Paris Cedex 05, FRANCE
Krzysztof M. GORSKI
   Department of Astrophysical Sciences, Princeton University
   Peyton Hall, Princeton NJ 08544, USA
Michel GUELIN
   Institut de Radioastronomie Millimétrique, Domaine Universitaire
   300 rue de la Piscine, F–38406 Saint-Martin-d'Hères Cedex, FRANCE
Michael A. JANSSEN
   Jet Propulsion Laboratory
   4800 Oak Grove Drive, Pasadena CA 91109, USA
Roman JUSZKIEWICZ
   Nicolaus Copernicus Astronomical Center
   Bartycka 18, 00–716 Warszawa, POLAND
Jean-Michel LAMARRE
   Institut d'Astrophysique Spatiale, Université Paris-Sud
   Bâtiment 120, F–91405 Orsay, FRANCE
Anthony LASENBY
   Mullard Radio Astronomy Observatory, Cavendish Laboratory
   Madingley Road, Cambridge CB3 0HE, UNITED KINGDOM

Vladimir LUKASH
    Space Research Institute, Academy of Sciences
    Profsojuznaja 84/32, 117810 Moscow, USSR

Enrique MARTINEZ-GONZALEZ
    Departamento da Fisica Moderna, Universidad de Cantabrià
    Avenida de Los Castros S/N, E–39005 Santander, SPAIN

Francesco MELCHIORRI
    Dipartimento di Fisica, Università "La Sapienza"
    Piazzale Aldo Moro 2, I–00185 Roma, ITALY

Georg RAFFELT
    Max-Planck Institut für Physik und Astrophysik
    Postfach 401212, D–8000 München, GERMANY

Pierre SALATI
    Laboratoire d'Annecy-le-Vieux de Physique des Particules
    Chemin de Bellevue, BP 909, F–74029 Annecy-le-Vieux Cedex, FRANCE

Norma SANCHEZ
    DEMIRM, Observatoire de Paris-Meudon
    F–92195 Meudon principal Cedex, FRANCE

Joseph SILK
    Astronomy Department, University of California at Berkeley
    Berkeley CA 94720, USA

George SMOOT
    Lawrence Berkeley Laboratory
    MS 50–232, Berkeley CA 94720, USA

Michael TURNER
    Fermi Laboratory, MS 209
    Box 500, Batavia IL 60510, USA

Gabriele VENEZIANO
    Theoretical Physics Division, CERN
    CH–1211 Genève 23, SWITZERLAND

Edward L. WRIGHT
    Department of Astronomy, University of California at Los Angeles
    405 Hilgard Avenue, Los Angeles CA 90024–1562, USA

PLATES OF COBE

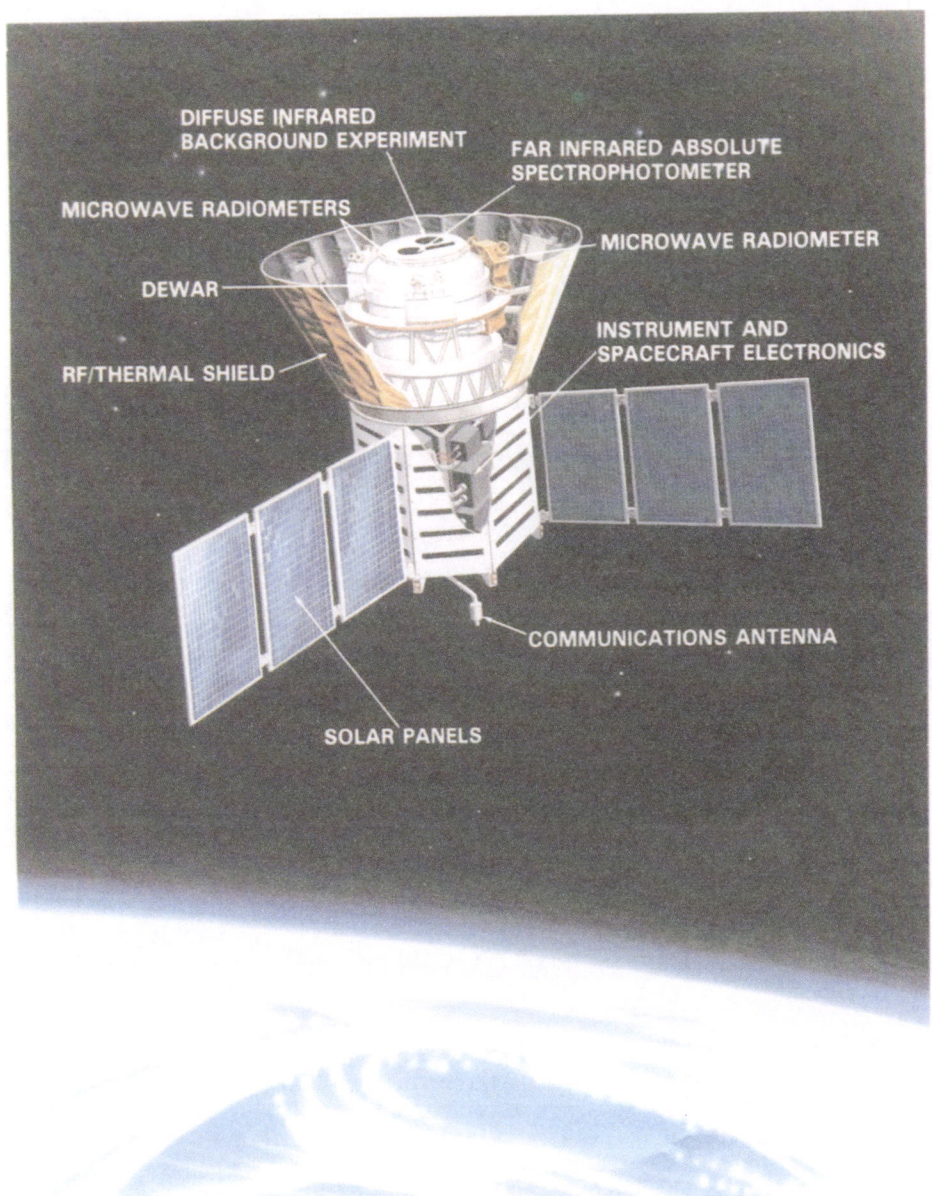

PLATE 1.

Artist's rendering of the Cosmic Background Explorer (*COBE*). The three panels of solar cells are shown deployed from the body of the spacecraft. The white area houses the power systems, telemetry command and data handling, and the bulk of the altitude control system. The upper section is surrounded by the deployed ground shield/sun shade shown in cut away. The large white structure in the centre is the superfluid helium dewar, operating at 1.5 K and containing the FIRAS and DIRBE instruments. The DMR is made of the three radiometer pairs located around the dewar.

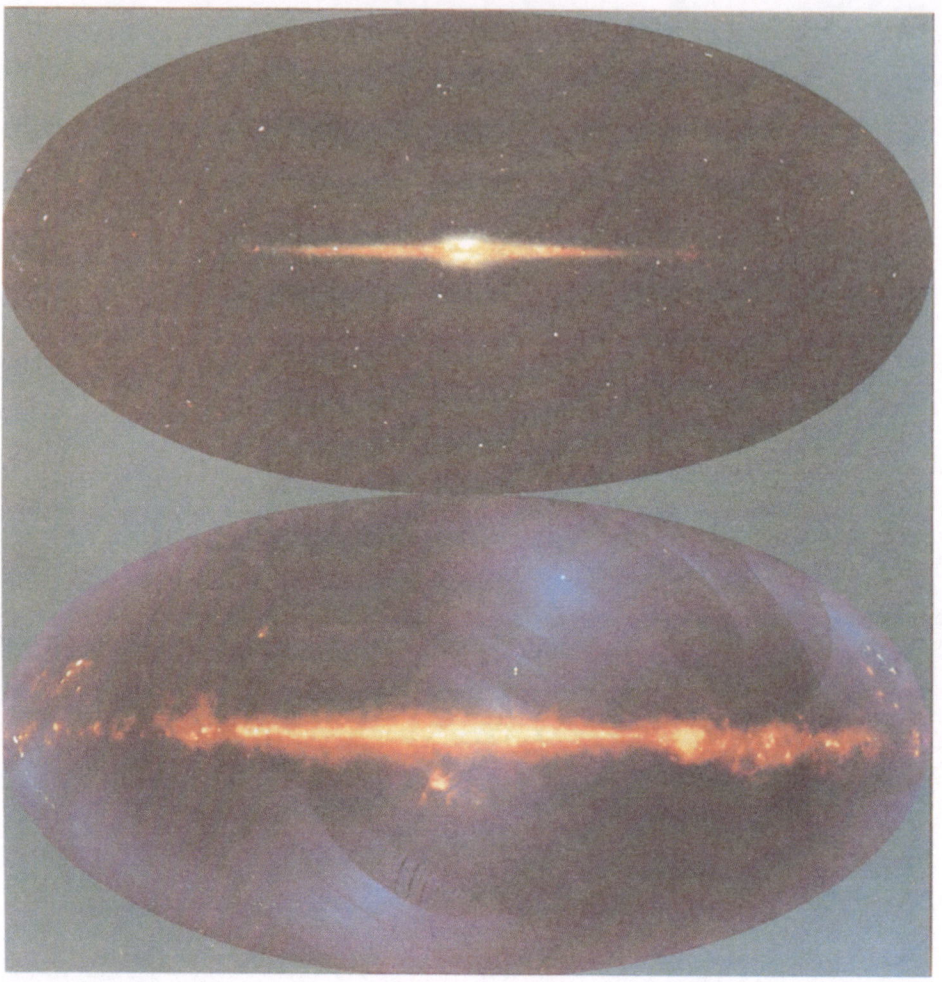

PLATE 2.

COBE DIRBE full sky maps in galactic coordinates. Upper portion is the combined 1.2, 2.2 and 3.4 micron maps shown in respective blue, green and red colors. The image shows both the thin disk and central bulge of the Milky Way. The lower portion shows the combined 25, 60 and 100 micron map in respective blue, green and red colors. Both maps are preliminary results of a full sky scan by the DIRBE instrument. The discontinuities apparent are due primarily to the change in zodiacal light, as the spacecraft and earth move around in the interplanetary dust but some part may be due to not having final instrument calibratiuon. The bright spots in the ecliptic are the moon and planets.

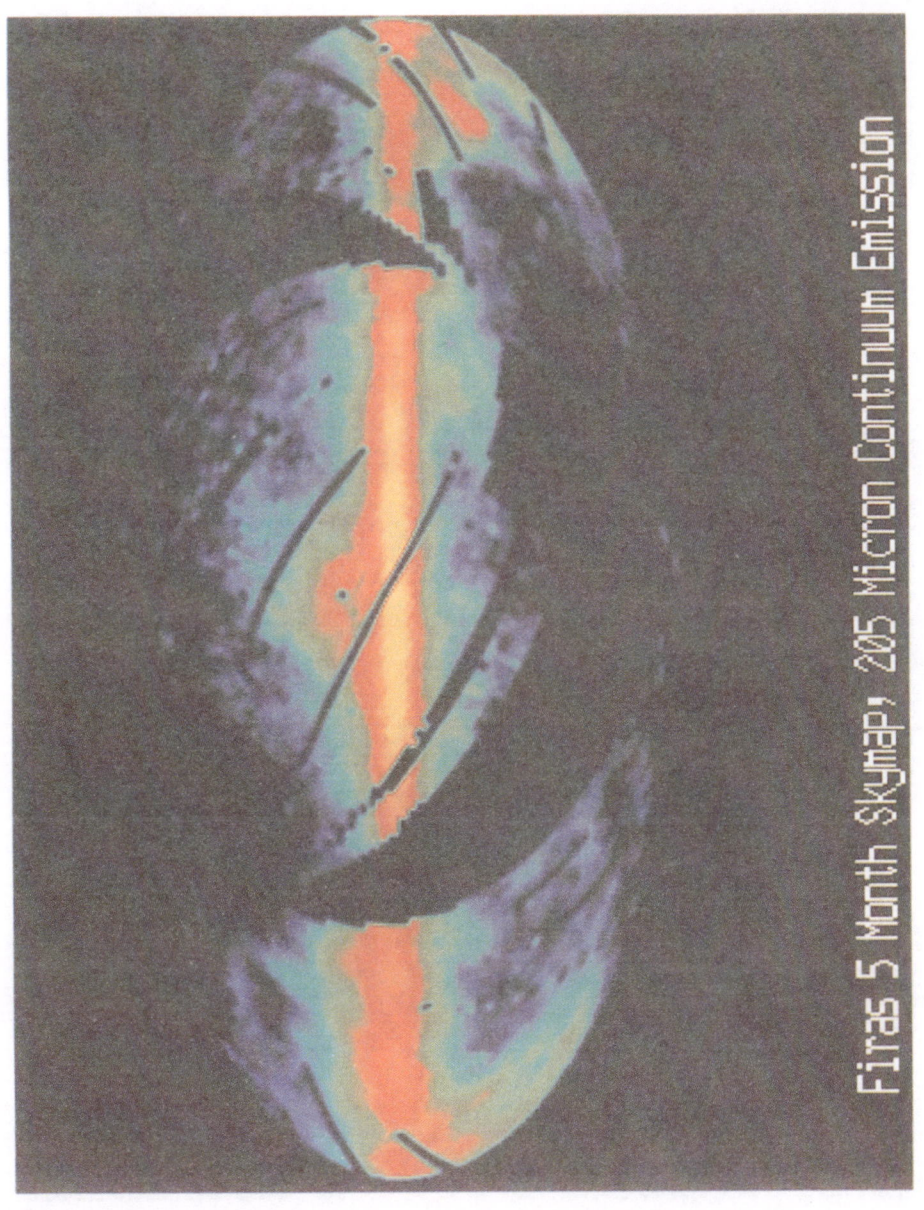

A)

PLATE 3.
 COBE FIRAS instrument maps of sky : A) dust (205 micron) ; B) [NII] (205 micron) ; C) [CII] (158 micron).

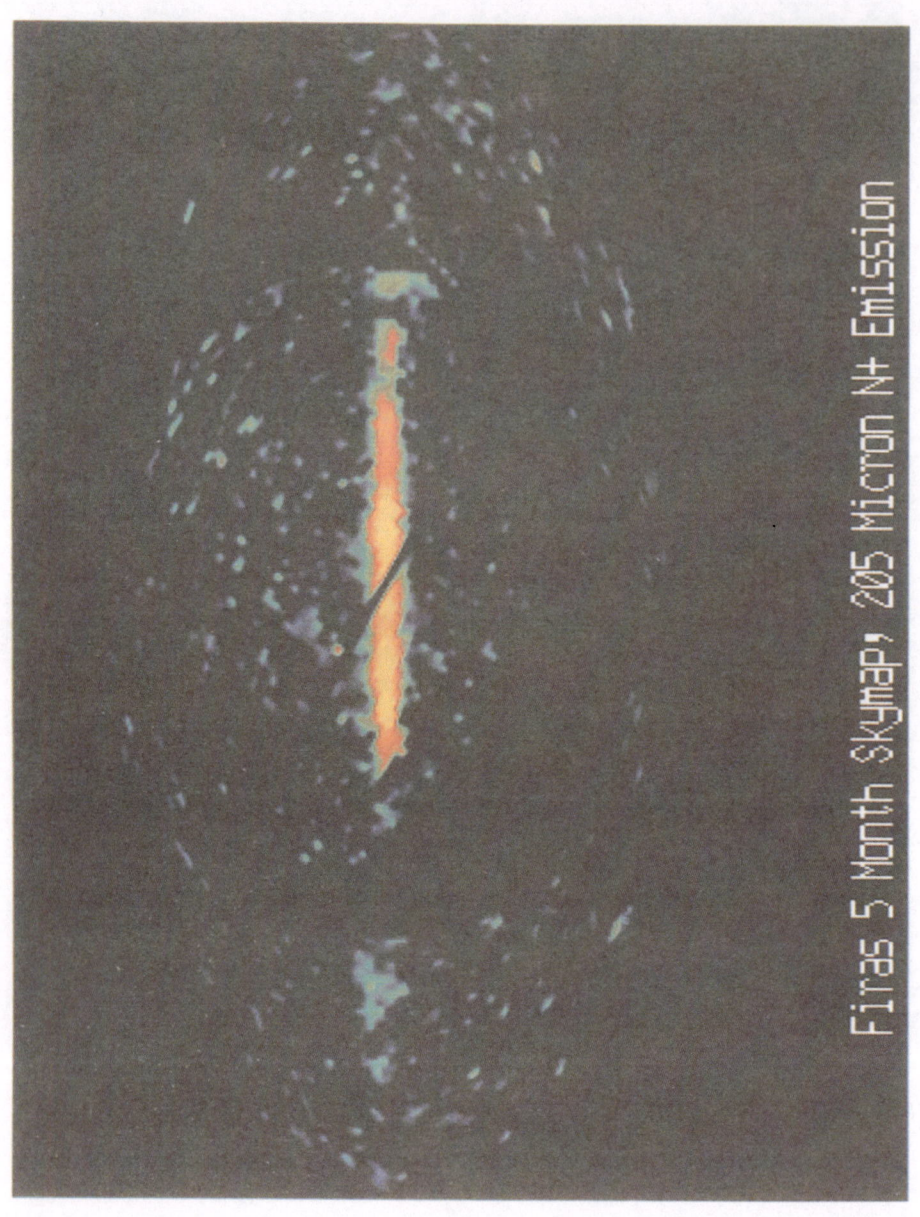

B)

PLATE 3.
COBE FIRAS instrument maps of sky : A) dust (205 micron) ; B) [NII] (205 micron) ; C) [CII] (158 micron).

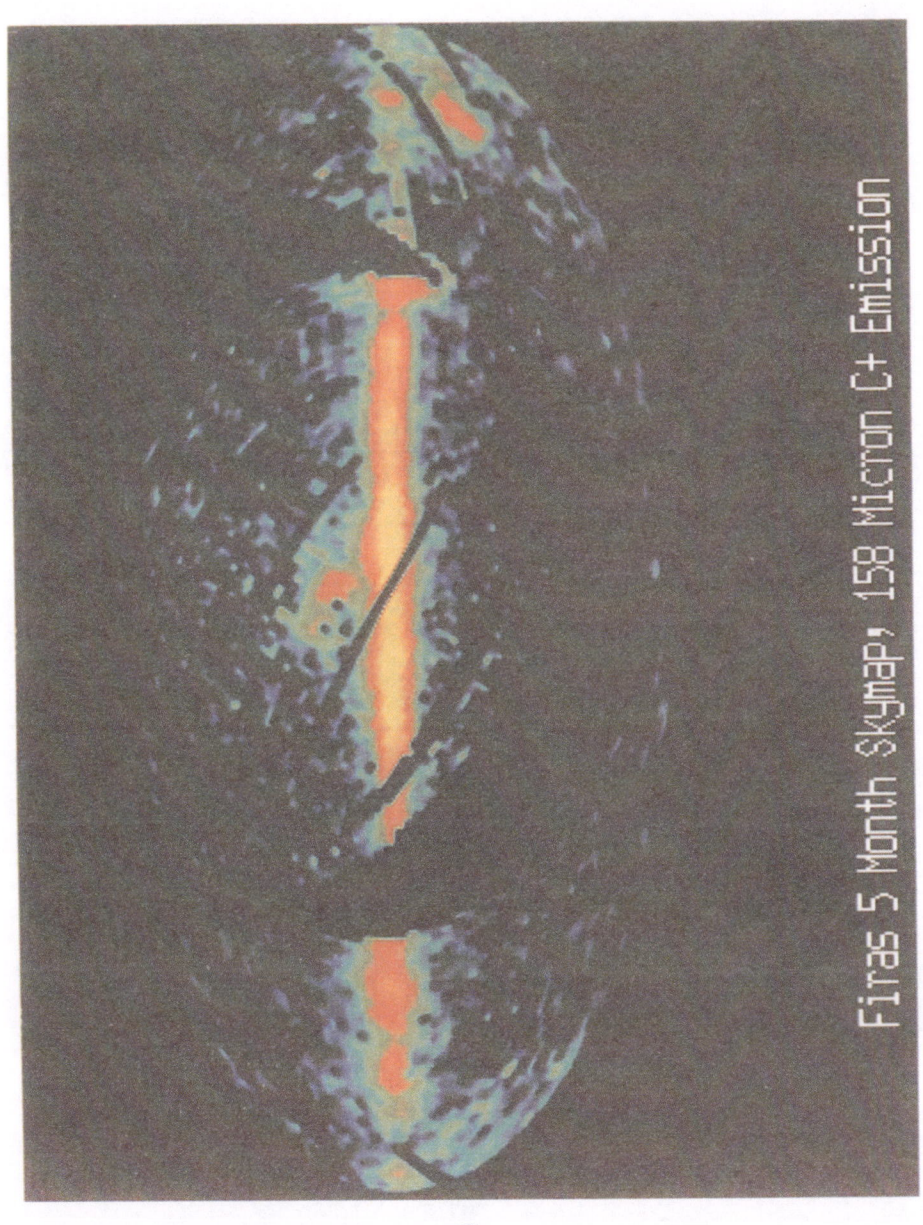

C)

PLATE 3.
   COBE FIRAS instrument maps of sky : A) dust (205 micron) ; B) [NII] (205 micron) ; C) [CII] (158 micron).

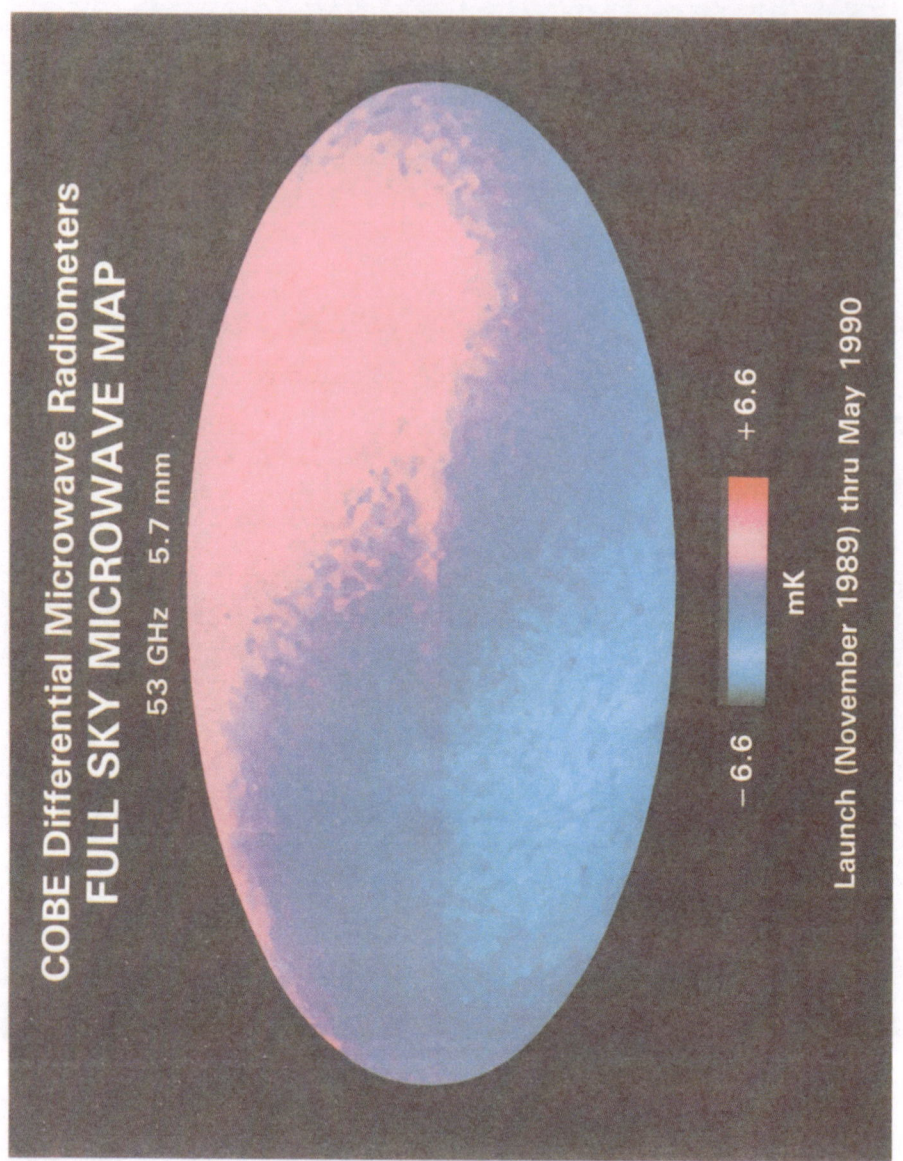

A)

PLATE 4.

COBE DMR full sky maps of the temperature of the sky at 53 GHz (wavelength 5.7 mm). The maps are in galactic coordinates and have been corrected to solar system bary-center. A) Relative sky brightness with mean removed and scaled to about –4 to +4 mK. The dipole anisotropy (± 0.1%) and the galactic plane are clearly visible. B) Relative sky brightness with monopole and dipole removed. The galactic plane emission is the only significant visible signal. The other structure evident is consistent with instrument observing noise.

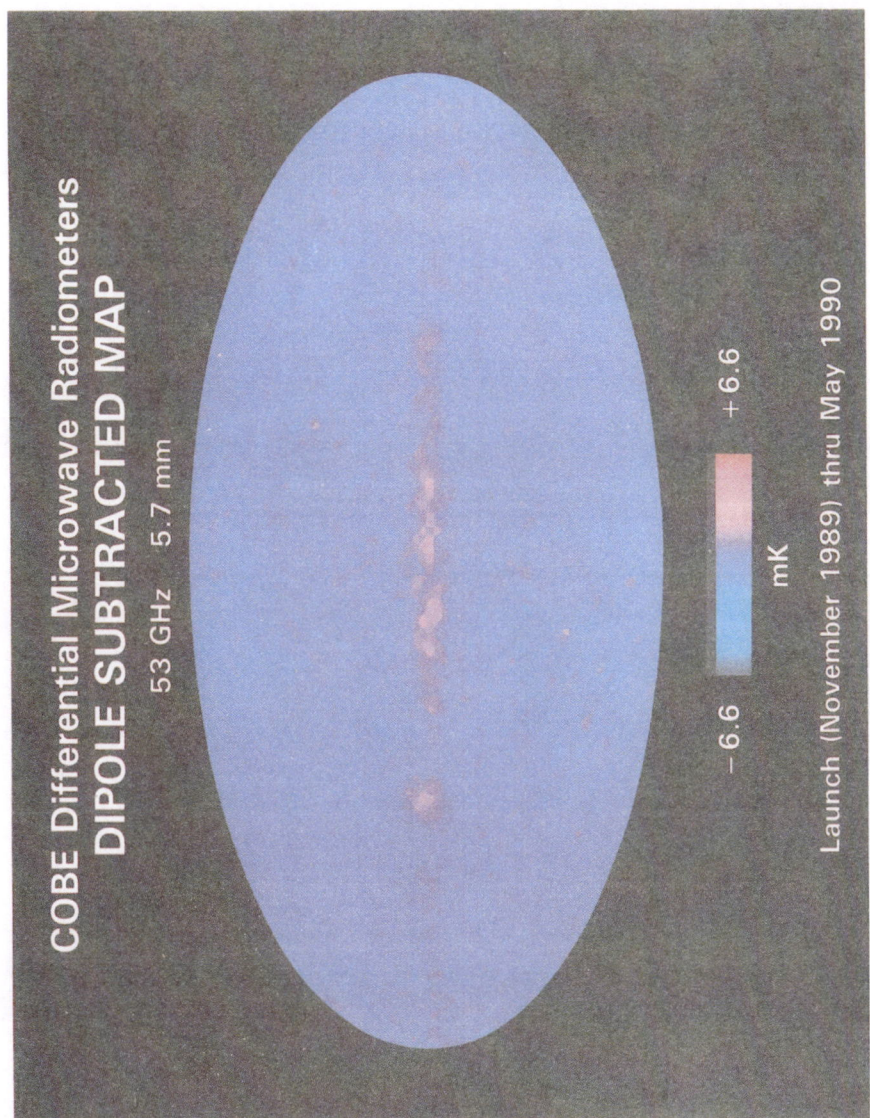

B)

PLATE 4.

COBE DMR full sky maps of the temperature of the sky at 53 GHz (wavelength 5.7 mm). The maps are in galactic coordinates and have been corrected to solar system barycenter. A) Relative sky brightness with mean removed and scaled to about −4 to +4 mK. The dipole anisotropy (± 0.1%) and the galactic plane are clearly visible. B) Relative sky brightness with monopole and dipole removed. The galactic plane emission is the only significant visible signal. The other structure evident is consistent with instrument observing noise.

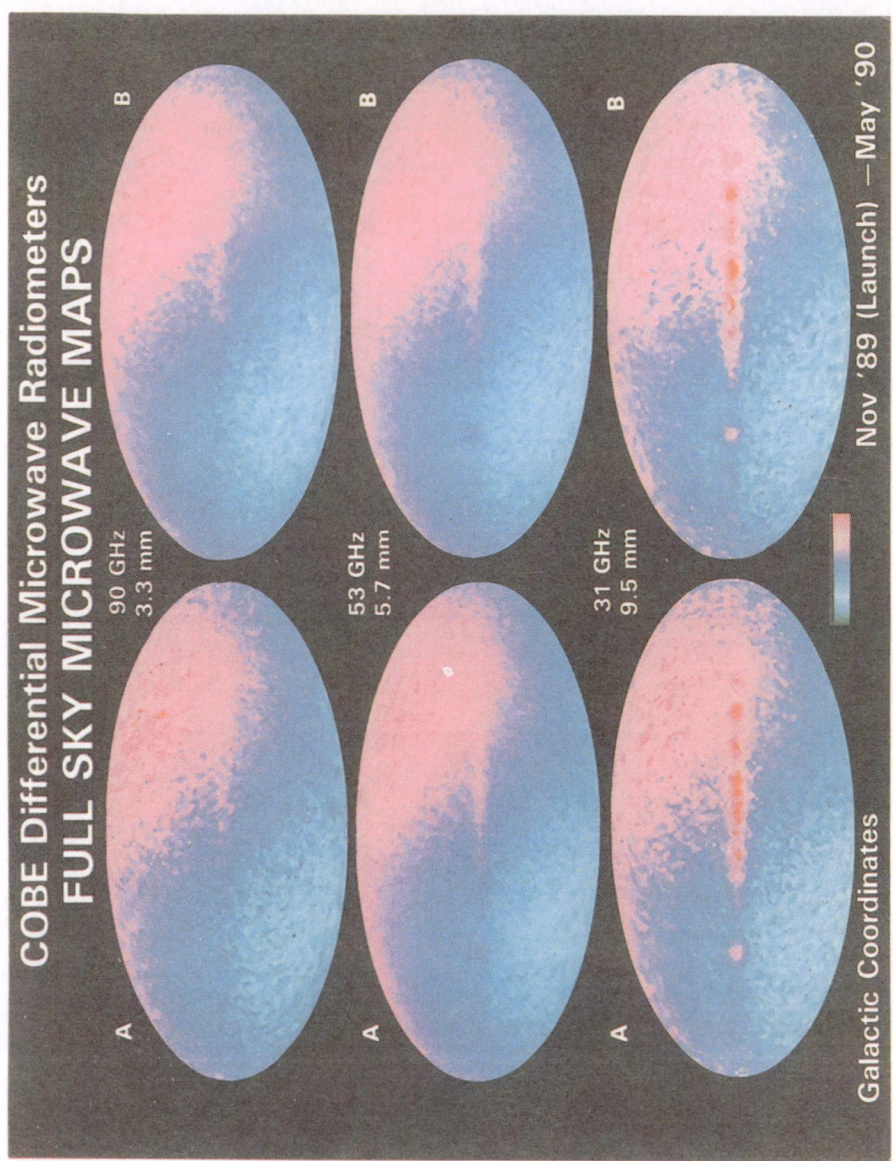

PLATE 5.

COBE DMR full sky maps of the relative temperature of the sky at frequencies 31.5, 53.0 and 90.0 GHz. The maps are in galactic coordinates and have been corrected to solar system barycenter. The maps show variations in received power and are insensitive to the mean temperature of about 2.735 K. Maps A and B correspond to two different channels at the three frequencies.

An Overview of the Cosmic Background Explorer (COBE) and Its Observations: New Sky Maps of the Early Universe

GEORGE F. SMOOT
Lawrence Berkeley Laboratory and Space Sciences Laboratory,
Bldg 50-352 - 1 Cyclotron Road
University of California, Berkeley, 94720

ABSTRACT. This paper[1] discusses the three instruments aboard NASA's Cosmic Background Explorer (*COBE*)[2] satellite and presents early results obtained from the first six months of observations. The three instruments (FIRAS, DMR, and DIRBE) have operated well and produced significant new results. The FIRAS measurement of the CMB spectrum supports the standard Big Bang model. The maps made from the DMR instrument measurements show a spatially smooth early universe. The maps of galactic and zodiacal emission produced by the DIRBE instrument are needed to identify the foreground emissions from extragalactic and thus to interpret its and the other *COBE* results in terms of events in the early universe.

1. INTRODUCTION

The standard cosmological model was established in 1965 with the discovery of the cosmic microwave background (CMB) and the recognition of its implications. The Friedmann-Robertson-Walker Big Bang model provided cosmologists with a theoretical framework for describing the development of the universe from a time of about one hundredth of a second through the present age of the universe. The most notable and precise successes of the Big Bang model were the calculation of the primordial nucleosynthesis of light elements (notably ^{4}He, ^{3}He, D, T, and ^{7}Li) and the existence of a thermal relic radiation (CMB) from this era. "The First Three Minutes" by Steve Weinberg (1977) and "Physical Cosmology" by Jim Peebles (1971) provide an excellent account of this model.

The development of the Standard Model of strong and electroweak interactions and ideas about grand unification provided a context and motivation for extending and expanding the original Big Bang model. Treating fundamental particles (quarks and leptons) as point-like particles with asymptotic freedom from strong interactions at high energies provides

---

[1] This is the introductory lecture of the Les Houches School on *the Infrared and Submillimetre Sky after COBE* held March 20-30, 1991.

[2] The National Aeronautics and Space Administration/Goddard Space Flight Center is responsible for the design, development, and operation of the Cosmic Background Explorer. GSFC is also responsible for the software development through to the final processing of the space data. The *COBE* program is supported by the Astrophysics division of NASA's Office of Space Science and Applications.

M. Signore and C. Dupraz (et al.), The Infrared and Submillimetre Sky after COBE, 1–14.
© 1992 *Kluwer Academic Publishers.*

a picture of the early universe as a dilute gas of weakly interacting quarks, leptons, and gauge bosons. This picture is more readily understandable than the old view of the epoch before 10^{-5} sec, when particles were crowded together with separations and horizons less than a typical particle size (e.g. a proton). Within this context it has been possible to develop a scenario of the universe from very early times (10^{-35} sec) including a phase transition induced period of inflation, baryogenesis, and other events treated as initial conditions in the original Big Bang model (e.g., the fluctuations that lead to formation of galaxies and other structures). I recommend "The Early Universe" by Rocky Kolb and Mike Turner (1990) as a description of this extended standard model.

The light elements and the cosmic microwave background are the most accessible relics from the era during which the universe was a relatively structureless plasma prior to its evolving to the highly-ordered state observed today. In standard models of cosmology, CMB photons have travelled unhindered from the surface of last scattering in the early universe to the present era; as such, the CMB maps the large scale structure of space-time in the early universe. Despite a quarter-century of effort, no intrinsic anisotropy in the CMB has been detected. Likewise, the search for spectral distortions reveal a featureless blackbody. These measurements can be used to constrain possible effects in the early universe, cosmology and particle physics theories.

The *COBE* (Cosmic Background Explorer) experiments are a precision test of the standard model of cosmology. A primary objective is to measure the fluxes and fluctuations of diffuse cosmological emission at millimeter and submillimeter wavelengths.

COBE was launched on November 18, 1989 into a 900-km circular, near-polar orbit (inclination 99°). The Earth's gravitational quadrupole moment precesses the orbit to follow the terminator, allowing the instruments to point away from the Earth and perpendicular to the Sun to avoid both solar and terrestrial radiation. Throughout most of the year, the satellite orbit provides an exceptionally stable environment (the exception is the two-month period surrounding summer solstice, when the Earth limb becomes visible over the instrument aperture plane and the satellite enters the Earth's shadow). Prior to depletion of the liquid helium cryogen on September 20, 1990, the DIRBE and FIRAS instruments mapped the sky 1.6 times. As the dewar warms, the DIRBE instrument continues to take data in its four shortest wavelength bands. The DMR instrument does not require cryogenic cooling and continues to operate normally.

The *COBE* Science Working Group (SWG), consisting of C. L. Bennett, N. W. Boggess, E. S. Cheng, E. Dwek, S. Gulkis, M. G. Hauser, M. Janssen, T. Kelsall, P. Lubin, J. C. Mather, S. Meyer, S. H. Moseley, T. L. Murdock, R. A. Shafer, R. F. Silverberg, G. F. Smoot, R. Weiss, D. T. Wilkinson, and E. L.Wright, is responsible for the scientific oversight. An artist's rendering of *COBE* is the first color plate in this book.

2. The DIRBE Instrument Description and Preliminary Results

The Diffuse Infrared Background Experiment (DIRBE) is designed to conduct a sensitive search for an isotropic cosmic infrared background (CIB) radiation over the spectral range from 1 to 300 micrometers. The cumulative emissions of pregalactic, protogalactic, and evolving galactic systems are expected to be recorded in this background. Since both the cosmic red-shift and reprocessing of short-wavelength radiation to longer wavelengths by dust act to shift the short-wavelength emissions of cosmic sources toward or into the infrared, the spectral range from 1 to 1000 micrometers is expected to contain much of the energy released since the formation of luminous objects, and might contain a total radiant energy density comparable to that of the CMB. The discovery and measurement of the CIB would provide new insight into the cosmic 'dark ages' following the decoupling of matter and the CMB and the low z universe.

Observing the CIB is a formidable task. Bright foregrounds from the Earth's atmosphere, from interplanetary dust scattering of sunlight and emission of absorbed sunlight, and from stellar and interstellar emissions of our own Galaxy dominate the diffuse sky brightness in the infrared. Even when measurements are made from space with cryogenically-cooled instruments, the local astrophysical foregrounds strongly constrain our ability to isolate and measure an extragalactic infrared background. Furthermore, since the absolute brightness of the CIB is of paramount interest for cosmology, such measurements must be done relative to a well-established absolute flux reference, with instruments which strongly exclude or permit discrimination of all stray sources of radiation or offset signals which could mimic a cosmic signal.

The DIRBE approach is to obtain absolute brightness maps of the full sky in 10 photometric bands (J[1.2], K[2.3], L[3.4], and M[4.9]; the four IRAS bands at 12, 25, 60, and 100 micrometers; and 120-200 and 200-300 micrometer bands). To facilitate determination of the bright foreground contribution from interplanetary dust, linear polarization is also measured in the J, K, and L bands, and all celestial directions are observed hundreds of times at all accessible solar elongation angles (depending upon ecliptic latitude) in the range 64° to 124°. The instrument is designed to achieve a sensitivity for each field of view of $\lambda I(\lambda) = 10^{-13}$ W cm^{-2} sr^{-1} (1 σ, 1 year). This level is well below most estimated CIB contributions. Extensive modeling of the foregrounds, just beginning, will be required to isolate and identify any extragalactic component.

The DIRBE instrument is an absolute radiometer, utilizing an off-axis Gregorian telescope with a 19-cm diameter primary mirror. The optical configuration is carefully designed for strong rejection of stray light from the Sun, Earth limb, Moon or other off-axis celestial radiation, or parts of *COBE*. Stray light rejection features include both a secondary field stop and a Lyot stop, super-polished primary and secondary mirrors, a reflective forebaffle, extensive internal baffling, and a complete light-tight enclosure of the instrument within the *COBE* dewar. Additional protection is provided by the Sun and Earth shade surrounding the *COBE* dewar, which prevents direct illumination of the dewar aperture by these strong local sources. The DIRBE instrument, which is maintained at a temperature below 2 K within the dewar, measures absolute brightness by chopping between the sky signal and a zero-flux internal reference at 32 Hz using a tuning fork chopper. Instrumental offsets are measured by closing a cold shutter located at the prime focus. All spectral bands view the same instantaneous field-of-view, 0.7° x 0.7°, oriented at 30° from the spacecraft spin axis. This allows the DIRBE to modulate solar elongation angles by 60° during each rotation, and to sample fully 50% of the celestial sphere each day. Four highly-reproducible internal radiative reference sources can be used to stimulate all detectors when the shutter is closed to monitor the stability and linearity of the instrument response. The highly redundant sky sampling and frequent response checks provide precise photometric closure over the sky for the duration of the mission. Calibration of the photometric scale is obtained from observations of isolated bright celestial sources. Careful measurements of the beam shape in pre-flight system testing and during the mission using scans across bright point sources allow conversion of point-source calibrations to surface brightness calibrations.

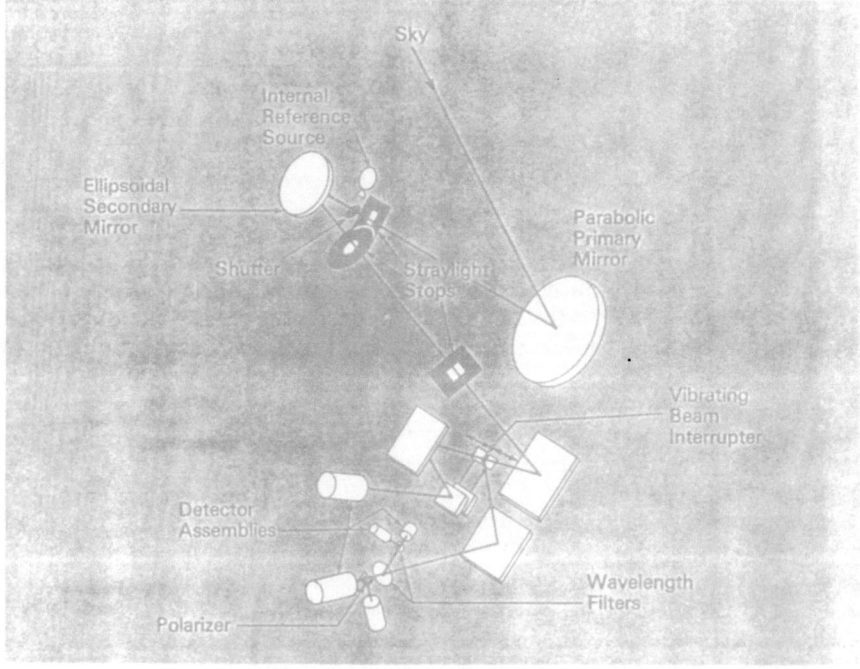

Figure 1. *COBE* DIRBE Instrument Schematic

Qualitatively, the initial DIRBE sky maps show the expected character of the infrared sky. For example, at 1.2 microns stellar emission from the galactic plane and from isolated high latitude stars is prominent. Zodiacal scattered light from interplanetary dust is also prominent. At fixed ecliptic latitude the zodiacal light decreases strongly with increasing solar elongation angle, and at fixed elongation angle it decreases with increasing ecliptic latitude. These two components continue to dominate out to 3.4 microns, though both become fainter as wavelength increases. The foreground emissions have relative minima at wavelengths of 3.4 microns and longward of 100 microns. A composite of the 1.2, 2.3, and 3.4 microns images is shown as blue, green, and red respectively in the second color plate. The Milky Way galactic bulge and disk stellar populations appear dramatically in this image because extinction at these wavelengths is far less than in visible light. A composite of the 25, 60, and 100 micron images using the same respective colors is in the same plate. At 12 and 25 microns, emission from the interplanetary dust dominates the sky brightness, again strongly dependent upon ecliptic latitude and elongation angle. At wavelengths of 60 microns and longer, emission from the interstellar medium dominates the galactic brightness, and the interplanetary dust emission becomes progressively less apparent. The patchy infrared cirrus noted in IRAS data is evident at all of these wavelengths. The discontinuities apparent are due primarily to the change in zodiacal light as the spacecraft and earth move around in the zodiacal dust.

The DIRBE flux limits (Hauser et al. 1991) have the expected spectral shape. They can be compared in detail with other measurements, e.g. IRAS, ZIP, Matsumoto et al., (1988A&B) and Noda et al. (1991). IRAS measurements toward the south ecliptic pole have been compared with DIRBE results. The two experiments are seen to agree reasonably well at 12 and 25 microns, but the DIRBE results are substantially fainter at 60 and 100 microns, reaching a factor of ~3 at 100 microns. The DIRBE and ZIP (Murdock & Price 1985) results at the north ecliptic pole agree to 10% at 10 and 20 microns. Comparison with Matsumoto et al. and Noda et al. shows fairly good agreement in the near infrared with a deeper minimum at 3.4 microns and reasonable agreement in the far infrared but not as sharp a bump at 160 microns. These results and interpretations are discussed in the lectures of Bernard Carr and François Boulanger and a paper by Bond, Carr, & Hogan (1991). The DIRBE (and FIRAS) data already rule out significant early VMO's and explosions and constrain galactic evolution models. The DIRBE data will clearly be a valuable new resource for studies of the interplanetary medium and Galaxy as well as the search for the CIB.

The DIRBE team is headed by the Principal Investigator Micheal Hauser and Deputy Principal Investigator Tom Kelsall. In addition to the COBE SWG participation the following people are working on DIRBE: S. Burdick, G. N. Toller, G. Berriman, B. B. Franz, K. J. Mitchell, M. Mitra, N. P. Odegard, F. S. Patt, W. J. Speisman, S.W. Stemwedel, J. A. Skard, A. Ahearn, W. Daffer, I. Freedman, S. Jong, A. Panitz, A. Ramsey, T. Roegner, V. Snowel, K. Tewari, H. T. Freudenreich, C. Lisse, J. Weiland.

3. FIRAS Instrument Description and CMB Spectrum Results

The purpose of the FIRAS is to compare the spectrum of the CMB with that of a precise blackbody, enabling the measurement of very small deviations from a Planckian spectrum. The FIRAS instrument covers two frequency ranges, a low frequency channel from 1 to 20 cm^{-1} and a high frequency channel from 20 to 100 cm^{-1}. It has a 7° diameter beam width, established by a non-imaging parabolic concentrator, which has a flared aperture to reduce diffractive sidelobe responses. The instrument is calibrated by a full beam, temperature-controlled external blackbody, which can be moved into the beam on command. The FIRAS is the first instrument to measure the background radiation and compare it with such an accurate external full-beam calibrator in flight. The spectral resolution is obtained with a polarizing Michelson interferometer, with separated input and output beams to permit fully symmetrical differential operation. One input beam views the sky or the full aperture calibrator, while the second input beam views an internal temperature controlled reference blackbody, with its own parabolic concentrator. Both input concentrators and both calibrators are temperature controlled and can be set by command to any temperature between 2 and 25 K. In standard operating condition the two concentrators and the internal reference body are commanded to match the sky temperature, thereby yielding a nearly nulled interferogram and reducing almost all instrumental gain errors to negligible values.

The external calibrator determines the accuracy of the instrument for broad band sources like the CMB. It is a re-entrant cone shaped like a trumpet mute, made of Eccosorb CR-110 iron-loaded epoxy. The angles at the point and groove are 25°, so that a ray reaching the detector has undergone 7 specular reflections from the calibrator. The calculated reflectance for this design, including diffraction and surface imperfections, is less than 10^{-4} from 2 to 20 cm^{-1}. Measurements of the reflectance of an identical calibrator in an identical antenna using coherent radiation at 1 cm^{-1} and 3 cm^{-1} frequencies confirm this calculation. The instrument is calibrated by measuring spectra with the calibrator in the sky horn while operating all other controllable sources within the instrument at a sequence of different temperatures.

The first results of the FIRAS (Mather *et al.* 1990) may be summarized as follows. The intensity of the background sky radiation is consistent with a blackbody at 2.735 ± 0.06 K. Deviations from this blackbody at the spectral resolution of the instrument are less than 1% of the peak brightness. The quoted temperature error is primarily due to an uncertainty in the thermometer calibration; we expect to reduce this uncertainty by additional tests. The measured spectrum is shown in Figure 2, and is converted to temperature units, where it is compared to previous measurements. The deviations can be fitted to a Bose-Einstein distribution, limiting the dimensionless chemical potential to $|\mu| < 10^{-2}$ (3σ). Fitting to the Sunyaev-Zel'dovich form for Comptonization (Zel'dovich and Sunyaev 1969), yields a limit of $|y| < 10^{-3}$ (3 σ). This strongly limits the existence of a smooth hot intergalactic medium: it can contribute less than 3% of the X-ray background radiation even at a reheating time as recent as $z=2$. One might think that the X-ray background could be produced by clumping the hot

IGM, since y is proportional to the electron density and the X-ray luminosity is proportional to the square of the electron density. However, such clumping would cause a net anisotropy. Following these arguments the size of the X-ray emitting regions must be less than a galactic diameter (Barcons, Fabian, & Rees 1991). There is no evidence of a spectral distortion, such as that reported by Matsumoto *et al.* 1988B. The measured temperature is consistent with many previous reports and the recent rocket result of Gush *et al.* (1990).

The variation of the spectrum with position in the sky as measured by the FIRAS is dominated by the dipole anisotropy of the CMB, plus a variation in the interstellar dust emission. A preliminary dipole anisotropy spectrum was determined by calculating the average spectra in two large circular regions of the sky each of angular diameter 60°, one centered at $(\alpha, \delta) =$ (11.1h, -6.3°) and the other at (23.1h, 6.3°), which lie along opposite ends of the dipole axis as determined by the DMR. The difference between these spectra is fit extremely well by the difference of two blackbodies, and is consistent with a peak dipole amplitude of 3.3 ± 0.3 mK and the assumed dipole direction (Cheng *et al.* 1990).

These precise limits to potential distortions covering the major portion of the CMB photons provide support for a key element in the models of Big Bang nucleosynthesis - the baryon to photon ratio and the interpretation of that number relative to the critical density. The FIRAS measurement actually counts the vast majority of photons and the precision of the measurement gives us the density:

$$\rho_\gamma = 4.722 \ (T/1K)^4 e^{4y} \ eV/cm^3 \sim 10^{-33} \ (To/2.7K)^4 e^{4y} \ gm/cm^3$$

$$n_\gamma = 20.286 \ (T/1K)^3 \ photons/cm^3 = 415 \ (To/2.735K)^3 \ photons/cm^3 \qquad (1)$$

FIRAS results decrease the error on n_γ from >20% to less than 5%. The limits on $|\mu|$ and $|y|$ in turn limit energy release in the early universe ($1+z < few \ 10^6$) to typically less than 1% of the energy in the CMB. This in turn limits cosmologically significant events such as primordial particle decay, superconducting cosmic strings, and turbulence.

The FIRAS instrument also measures the galactic emission and has made a survey of the dust and line emission in its bands. The dust emission is strong; the FIRAS provides good spectral information on that dust and has found some ten emission lines (Wright *et al.* 1991). In addition FIRAS has produced maps of the dust emission (205 μm for instance), the 205 μm NII line and 158 μm CII line. Ned Wright discusses the dust spectrum and maps in his lecture.

The FIRAS team is headed by the Principal Investigator John Mather, Deputy Principal Investigator Rick Shafer and Deputy Project Scientist for Data Ed Cheng. In addition to the *COBE* SWG participation the following people are working on FIRAS: Dale Fixsen, Gene Eplee, Joel Gales, Rich Isaacman, Snehevadan Macwan, Derck Massa, Muriel Taylor, Alice Trenholme, Dave Wynne, Steve Alexander, Dave Bouler, Nilo Gonzalez, Ken Jensen, Shirley Read, Larry Rosen, Fred Shuman, Frank Varosi, Harte Wang, Will Daffer, Jim Krise, Nick Iascone, and Courtney Scott.

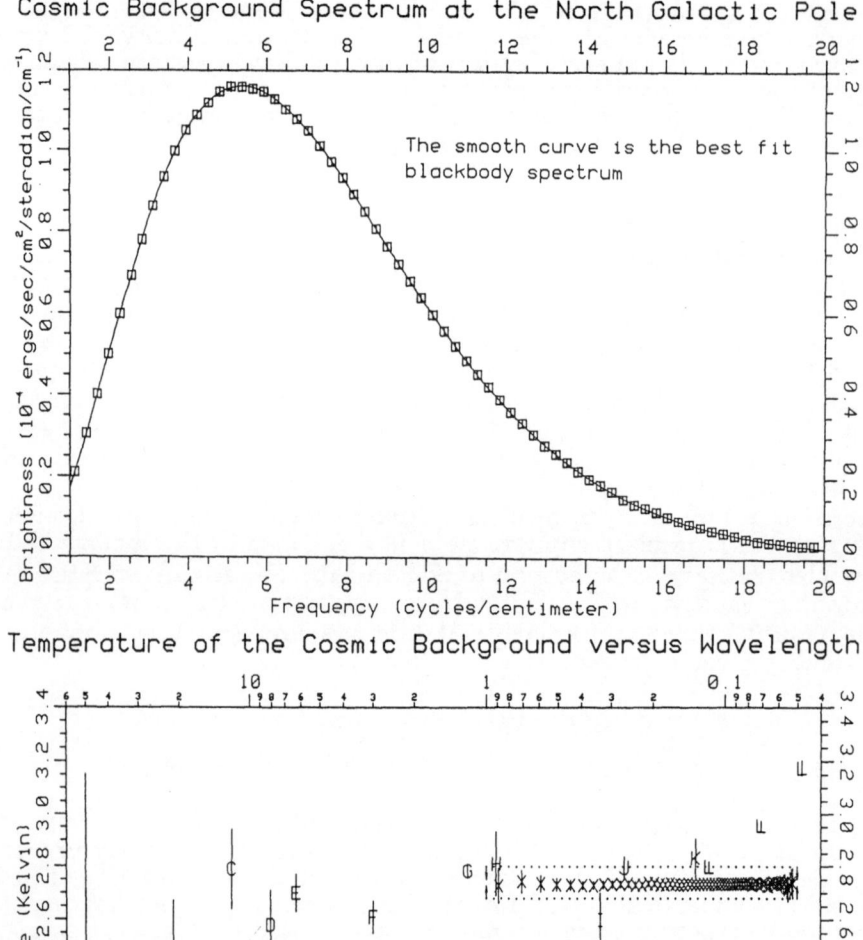

Figure 2. A. *COBE* FIRAS spectrum measurement results. The boxes are the data points and the line the best fit Planckian (2.735 K). The data are from near the north galactic pole. B. Plot of Temperature versus wavelength of recent CMB spectrum measurements.

4. DMR

4.1 DMR INSTRUMENT DESCRIPTION AND OPERATION

The *COBE* Differential Microwave Radiometers (DMR) instrument is making precise maps of the microwave sky on large angular scales. The DMR consists of six differential microwave radiometers, two independent radiometers at each of three frequencies: 31.5, 53, and 90 GHz (wavelengths 9.5, 5.7, and 3.3 mm). At these frequencies the CMB exceeds the foreground galactic emission by at least a factor of about 1000. The multiple frequencies allow subtraction of galactic emission using its spectral signature. Each radiometer measures the difference in microwave power between two regions of the sky separated by 60°. The combined motions of spacecraft spin (75 s period), orbit (103 minute period), and orbital precession (~1 degree per day) allow each sky position to be compared to all others through a highly redundant set of all possible difference measurements spaced 60° apart. Each radiometer is a heterodyne receiver whose input is switched at 100 Hz between two identical corrugated horn antennas. The compact low-sidelobe antennas' main beam profile is well described by a Gaussian of 7° FWHM and point 60° apart, 30° to either side of the spacecraft spin axis (Toral *et al.* 1989). The two channels at 31.5 GHz share a single antenna pair with an orthomode transducer splitting the input into opposite circular polarizations. Both channels share a common enclosure and thermal regulation system. The 53 and 90 GHz radiometers are similar but have two antenna pairs at each frequency, each with identical linear polarization response. A detailed description of the DMR instrument may be found in Smoot *et al.* (1990).

Three independent techniques are used to determine the radiometer calibration. Solid-state noise sources provide in-flight calibration by injecting broad-band microwave power into the front end of each radiometer every two hours. Prior to launch the noise source signals were calibrated by comparing to the signal produced by targets of known, dissimilar temperatures (approximately 300 K and 77 K) filling the antenna apertures. The DMR observes the Moon for a fraction of an orbit during two weeks of every month providing an independent determination of the calibration factor. The Earth's ~30 km s^{-1} motion about the solar system barycenter produces a Doppler-shift dipole of known magnitude (~0.3 mK) and direction. The DMR measures this dipole, but at low signal to noise. Given sufficient observing time (more than one year), this method is expected to produce the most accurate determination of the absolute calibration of the DMR instrument. A forthcoming paper (Bennett *et al.* 1991) discusses the DMR calibration more fully.

The DMR has been operating well since about a week after the *COBE* launch (November 1989) and is approved for two years of operation. Two additional years of operation have been requested.

4.2 DMR DATA REDUCTION AND ANALYSIS

Although the experiment has been designed to minimize or avoid systematic uncertainty, the sky maps may be contaminated by foreground sources, instrumental effects and the data processing. The data reduction process must distinguish cosmological signals from a variety of potential systematic effects. The most obvious source of non-cosmological signals is the presence in the sky of foreground microwave sources. Although the DMR instrument is largely shielded from such sources, their residual or intermittent effect must be considered. A second class of potential systematics is the effect of the changing orbital environment. Instrument performance varies slightly with changes in temperature, voltage, and local magnetic field, each of which is modulated by the *COBE* orbit. Longer-term drifts can also affect the data. Finally, the data reduction process itself may introduce or mask features in the data. The DMR data are differential; the map making algorithm is subject to concerns of both coverage (closure) and solution stability. Other data reduction processes, particularly the calibration and baseline subtraction, are also a source of potential artifacts. All potential sources of systematic error must be identified and their effects measured or limited before maps with reliable uncertainties can be produced.

Currently the absolute calibration of the instrument is uncertain to ~5%. This uncertainty does not create artifacts in the maps but affects the calculated amplitude of the dipole. We continue to acquire and analyze calibration information and anticipate improved calibration in the future. The next largest effect is the modulation of the instrument output by the Earth's magnetic field. Magnetic effects will be modelled and removed in future analysis. The largest potential effects upon the quadrupole and higher-order coefficients are Earth emission and the possible undetected calibration drifts. The current 95% C.L. upper limits to combined systematic errors in the DMR maps are $\Delta T/T_0 < 8x10^{-5}$ for the dipole anisotropy and $\Delta T/T_0 < 3x10^{-5}$ for the quadrupole and higher-order terms.

As the DMR gathers redundant sky coverage and as analysis continues, we anticipate refined estimates of, or limits on, these effects. It is important to note that the DMR is free from some of the systematics of previous large-scale sky surveys. The multiple differences generated by various chopping frequencies (spin, orbit, and precession periods) allow separation of instrumental from celestial signals. Two independent full-sky maps, with matched beams at each of three frequencies, provide a powerful tool for analysis and removal of possible systematic effects.

4.3 DMR RESULTS

The main results from the DMR at present are the maps and their expansion in spherical harmonics (Smoot et al. 1991A&B). The color plate shows the microwave sky on a linear scale. The most noticeable effect is the extreme uniformity of the CMB. The next color plate shows preliminary maps of the microwave sky for each of the six DMR channels. The independent maps at each frequency enable celestial signals to be distinguished from noise or spurious features: a celestial source will appear at identical amplitude in both maps. The three frequencies allow separation of cosmological signals from galactic foregrounds based on spectral signatures. The maps are corrected to solar-system barycenter and do not include data with the Moon within 25° of an antenna; no other systematic corrections have been made. All six maps clearly show the dipole anisotropy and galactic emission. The dipole appears at similar amplitudes in all maps while galactic emission decreases sharply at higher frequencies, in accord with the expected spectral behavior.

An observer moving with velocity β = v/c relative to an isotropic radiation field of temperature T_0 observes a Doppler-shifted temperature

$$T(\theta) = T_0 \frac{(1-\beta^2)^{1/2}}{1-\beta\cos(\theta)} = T_0 [1 + \beta\cos(\theta) + \frac{1}{2}\beta^2 \cos(2\theta) + O(\beta^3)] \qquad (2)$$

The first term is the monopole CMB temperature without a Doppler shift. The second term, proportional to β, is a dipole distribution, varying as the cosine of the angle between the velocity and the direction of observation. The term proportional to β^2 is a quadrupole with amplitude reduced by 1/2 β from the dipole amplitude. The DMR maps clearly show a dipole distribution consistent with a Doppler-shifted thermal spectrum, implying a velocity for the solar system barycenter of β=0.00122 ± 0.00003 (68% CL), or v = 365±18 km s^{-1} toward (l,b) = (265°±2°, 48°±2°), where we assume a value T_0=2.735 K. The solar system velocity with respect to the local standard of rest is estimated at 20 km s^{-1} toward (57°,23°), while galactic rotation moves the the local standard of rest at 220 km s^{-1} toward (90°,0°) (Kerr & Lyndon-Bell 1990, Fich, Blitz, & Stark, 1989). The DMR results thus imply a peculiar velocity for the Galaxy of v_g = 547 ± 17 km s^{-1} in the direction (277° ± 2°, 29° ± 2°) and a peculiar velocity of the local group, v_{lg} = 622 ± 10 km s^{-1} toward (277°± 2°, 30° ± 2°) using local group velocity of 308 km s^{-1} toward l = 107° and b = -7° (Yahil, Tamman, & Sandage 1977, Yahil *et al.* 1986).

The angular distribution in the sky maps is consistent with a dipole anisotropy over the full sky. The spectrum of all published dipole parameters, including those from the FIRAS experiment (Cheng *et al.* 1990), is consistent with a Doppler-shifted blackbody origin. With high probability the dipole comes from our peculiar motion and is not due to an intrinsic dipole from a large scale gradient (Paczynski and Piran 1990) or a net velocity of the CMB relative to the comoving frame (Collins & Hawking 1973). The dipole anisotropy is more than 30 times larger than any other large angular scale anisotropy. It is very difficult to establish a universe with a large intrinsic dipole anisotropy without comparable quadrupolar and higher order moment anisotropies. Evidence for a Doppler origin from our peculiar velocity relative to the comoving CMB frame can be found by comparing the velocity inferred from the dipole anisotropy with the velocity expected from gravitational perturbations in the region 100 Mpc around our location. Our peculiar velocity can be estimated either by adding up observed sources such as *IRAS* galaxies (light⇒mass⇒potential), or by inferring potential from large scale velocity flows (the "Great Attractor"). One readily obtains 80% of the CMB dipole implied velocity with errors on the order of 20% or greater. We adopt the working hypothesis that the dipole is generated by the Doppler shift resulting from the galactic peculiar velocity relative to the comoving frame. The Doppler shift origin can in principle be tested by looking for the predicted kinematic quadrupole illustrated in equation (2).

The last color plate shows the DMR maps with this dipole removed from the data. The only large-scale feature remaining is galactic emission, confined to the plane of the galaxy. This emission is present at roughly the level expected before flight and is consistent with emission from electrons (synchrotron and HII) and dust within the galaxy. The ratio of the dipole anisotropy (the largest cosmological feature in the maps) to the Galactic foreground reaches a maximum in the frequency range 60—90 GHz. There is no evidence of any other emission features.

We have made a series of spherical harmonic fits to the data, excluding data within several ranges of galactic latitude. The only large-scale anisotropy detected to date is the dipole. The quadrupole anisotropy is limited $\Delta T/T < 3 \; x \; 10^{-5}$ and higher-order terms are limited to amplitude $\Delta T/T < 10^{-4}$. Similarly, a search for Gaussian or non-Gaussian fluctuations on the sky showed no features to limit $\Delta T/T < 1.5 \; x \; 10^{-4}$. The reported uncertainties are 95% confidence level unless otherwise stated, and include the effects of systematics. These results reflect "quick look" data and are still limited by uncertainties in estimates of many potential systematic errors that may be removed in further data processing.

One important effect that may ultimately limit the detectable cosmological quadrupole and other large angular scale anisotropies is confusion from galactic emission. Galactic emission (including the galactic plane) produces a quadrupole whose amplitude decreases with increasing frequency until dust emission becomes important. However, since the CMB amplitude also decreases with frequency, the DMR 53 and 90 GHz channels bracket the apparent best observing frequency range. Including the galactic plane produces a quadrupole amplitude with a minimum of 50 μK. To obtain the ultimate sensitivity we need a level 20 to 30 times lower. How much does simply cutting the galactic plane from the fitting achieve? Tests using the 408 MHz and *IRAS* 100 micron maps indicate that reasonable galactic latitude cuts gain roughly a factor of 5 to 10. We will have to obtain the remaining factor of 2 to 4 from careful modeling.

4.4 DMR DISCUSSION

The DMR limits to CMB anisotropies provide significant new limits to the dynamics and physical processes in the early universe. The dipole anisotropy provides a precise measure of the Earth's peculiar velocity with respect to the co-moving frame. I will discuss these in more detail in my second lecture.

The DMR team is headed by the Principal Investigator George F. Smoot and Deputy Principal Investigator Charles Bennett. In addition to the *COBE* SWG participation the following people are working on DMR: Alan Kogut, Jon Aymon, Charles Backus, Giovanni De Amici, Kevin Galuk, Gary Hinshaw, Peter D. Jackson, Phil Keegstra, Robert Kummerer, Charles Lineweaver, Laurie Rokke, Luis Tenorio, Richard Mills, Jairo Santana, and Peter Young.

5. CONCLUSIONS

The *COBE* instruments have produced exceptionally good maps of the full sky from 1 micron to 1 cm wavelength. DIRBE has mapped the sky from 1 to 300 micron sky and produced preliminary estimates of dark sky fluxes. FIRAS has measured the CMB spectrum quite precisely and shown no deviations from a blackbody spectrum. Future improvements and maps are expected. The FIRAS and DMR data show the dipole anisotropy, consistent with a Doppler-shifted thermal spectrum. The DMR maps show galactic emission in the plane of the galaxy; there is no evidence for any other large-scale feature. The results are currently limited by instrument noise and upper limits to potential sources of systematic error.

ACKNOWLEDGEMENTS

The *COBE* instruments and satellite have been working well and continue to collect data due to the excellent work by the staff and management of the *COBE* Project. It is a pleasure to thank and acknowledge the many people who worked on the project and contributed to its great success. The *COBE* research is a team effort including the 19 original Science Working Group members and other scientists and professionals (see for example, Smoot *et al.*). This work is supported by NASA and in part by the Director, Office of Energy Research, Office of High Energy and Nuclear Physics, Division of High Energy Physics of the U.S. Department of Energy under Contract No. DE-AC03-76SF00098. We thank Monique Signore, Christophe Dupraz, and the Les Houches staff for the delightful and enlightful school.

6. REFERENCES

Barcons, X., Fabian, A.C., & Rees, M.J. 1991, **Nature**,
Bennett, C.L., *et al.*, 1991 in preparation.
Bond, R., Carr, B. & Hogan, C 1991, *Ap. J.*, **367**,420-454.
Cheng *et al.* 1990, *Bull. APS*, **35**, 937.
Collins, C.B., and Hawking, S.W., *MNRAS*, **162**, 307 (1973).
Fich, M., Blitz, L., and Stark, A., *Ap. J.*, **342**, 272 (1989).
Hauser, M. G. *et al.* 1991, "After the First Three Minutes", AIP, Holt, Bennett, & Trimble
Gush, H.P., Halpern, M. & Wishnow, E. (1990) *Phys. Rev. Lett.*, **35**, 937.
Kerr, F.J., and Lyndon-Bell, D., *MNRAS*, **221**, 1023 (1990).

14

Kolb, E.W. & Turner, M. 1990,"The Early Universe", Addison-Wesley.
Mather *et al.* 1990, *Ap. J.*, **354**, L37-L41.
Matsumoto, T., Akiba, M., & Murakami, H, 1988A, *Ap. J.*, **332**, 575.
Matsumoto *et al.* 1988B, *Ap. J.*, **329**, 567-571.
Murdock, T. L., & Price, S. D. 1985, *Astr. J.*, **90**, 375.
Noda *et al.* (1991) preprint
Paczynski and Piran 1990, *Ap. J.*, **364**, 341-348.
Peebles, P.J.E., 1971, "Physical Cosmology", Princeton.
Smoot, G.F., *et al.*, *Ap. J.*, **360**, 685 (1990).
Smoot, G. F., *et al.*, *Ap. J. Let.*, **371**, L1-L5 1991A
Smoot, G. F., *et al.*, *Adv. Space Res.* **11**, No2, pp.(2)193-(2)205, 1991B.
Toral, M.A., *et al.*, *IEEE Transactions on Antennas and Propagation*, **37**, 171 (1989).
Wright, E.L., *et al.* 1991 accepted *Ap. J.*
Weinberg, S. (1972) "Gravitation and Cosmology", Wiley.
Weinberg, S. (1977) "The First Three Minutes", Basic Books
Yahil, A., Tamman, G., and Sandage, A., *Ap. J.*, **217**, 903 (1977).
Yahil *et al.* 1986, *Ap.J.*, **301**, L1-L5.
Zel'dovich, Ya. B. and Sunyaev, R. A. 1969, *Ap. Space Sciences*, **4**, 301.

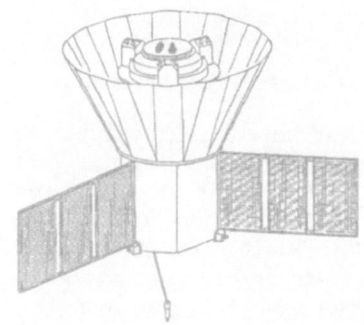

STRINGS, GRAVITY, AND THE CONSTANTS OF NATURE

G. Veneziano
CERN, Geneva
Switzerland

ABSTRACT. After discussing some non-trivial properties of classical string motions in gravitational backgrounds, we turn to the standard model and, in particular, to its (usually understated) shortcomings. We then discuss how quantum strings might resolve the SM difficulties by providing a finite theory of all interactions, including gravity. An outcome of this optimistic scenario is the calculability-in-principle of the fundamental constants of Nature, which simply appear, in string theory, as vacuum parameters.

1. OUTLINE

The subject of this talk is the fascinating, highly non-trivial relationship between strings and gravity. Such a relation is not just interesting: it is the very "raison d'être" of string theory as a (candidate) consistent theory of classical and quantum gravity.

But, of course, string theory is not limited to gravitational phenomena. The same mechanism that presumably renders quantum string gravity (QSG) consistent should also free gauge interactions from their (in)famous, though by now accepted, ultraviolet (UV) divergences.

We shall argue that there is a very deep consequence stemming from the UV finiteness of string theory: unlike in the standard model (SM), where many free parameters inevitably occur, the fundamental constants of Nature are in principle calculable in quantum string theory (QST). As we shall argue, they are nothing but vacuum expectation values (VEVs) to be fixed, hopefully, by some non-perturbative dynamics.

We shall start by reviewing some recent work dealing with classical strings in gravitational backgrounds. Then, after recalling the pluses and minuses of the standard model, we shall outline and contrast the way the constants of Nature originate in a renormalized theory (such as the SM) and in a finite theory (such as QST).

2. CLASSICAL STRINGS

As elementary, classical systems strings come next to points as the least complicated objects. Denoting by d the intrinsic dimensions of the system, we have $d = 0$ for points, $d = 1$ for strings (which can be open or closed), $d = 2$ for membranes (open or

15

M. Signore and C. Dupraz (et al.), The Infrared and Submillimetre Sky after COBE, 15–33.

closed, again), etc. Each one of these systems can be embedded in a D-dimensional space-time (or target space) in which it moves. If we endow space-time with a metric $g_{\mu\nu}$ ($\mu, \nu = 0, 1, \ldots, D-1$), i.e., with a gravitational field, the classical motion of each elementary object will be given by a generalized geodesic: a geodetic line for the point, a geodetic surface for a string, a geodetic volume for a membrane, etc., corresponding to a minimal length, area, volume, respectively.

What is then so special about $d = 1$, i.e., about strings? Apparently nothing! Points, if any, look special. Indeed, there are questions that we may only ask to extended objects ($d \geq 1$) not to point-particles.

Consider for instance the metric $g_{\mu\nu}$ to be of a cosmological type, e.g., describing an "expanding" Universe

$$ds^2 = dt^2 - R^2(t)d\vec{r}^2; \quad dR/dt > 0 \tag{2.1}$$

We may ask to a string (or to a membrane):

 i) How does the size of the string change with t?
 ii) Is the expansion real in the sense that the ratio (distance between strings)/(size of each) grows. If yes, does it grow like $R(t)$?
 iii) Can we have new phenomena occur because of the extended nature of strings?

A basic (and trivial) point to remember before answering these questions is the following.

Classical strings have **no** characteristic size and, in particular, **no minimal** size. A corollary is that the size L of a classical string, being the only scale in the problem, is irrelevant in flat spacetime.

This is not so, however, in a cosmological background the reason being that such a metric has a scale of its own, $H = \frac{1}{R}\frac{dR}{dt}$. The dimensionless combination LH does matter and, indeed, one finds that

If **LH ≪ 1**, i.e., if the string is "small", it behaves like a physical point, its size keeps constant during the motion and the ratio distance/size grows like R. A solution can be constructed in terms of a systematic expansion [1] around the point-particle geodesic.

If **LH ≫ 1**, i.e., if the string is "large", its behaviour is markedly different from that of a point. In particular, if $\ddot{R} > 0$ (inflation) the string develops "Jeans" instabilities [2] and its size starts to grow with R. In the limiting case $LH \gg 1$ a new expansion is valid [3] around an asymptotic (large R) solution in which the comoving size of the string gets frozen, the oscillatory behaviour stops and proper sizes grow like the scale factor (so that distance/size → constant).

An interesting question concerning this latter regime of high instability arises: if fast inflation can give rise to string instabilities is it possible that, vice versa, highly unstable strings may provide a suitable source for inflation? Scenarios in which the answer to this question is in the affirmative have been given the name of "self-sustained inflation" (in that no cosmological constant is needed to drive the inflationary process) and examples have been provided [4].

In the asymptotic expansion of Refs. [3] it is easy to see that, in the perfect fluid approximation, the equation of state of unstable strings is of the type

$$p = -\rho/(D-1) + (\text{positive non-leading terms}) \tag{2.2}$$

Thus the pressure, though opposite to that of point particles, is not sufficiently negative to drive inflation for $D \geq 4$. The statement follows from the D-dimensional Einstein equations which imply:

$$(D-1)(D-2)\frac{d^2R/dt^2}{R} = -8\pi G_N[\rho(D-3)+p(D-1)]$$
$$= -8\pi G_N\rho(D-4) + \text{neg. terms} \tag{2.3}$$

For $\rho > 0, D \geq 4$ this implies $\ddot{R} < 0$, i.e., no inflation.

The appealing idea of a self-sustained inflation is not necessarily dead, however. There are several possible ways out worth exploring:

a) Add a "viscosity" term to the equation of state in order to reflect quantum creation of strings . This can be made to work, albeit at the cost of choosing an "ad hoc" viscosity [5].

b) Start from $D > 4$ (after all that is what one has to do, at the quantum level, with string theory, see below) and look for solutions describing anisotropic self-sustained inflation of three spatial dimensions accompanied by the contraction of $(D-4)$ internal dimensions. Solutions of this kind have been recently found [6]. Certain initial conditions have to be fulfilled in order to obtain sufficient inflation (for solving the usual difficulties of standard cosmology).

c) Take into account the string modifications of Einstein's equations (see below), in particular the presence of another massless field, the dilaton ϕ. It has been pointed out recently [7,8] that the string-modified Fiedmann-Einstein equations exhibit an interesting symmetry under inversion of the scale factor (accompanied by a suitable change in ϕ) and possess generically cosmological solutions of the inflationary type even in the absence of classical stringy sources. This possibility is being further analyzed at present.

At this point we shall leave strings momentarily and discuss an apparently unrelated subject: the standard model.

3. PLUSES AND MINUSES OF THE STANDARD MODEL

The SM is our state-of-the-art theory of three of the fundamental forces of Nature: thus, in the SM, electromagnetic interactions are described, at low energy, by the usual $U(1)$ gauge theory but get unified at higher energies with the weak interactions within an $SU(2) \times U(1)$ gauge group. Finally, strong interactions are added in through an extra factor $SU(3)$, to yield the full SM gauge group: $SU(3) \times SU(2) \times U(1)$. Like any theory, the SM too has its own pluses and minuses:

A) SM PLUSES

These are many, not the least one being that the SM seems to explain the data (and maybe I should stop there)! But, also, from the theorist's viewpoint, it reduces *three* fundamental forces to *a single* physical principle :

GAUGE INVARIANCE

It so happens that, precisely in D=3+1 dimensions of space-time, gauge theories are just "renormalizable". Let me explain in simple terms what is meant by this and why it is so important:

The classical Lagrangian of the standard model has the schematic form:

$$L_{class} = F_{\mu\nu}^2 + \bar{\psi}\gamma D\psi + L_{Higgs}. \qquad (3.1)$$

However, experimentally, what one measures is the full, quantum-corrected, so-called effective Lagrangian:

$$L_{eff} = L_{class} + L_{quantum} = Z_1 F_{\mu\nu}^2 + Z_2 \bar{\psi}\gamma D\psi + Z_3 \ldots + L_{finite}. \qquad (3.2)$$

A renormalizable theory is defined by the property that, although (as opposed to a finite theory) it does have quantum infinities, these can all be lumped into a *finite* number of constants Z_i which multiply precisely the terms appearing in L_{class}. Thus, even if someone had given us the parameters appearing in L_{class}, there would still be no hope of computing those appearing in the *measured* L_{eff}. Rather, one fixes the "renormalized" parameters (masses, couplings) from the experiments while everything else is determined, in principle, by the theory.

To summarize:

IN A RENORMALIZABLE THEORY L_{eff} IS A *FINITE* FUNCTION OF A *FINITE* NUMBER OF *ARBITRARY* PARAMETERS.

The above statement shows, at the same time, the strength and weakness of renormalizable theories: it is good to have a *finite* number of arbitrary parameters; yet it would have been even better if they were not *arbitrary*.

B) SM MINUSES

1. There are "many" (15 or so) arbitrary parameters:

α, G_F, $\sin^2 \theta_W$, Λ_{QCD} ... but also m_q, m_l, m_H, $\sin\theta_C$ ...

2. The Higgs sector is "ugly" with

$m_H^2 = m_H^2(class) + O(\Lambda^2)$, where Λ is the ultraviolet cut-off, so that a huge fine-tuning is needed in order to avoid that $m_H \gg m_W$.

3. Some sectors of the theory may be sick and/or trivial, e.g. the $U(1)$ and Higgs sectors.

What is meant by this is the so-called Landau pole phenomenon by which, in non-asymptotically free theories, the effective coupling becomes infinite at a finite energy scale which is to be interpreted as an ultraviolet cut-off. Alternatively, if the cut-off is sent to infinity, the effective coupling goes to zero at any finite scale (triviality). This is what led Landau to say, in 1955:

"The Lagrangian is dead. It should be buried, with all due honour, of course."

Today we know of some possible remedies to the above problems: they go under the name of supersymmetry, grand unified theories (GUTs), technicolour and all possible combinations thereof. Yet none of this appears to date to solve another fundamental problem with the SM:

THE FRAMEWORK OF QUANTUM FIELD THEORY (ON WHICH THE SM RESTS) DOES NOT WORK FOR THE 4TH FUNDAMENTAL FORCE OF NATURE: GRAVITY.

Why is this so? It has to do again with quantum corrections to L_{class}.

Let us compare the correction due to the emission and reabsorption of a photon by an electron with those of a graviton. Let us call x the distance between the point of emission and that of absorption. An elementary calculation shows that the corresponding probability amplitudes (to be integrated over x) have the approximate form:

$$dP(x) \sim \alpha \frac{dx}{x} \rightarrow P \sim \alpha ln \frac{l}{\epsilon} \tag{3.3}$$

for the photon and

$$dP(x) \sim G_N \hbar \frac{dx}{x^3} \rightarrow P \sim \frac{\lambda_P^2}{\epsilon^2} \tag{3.4}$$

for the graviton, where α is the fine-structure contant, $\lambda_P = \sqrt{G_N \hbar}$ is the Planck length, or about $10^{-33} cm.$, l is the typical scale involved in the process, and we have introduced an ultraviolet (short-distance) cut-off ϵ.

We see that the divergence as $\epsilon \rightarrow 0$ is much more violent for the graviton than for the photon, but this perse would not be a disaster (after all we have already seen the quadratically divergent renormalization of the Higgs mass in the SM). What is lethal though is that "a singularity may hide another singularity". A careful computation reveals indeed that there is also a logarithmically divergent contribution for the graviton case which, unlike the quadratic one, cannot be reabsorbed into a redefinition of the electron mass or of the Newton constant. In formulae we find:

$$L_{eff} = L_{class} + L_{quantum} = (16\pi G_N)^{-1} R + (ln\epsilon) R^2 + \dots \tag{3.5}$$

and it gets worse and worse as we go to higher and higher loops! At each order infinities appear multiplying more and more complicated terms none of which existed in L_{class}. Conventional quantum gravity is a *non − renormalizable* theory, i.e. one in which infinities multiply an *infinite* number of (theoretically unpredictable) terms. Put differently:

NO PREDICTIVITY IS LEFT FOR NON-RENORMALIZABLE THEORIES AND, UNFURTUNATELY, STANDARD QUANTUM GRAVITY IS ONE OF THEM.

We shall now take a little step backward in time and ask:

1. How did we get to the SM?

2. Where might we go from the SM?

We have seen that renormalizability was a crucial issue, being related to predictivity and thus, ultimately, to the number of free parameters, or fundamental constants, of the theory.

We shall thus try to answer the above questions by addressing a third one:

HOW DID (WILL) OUR LIST OF FUNDAMENTAL CONSTANTS EVOLVE?

I shall start this discussion with a quotation from Weinberg [9] which, I think, explains very well what he and I mean, in practice, by a "fundamental constant of Nature". Weinberg says:

"The list of fundamental constants depends on who (and I would add on when one) is compiling the list... A hydrodynamicist would put in the list the density and viscosity of water, an atomic physicist would put the mass of the proton and the charge of the electron". Weinberg goes on to specify what goes in *his list:*

"....a list of constants whose value we cannot calculate with precision in terms of more fundamental constants, not just because the calculation is too complicated, but because we do not know of anything more fundamental. .... The membership of such a list thus reflects our present understanding of fundamental physics...."

Clearly, our understanding of physics has evolved considerably in the last century. And, indeed, the same has happened to the above list. Let me take up things not a century ago but, for brevity, about 20-25 years ago, in the late 60's.

By then, of course, the two big "revolutions" of scientific thought, relativity and quantum mechanics, had been absorbed. Each had left a very conspicuous trace in our list, respectively:

c = the speed of light in vacuum

$\hbar$ = the minimum quantum uncertainty.

Besides these two privileged universal constants, there were others which referred, specifically, to each one of the four fundamental forces. Thus the list continued roughly as follows:

Table I. Fundamental Constants ca 1968

FORCE	CONSTANTS
ELECTROMAGNETIC	$\alpha = e^2/4\pi\hbar c;\quad m_e, m_\mu$
WEAK	$G_F, sin\theta_C;\quad m_\nu(= 0?), g_A/g_V, \cdots$
STRONG	$G_{\pi NN}, G_{\pi\pi\rho};\quad m_{p,n}, m_\pi, m_\rho, m_\Delta \cdots$
GRAVITATIONAL	$G_N, \Lambda_{cosm}(= 0?)$

The following observations about the above list come easily to mind:

a) Electromagnetic interactions are "economical" in terms of the number of constants. They are the best understood (QED is their "theory");

b) same is true for gravity, which, however, remains unquantized;

c) Strong interactions are the most "abundant and redundant" in the number of parameters. The old Dual Model was an attempt to drastically reduce that number;

d) Weak interactions are somewhere in the middle: they have a moderate number of parameters but, unlike the EM interactions, their quantum theory is still affected by problems (infinities).

Let us now look at the situation fifteen years or so later, in 1983, when Weinberg wrote the quoted article. The situation had drastically evolved and is summarized in a new table containing, besides c and $\hbar$:

Table II. Fundamental Constants ca 1983

FORCE	CONSTANTS		
ELECTROWEAK AND STRONG: $SU(3) \times SU(2) \times U(1)$	$\alpha \to \alpha$ $\alpha_W' \to sin^2\theta_W$ $\alpha_s \to \Lambda_{QCD}$	Yukawa coupl. $\to m_i$, KM mat. Higgs mass & self-coupling $\theta_{QCD}(= 0?)$	
GRAVITATIONAL:	G_N	$\Lambda_{cosm}(= 0?)$	

The enormous progress made in those fifteen years is clearly apparent in the comparison of the two tables. Yet, apart from the "minor", "aesthetic" problems mentioned above, the shocking thing about the comparison is that absolutely nothing had happened to the last entry: gravity is treated in 1983 as in 1966, i.e. through classical general relativity (CGR). As already mentioned, this reflects the failure of all attempts to construct a consistent theory of quantum gravity. It is in this context that string theory made an unexpected come-back in 1984. I shall try to explain why.

4. QUANTUM STRING THEORY FOR PEDESTRIANS

Let us recall the form of the action for a system of points and for the string:

$$S_{points} = \sum_i m_i c \int d\ell_i + \text{interactions}$$
$$S_{string} = T \int d\sum$$

(4.1)

The first is proportional to the sum of the length of the paths described by the points, each one weighted by a different constant, its mass. Interactions modify the action and are quite arbitrary. The second is given by a single term proportional to the area swept by the string. A single constant appears in place of the masses (the string tension T) and, most remarkably, interactions will come out naturally and uniquely.

The area is swept in space-time and therefore space and time have to be measured in the same units. A constant c (later to be recognized as the speed of light), is implicitly multiplying time intervals. Having c in our list, we proceed without writing it explicitly each time.

In order for the action to have its usual dimensions, T has to have dimensions energy/length as appropriate for a tension. However, as long as we work at the classical level, T is completely irrelevant (for points, analogously, only mass ratios are important). This is because, classically, only the stationary points of S do matter and these are invariant under a rescaling of the action.

Still classically, free points move along straight lines, while the string has a rich variety of motions all of which, however, satisfy the constraint:

$$M^2 \geq 2\pi T J.$$

(4.2)

The mass-squared of the string is bound from below by a multiple of its angular momentum, the equality sign being reached for a rigid, rotating stick whose end points (for an open string) move at the speed of light.

At the quantum level the difference between points and strings becomes even bigger. Of course we know that angular momentum becomes quantized in units of $\hbar$:

$$J = n\hbar$$

(4.3)

Not surprisingly because of eq.(4.2), $M^2(2\pi T)^{-1}$ also becomes quantized in units of $\hbar$:

$$M^2(2\pi T)^{-1} = m\hbar. \qquad (4.4)$$

After all the string is a collection of harmonic oscillators!

What becomes of the classical inequality (4.2)? Here comes the first surprise . Instead of a naive $m \geq n$ one finds:

$$m \geq n - a_0 \text{ with } a_0 = +1(+2) \text{ for the open (closed) string.} \qquad (4.5)$$

The origin of a_0 is that of a zero-point energy (or normal-ordering constant) analogous to the famous $1/2$ of the harmonic-oscillator levels $(n + 1/2)\hbar\omega$. This normal ordering constant is of paramount importance for string theory in that it allows the existence of classically forbidden, massless spin 1 and spin 2 states (for open and closed strings, respectively). These objects are believed to exist in Nature, in the form of gauge bosons and gravitons respectively, and to mediate all known forces, including gravity. That's a good start indeed for *quantum* string theory!

Quantization is responsible for another important property of strings: strings acquire a typical, better, a *minimal* size. Recall again that position and momentum uncertainty behave rather symmetrically for the harmonic oscillator, both scaling like the square root of the Planck constant. The same is true for the string where we get:

$$\Delta x \simeq (\hbar/T)^{1/2} \quad ; \quad \Delta p \simeq (\hbar T)^{1/2}. \qquad (4.6)$$

So far we have introduced three fundamental constants:

$$c \quad , \quad T \quad , \quad \hbar$$

Classically, there was no $\hbar$, of course, and T was irrelevant. Quantum mechanically it is only a combination of $\hbar$ and T that appears. Since the quantum theory depends on S only through the pure number, $S/\hbar$, the combination that survives is obviously:

$$T/\hbar = \lambda_s^{-2}. \qquad (4.7)$$

which defines a fundamental length, λ_s. It looks funny, at first sight, that we managed to get rid of $\hbar$ and T in favour of a single quantity, λ_s. Actually, in order to do so, we have implicitly changed units of energy. The natural unit of energy in string theory is *length*, with T providing the *conversion* factor to and from c.g.s. (or normal particle theory) units:

$$E \to \overline{E} \equiv E/T \text{ which is a length.} \qquad (4.8)$$

In other words, using the string tension, I can give the energy of a string by giving its length: energy is length and length is energy in string theory! If natural string units are consistently used, only two dimensionful constants are needed for the relativistic and quantum nature of the theory [10].

It looks like no achievent at all to have replaced c and $\hbar$ by c and λ_s through a change of units. But, actually, we have gained a lot:

THE PLANCK CONSTANT λ_s IS ALSO, IN STRING THEORY, THE SHORT DISTANCE CUT-OFF.

This is so because, as we have seen, strings acquire, through quantization, a finite, minimal size $O(\lambda_s)$. The finite size induces in turn a cut-off (in the way of a form factor) on large virtual momenta. Incidentally, this is what makes sense of quantum string gravity. A typical one-loop (quantum gravity) correction normalized to the tree (classical) value has the form given in eq.(3.4). With the string cut-off $(\hbar T)^{1/2}$ on the virtual momentum one gets a finite correction of order λ_P^2/λ_s^2. Quantum gravity corrections are (typically) small if the string length parameter λ_s is somewhat larger than the Planck length λ_P!

5. ORIGIN OF FUNDAMENTAL CONSTANTS IN QUANTUM FIELD AND QUANTUM STRING THEORY

QFT

As mentioned in the introduction the basic object of QFT is the so-called effective Lagrangian (or its space-time integral, the effective action S_{eff}) of eq. (3.2). Let us discuss some of its basic properties:

1. S_{eff} possesses all the symmetries of the quantum theory. In general these are fewer than those of the classical theory because of anomalies);

2. The stationary points of S_{eff} correspond to the possible ground states of the theory (at the quantum level, of course). A most famous example is the so-called Coleman-Weinberg mechanism [11] in which, typically, the classical action has a symmetric ground state (say $\phi = 0$) while the effective action has a symmetry-breaking ground state (say $\phi \neq 0$);

3. Ground state expectation values are obtained by setting to 0 the first derivatives of S_{eff}. Higher derivatives provide other physical quantities such as masses, coupling constants, scattering and decay amplitudes. Thus we can say that it is S_{eff} that is being tested at LEP!

4. S_{eff} depends on its arguments (the renormalized fields) through a finite number of (finite) parameters λ_j:

$$S_{eff} = S_{eff}(\phi_i; \lambda_j), \qquad (5.1)$$

which are nothing but the uncalculable fundamental constants that a renormalizable field theory has to live with. These are QFT's fundamental constants!

5. Last, but not least, there must be a recipe for computing S_{eff} from S_{class}. It can be expressed in a deceptively simple form (which took about 30 years to be defined):

$$\exp(-S_{eff}(\phi_i^c;\lambda_j)/\hbar) \;\Leftrightarrow\; \int D\phi_i(x)\exp(-\tilde{S}(\phi_i,\phi_i^c;\lambda_j)/\hbar). \qquad (5.2)$$

Here ϕ^c is called the "classical" field as opposed to ϕ, the quantum field on which one "integrates". $\tilde{S}$ is a functional that we need not specify. At the end of the calculation one will get a (generally non-local) functional of ϕ^c which also depends on the fundamental constants λ_i. Then ϕ^c is simply renamed ϕ.

QST

1. String physics is also encoded in some S_{eff}. However, the recipe to compute S_{eff} is now

$$\exp(-S_{eff}(\phi_i^c)/\hbar) \;\Leftrightarrow\; \int DX^\mu(\sigma,\tau)\exp(-\tilde{S}(X^\mu;\phi_i^c)/\hbar) \qquad (5.3)$$

indicating some major differences with respect to QFT.

2. There is only *one* set of $\phi's$ and it's clearly the analogue of ϕ^c (it is not integrated over). Instead, the role of the quantum fields is now played by the string coordinate $X^\mu(\sigma,\tau)$ (and, in canonical quantization, by its conjugate momentum $P_\mu(\sigma,\tau)$). In other words we are "back", in QST, to *first* quantization! Recall, incidentally, how things went for relativistic quantum particles. There were problems with the relativistic generalizations of Schroedinger's equation and thus people preferred to take them as *classical* field equations to be quantized again (hence the name of second quantization). Somehow, all this is not needed (and might even be wrong) in string theory!

3. Finally, $\tilde{S}$ contains no arbitrary parameters since it is possible to absorb them in the redefinition of the classical fields. Furthermore, the ultraviolet finiteness of loop calculations guarantees that no "renormalization" is needed in order to compute S_{eff} through eq.(5.3) and therefore that no new arbitrary parameters are generated from the loops. We conclude that, in string theory:

S_{eff} DOES NOT CONTAIN ARBITRARY CONSTANTS.

But then where do the constants of Nature come from in QST? The (obvious?) answer is [12]:

THE FUNDAMENTAL CONSTANTS OF QST ARE VACUUM EXPECTATION VALUES!

We shall illustrate this point with some examples concentrating our attention on two $\phi's$ which are of capital importance in string theory:

The so-called "metric" : $G_{\mu\nu}$

The so-called "dilaton" : ϕ

These two fields are defined by the way they enter $\tilde{S}$ of eq. (5.3) [13]:

$$\tilde{S}/\hbar = -1/2 \int d^2\xi g^{1/2} \{g^{\alpha\beta}\partial_\alpha X^\mu \partial_\beta X^\nu G_{\mu\nu}(X) + {}^{(2)}R(\xi)\phi(X)/4\pi\}. \tag{5.4}$$

Here, as usual, X^μ is the string coordinate, $g_{\alpha\beta}$ is the metric on the world sheet whose coordinates σ and τ are collectively denoted by ξ. ${}^{(2)}R(\xi)$ is the curvature scalar on the world sheet which satisfies the well-known formula:

$$(2 - 2g) = \int d^2\xi g^{1/2} {}^{(2)}R(\xi)/4\pi, \tag{5.5}$$

where g is the genus of the two-dimensional Riemann surface described by the metric $g_{\alpha\beta}$ (thus $g = 0$ for a sphere, $g = 1$ for a torus, $g = 2$ for a sphere with two handles etc). In string theory, the sum over g implicit in the integral over all metrics in eq. (5.3), corresponds to the sum over loops of ordinary quantum field theory.

Note that, in order for eq.(5.4) to define a dimensionless quantity, we have to attribute to $G_{\mu\nu}$ dimensions of $(length)^{-2}$ once we call $(length)$ the dimension of X. By contrast, ϕ is dimensionless.

What is the general structure of S_{eff}? First of all, it can be written as the sum of contributions from different Riemann surfaces (string-loop expansion):

$$S_{eff} = \sum S_{eff}^g = S_{eff}^0 + S_{eff}^1 + S_{eff}^2 + \ldots \tag{5.6}$$

Next, if we look at the lowest order ($g = 0$) term, we find:

$$S_{eff}^{g=0} = \int d^D x \sqrt{-det G_{\mu\nu}} e^{2\phi}[\frac{D - 10}{3} + R(G) + 4\partial_\mu\phi\partial^\mu\phi + F_{\mu\nu}^2 + \ldots] \tag{5.7}$$

where D is the number of components of X, i.e. the number of dimensions of space-time, and, as promised, no fundamental constant (not even c, or λ_s of our Sect.2) appears! The dots indicate here terms with more than two derivatives acting on the fields.

We have now to put to zero the first derivatives of S_{eff} i.e.

$$\delta\Gamma/\delta G_{\mu\nu} = \delta\Gamma/\delta\phi = \ldots = 0 \tag{5.8}$$

If we look for a slowly varying solution of these equations we are entitled to neglect the dots in (5.7). More doubiously, we shall also neglect the higher genus contributions a point we shall come back to. We then find that a solution of this kind exists only in $D = 10$ and has the form:

$$G_{\mu\nu} = \lambda_s^{-2}\eta_{\mu\nu} \quad , \quad \eta_{\mu\nu} = \text{diag}(-c^2, +1, +1, \ldots +1)$$

$$\phi = constant \tag{5.9}$$

We have finally recovered our fundamental constants c and λ_s as VEV's. They came from our desire of finding a flat-metric vacuum and from the unconventional dimensions of $G_{\mu\nu}$. Precisely the same dimensions for G have been advocated for different reasons in [14].

It can be argued that the role of Einstein's metric $g_{\mu\nu}^{Einst}$ is played, in string theory, by $\lambda_s^2 G_{\mu\nu}$ so that

$$\lambda_s^2 (ds^2)_{GR} = G_{\mu\nu} dx^\mu dx^\nu \equiv (ds^2)_{String} \tag{5.10}$$

String distance larger or smaller than 1 defines, in string theory what we mean by large or short distance.

What determines the actual values of c and λ_s? Let us distinguish two cases:

a) If the dimension we are looking at is non-compact (the case we believe to be true for our 3+1 dimensional world) the actual value of these constants is irrelevant. What the theory should determine, in this case, are the ratios of c and λ_s to some physical speed and length that we call cm./sec. and cm. Thus, in principle, the theory allows to compute c, λ_s in centimetres and seconds.

b) If a dimension is instead compact (those in excess of 4 should, physically), say a circle of radius R, the ratio

$$\rho = R/\lambda_s$$

i.e. how many Planck constants is the dimension large is relevant. Actually, a very non-trivial property of (closed) string theories says that ρ and $1/\rho$ define the same theory [15].

The case $\rho = 1$ is of course special: it is the fixed point of the "duality" transformation. One finds that, at this special value, the gauge symmetry generated from compactification (a la Kaluza-Klein) is enhanced. It has been argued [10,16] that this special value (a radius equal to the fundamental Planck constant) will represent the preferred value for the extra dimensions (thus stabilizing the conventional KK picture) and [17] some minimal value of the scale factor of the Universe at the big bang (equivalently a maximal temperature) thus avoiding the initial singularity.

Let us now suppose that, in the early universe, the metric was not really flat (because of high matter density or of an effective cosmological constant, see my previous lecture) but just conformally flat i.e.

$$dx^\mu dx^\nu \lambda_s^2 G_{\mu\nu} = g_{\mu\nu} dx^\mu dx^\nu = -dt^2 + a^2(t) d\bar{x}^2 \tag{5.11}$$

The conventional interpretation of $a(t)$ is that it represents the scale factor of the universe. The distance of any two points expands (for a an increasing function of t) by a factor $a(t)/a(0)$ as we go from time 0 to time t. Quantum String Theory affords an amusing reinterpretation of this expansion. Because of the way the metric enters the action, we may reinterpret a varying scale factor $a(t)$ as a varying λ_s

$$\lambda_s \rightarrow \lambda_s(t) = \lambda_s/a(t) \tag{5.12}$$

For an increasing $a(t)$, the Planck constant of string theory is becoming smaller with time... and the Universe is becoming more and more classical! The Universe expands because our rods, made of atoms, shrink as a consequence of (5.12). The light that reaches us today from a distant star was emitted a long time ago i.e. when the Planck constant was larger by a factor $a(now)/a(t_{emission})$. All other fundamental constants being equal, its wavelength was larger precisely by the same factor: the red shift formula is thus perfectly recovered. Incidentally, a similar description of the red shift was given by Dirac in a different context, that of his Large Number Hypothesis [18].

After having discussed the dimensionful constants we now turn to the dimensionless ones, typically the fine structure constant(s) α_i where i labels the various gauge interactions.

We find again, and not surprisingly, that these are again related to the Vacuum Expectation Values (VEV) of certain fields. The most fundamental dimensionless coupling is related, in string theory, to the dilaton field ϕ appearing in eq (5.4). The reason why ϕ is related to a coupling constant was first pointed out by Witten [19]. It is related to eq. (5.5) which implies that a definite dependence of S^g_{eff} of eq. (5.6) from ϕ :

$$S^g_{eff} \simeq \exp(2\phi(1-g)) \tag{5.13}$$

But this means a factor $(\alpha_{SL})^g$ per loop where:

$$\alpha_{SL} = \exp(-2\phi) \tag{5.14}$$

Thus, indeed, α_{SL} is the loop-expansion parameter, the analog of the fine structure constant in QED. The identification can also be seen by noting that, for $g = 0$, we get from (5.7) a factor $1/\alpha_{SL}$ sitting in front of the tree level action. The latter has the form of a gauge-plus-gravity action where the fields are already rescaled so that the appropriate powers of the gauge and Newton couplings multiply simply the action. The above identification yields the two fundamental relations of (closed or heterotic) string theory:

$$\alpha_{GUT} = \alpha_{SL}\lambda_s^{D-4}$$
$$\lambda_P^{D-2} \equiv G_N\hbar = \alpha_{SL}\lambda_s^{D-2} \tag{5.15}$$

where α_{GUT} is the grand unification coupling at the grand unification scale and D is the number of space-time dimensions. The string loop expansion parameter (the dilaton VEV) thus gives the ratio between the Planck length and the fundamental length of the theory! This means, unfortunately, that the string scale cannot be a lot larger than the Planck scale and that, consequently, the new physics implied by the string will only be felt at energies of the order of the Planck mass.

From (5.15) we immediately get the relation:

$$G_N \hbar = \alpha_{GUT} \lambda_s^2 \tag{5.16}$$

Compactification of string theory from D=26 or D=10 to D=4 affects the relations (5.15) whithout changing (5.16). The physical meaning of the latter equation is striking. It says that:

HEAVY, CHARGED STRINGS HAVE IDENTICAL (UP TO CLEBSHES) GAUGE AND GRAVITATIONAL INTERACTIONS

a property I am tempted to call (GRAND UNIFICATION)[2]. Of course neutral and heavy, or light and charged (e.g. the electron) strings are the exceptions confirming this rule.

This unification of all forces works fine at tree level. What about loops? Obviously α_{GUT} controls gauge loops. For gravitational loops let us go back to the estimate (3.4) and insert eq. (5.16). We find, not without satisfaction, that quantum gravity corrections are also controlled by α_{GUT}. Thus string unification seems to persist beyond tree level. At a closer look this turns out to be only "logarithmically true" i.e. modulo running. Only the short distance couplings are unified while the low energy can be substantially different (even for heavy strings).

Finally I shall mention two important aspects of string unification. Because of the UV finiteness of the theory, the tree level, unified couplings are not bare, unobservable quantities: they are physical and observable, in principle, by a short distance experiment. As a corollary, non-asymptotically free theories make perfect sense in string theory. Their couplings, being finite at tree level, will just decrease towards lower energies as a result of loops (screening). Landau poles and similar deseases of field theories will be pushed to scales where the field theory approximation is certainly invalid.

6. WHAT COULD SIT BEHIND THE MAGIC OF STRINGS?

I shall end up this talk with a couple of (wild) guesses on what could lie behind all these theoretical miracles of string theory.

a. AN ENLARGED UNCERTAINTY PRINCIPLE ?

A naive guess on how the usual uncertainty principle could be enlarged in string theory comes from the idea that, in the presense of a large momentum transfer Δp, the system exchanged, being itself necessarily stringy, has to have a spacial extent $O(\Delta p/T)$. One would thus expect something like [20]:

$$\Delta x = \hbar/\Delta p + \Delta p/T \tag{6.1}$$

Actual studies of Planckian energy superstring collisions [21] have shown that eq.(6.1) holds only if we interpret Δp as the average momentum transfer per loop i.e., instead of (6.1), one finds:

$$\Delta x = \hbar/\Delta p_s + \Delta p_s/T \quad , \quad \Delta p_s = \Delta p/ < 1 + g > \qquad (6.2)$$

where $< g >$ is the average genus (loop order) contributing to the process in the particular kinematical regime. The first term in (6.2) looks like the usual uncertainty principle apart from the fact that the momentum transfer per loop Δp_s replaces the overall momentum transfer. This can be seen to reproduce, in the appropriate kinematical regime, the expected classical dependence of the impact parameter of the collision (identified with Δx) from the energy and scattering angle. The second term is a typical string effect. The $x + 1/x$ structure of (6.2) yields a lower bound on Δx:

$$\Delta x > \lambda_s \qquad (6.3)$$

which is nearly saturated [21] in an appropriate kinematical region. This is the enlargement of the uncertainty principle caused by the string length parameter!

b. AN ENLARGED EQUIVALENCE PRINCIPLE?

In ref. [22] one has tried to give an explicit argument in favour of the conjecture [23] that an enlarged equivalence principle lies behind string theory. The basic idea is simple: we know that string theory contains general relativity and, hence, that it obeys, at least at large scales, Einstein's equivalence principle. Mathematically this statement can be expressed as the invariance of S_{eff} of eq. (5.3) under a change of the space-time metric $G_{\mu\nu}$ which corresponds to a General Coordinate Transformation (GCT).

The question is whether or not S_{eff} is invariant under a larger class of transformations which affect the metric non-trivally but only on very short scales, i.e. on scales much shorter than λ_s. In order to see if this is the case we have considered [22] the scattering of point-like and of string-like particles by massless, gravitational shock waves (which are precisely those relevant for High Energy superstring collisions) before and after adding ripples in the metric that live on scales much shorter than λ_s. We have found that, for point particles, the ripples influence the scattering process, while strings appear to be "blind" to such modifications. We thus have a confirmation that distance degrees of freedom are irrelevant in string theory.

The above conclusion fits very well with other arguments in favour of a minimal length observable in string theory based on "duality" [15-17] , on the study of the free energy at very high temperatures [24] and of discretized versions of string theory [25].

7. CONCLUSIONS

My main conclusions are briefly summarized below:

1. Gravity needs (something like) string theory...in order to cure its problems (classical singularities and quantum infinities).

2. String theory need gravity....as its testing grounds. Is for instance the dilaton a "string killer"? How about inflation and the cosmological constant in string theory? It is in these questions that, presumably, the future of string theory will be decided.

3. Quantum Strings have few (maybe too few!) fundamental constants: all of them are vacuum parameters.

4. Had we been lucky we would have found that eqs. (5.8) already determine a unique solution which looks like the real world. Unfortunately, this is not (and by far) the case at least order by order in the expansion (5.6). In perturbation theory there are too many degenerate, inequivalent vacua corresponding to different numbers of uncompactified dimensions, different gauge symmetries, different numbers of generations etc. Hopefully, non-perturbative phenomena of the kind known from supersymmetric gauge theories will resolve this huge degeneracy.

But, whether or not there will be a unique vacuum and thus a fully determined set of fundamental constants, or whether a residual degeneracy will force us to choose some of them arbitrarily, it will still be true that, in the same way as Sid Coleman used to say:

THE SYMMETRIES OF THE VACUUM ARE THE SYMMETRIES OF THE WORLD,

we are able to claim that, in String Theory:

THE CONSTANTS OF THE VACUUM ARE THE CONSTANTS OF THE WORLD.

REFERENCES

1. H. De Vega and N. Sànchez, *Phys.Lett.* **B197** (1987) 320.

2. N. Turok and P. Bhattacharjee, *Phys.Rev.* **D29** (1984) 1557; N. Sànchez and G. Veneziano, *Nucl.Phys.* **B333** (1990) 253.

3. M. Gasperini, N. Sànchez and G. Veneziano, Highly Unstable Fundamental Strings in Inflationary Cosmologies, CERN Preprint TH. 5893/90 (1990), to appear in Int.J.Mod.Phys. ; M. Gasperini, Kinematic Interpretation of String Instability in a Background Gravitational Field, Torino University Preprint DFTT-38/90 (to appear in Phys.Lett.B); Nguyen Suan Han and G. Veneziano, Mod. Phys. Lett. **6** (1991) 1993.

4. Y. Aharonov, F. Englert and J. Orloff *Phys.Lett.* **B199** (1987) 366; N. Turok, *Phys.Rev.Lett.* **60** (1988) 549..

5. J.D. Barrow, *Nucl.Phys.* **B310** (1988) 743.

6. M. Gasperini, N. Sànchez and G. Veneziano, Self-Sustained Inflation and Dimensional Reduction from Fundamental Strings, CERN preprint TH-6010/91.

7. G. Veneziano, Scale Factor Duality for Classical and Quantum Strings, CERN preprint TH-6077/91.

8. K. A. Meissner and G. Veneziano, Symmetries of Cosmological String Vacua, CERN preprint TH-6138/91.

9. S. Weinberg in "The Constants of Physics", Phil. Trans. R. Soc. Lon. **A310** (1983) 249.

10. G. Veneziano, Europhysics Lett. **2** (1986) 133.

11. S. Coleman and E. Weinberg, Phys. Rev. D7 (1973) 1888.

12. G. Veneziano, in "The Challenging Questions", Erice 1989 (A. Zichichi editor, Plenum Press, N.Y. and London) p. 171.

13. A. M. Polyakov, Phys. Lett. **103B** (1981) 207, 211; see also M. Ademollo et al., Nuovo Cim. **21A** (1974) 77; E.S. Fradkin and A.A. Tseytlin, Phys. Lett. **158B** (1985) 316.

14. V. de Alfaro, S. Fubini and G. Furlan, Nuovo Cim. **A50** (1979) 523; ibid. **B57** (1980) 227.

15. K. Kikkawa and M. Yamasaki, Phys. Lett. **149B** (1984) 357; N. Sakai and I. Senda, Prog. Theor. Phys. **75** (1986) 692.

16. T.R. Taylor and G. Veneziano, Phys. Lett. **212B** (1988) 147.

17. R. Brandemberger and C. Vafa, Nucl. Phys. **B316** (1988) 391.

18. P. A. M. Dirac, Nature **139** (1937) 323.

19. E. Witten, Phys. Lett. **149B** (1984) 351.

20. G. Veneziano, invited talk at the annual meeting of the Italian Phys. Soc. (Naples, Oct. 1987); D.J. Gross, Proc. XXIV Int. Conf. on High Energy Physics, Munich, Aug. 1988 (R. Kotthaus and J.H. Kühn Eds., Springer-Verlag Publ. Co.) p 310.

21. D. Amati, M. Ciafaloni and G. Veneziano, Int. Journ. Mod. Phys. **3A** (1988) 1615; Phys. Lett. **B216** (1989) 41; G. Veneziano, Proc. of Superstring '89 workshop, Texas A and M University, R. Arnowitt et al. eds. (World Scientific, Singapore, 1990) p. 86.

22. M. Fabbrichesi and G. Veneziano, Phys. Lett. **233B** (1989) 135.

23. G. Veneziano, Proc of the 5th Marcel Grossmann meeting,(Perth, Aug.1988), D.G. Blair and M.J. Buckingham eds. (World Scientific, Singapore, 1989) p. 173.

24. J.J. Atick and E. Witten, Nucl. Phys. **B310** (1988) 291; see also: Ya I. Kogan, JETP Lett. **45** (1987) 709; B. Sathiaplan, Phys. Rev. **D35** (1987) 3277.

25. M. Karliner, I. Klebanov and L. Susskind, Int. Journ. Mod. Phys. **A3** (1988)1981; T. Yoneya, "On the Interpretation of Minimal Length in String Theory", Univ. of Tokyo preprint (1989); K. Konishi, G. Paffuti and P. Provero, "Minimum Physical Length and the Generalized Uncertainty Principle in String Theory", Univ. of Pisa preprint, IFUP-TH 46/89.

FIRST-ORDER INFLATION

MICHAEL S. TURNER

NASA/Fermilab Astrophysics Center
Fermi National Accelerator Laboratory
Batavia, IL 60510-0500
and
Departments of Physics and Astronomy & Astrophysics
Enrico Fermi Institute
The University of Chicago
Chicago, IL 60637-1433

I discuss the most recent—and in my opinion—most attractive model of inflation. In first-order inflation the inflationary epoch is associated with a first-order phase transition, with the most likely candidate being GUT symmetry breaking. The transition from the false-vacuum, inflationary phase to the true-vacuum, radiation-dominated phase proceeds through the nucleation and percolation of true-vacuum bubbles. The first successful—and simplest—model of first-order inflation, extended inflation, is discussed in some detail: evolution of the cosmic-scale factor, reheating, density perturbations, and the production of gravitational waves both from quantum fluctuations and bubble collisions. Particular attention is paid to the most critical issue in any model of first-order inflation: the requirements on the nucleation rate to ensure a graceful transition from the inflationary phase to the radiation-dominated phase.

M. Signore and C. Dupraz (et al.), The Infrared and Submillimetre Sky after COBE, 35–74.
© 1992 Kluwer Academic Publishers.

I. Introduction

The virtues of inflation are well known: Inflation greatly lessens the dependence of the present state of the Universe upon its initial state; in particular, it has the potential to account for the large-scale isotropy and homogeneity, and the flatness of the Universe, the origin of the primeval density inhomogeneities necessity to seed structure formation, as well as providing a solution to the monopole problem. While other ideas have been suggested to account for these features of the present Universe, inflation is particularly attractive because it is based upon well defined—albeit speculative—particle physics at energy scales well below the Planck scale (where quantum gravitational effects should not be of concern) [1].

The basic idea underlying inflation is an epoch where the energy density of the Universe is dominated by vacuum energy, so that the cosmic-scale factor $R(t)$ grows exponentially. Provided this epoch lasts long enough so that the cosmic-scale factor grows by 60 or so e-folds, a small, subHubble-sized patch of the Universe can grow to a size that is large enough to encompass all that we see today.

There are several huddles that an inflationary model must clear before it can be called successful: (i) sufficient inflation, 60 or so e-foldings; (ii) graceful transition from the vacuum-energy dominated de Sitter phase back to a radiation-dominated FRW phase; (iii) sufficiently high "reheat temperature" to permit baryogenesis (10^{14} GeV?; 10^{10} GeV?; 1 GeV?) and to ensure that the Universe is radiation dominated before the epoch of primordial nucleosynthesis ($T_{BBN} \sim$ MeV); (iv) density perturbations of appropriate amplitude, $(\delta\rho/\rho)_{HOR}$ of order 10^{-5} are required to initiate structure formation and less than order 10^{-4} to be consistent with the current upper limit to the quadrupole anisotropy; (iv) no overproduction of gravitons, other light particles produced as quantum fluctuations, or monopoles.

Inflationary models can be grouped into two broad categories: (1) slow rollover [2]; and (2) first order, where these descriptives refer to how the transition from the de Sitter to the radiation-dominated epoch is accomplished. In slow-rollover inflation, the Universe is vacuum dominated because some (weakly coupled) scalar field is displaced from the minimum of its potential; owing to the flatness of its potential this scalar field, often called the "inflaton," very slowly rolls to the minimum of its potential, during which time the Universe "inflates" (grows in size exponentially); see Fig. 1. When the inflaton reaches the minimum of its potential it oscillates about the minimum; these oscillations correspond to zero-momentum inflaton particles. The decay of these particles, into which the vacuum energy has been converted, leads to the reheating of the Universe. Slow, but inevitable, rolling to the true vacuum accomplishes the transition from the inflationary epoch to the radiation-dominated epoch.

The production of density perturbations of appropriate amplitude requires that the inflaton be extremely weakly coupled: self coupling λ of order 10^{-15} or so. Assuming that the underlying particle-physics model is to satisfy the criterion of "weak naturalness" (couplings not spoiled by radiative corrections, or so-called set it and forget it naturalness), all the couplings of the inflaton must be very small. Because of this the inflaton must be a gauge singlet, and the reheating temperature, which is determined by its decay width: $T_{RH} \sim \sqrt{\Gamma m_{Pl}}$, is typically small ($\ll 10^{14}$ GeV) compared to the scale of the vacuum energy which drives inflation ($\rho_{VAC} \sim V^{1/4} \sim 10^{14}$ GeV).

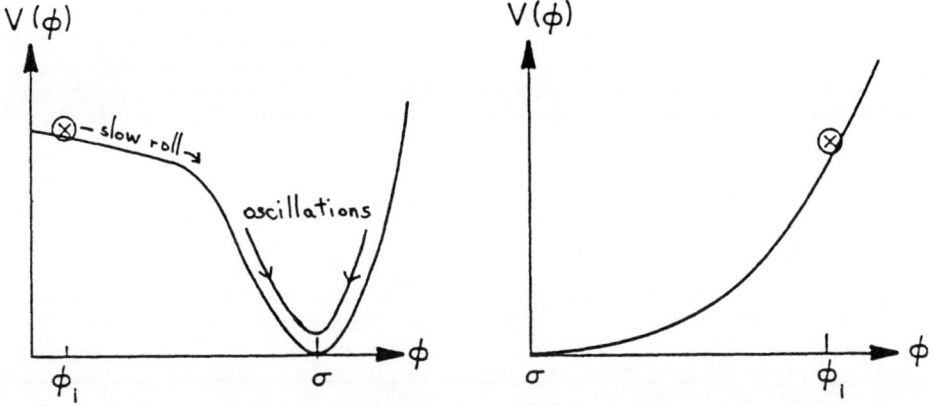

Figure 1

Slow-rollover inflation solves the graceful-exit problem that made Guth's original model of inflation a "no go" [3]. However, the consideration of density perturbations implies that the inflaton field must be weakly coupled [4], essentially decoupled from everything else in theory. While there are other very small dimensionless numbers in physics that must be explained—ratio of the weak scale to the Planck scale, ratio of the top-quark mass to that of the electron, and so—and one could hope that the inflaton's very tiny coupling might be related in some way to these other small numbers (and in at least one model is; see Ref. [5]), the very small self coupling required for the inflaton gives one pause, and even begs one to look for a better solution.

First-order inflation, as its name indicates, is associated with a first-order (symmetry-breaking) phase transition: The scalar (Higgs) field responsible for symmetry breaking (e.g., GUT SSB) becomes trapped in a metastable minimum of its potential (false-vacuum state); see Fig. 2. While the scalar field is trapped in the false vacuum the Universe expands exponentially. The trick in first-order inflation is the transition to the true vacuum and "graceful exit" from the de Sitter phase: Because of the barrier that separates the zero-energy minimum of the potential (true-vacuum state) from the false vacuum, the transition must occur via the nucleation of bubbles of true vacuum [6].

(A vacuum bubbles is a spherical region of space where the Higgs field is in the true vacuum; outside a vacuum bubble space is still in the false vacuum; the bubble wall, which is the transition region, expands at near the speed of light. The nucleation of a vacuum bubble is a quantum event; the nucleation rate (per unit volume per unit time) is typically of the form: $\Gamma = Ce^{-A}$, where C is a constant whose dimensions are (energy)^4, and A is the tunneling action [6].)

First-order inflation is inherently attractive because it is associated with spontaneous symmetry-breaking, a phenomenon for which there is at least circumstantial evidence (the electroweak model). However, the graceful-exit problem is a difficult one to overcome: Guth's original model of inflation ("old inflation") was not viable because the transition from the inflationary epoch back to a radiation-dominated Universe left the Universe terribly "scarred" [3,7]. The reason for this is simple to see.

Once a bubble is nucleated in de Sitter space its comoving size only grows until it becomes Hubble-sized; thereafter its size just grows as the scale factor. Because of this fact the effectiveness of bubble nucleation is naturally measured by the bubble nucleation rate per Hubble time per Hubble volume: $\epsilon \equiv \Gamma/H^4$. If ϵ is much greater than one, then many bubbles are nucleated per Hubble time per Hubble volume, they collide before they cease growing, and rapidly convert all of space into the true vacuum, completing the phase transition. Reheating is very efficient: The vacuum energy is converted into the kinetic energy of the bubble walls, and then into radiation when the walls collide [8]; essentially all of the vacuum energy is converted into radiation and $T_{\rm RH} \sim \rho_{\rm VAC}^{1/4}$.

On the other hand, if ϵ is much less than one, bubble nucleation is rare, bubbles do not collide before they cease growing, and "most" of the Universe remains trapped in the false-vacuum state.

The "formula" for successful first-order inflation is clear: At early times ϵ should be small, so that the Universe remains trapped in the false vacuum and inflates; after sufficient inflation ϵ should become much larger than one so that the Universe completes the phase transition to the true vacuum and becomes radiation dominated; see Fig. 3. *But,*

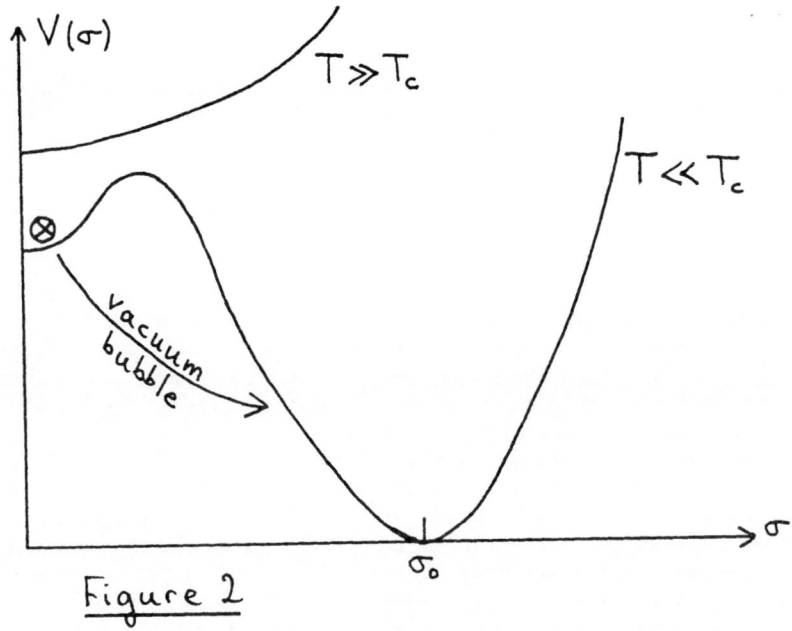

$V(\sigma)$

$T \gg T_c$

$T \ll T_c$

vacuum bubble

σ

σ_0

Figure 2

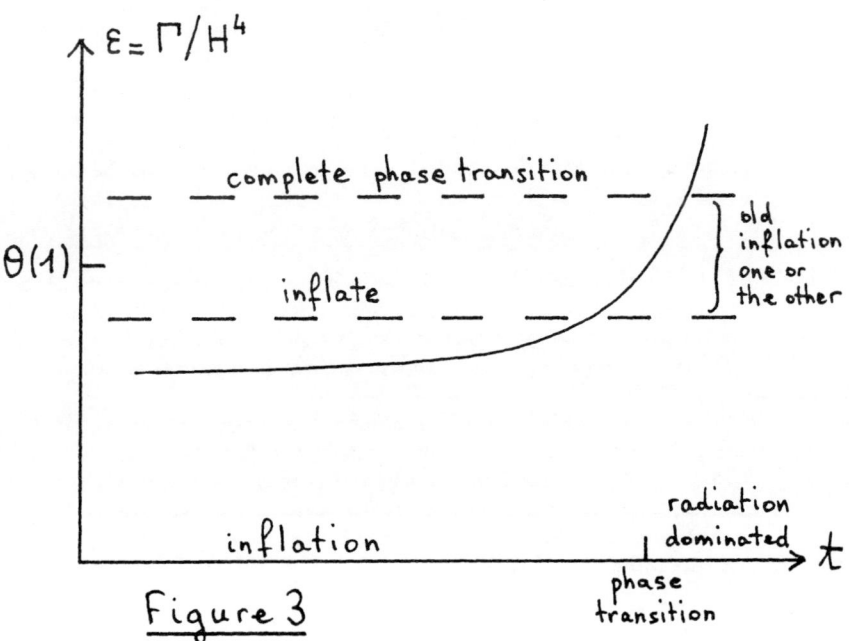

$\varepsilon = \Gamma / H^4$

complete phase transition

$\theta(1)$

old inflation one or the other

inflate

inflation

radiation dominated

t

phase transition

Figure 3

once the Universe becomes trapped in the false vacuum there is no reason that Γ or H should evolve—de Sitter space is time-translation invariant—and thus ϵ remains constant: ϵ can either be less than one or greater than one—but not both. However, a successful inflationary scenario requires that ϵ begin less than one and become greater than one. This was the downfall of old inflation: the graceful-exit problem [3,7].

Recent models of first-order inflation, beginning with extended inflation [9], have solved the graceful-exit problem, by cleverly arranging for Γ and/or H to evolve with time so that ϵ can be less than unity at early times and greater than unity at late times; see Fig. 3. How is this accomplished? In the first successful model of first-order inflation it was H that varied. In this model (extended inflation), gravity is described by Brans-Dicke-Jordan theory; during the false-vacuum epoch the Brans-Dicke (or dilaton) field Φ, which controls the strength of gravity, evolves, so that the expansion is not exponential, but power law; thus H varies as const$/t$ and ϵ increases as t^4. This is the model that we shall analyze in detail. Since then other models have been put forth, wherein Γ evolves because the coupling of the Higgs field to the dilaton or other fields that are evolving causes the action A for bubble nucleation to vary [10].

It should be clear that the key issue in first-order inflation is physics of the completion of the phase transition, which depends upon a competition between the bubble-nucleation rate and the expansion rate. After briefly describing the essentials of extended inflation we will go on to discuss the requirements on the bubble-nucleation rate to ensure a graceful exit and successful inflationary model.

We will not discuss the basics of inflation here—motivation, kinematics, production of density perturbations, and so on—and refer the reader to Refs. [1] for such.

II. Extended Inflation in Brief

(a) The basics

The model is based upon the Brans-Dicke-Jordan theory of gravity, derived from the action

$$S = \int d^4x \sqrt{-g} \left\{ -\frac{\mathcal{R}\Phi}{16\pi} + \frac{\omega}{16\pi} \frac{(\partial_\mu \Phi)^2}{\Phi} + \mathcal{L}_{\text{matter}} \right\}, \tag{2.1}$$

where $\mathcal{L}_{\text{matter}}$ includes the Higgs field σ and all the other matter fields in the theory,

$$\mathcal{L}_{\text{matter}} = \frac{1}{2}(\partial_\mu \sigma)^2 - V(\sigma) + \cdots.$$

Note that the effective gravitational "constant" is set by the value of the Brans-Dicke field Φ (which is a dynamical field and has dimensions of energy^2): $G_{\text{eff}} = \Phi^{-1}$. In the limit that $\omega \rightarrow \infty$, the theory becomes general relativity, because the coefficient in front of the kinetic term for Φ becomes infinite, so that the Brans-Dicke field freezes out (i.e., becomes nondynamical).

We denote the energy scale of the scalar potential by m_U: $V(\sigma = 0) \equiv m_U^4$. At high temperatures, $T \gg T_C \sim m_U$, the minimum of the finite-temperature, effective potential $V_T(\sigma)$ is $\sigma = 0$, corresponding to the state when the symmetry is restored. At low temperatures, $T \ll T_C$, the minimum of the scalar potential obtains for a value $\sigma = \sigma_0 \neq 0$, signaling spontaneous symmetry breaking (SSB); see Fig. 2. For GUT SSB one might expect m_U to be $\mathcal{O}(10^{15}\text{ GeV})$.

We are interested in the case where the SSB phase transition is first-order, so that at temperatures below $T_C \sim m_U$, a barrier separates $\sigma = 0$ from $\sigma = \sigma_0$. In this case the Higgs field becomes trapped in the metastable, symmetric state (false vacuum) and must tunnel to the true vacuum (symmetry-broken state). Once the temperature of the Universe drops below T_C, the energy density of the Universe becomes dominated by the energy density associated with the false-vacuum state, which drives a rapid expansion, causing the Universe to supercool in the metastable, false-vacuum state since $T \propto R^{-1}$. This is where our analysis begins: We assume that the temperature of the Universe is much less than T_C, so the radiation energy density is negligible and the zero-temperature scalar potential is appropriate to use. We will denote the bubble-nucleation rate by Γ, which in the "Jordan" frame (the frame where the action is given by Eq. (2.1)) is constant.

During inflation the Higgs field σ remains trapped in the false-vacuum, and we will not consider its evolution. (Inflation is ended when the Higgs field finally tunnels to the true vacuum, the details of which we will discuss in the next Sections.) The equations of motion for the cosmic-scale factor $R(t)$ and the Brans-Dicke field Φ are given by:

$$\ddot{\Phi} + 3H\dot{\Phi} - \frac{1}{R^2}\nabla^2\Phi = \frac{8\pi}{2\omega+3}(\rho - 3p) \simeq \frac{32\pi}{2\omega+3}m_U^4; \tag{2.2a}$$

$$H^2 \equiv \frac{\dot{R}^2}{R^2} = \frac{8\pi\rho}{3\Phi} + \frac{\omega}{6}\frac{\dot{\Phi}^2}{\Phi^2} - H\frac{\dot{\Phi}}{\Phi}; \tag{2.2b}$$

where during inflation the energy density of the Universe $\rho \simeq m_U^4$, the pressure $p \simeq -\rho$, and R and Φ evolve as

$$R(t) \propto t^{\omega+1/2}; \qquad \Phi \propto t^2. \tag{2.3}$$

For simplicity we have assumed that Φ and σ are spatially constant in the region of interest so that the ∇^2 terms can be ignored; in any case once the rapid expansion begins, these terms decrease quickly. After inflation the Brans-Dicke field evolves very little; thus its value at the end of inflation ($t = t_*$) must be equal to the Planck-mass squared, $\Phi(t_*) = m_{\rm Pl}^2$.

Because of the evolution of the Brans-Dicke field, gravity weakens during inflation and the "effective energy density" that drives inflation, ρ/Φ, decreases; this results in the growth of the scale factor being power-law, rather than exponential (approaching exponential in the $\omega \to \infty$, general relativity, limit).* Thus, the expansion rate H is not constant; rather it decreases with time: $H \propto (\omega + 1/2)/t$. The all-important quantity $\epsilon = \Gamma/H^4$ varies as t^4, which is just what the doctor ordered: ϵ can be small at early times, keeping the Universe trapped in the false vacuum, and large at late times, ensuring that the phase transition is completed gracefully. Clearly, the requirement of a graceful exit must impose restrictions on ω, as in the limit $\omega \to \infty$ extended inflation becomes "old inflation." As we shall discuss later, ensuring that the Universe recovers from extended inflation without "noticeable scars" requires that $\omega \lesssim \mathcal{O}(20)$.

(b) Density perturbations†

---

* Solving the horizon and flatness problems only requires that the cosmic-scale factor grow faster than t, which is achieved for $\omega \geq 1/2$.

† The discussion of density perturbations follows that of Ref. [11].

The calculation of density perturbations is most easily and clearly treated by making a scale (conformal or Weyl) transformation, so that the re-scaled theory closely resembles general relativity. In the re-scaled, or Einstein, frame, extended inflation takes on the appearance of slow-rollover inflation with an exponential potential, and density perturbations are simple to compute. But first, we need some simple kinematical relations.

In computing density perturbations we need to know when a given comoving scale λ "went outside the horizon;" more precisely, when its physical size $\lambda_{\rm phys} = R\lambda$ became larger than the Hubble radius H^{-1} at that epoch; see Fig. 4. Normalizing the scale factor so that its current value is unity, it is a simple exercise to show that the comoving scale λ crossed outside the Hubble radius at a time

$$t/t_* = 10^{25/(\omega-1/2)} \left[\frac{m_U}{m_{\rm Pl}}\right]^{1/(\omega-1/2)} \lambda_{\rm Mpc}^{1/(\omega-1/2)}; \qquad (2.4)$$

when the value of the Brans-Dicke field was

$$\frac{m_{\rm Pl}^2}{\Phi} = \left[10^{25}\frac{m_U}{m_{\rm Pl}}\lambda_{\rm Mpc}\right]^{2/(\omega-1/2)} \qquad (2.5)$$

Here, $\lambda_{\rm Mpc} = \lambda/{\rm Mpc}$ and the time t_* corresponds to the end of inflation.

We are now ready to re-scale the theory; we will denote quantities in the re-scaled frame with an overbar. The conformal transformation is given by

$$\bar{g}_{\mu\nu} = \Omega^{-2}(t)g_{\mu\nu}, \qquad \Omega^2(t) = m_{\rm Pl}^2/\Phi, \qquad \Psi = \Psi_0 \ln(\Phi/m_{\rm Pl}^2),$$

where $\Psi_0^2 = (2\omega+3)m_{\rm Pl}^2/16\pi$. Note that at late times the conformal factor $\Omega^2(t) \rightarrow 1$ as $\Phi = m_{\rm Pl}^2$. In the re-scaled (Einstein) frame the theory is defined by the action

$$\bar{S} = \int d^4\bar{x}\sqrt{-\bar{g}}\left\{-\frac{\bar{\mathcal{R}}}{16\pi G} + \frac{(\partial_\mu\Psi)^2}{2} - e^{-2\Psi/\Psi_0}V(\sigma) + e^{-\Psi/\Psi_0}(\partial_\mu\sigma)^2/2 + \cdots\right\}, \qquad (2.6)$$

where $G = m_{\rm Pl}^{-2}$ is Newton's constant.

So long as the Higgs field is anchored in the false vacuum, its kinetic term is irrelevant and $V(\sigma) = m_U^4$. The only dynamical field is the re-scaled Brans-Dicke field Ψ, which in the Einstein frame has the canonical kinetic-energy term and an exponential potential. Its equation of motion is just:

$$\ddot{\Psi} + 3\bar{H}\dot{\Psi} + dV(\Psi)/dt = 0, \qquad (2.7)$$

where $V(\Psi) = m_U^4 \exp(-2\Psi/\Psi_0)$ and overdot here denotes $d/d\bar{t}$. The equation of motion for the scale factor $\bar{R}$ is just the usual Friedmann equation,

$$\bar{H}^2 = \left(\frac{\dot{\bar{R}}}{\bar{R}}\right)^2 = \frac{8\pi}{3m_{\rm Pl}^2}V(\Psi). \qquad (2.8)$$

These equations are just those for an inflaton with an exponential potential. *Extended inflation, when viewed in the Einstein frame, resembles slow-rollover inflation, with the re-scaled Brans-Dicke field assuming the role of the inflaton!*

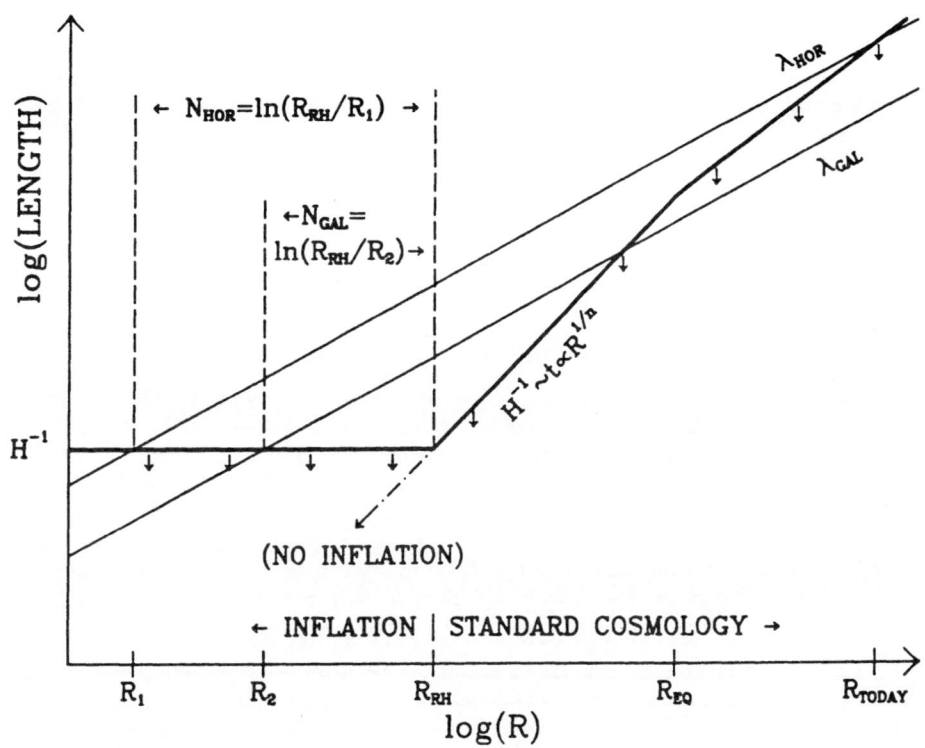

Fig. 4: The evolution of the physical size of the comoving scale, λ, and of the Hubble radius, H^{-1}, in the standard and the inflationary cosmologies. In the standard cosmology (i.e., no inflation) a given scale crosses the horizon but once; while in the inflationary cosmology all scales begin sub-horizon sized, cross outside the horizon ("good bye") during inflation, and re-enter again ("hello again") during the post-inflationary epoch. Note that the largest scales cross outside the horizon first and re-enter last. The growth in the scale factor $[N = \ln(R_{RH}/R)]$ between the time a scale crosses outside the horizon during inflation and the end of inflation is also indicated. For a galaxy, $N_{GAL} = \ln(R_{RH}/R_1) \sim 45$, and for the present horizon scale, $N_{HOR} = \ln(R_{RH}/R_2) \sim 53$. Causal microphysics operates only on scales less than H^{-1} (indicated by arrows). During inflation $H^{-1} \equiv H_I^{-1} = const$, and in the post-inflation era, $H^{-1} \sim t \propto R^{1/n}$ ($n = 1/2$—radiation dominated, $n = 2/3$—matter dominated).

We know how to analyze the production of density perturbations in slow-rollover inflation, and so we can apply that knowledge here. In the slow-rollover approximation, $\dot{\Psi} \simeq -(dV\Psi/d\Psi)/3\bar{H}$,* and the amplitude of density perturbations when they cross back inside the horizon during the post-inflation FRW phase is just:

$$\left(\frac{\delta\rho}{\rho}\right)_{\mathrm{HOR}} \simeq \frac{\bar{H}^2}{d\Psi/d\tilde{t}} \simeq -\frac{3\bar{H}^3}{dV/d\Psi}, \tag{2.9}$$

where as usual the quantities on the right-hand side of the equation are to be evaluated when the scale in question crosses outside the horizon during inflation.

Bringing it all together, we have

$$\left(\frac{\delta\rho}{\rho}\right)_{\mathrm{HOR}} \simeq 4\pi\sqrt{(\omega/3+1/2)}\cdot 10^{50/(\omega-1/2)}(m_U/m_{\mathrm{Pl}})^{(2\omega+1)/(\omega-1/2)}\lambda_{\mathrm{Mpc}}^{2/(\omega-1/2)}. \tag{2.10}$$

Note that, (i) the amplitude is controlled by the ratio of two energy scales, m_U and m_{Pl}, squared; (ii) the spectrum is not quite scale invariant: $(\delta\rho/\rho)_{\mathrm{HOR}} \propto \lambda^{2/(\omega-1/2)}$; (iii) the couplings of the Higgs field σ never came into play; and (iv) for an interesting value of $m_U \sim 10^{15}$ GeV and $\omega \sim 10$, $(\delta\rho/\rho)$ is in the right ballpark, around 10^{-4}. This is in sharp contrast to slow-rollover inflation where the amplitude of the density perturbations typically depends upon the self coupling of the inflaton field λ: $(\delta\rho/\rho)_{\mathrm{HOR}} \sim 10^3\lambda^{1/2}$, which requires a tiny value of λ to achieve density fluctuations of an appropriate size.

(c) Graviton production from quantum fluctuations

The graviton field is excited during inflation by de Sitter space quantum fluctuations. The dimensionless amplitude of a gravitational-wave perturbation is given by $\bar{h}_\lambda \sim \bar{H}/m_{\mathrm{Pl}}$ as the scale λ goes outside the horizon during inflation. While the mode is outside the horizon $\bar{h}_\lambda$ remains constant. Since the Einstein and Jordan frames coincide after inflation the amplitude of the gravity-wave perturbation when it crosses back inside the horizon in the post-inflation FRW epoch is

$$h_\lambda \simeq \frac{\bar{H}}{m_{\mathrm{Pl}}} \simeq \sqrt{8\pi/3}\cdot 10^{50/(\omega-1/2)}(m_U/m_{\mathrm{Pl}})^{(2\omega+1)/(\omega-1/2)}\lambda_{\mathrm{Mpc}}^{2/(\omega-1/2)} \tag{2.11}$$

Like density perturbations, the amplitude of gravity-wave perturbations is controlled by m_U/m_{Pl} and the spectrum is only scale-invariant in the $\omega \to \infty$ limit. From the dimensionless amplitude it is straightforward to obtain the present energy density in relic gravitational waves; on the present horizon scale it is

$$\Omega_{\lambda\sim 3000\,\mathrm{Mpc}} = \frac{\lambda d\rho_{\mathrm{GW}}/d\lambda}{\rho_{\mathrm{crit}}} \simeq 10^{114/(\omega-1/2)}(m_U/m_{\mathrm{Pl}})^{2(2\omega+1)/(\omega-1/2)} \tag{2.12}$$

For $\omega \sim 10$ and $m_U \sim 10^{15}$ GeV, $\Omega \sim 10^{-10}$. On scales smaller than the current Hubble scale, Ω_λ scales as

$$\Omega_\lambda \propto \lambda^{2+4/(\omega-1/2)}; \qquad 13\,\mathrm{Mpc} \lesssim \lambda \lesssim 3000\,\mathrm{Mpc} \tag{2.13a}$$

---

* It is simple to show that for $\omega \gg 1$, the slow-roll approximation is valid.

$$\Omega_\lambda \propto \lambda^{4/(\omega-1/2)}; \qquad \lambda \lesssim 13\,\mathrm{Mpc} \tag{2.13b}$$

(d) Gravity waves from "vacuum popping"

In first-order inflation there is another, even more important source of gravitational waves [12,13]: Those that are produced by the collisions of bubbles of true vacuum when the phase transition is being completed. That this is an important source is easy to understand: At the end of inflation the entire energy density of the Universe resides in the kinetic energy of colliding bubble walls.

It is straightforward to estimate the fraction of the total energy density of the Universe that goes into gravitational waves. Consider gravitational wave production from the collision of two vacuum bubbles whose size at the end of inflation is d. The power emitted in gravitational waves is given roughly by the quadrupole formula:

$$P_{\mathrm{GW}} \sim G\left[d^3 Q/dt^3\right]^2; \qquad \Rightarrow E_{\mathrm{GW}} \sim P_{\mathrm{GW}}\Delta t \sim G\Delta t\left[d^3 Q/dt^3\right]^2, \tag{2.14}$$

where $Q \sim d^5 \rho_{\mathrm{VAC}}$ is the quadrupole moment of the energy distribution of the colliding bubbles. If we suppose that the size of the bubbles when they collide sets the length/time scale for this problem (which turns out to be a very good approximation [13]), it then follows that

$$E_{\mathrm{GW}} \sim G\rho_{\mathrm{VAC}}^2 d^5. \tag{2.15}$$

The vacuum energy liberated by the collision of these two colliding bubbles is $E_{\mathrm{VAC}} \sim \rho_{\mathrm{VAC}} d^3$, thus the fraction of energy liberated that goes into gravitational waves is

$$\frac{E_{\mathrm{GW}}}{E_{\mathrm{VAC}}} \sim G\rho_{\mathrm{VAC}} d^2 \sim \left(\frac{d}{H^{-1}}\right)^2, \tag{2.16}$$

where the final expression follows since $H^2 \sim G\rho_{\mathrm{VAC}}$. This expression is generally valid in first-order inflation (or in any strongly first-order phase transition). The import of this result is clear: The fraction of vacuum energy that is radiated in gravitational waves is proportional to the square of the typical bubble size relative to the Hubble length. As we shall see when we discuss bubble nucleation in more detail, d/H^{-1} is a number that is expected to be in the range of few$\times 10^{-2}$ to 1.

Next consider the ratio of the energy density in gravitational waves to that in photons; just after inflation

$$\frac{\rho_{\mathrm{GW}}}{\rho_\gamma} \sim g_{*\mathrm{RH}}\left(\frac{d}{H^{-1}}\right)^2, \tag{2.17}$$

where $g_{*\mathrm{RH}}$ ($\sim 100 - 1000?$) is the number of ultrarelativistic degrees of freedom at temperature T_{RH}, and therefore $\rho_\gamma \sim \rho_{\mathrm{total}}/g_{*\mathrm{RH}}$. Since photons and gravitons remain relativistic until the present the ratio of their energy densities would remain constant except for the fact that various species become nonrelativistic and transfer their entropy to the photons. Taking this into account, it follows that today the ratio of energy densities in gravitons to that in photons is

$$\left(\frac{\rho_{\mathrm{GW}}}{\rho_\gamma}\right)_{\mathrm{today}} \sim g_{*\mathrm{RH}}^{-1/3}\left(\frac{d}{H^{-1}}\right)^2 \sim 0.1\left(\frac{d}{H^{-1}}\right)^2. \tag{2.18}$$

At the end of the phase transition the characteristic wavelength of these gravitons was $\lambda \sim d$; since then they have been red shifted by a factor of $T_{RH}/3$ K, and their wavelengths today are

$$\lambda_{GW} \sim 10^4 \text{ cm} \left[\frac{(d/H^{-1})}{(T_{RH}/10^{14} \text{ GeV})} \right]. \tag{2.19}$$

Since the fraction of critical density today contributed by photons is $\mathcal{O}(10^{-4})$, that contributed by gravitons from "vacuum popping" could be as large as 10^{-5} or so. Moreover, if detected their characteristic wavelength would provide information about the energy scale associated with inflation.

The energy spectrum of gravitons produced by "vacuum popping" and as quantum fluctuations is shown in Fig. 5. The spectrum of quantum-fluctuation-produced gravitons extends over many orders of magnitude in wavelength; however, the overall scale of the spectrum is tightly constrained by the gravitational-wave mode that is just re-entering the horizon today, which leads to a quadrupole anisotropy of the CMBR: $3 \times 10^{-5} \gtrsim \delta T/T \simeq h(\lambda \sim 3000 \text{ Mpc})$. This in turn constrains the long plateau of the spectrum: $\Omega_{GW} \lesssim 10^{-13}$ or so. In contrast the gravitational waves produced from vacuum popping in first-order inflation have a characteristic wavelength and Ω_{GW} can be as large as 10^{-5} or so.

III. Bubble Nucleation in First-order Inflation†

The analysis of the most general model of first-order inflation involves, as one might appreciate, numerous subtleties. However, much can be learned by considering a subset of models that gracefully exit with ease, so to speak. While this subset of models certainly does not encompass all "successful" models of first-order inflation, we believe the subset is quite general and of great pedagogical use. These simple models are characterized by two important features: (i) completion of the phase transition in less than a Hubble time or so; and (ii) simple functional dependence for Γ, either exponential or linear in time (see Fig. 6). The first assumption allows us to ignore most of the effects of the expansion, and the first together with the second allows one to express all quantities of interest in simple analytical forms.

The exponential form for Γ is motivated by models where the tunneling action is time dependent, e.g., because of the evolution of another field that affects the tunneling action or the coupling of the tunneling field to the dilaton field which itself is evolving. The second form for Γ is motivated by the idea of modeling Γ around the time the phase transition by a linear fit. (Likewise, the exponential form can in many circumstances be "fit" to the desired functional form for the nucleation rate over the time interval of interest.)

After we consider both of these simple models for $\Gamma(t)$ we shall turn to the issue of a class of unwanted relics (first discussed in [15]): big bubbles, and we shall discuss the constraints on Γ/H^4 during the course of first-order inflation that follow from requiring that "big bubbles" do not scar the post-inflationary Universe. After that we will analyze more complicated models, beginning with extended inflation.

(a) Preliminaries

A crucial quantity in discussing the completion of a phase transition is the probability $p(t)$ that a given point in space remains in the false vacuum at time t. It is straightforward

---

† Sections III through VI closely follow Ref. [14]

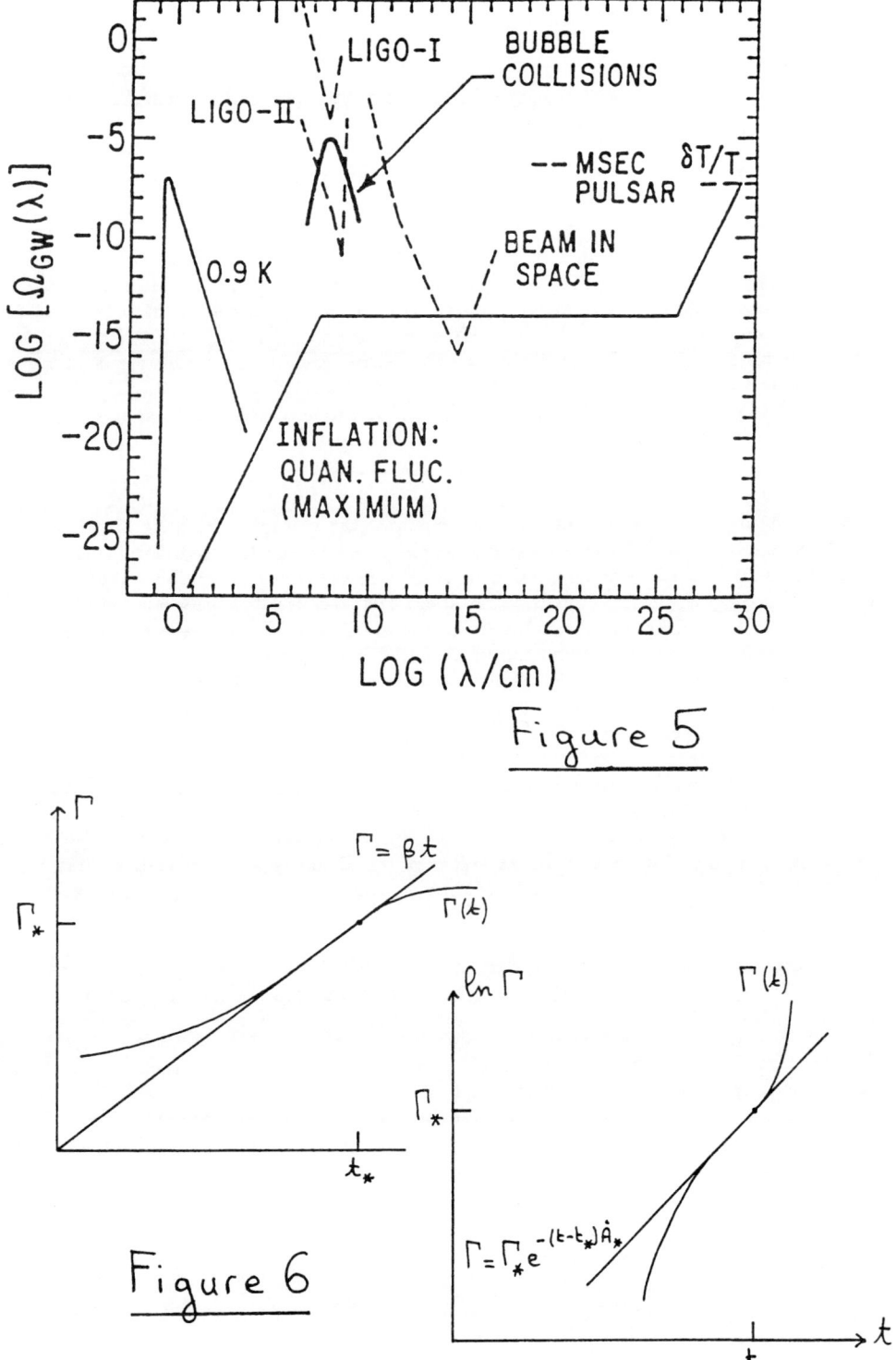

Figure 5

Figure 6

48

to show that

$$p(t) = \exp[-I(t)]; \qquad I(t) = \int_{t_0}^{t} \Gamma(t')V(t,t')dt'; \qquad (3.1)$$

where $V(t,t')$ is the volume occupied by a bubble that was nucleated at time t' at time t and t_0 is the time when the phase transition begins [16].* If we further assume that at nucleation bubbles are of negligible size (in any case, at early times the size is indeterminate since bubble nucleation is a quantum event) and ignore the expansion of the Universe, then

$$V(t,t') \simeq \frac{4\pi}{3}(t-t')^3.$$

Also of interest is the distribution of bubble sizes. If we again ignore bubble collisions and the expansion of the Universe, and assume that at nucleation bubbles are of negligible size, then at time t the number of bubbles dN of radius r to $r + dr$ is just equal to the number of bubbles nucleated between time $t - r$ and $t - r - dr$. That number is just proportional to the nucleation rate at time $t - r$ times the fraction of space still in the false-vacuum state at that time; thus the distribution of bubble sizes at time t is given by

$$\frac{dN(t)}{dr} \propto \Gamma(t-r)\exp[-I(t-r)]. \qquad (3.2)$$

How is one to judge the success, or nonsuccess, of the transition from de Sitter to radiation-dominated FRW? There are several crucial issues to be addressed. First, can a FRW frame be established?—which is essential since we have very good evidence that the Universe is described by an FRW model from about 10^{-2} sec after the bang onward. Because of the $O(3,1)$ invariance of the de Sitter space-time associated with the false-vacuum state there are an infinity of equivalent candidate FRW frames, related by Lorentz boosts. How is the FRW frame to be chosen? This was one of the the fundamental difficulties of old inflation: The distribution of bubbles in old inflation was $O(3,1)$ invariant,* and thus the only hope of defining an FRW frame was to use the interior of a single bubble. Unfortunately, the entropy contained within a single bubble (about 10^{12}) was far too small to account for the large entropy within our current Hubble volume, about 10^{88} [17]. If the bubble nucleation rate varies with time the situation changes dramatically: The surfaces of constant Γ define a unique frame, which is the natural candidate for the post-inflation FRW frame. Likewise, if the expansion rate $H = H(t)$. In many recent models of first-order inflation Γ and/or H are a function of time because one or both depend upon the evolution of a scalar field other than the Higgs field that becomes trapped in the false vacuum.

Next, there is the issue of the completion of the phase transition. Certainly one wants the fraction of comoving (coordinate) space remaining in the false vacuum to diminish to

---

* The value of t_0 will for most of our purposes be irrelevant; we will only use the fact that by the end of the phase transition the Universe has been in the false vacuum state for many e-foldings of the cosmic-scale factor: i.e., $(t - t_0) \gg H^{-1}$.

* In old inflation the distribution of bubble size was given by $dN/dr \propto r^{-4}$, or $r^3 dN/d\ln(r) \propto$ const, where the physical significance of the second form is manifest: constant volume distribution of bubbles per logarithmic interval [17].

zero. However, since those regions of space that remain in the false vacuum continue to expand exponentially, while the regions in the true vacuum only expand as $\sqrt{t}$, the volume of physical space in the false vacuum, which is proportional to $p(t)e^{3Ht}$, can continue to grow even as $p(t) \to 0$. This was the "Swiss-cheese" problem associated with old inflation [17]. It is clear that the phase transition cannot be viewed as having been completed until the volume of physical space in the false vacuum decreases to zero—if not everywhere, at least in a region that can encompass the currently observable Universe.

Finally, there is the issue of the distribution of bubble sizes. If the distribution extends over a very wide range of scales, the level of inhomogeneity will be inconsistent with the isotropy observed today; in addition, it could well make it impossible to define the FRW frame. Clearly the distribution of bubble sizes depends upon the duration of the phase transition: If the phase transition occurs quickly—say in a time of order the Hubble time—then the bubble sizes will have a very limited range of sizes. On the other hand, if the phase transition takes a long time—many, many Hubble times—then because of the conformal stretching of bubbles nucleated early by the expansion of the Universe the range of bubble sizes will be very great. The precise dividing line between an acceptable and an unacceptable distribution of bubble sizes remains to be seen—and will be discussed later. Note too, that the distribution of bubbles is not unrelated to the previous issue, the duration of the phase transition.

Doubtless, there are other issues to be considered. For the moment, these are the ones that will occupy us. The simple models that we will discuss shortly will answer these concerns because the phase transition occurs rapidly: time of order a Hubble time or less. In order to illustrate quantitatively how our simple models avoid the pitfalls mentioned above it is useful to define some "milestones" during the phase transition. They are:

- The time t_* when essentially all of space is in the true-vacuum state. Specifically we define t_* by the condition: $p(t_*) = e^{-M}$, where $I(t_*) = M$ is a large number, say 10 to 100. The fraction of space remaining in the false vacuum at time t_* is exponentially small.

- The duration of the phase transition δt. For these simple models we define δt to be the time it takes for $p(t)$ to go from being nearly unity (all of space in the false vacuum) to being nearly zero (all of space in the true vacuum). To be specific, $\delta t \equiv t_* - t_m$, where t_m is defined by $p(t_m) = e^{-m}$ and $I(t_m) = m$ is a small number, say 0.1 to 1.

- The time t_e when the volume of physical space in the false-vacuum state begins to decrease. Since regions of space that remain in the false vacuum continue to expand exponentially even as $p(t)$ becomes very small, the volume of "physical space" still in the false vacuum, $\propto e^{3Ht}p(t)$, can continue to grow. (In fact, this was a major defect of "old" inflation [17].) If the phase transition is to successfully complete, then the volume of physical space remaining in the false vacuum must ultimately decrease to zero (at least within our Hubble volume, if not everywhere). We define t_e to be the time when $V_{\text{phys}} \propto e^{3Ht}p(t)$ begins to decrease; that is,

$$V_{\text{phys}}^{-1} \frac{dV_{\text{phys}}}{dt} = (3H - dI/dt) \leq 0 \qquad \text{(for } t \geq t_e\text{)}. \tag{3.3}$$

The following relations are satisfied for our "fast" and "foolproof" simple models: $\delta t \lesssim H^{-1}$ and $(t_* - t_e) \lesssim H^{-1}$. These two constraints ensure a rapid phase transition, where the range of bubble sizes is limited. While they are certainly *sufficient* conditions for a successful phase transition, they are not *necessary* conditions: More complicated scenarios may indeed successfully implement first-order inflation without satisfying these requirements for speed. For example, $p(t) \rightarrow 0$ could occur long before V_{phys} begins to decrease; thus large regions of physical space would still be in the false vacuum as $p(t) \rightarrow 0$. A rapid increase in $I(t)$, concurrent with the onset of the decrease of V_{phys}, could allow a large region, still in the false vacuum, to undergo a graceful exit, creating a smooth patch much larger in size that our current Hubble volume, in spite of the fact that much (most?) of the Universe would remain unsuitable for "our purposes." We will leave the discussion of scenarios of this ilk for later, after our intuition is better developed. Now, on to the analysis of our two models that gracefully exit with ease.

IV. Two Simple Models

(a) Model 1: $\Gamma = C \exp[-A(t)]$

As mentioned above, this form for the bubble-nucleation rate Γ is motivated by a time-dependent action $A(t)$ for bubble nucleation. Of course the constant C—whose dimensions are (energy)^4—may also be time dependent, but we can ignore that time dependence as all "the real action" is in the action.*

First, it is useful to expand $A(t)$ about $t_\bullet$, the time when the phase transition is complete:

$$A(t) = A_\bullet + (t - t_\bullet)\dot{A}_\bullet + \cdots ;$$

where $A_\bullet = A(t_\bullet)$, $\dot{A}_\bullet = \dot{A}(t_\bullet)$, and we will only keep the first two terms. We note that $A_\bullet$ should be greater than order unity (probably much greater), since the usual treatment of bubble nucleation breaks down for $A \lesssim \mathcal{O}(1)$. (If the tunneling action is of order unity or smaller, the phase transition proceeds via spinodal decomposition—formation of very irregularly shaped fluctuation regions—rather than via the nucleation of bubbles of true vacuum.) Further, $\dot{A}_\bullet$ is negative—the action must be decreasing—and it is reasonable to expect that $|\dot{A}_\bullet| \sim A_\bullet/t_\bullet$—as would follow if $A_\bullet \propto t^q$, $q < 0$.

It is now straightforward to evaluate $I(t)$ for times not too different from $t_\bullet$:

$$I(t) = \frac{4\pi C}{3} e^{-A(t)} \int_0^t u^3 \exp(\dot{A}_\bullet u) du \simeq \frac{8\pi C}{\dot{A}_\bullet^4} e^{-A(t)} \left(= \frac{8\pi}{\dot{A}_\bullet^4} \Gamma(t) \right). \tag{4.1}$$

In evaluating the integral in Eq. (4.1) we have made use of the fact that $|\dot{A}_\bullet|(t - t_0) \gtrsim A_\bullet \gg 1$, so that the integral is just $3!/\dot{A}_\bullet^4$.

Recall that $t_\bullet$, the time when essentially all of space is in the true vacuum, is defined by $I(t_\bullet) = M$ ($M = 10 - 100$), and that the duration of the phase transition $\delta t = (t_\bullet - t_m)$, where $I(t_m) = m$ ($m = 0.1 - 1$). And further, that t_e is the time when the volume of physical space in the false-vacuum state begins to decrease: $dV_{\text{phys}}/dt \propto (3H - dI/dt) \leq 0$ for $t \geq t_e$. Using Eq. (4.1) it is simple to compute δt and t_e:

$$\delta t = \frac{\ln(M/m)}{|\dot{A}_\bullet|} \simeq \frac{2}{|\dot{A}_\bullet|}; \qquad |\dot{A}_\bullet| \equiv H/\alpha \ \Rightarrow \delta t = 2\alpha H^{-1}; \qquad t_e = t_\bullet - \ln[M/3\alpha]\frac{\delta t}{2}.$$

---

* Much of our analysis here follows that of Ref. [18], though we will try to be more quantitative. Our more quantitative analysis also bears on the main issue of Ref. [18]: the distribution of bubble sizes in a noninflationary, first-order phase transition.

(The actual value of t_* is irrelevant, although one expects it to be of order H^{-1}.) We see that the duration of the phase transition δt ($= 2\alpha H^{-1}$) is less than a few Hubble times provided that $|\dot{A}_*| \gtrsim H$, and further that if this condition is satisfied, $(t_* - t_e)$ is also less than order a Hubble time, and that t_e precedes, or occurs at about the same time, as t_m.

By using Eq. (4.1) and $\delta t = 2\alpha H^{-1}$ we can solve for the value of Γ/H^4 at $t = t_*$:

$$\left(\frac{\Gamma}{H^4}\right)_{t_*} = \frac{M}{8\pi\alpha^4}. \tag{4.2}$$

If the transition lasts a Hubble time, i.e., $\alpha \sim 1$, then Γ/H^4 is order unity at $t = t_*$; if it is much quicker, i.e., $\alpha \ll 1$, then Γ/H^4 is much greater than unity at $t = t_*$. We can turn Eq. (4.2) around to solve for A_*:

$$A_* = \ln\left[\frac{32\pi\alpha^4}{M}\frac{C}{H^4}\right] \sim 4\ln\left[\frac{m_{\rm Pl}}{m_U}\right]. \tag{4.3}$$

Here we have written $C = m_U^4$; if m_U is of order the unification scale—say 10^{15} GeV—as one might expect for inflation, then A_* is of order 100. Conceivably C could be of order H^4, in which case A_* is of order unity or so; as noted earlier this makes the tunneling approximation marginal at best.

If we define $|\dot{A}_*| \equiv \beta A_* H$ (as discussed earlier β is expected to of order unity), we can then solve for the duration of the phase transition δt in terms of β and m_U:

$$\delta t \simeq \frac{2}{|\dot{A}_*|} = \frac{2H^{-1}}{\beta A_*} \simeq \frac{H^{-1}}{2\beta\ln(m_{\rm Pl}/m_U)} \quad \text{or} \quad \alpha = \frac{1}{4\beta\ln(m_{\rm Pl}/m_U)}.$$

This means that if C is order the unification scale to the fourth power, $\delta t \sim 10^{-2}\beta^{-1}H^{-1}$. On the other hand, if C is significantly smaller than the fourth power of the unification scale, or if $\beta \ll 1$, δt could be of order H^{-1}.

Now to the distribution of bubble sizes. Starting with Eq. (3.2) and using the facts that $I(t_* - r) = M\exp(r\dot{A}_*)$ and $\Gamma(t_* - r) = \Gamma_*\exp(r\dot{A}_*)$ it follows that the distribution of bubble sizes when the transition is complete $(t = t_*)$ is

$$\frac{dN}{dr} \propto \exp\left(-2r/\delta t - Me^{-2r/\delta t}\right); \tag{4.4}$$

where we have used $\delta t \simeq 2/|\dot{A}_*|$. Writing dN/dr as $\exp[-f(r)]$ and expanding $f(r) = 2r/\delta t + M\exp(-2r/\delta t)$ in a Taylor series,

$$f(r) = 3 + 2(r - r_0)^2/r_0^2,$$

where $r_0 = \ln(M)\delta t/2 \simeq \delta t$ we arrive at the following for the distribution of bubble sizes at $t = t_*$:

$$\frac{dN}{dr} \propto \exp\left[-2(r - r_0)^2/r_0^2\right]. \tag{4.4'}$$

The distribution of bubble sizes is gaussian, peaks at $r = r_0 \simeq \delta t$, and has a width of order $\delta t/2$; see Fig. 7a. This is in sharp contrast to old inflation where $dN/dr \propto r^{-4}$. Recalling

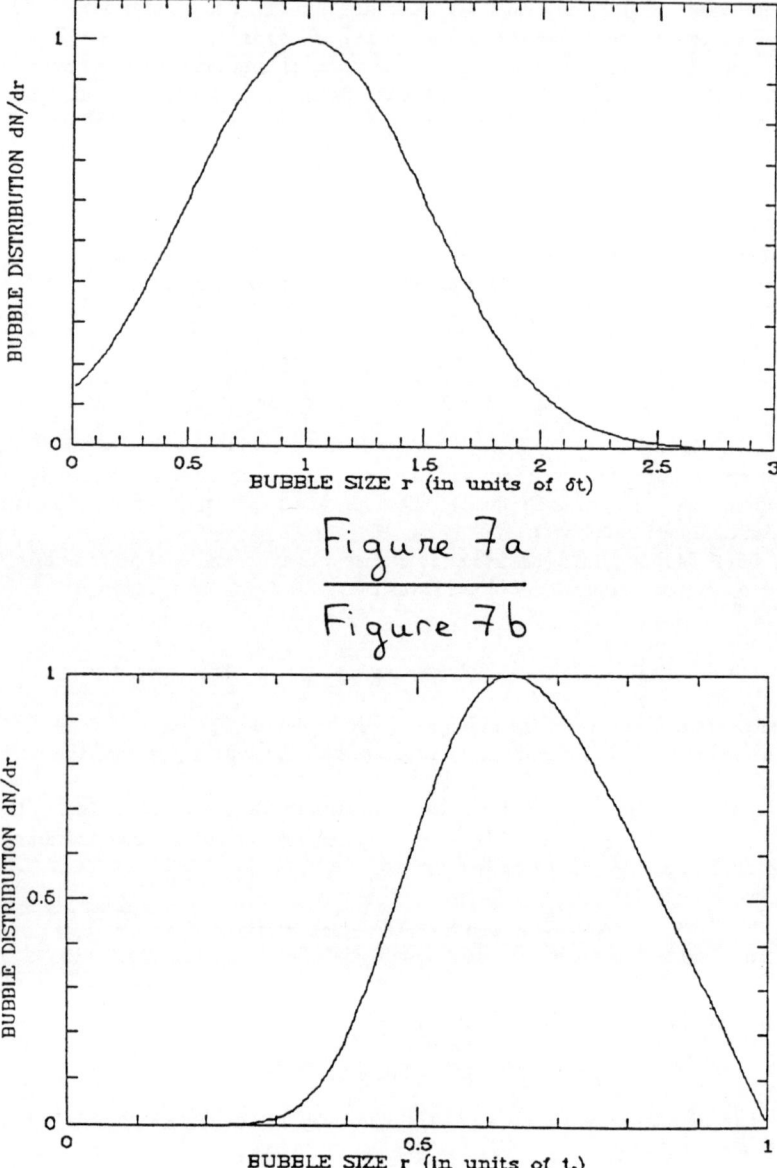

Figure 7a

Figure 7b

our expectations for δt—from order $10^{-2}H^{-1}$ to order H^{-1}—we see that the characteristic bubble size is between 1% and 100% of the Hubble radius.

(b) Model 2: $\Gamma = \beta t$

Next consider a slightly different model where Γ increases linearly with time.* Two remarks are in order: (i) Since we don't want Γ to be negative we should actually write $\Gamma = \beta\theta(t)t$, where $\theta(t)$ is the Heaviside function; (ii) The apparent absence of a constant term merely corresponds to our definition of the origin of time. The essence of this simple model is that we are using the form $\Gamma = \beta t$ to fit: (i) the rate of change of Γ around the end of the phase transition, quantified by β; and (ii) the value of Γ at the end of the phase transition, $\Gamma_* = \beta t_*$ (see Fig. 6).

With $\Gamma = \beta t$ it is straightforward to compute $I(t)$:

$$I(t) = \frac{\pi}{15}\beta t^5.$$

Once we have $I(t)$ we can compute everything else:

$$t_* = \left(\frac{15M}{\pi}\right)^{1/5}\beta^{-1/5} \simeq 2\beta^{-1/5}; \qquad \delta t = \left[1 - \left(\frac{m}{M}\right)^{1/5}\right]t_* \simeq t_*.$$

Note that the timescale for the duration of the phase transition is controlled by $\beta^{-1/5}$. Further, if we write $\beta \equiv \zeta^5 H^5$, then

$$\delta t \simeq t_* \simeq \frac{2}{\zeta}H^{-1}.$$

The condition that phase transition occur in less than or of order of a Hubble time corresponds to $\zeta \gtrsim 1$, or $\beta \gtrsim H^5$. Consider the value of Γ/H^4 at the end of the phase transition ($t = t_*$):

$$\left(\frac{\Gamma}{H^4}\right)_{t_*} = \frac{\beta t_*}{H^4} = 2\zeta^4;$$

if $\zeta \gtrsim 1$ the value of $(\Gamma/H^4)_*$ is greater than unity.

Note that if β were of order m_U^5, then $\zeta \sim m_{Pl}/m_U \sim 10^4$ or so. In this case the phase transition would occur very rapidly, $\delta t \sim 10^{-4}H^{-1}$. We remind the reader that our origin of time was chosen so that the form $\Gamma = \beta t$ describes both the rate of change of Γ around $t = t_*$ and the value of Γ at $t = t_*$. Thus the fact that $t_* \ll H^{-1}$ does not *necessarily* mean that the phase transition occurs too soon to permit sufficient inflation.

The condition that the physical volume in the false-vacuum decrease, $V_{\text{phys}}^{-1}dV_{\text{phys}}/dt = (3H - \pi\beta t^4/3) \le 0$, is satisfied for times

$$t \ge t_e = \left(\frac{9}{\pi}\right)^{1/4}\zeta^{-5/4}H^{-1} \simeq \left(\frac{9}{16\pi\zeta}\right)^{1/4}t_*.$$

---

* Note, "this β" is not related to the β of the previous model; here β has dimensions of (energy)^5.

Note that for $\zeta \gtrsim 1$ the physical volume of space remaining in the false-vacuum state begins to decrease just before $t = t_m$, the time when fraction of space in the false vacuum is order e^{-m}.

Finally, consider the distribution of bubble sizes r at the end of the phase transition ($t = t_*$):

$$\frac{dN}{dr} \propto \Gamma(t_* - r) \exp[-I(t_* - r)] \propto \left(1 - \frac{r}{t_*}\right) \exp\left[-M\left(1 - \frac{r}{t_*}\right)^5\right]; \qquad (4.5)$$

valid for $r \leq t_*$. The distribution of bubble sizes is sharply peaked at the characteristic size $r \simeq \delta t$, with a width of order $t_*/M^{1/5}$; see Fig. 7b. For $\beta \sim H^5$ ($\zeta \sim 1$), the typical bubble size is comparable to the Hubble radius.

These two simple—but well motivated—models serve to illustrate how first-order inflation can successfully terminate. However, we should again emphasize that the conditions on Γ and H derived are only sufficient, not necessary. We will return to discuss models where the success is "more marginal." But first, we will briefly discuss an important generic constraint on first-order inflation: the absence of large, deleterious bubbles.

(c) Distribution of "colliding-bubble" sizes

It is interesting to ask, What is the distribution of the sizes of bubbles that collide with a given bubble? In old inflation, the distribution was skewed towards bubbles of much smaller size than the reference bubble, which was indicative of the nonsuccess of old inflation [17]. In the limit of a fast transition, where we can ignore the effects of the expansion, it is very simple to compute this distribution, which we denote by dN_C/dr_C.

To begin, let us assume that the reference bubble was nucleated at time t at position $r = 0$, and that we want the distribution of colliding-bubble sizes at the end of the phase transition ($t = t_*$). (Of course, we can calculate it at any arbitrary time.) The size of the reference bubble at time t_* is: $r_0 = t_* - t$. First consider "colliding bubbles" that are nucleated *after* the reference bubble is nucleated: nucleation time $t' \geq t$. There are two conditions on the radial distance r from the reference bubble where a "colliding bubble" can be nucleated: (i) it must be nucleated close enough so that it collides with the reference bubble by time t_*: $r < r_0 + r_C$, where $r_C = t_* - t'$ is the size of the "colliding bubble" at time t_*; and (ii) it cannot be nucleated within the reference bubble itself: $r > t' - t = r_0 - r_C$.

Thus the volume within which the colliding bubble can be nucleated is given by

$$V = \frac{4\pi}{3}\left[(r_0 + r_C)^3 - (r_0 - r_C)^3\right] = 8\pi(r_0^2 r_C + r_C^3/3).$$

The fraction of this volume V that is still in the false vacuum at time t' is $p(t')V$. Therefore, the number of colliding bubbles that have size $r_C \leq r_0$ at the end of the phase transition is just

$$\frac{dN_C}{dr_C} = 8\pi(r_0^2 r_C + r_C^3/3)\, p(t_* - r_C)\Gamma(t_* - r_C) \propto 8\pi(r_0^2 r_C + r_C^3/3)\frac{dN(r_C)}{dr};$$

valid for $r_C \leq r_0$, where we have recognized the fact that $p(t_* - r_C)\Gamma(t_* - r_C)$ is just the distribution of bubble sizes at the end of the phase transition: The distribution of colliding bubbles is just proportional to the distribution bubbles of size r_C times a geometrical factor.

Now consider colliding bubbles that are nucleated *before* the reference bubble is nucleated: that is, $t' < t$, which implies that $r_C > r_0$. The first condition on the nucleation position r is the same as before. The second condition is different: now it is the reference bubble that cannot be nucleated inside the colliding bubble, which implies that $r > r_C - r_0$. It then follows that

$$\frac{dN_C}{dr_C} \propto 8\pi(r_C^2 r_0 + r_0^3/3)\frac{dN(r_C)}{dr};$$

valid for $r_C > r_0$. Bringing the two formulae together we have

$$\frac{dN_C}{dr_C} \propto \left[\max(r_0, r_C)^2 \min(r_0, r_C) + \frac{1}{3}\min(r_0, r_C)^3\right]\frac{dN(r_C)}{dr}. \tag{4.6}$$

V. Big Bubbles are Bad!

(a) Preliminaries

As Guth and Weinberg [17] and Weinberg [15] have emphasized big bubbles are a potentially dangerous relic of first-order inflation. The absence of "scars" in the post-inflationary Universe from big bubbles can be used to sharply constrain the value of Γ/H^4 at various times during the inflationary epoch, well before the end of inflation. Although some had expressed hope that the production of a few astrophysical-size bubbles might play a role in accounting for the voids seen in the distribution of galaxies, we will see that the "big-bubble" constraints dash this hope.

The bubbles that we are concerned with here are those rare bubbles that are nucleated long before the end of inflation. It is useful to characterize such bubbles by N, the number of e-foldings ($\equiv \int H \, dt$) which they were nucleated before the end of inflation. The reason for this is simple: About a Hubble time after a bubble is nucleated it crosses outside the "horizon" (i.e., its physical size becomes larger than H^{-1}); thereafter it conformally stretches with the expansion* (i.e., physical size grows as the cosmic-scale factor R). Thus at the end of inflation a "big bubble" is about a factor of e^N larger than the Hubble radius, which sets the scale for the the typical size of bubbles being nucleated as the phase transition is being completed. (We assume here that the expansion rate does not change greatly; if it does, our results change quantitatively, but not qualitatively; see below.) If the bubble continued to grow conformally until the present, its size D would be related to N by

$$\ln(D/\text{Mpc}) \simeq (N - 48) - \frac{2}{3}\ln\left(T_{\text{RH}}/10^{14}\,\text{GeV}\right), \tag{5.1}$$

where T_{RH} is the temperature to which the Universe is reheated after inflation. (If we normalize the cosmic-scale factor to be unity at the present, then D is also the comoving size of the bubble as long as it is being conformally stretched.) An equally, if not more important, issue is when the bubble crosses back inside the horizon; the temperature of the Universe at horizon crossing is related to D by

$$T_{\text{HOR}} \sim \frac{300\,\text{eV}}{D/\text{Mpc}} \sim \frac{300\,\text{eV}}{\exp(N - 48)} \qquad \text{for } D \lesssim 13\,\text{Mpc} \ (N \lesssim 50); \tag{5.2a}$$

---

* The rigorous proof and discussion of this issue is relegated to an Appendix.

$$T_{\text{HOR}} \sim \frac{1000\,\text{eV}}{(D/\text{Mpc})^2} \sim \frac{1000\,\text{eV}}{[\exp(N-48)]^2} \qquad \text{for } D \gtrsim 13\,\text{Mpc} \ (N \gtrsim 50); \qquad (5.2b)$$

where in relating D to N we have ignored the "ln term" in Eq. (5.2).

[In Eqs. (5.1, 5.2) we have assumed that H is constant during inflation. Suppose that it varies: $H = p/t$, corresponding to the cosmic-scale factor growing as $R \propto t^p$ (to solve the horizon problem p must be greater than 1). Keeping the same definition for $N \equiv \int H dt$, it follows that the value of the Hubble parameter N e-foldings before the end of inflation, H_N, is related to its value at the end of inflation, H_*, by

$$\frac{H_N}{H_*} = \exp(N/p);$$

and further that the analogues of Eqs. (5.1, 5.2) are obtained by replacing N with $N' \equiv N - N/p$.]

At early times the interior of a bubble of true vacuum is empty; the false-vacuum energy liberated by the bubble resides in the "kinetic energy" of the bubble wall (both spatial and temporal gradients of the scalar field). When two bubbles collide the kinetic energy of their walls is converted into massive scalar particles which eventually decay (on a timescale τ controlled by their mass and couplings to other fields: $\tau \sim m_\phi^{-1}$) into relativistic particles that quickly thermalize [19]. For the Hubble-sized (or smaller) bubbles that nucleate and collide at the end of inflation (and thereby terminate the phase transition) the story is simple: Because the bubble interiors which are initially devoid of radiation are subhorizon sized and the Hubble radius is growing relative to them, the thermalized false-vacuum energy quickly diffuses and fills the former empty bubble interiors, so that the Universe quickly becomes homogeneous. The potential scars from this process are on the scale of the Hubble radius at the end of inflation, which corresponds to a comoving scale of only about $10^{-21}\,\text{Mpc}$.

The story for large bubbles is very different. When the phase transition is ending via the rapid nucleation of Hubble-sized bubbles, "big bubbles" are very much superhorizon sized: sizes of order $e^N H^{-1}$. The energy carried by the expanding wall of a big bubble is thermalized through collisions with the Hubble-sized bubbles being nucleated at the end of inflation. These collisions look very different than the ones just discussed above: a bubble of size H^{-1} collides with one of size $e^N H^{-1}$. However, when viewed in a Lorentz-boosted frame along their bubble axis with Lorentz factor $\gamma_{\text{big}} = e^{N/2}/2$ the bubbles are of equal size. In the "equal-bubble" frame the conversion of wall energy into particles proceeds as before and takes a time of order τ. However, seen in the FRW frame (rest frame of the Universe) it takes a factor of γ_{big} longer. (Since τ is expected to be of order the unification time scale, say $(10^{15}\,\text{GeV})^{-1} \sim 10^{-40}\,\text{sec}$, and γ_{big} is at most 10^{11} or so, the conversion time is still very short on an absolute scale.) Moreover, if the particles produced in the "equal-bubble" frame are characterized by energy m_U, in the FRW frame they have energy $\gamma_{\text{big}} m_U$ and essentially all move in the direction of the motion of the big-bubble wall, thereby comprising a rapidly expanding, narrow shell of relativistic particles. (The energies of these particles are quickly degraded by collisions with the ambient thermal bath of particles in the Universe.) Most importantly, the interior of a big bubble is empty—and on the basis of causality alone cannot be filled with thermal radiation until the bubble "crosses back inside the horizon." Therein lies the danger of very big bubbles: They represent highly inhomogeneous regions of the Universe that cannot be homogenized until they cross back inside of the horizon, at a relatively late epoch, cf., Eqs. (5.1, 5.2).

Before discussing the many potential dangers of big bubbles it is useful to estimate the number of bubbles of a given size that one expects to find within our current Hubble volume. Since shortly after a big bubble is nucleated its size simply conformally stretches as the Universe expands, the ratio of its size to that of the comoving scale that corresponds to the currently observable Universe ($\equiv \lambda_U$) remains fixed (see Fig. 8). Thus, when a bubble that was nucleated N e-folds before the end of inflation came into existence, the scale λ_U had a physical size $H_0^{-1}/D(N)$ larger than the Hubble radius H^{-1} then ($H_0^{-1} \simeq 3000h^{-1}$ Mpc is the present Hubble radius). Therefore, the expected number of big bubbles (characterized by N) within the currently observable Universe is simply:

$$\mathcal{N} \simeq \left(\frac{\Gamma}{H^4}\right)_N \left(\frac{H_0^{-1}}{D(N)}\right)^3 ; \qquad (5.3)$$

where subscript N indicates that Γ/H^4 is to be evaluated N e-foldings before the end of inflation. It is also useful to compute the volume fraction f_D of the Universe that is comprised of big bubbles of comoving size D; it is given by

$$f = \frac{\mathcal{N}D(N)^3}{H_0^{-3}} \simeq \left(\frac{\Gamma}{H^4}\right)_N ; \qquad (5.4)$$

where this expression is valid so long as these bubbles are conformally stretching with the expansion. Conformal stretching of big bubbles must continue at least until they become subhorizon sized.

(b) Primordial nucleosynthesis

The standard scenario of primordial nucleosynthesis is remarkably successful, accounting for the primeval abundances of D, ^{3}He, ^{4}He, and ^{7}Li, provided that the baryon-to-photon ratio η lies in the narrow interval, 3×10^{-10} to 5×10^{-10}, and the effective number of light neutrino species is less than 3.4 [20]. Moreover, deviations from the standard scenario are remarkably unsuccessful; e.g., inhomogeneous nucleosynthesis: When the level of inhomogeneity in the baryon-to-photon ratio, e.g., due to the quark/hadron transition, is raised to the level that it affects nucleosynthesis the concordance of the predictions of primordial nucleosynthesis disappears [21].

Bubbles that are Hubble-sized (or larger) at the epoch of nucleosynthesis, $T \sim$ MeV and $t \sim$ sec—and therefore not yet homogenized—have the potential to upset the successful predictions of nucleosynthesis. We can be certain that nucleosynthesis in these unhomogenized regions proceeds radically differently; however, precise statements are beyond the scope of the present work. The potential to overproduce D, ^{3}He, or ^{7}Li is great, since they are produced in such small quantities in the standard scenario: a few parts in 10^5 for ^{3}He and D, and a few parts in 10^{10} for ^{7}Li. To be absolutely ^{7}Li-safe one would have to require that f_{BBN} be less than about 10^{-10}, so that even if in a big bubble region the ^{7}Li production were 100%, after mixing ^{7}Li would be diluted to a safe level; for ^{3}He and D, the corresponding absolutely safe limit would be $f_{\text{BBN}} \lesssim 10^{-5}$.

No doubt $f_{\text{BBN}} \lesssim 10^{-10}$ or even $\lesssim 10^{-5}$ is far too stringent a constraint; however, one might prudently require $f_{\text{BBN}} \lesssim 10^{-3}$ or so. Taking $T_{\text{RH}} \sim 10^{14}$ GeV and $T_{\text{BBN}} \sim 1$ MeV,

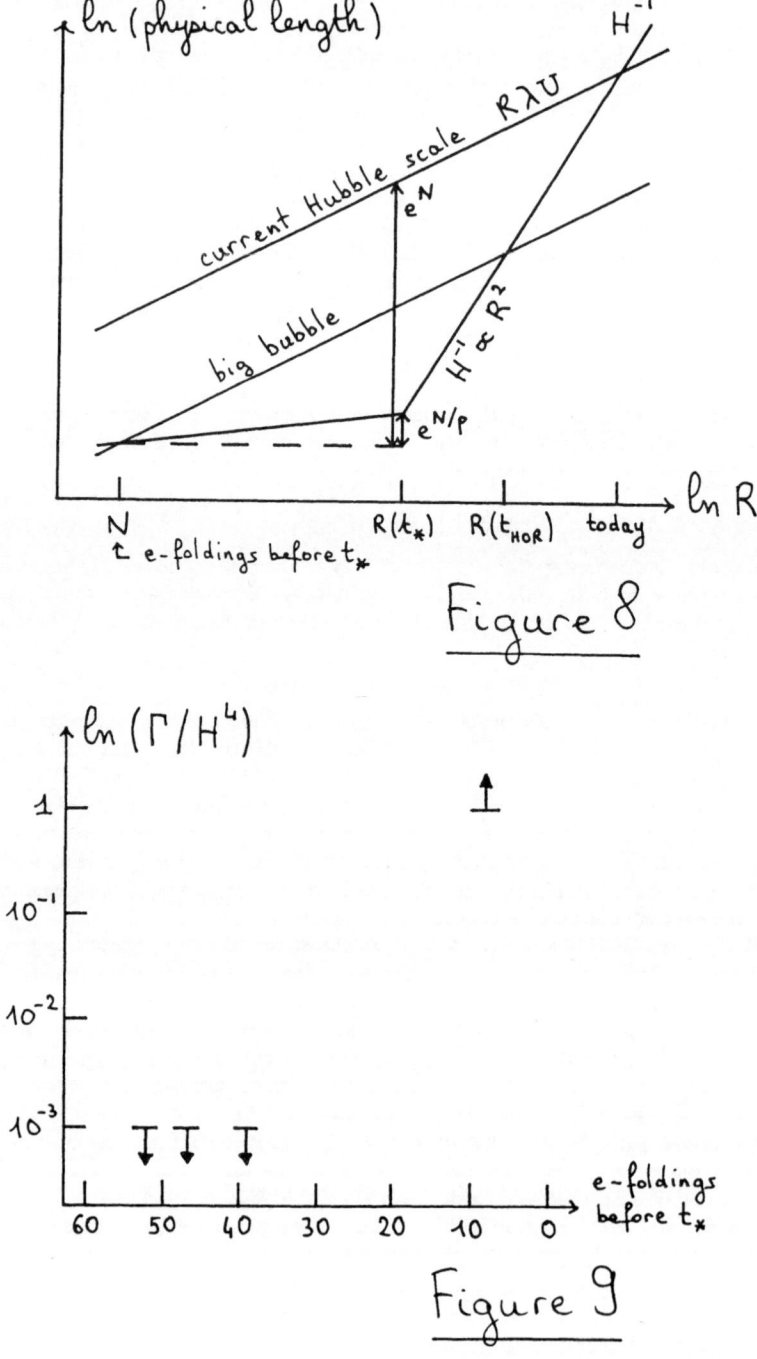

Figure 8

Figure 9

this constrains the value of (Γ/H^4) $N \simeq 39$ e-foldings before the end of inflation:

$$\left(\frac{\Gamma}{H^4}\right)_{N\simeq39} \lesssim 10^{-3}. \tag{5.5}$$

(c) CMBR distortions

The decoupling of (baryonic) matter and radiation occurs at a temperature of order $T_{DEC} \sim 0.3\,eV$. Needless to say the presence of an unhomogenized bubble near the last scattering surface would lead to a massive distortion of the Cosmic Microwave Background Radiation (CMBR) temperature. The CMBR temperature is spatially uniform to almost a part in 10^5 on angular scales from less than $1°$ to $90°$ (aside from the dipole anisotropy, which is presumably due to our motion with respect to the cosmic rest frame), and the CMBR spectrum is consistent with that of a black body at temperature 2.74 ± 0.01 K over almost three and one-half decades in wavelength ($\lambda \sim 100\,cm - 0.03\,cm$) [22]. Thus we can be certain that there could have been no Hubble-sized bubble within a "decoupling Hubble radius" (comoving length scale of about $H_{DEC}^{-1} \sim 100\,Mpc$) of the last-scattering surface. Since that volume corresponds to a fraction $H_{DEC}^{-1}H_0^{-2}/H_0^{-3} \sim 3 \times 10^{-2}$ of the currently observable Universe and the expected number of "decoupling-sized" bubbles within the present Hubble volume is order $(\Gamma/H^4)_N(H_0^{-1}/H_{DEC}^{-1})^3 \sim 3 \times 10^4$, the constraint that follows is

$$\left(\frac{\Gamma}{H^4}\right)_{N\simeq53} \lesssim 10^{-3}, \tag{5.6}$$

where we have again for simplicity taken $T_{RH} \sim 10^{14}$ GeV. (This is in essence the constraint derived in Ref. [15].)

Finally consider the so-called μ-distortion: That is, that the CMBR is characterized by a Bose-Einstein distribution with chemical potential μT, $n(p) = 1/[\exp(\mu + p/T) - 1]$, rather than a Planck distribution, $n(p) = 1/[\exp(p/T) - 1]$. When the temperature of the Universe drops below about a keV, photon-number changing processes, e.g., the "double-Compton" process $e^- + \gamma \leftrightarrow e^- + \gamma + \gamma$, become ineffective (interaction rate per photon less than the expansion rate); thus any energy injected after this epoch will lead to a Bose-Einstein, rather than Planck, distribution. COBE and COBRA observations constrain μ to be less than about 10^{-3} [22].

Bubbles that are still unhomogenized at this epoch ($T \sim$ keV) will lead to a μ-distortion. Again, a precise calculation of the size of such a distortion is beyond the scope of this paper. If we suppose that the potential μ-distortion within a Hubble-sized bubble at this epoch is order unity, we might estimate the μ-distortion after such bubbles have been homogenized as the fraction of our Hubble volume occupied by such bubbles, or $(\Gamma/H^4)_N$. Requiring that $\mu \lesssim 10^{-3}$ results in the constraint

$$\left(\frac{\Gamma}{H^4}\right)_{N\simeq47} \lesssim 10^{-3}, \tag{5.7}$$

again taking $T_{RH} \sim 10^{14}$ GeV.

The three constraints to $(\Gamma/H^4)_N$ just discussed, as well as the value of Γ/H^4 at the end of inflation required to ensure proper completion of the phase transition (for our two simple models) are illustrated in Fig. 9. Roughly speaking, Γ/H^4 must be of order unity at the end of inflation, and less than about 10^{-3} at 39 e-foldings before the end of inflation (primordial nucleosynthesis), at 43 e-foldings before the end of inflation (μ-distortion), and 53 e-foldings before the end of inflation (CMBR temperature). If we understood the physics of the early Universe at temperatures well above that of primordial nucleosynthesis, e.g., the electroweak phase transition ($N \sim 27$), baryogenesis ($N \sim$ few), and so on, we could in principle place constraints on $(\Gamma/H^4)_N$ even closer to the end of inflation.

The constraints just derived imply that the fraction of space occupied by big bubbles of astrophysically interesting size—say greater than of order 1 Mpc—is less than order 10^{-3}. To be more specific, consider the number of big bubbles of comoving size of order 30 Mpc expected within 100 Mpc of the Milky Way (a volume about the size of the CfA$_2$ red-shift survey): The expected number is only about 3×10^{-2}! The stringent constraints to the number of big bubbles that follow from nucleosynthesis and from the uniformity of the CMBR make it very unlikely that big bubbles produced by first-order inflation have anything to do with the observed structure today on scales of order 30 Mpc (e.g., large voids, walls, and so on). On the other hand, whether or not such structures need explanation beyond that of inflation-produced curvature perturbations still remains to be seen.

(d) $\Gamma = Ce^{-A(t)}$

Before we go on, we can apply our "big-bubble" constraint to the model of bubble nucleation where $\Gamma(t) = C\exp[-A(t)]$. (Because the origin of time is arbitrary in the $\Gamma(t) = \beta t$ model, one cannot apply the "big-bubble" constraint to this model without further knowledge of $\Gamma(t)$ for $t \ll t_*$.) Assuming that the expansion is exponential, epoch characterized by "N e-foldings" before the end of inflation occurs at a time, $t_{bb} \simeq t_* - NH^{-1}$. The value of (Γ/H^4) at this epoch must be less than about 10^{-3}. Using the value of (Γ/H^4) at the end of inflation, and using the Taylor expansion for $A(t)$, it follows that:

$$\left(\frac{\Gamma}{H^4}\right)_{t_{bb}} \simeq e^{-NH^{-1}|\dot{A}_*|}\left(\frac{\Gamma}{H^4}\right)_{t_*} \simeq \mathcal{O}(2/\alpha^4)\exp(-N/\alpha) \lesssim 10^{-3}.$$

Recall, the quantity α is related to the duration of the phase transition: $\delta t \equiv 2\alpha H^{-1}$, [and $\dot{A}_*$: $|\dot{A}_*| = H/\alpha$]. Provided that α is less than $\mathcal{O}(10)$, this constraint is easily satisfied.

We should caution that the use of only the first two terms in the Taylor expansion of $A(t)$ to approximate $A(t_{bb})$ may not be valid. However, if $A(t)$ does vary exponentially, it seems likely that at early times Γ/H^4 will be sufficiently small to satisfy the big bubble constraint. Further, the results that we have used from Section IVa are only strictly valid for $\alpha \lesssim 1$; in Section VIb we will treat the case of $\alpha \gtrsim 1$ (slow transition), and we will show that the big-bubble constraint is indeed $\alpha \lesssim \mathcal{O}(10)$.

VI. More-complicated Models

(a) Extended inflation

As a warm-up exercise before we go on to study more exotic models, it is worth analyzing the first (and almost) successful model of first-order inflation: extended inflation [9]. This model is based upon the Brans-Dicke-Jordan theory of gravity. In the so-called Jordan frame, Γ and ρ_{VAC} are constant; however, the effective gravitational constant, which is set by the inverse of the Brans-Dicke field Φ, evolves as t^{-2}. Owing to the variation of the gravitational constant, the scale factor only grow as a power, $R \propto t^{\omega+1/2}$, so that $H = (\omega + 1/2)/t$. The quantity ω is the coefficient in front of the kinetic energy term for the Brans-Dicke field; in the $\omega \to \infty$ limit the Brans-Dicke-Jordan theory becomes general relativity. [In the Einstein frame, where gravity is just that of general relativity, the Higgs field ϕ has nontrivial couplings to the Brans-Dicke field Φ; in this frame $\Gamma \propto \bar{t}^{-2}$ and $\bar{H} \propto \bar{t}^{-1}$, where $\bar{t} \propto t^2$ is the time variable in the Einstein frame (see Ref. [23] for details).]

For this model the duration of the phase transition is order several Hubble times, and it is not a good approximation to neglect the expansion of the Universe during the phase transition. The probability that a point remains in the false vacuum is still given by $\exp[-I(t)]$; however, now [16]

$$I(t) = \int_{t_0}^{t} \Gamma(t')R(t')^3 V(t,t')dt'; \qquad V(t,t') = \frac{4\pi}{3}r(t,t')^3. \qquad (6.1)$$

Here, $r(t,t')$ is the comoving radius of a bubble nucleated at time t', at time t; if once again we assume that the bubble is nucleated with negligible size and expands at the speed of light, then

$$r(t,t') = \int_{t'}^{t} du/R(u) = \frac{1}{\omega - 1/2}\left(\frac{1}{t'^{\omega - 1/2}} - \frac{1}{t^{\omega - 1/2}}\right). \qquad (6.2)$$

From this it follows that

$$I(t) = \frac{\pi \Gamma t^4}{3\omega^3}g(\omega); \qquad (6.3a)$$

$$g(\omega) = \frac{\omega^3}{(\omega - 1/2)^3}\left[1 - \frac{24}{2\omega + 7} + \frac{12}{2\omega + 3} - \frac{8}{6\omega + 5}\right]; \qquad (6.3b)$$

note that for $\omega \gg 1$, $g(\omega) \to 1$. Had we ignored the expansion, $I(t) = \pi \Gamma t^3/3$.

We can now compute t_* and δt:

$$t_* = \left(\frac{3M}{\pi g(\omega)}\right)^{1/4} \omega^{3/4}\Gamma^{-1/4} \simeq 2\omega^{3/4}\Gamma^{-1/4}; \qquad \delta t = \left(1 - (m/M)^{1/4}\right)t_* \simeq t_*.$$

The condition that the physical volume of space remaining in the false vacuum decreases, $V_{\text{phys}}^{-1}dV_{\text{phys}}/dt = 3(\omega + 1/2)/t - I'(t) \leq 0$, is satisfied for $t \geq t_e$, where

$$t_e = \left(\frac{9(\omega + 1/2)}{4\pi\omega g(\omega)}\right)^{1/4} \omega\Gamma^{-1/4} \simeq \omega\Gamma^{-1/4} \simeq \frac{\omega^{1/4}}{2}t_*. \qquad (6.4)$$

Note that the timescale of the phase transition is set by $\omega\Gamma^{-1/4}$, and that $\delta t \sim t_e \sim t_*$. The requirement that the phase transition last at most of order a Hubble time or so is:

$$\delta t \simeq t_* \lesssim H_*^{-1} \simeq t_*/(\omega - 1/2),$$

implies that ω should not be too much larger than unity. As $\omega \to \infty$, the general relativity limit, the duration of the phase transition becomes an infinite number of Hubble times, the disease of old inflation. As we shall see shortly, the "big-bubble constraint" also argues for small ω. Note too, since $H \simeq \omega/t$, t_* and H_*^{-1} are automatically of the same order, modulo ω.

Next, consider the distribution of bubbles sizes at the end of inflation. Because the duration of the phase transition is several Hubble times we cannot neglect the expansion and Eq. (3.2) for the distribution of bubble sizes must be modified. The number of bubbles nucleated from time t to time $t + dt$ is just

$$dN \propto \Gamma(t)R(t)^3 p(t)dt,$$

where $R(t)$ is the value of the cosmic-scale factor, and the amount of physical space still in the false vacuum at time t is $\propto R(t)^3 p(t)$. The comoving radius of a bubble nucleated at time t at the end of the phase transition $t = t_*$ is:

$$r(t_*, t) \equiv \int_t^{t_*} du/R(u),$$

where as usual we assume that the size of the bubble at nucleation is negligible and that the bubble expands at the speed of light. It is now a simple matter to change variables to obtain $dN(r)$, the number of bubbles at time t_* with comoving radius between r and $r + dr$:

$$\frac{dN}{dr} \propto \{\Gamma(t)R(t)^4 p(t)\}_{t=t(r)}, \tag{6.5}$$

where $r(t_*, t)$ must be inverted to obtain $t(r)$ (of course, the functional form of $t(r)$ depends upon t_*). Note too that the time at which the distribution is evaluated, here $t = t_*$, is arbitrary; the bubble distribution can be obtained at any time. In deriving Eq. (6.5) we have ignored bubble overlaps (collisions).

We can apply Eq. (6.5) to extended inflation. To do so write $R(t) = t^{\omega+1/2}$; then

$$t = t_* \left(\frac{1}{1 + r/r_*}\right)^{1/(\omega-1/2)},$$

where r_* is the comoving length that corresponds to the Hubble radius at $t = t_*$: that is, $r_* = H_*^{-1}/R(t_*) = t_*^{-\omega+1/2}/\omega$. From this it is straightforward to derive the distribution of bubble sizes at $t = t_*$:

$$\frac{dN}{dx} \propto \left(\frac{1}{1+x}\right)^{4+4/\omega} \exp\left\{-\frac{\pi}{3}\omega(1+x)^{-4/\omega}\right\}, \tag{6.6}$$

where we have taken the large ω limit and $x = r_{phys}/H_*^{-1} \simeq r_{phys}/\Gamma^{-1/4}$ (r_{phys} is the physical size of the bubble at time t_*); see Fig. 7c. This expression agrees with a similar result derived in Ref. [15]. Note that in the large-ω limit, the distribution becomes identical to that of old inflation: $dN/dr \propto r^{-4}$.

Distribution (22) is very flat. To see this write $dN/dx = \exp[-f(x)]$, where $f(x) = (4 + 4/\omega)\ln(1 + x) + \pi\omega(1 + x)^{-4/\omega}/3$ and expand $f(x)$ in a Taylor series:

$$f(x) = f(x_0) + \frac{(x - x_0)^2}{2!}f''(x_0) + \cdots \simeq \frac{\pi\omega}{3} + \frac{8}{\omega}\left(\frac{3 + 3/\omega}{\pi}\right)^{\omega/2} x^2;$$

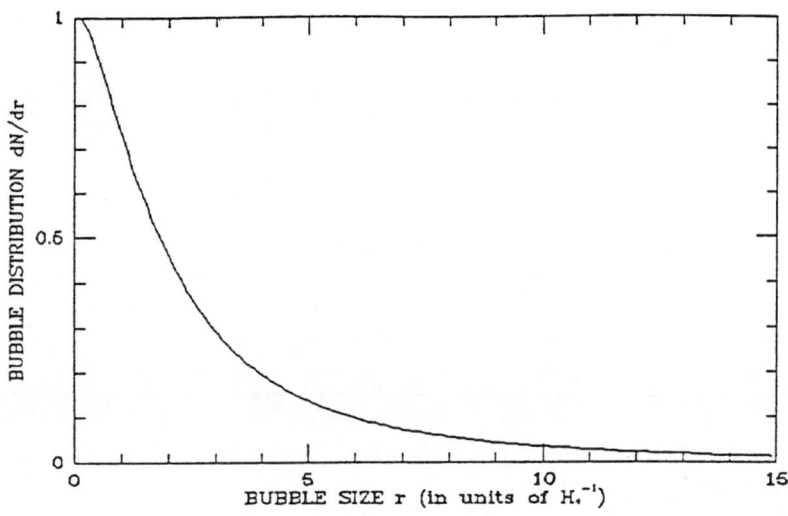

Figure 7c

where $f'(x_0) = 0$ and $x_0 = [\pi/(3 + 3/\omega)]^{\omega/4} - 1 \simeq 0$. We can then approximate the distribution of bubble sizes as:

$$\frac{dN}{dr} \propto \exp(-x^2/2\sigma^2); \qquad \sigma^2 = \left(\frac{\pi}{3 + 3/\omega}\right)^{\omega/2} \frac{\omega}{16}; \qquad (6.7)$$

This is a (truncated) gaussian with width $\sigma \sim \sqrt{\omega}/4$, corresponding to a flat distribution of physical bubble sizes from zero to $r_{phys} \simeq \sqrt{\omega} H_*^{-1}/4$. Not unexpectedly, in the limit $\omega \to \infty$ (old inflation), the width of the distribution becomes infinite.

Now let's consider the big-bubble constraint. We could in principle use distribution (22) to derive the constraint; however, it is more useful to implement it in the form: $(\Gamma/H^4)_N \lesssim 10^{-3}$ for $N \simeq 30 - 40$. During extended inflation Γ is constant; H varies. As discussed in the previous Section, $H_N = \exp[N/(\omega + 1/2)]H_*$, so that

$$\left(\frac{\Gamma}{H^4}\right)_N = e^{-4N/(\omega+1/2)} \left(\frac{\Gamma}{H^4}\right)_{t_*}.$$

The end of inflation occurs at t_e or t_*, depending which is larger. The values of Γ/H^4 at times t_e and t_* are:

$$\left(\frac{\Gamma}{H^4}\right)_{t_e} = \frac{\omega^4}{(\omega + 1/2)^4} \simeq 1; \qquad \left(\frac{\Gamma}{H^4}\right)_{t_*} = \frac{2^4}{\omega}.$$

Depending upon the value of ω, t_e may precede ($\omega \sim \mathcal{O}(1)$) or follow ($\omega \gg 1$) t_*; for the values of ω of interest, $\omega \sim \mathcal{O}(10)$, these two times are about the same, and the value of Γ/H^4 at the end of the phase transition is approximately unity. The big-bubble constraint, $(\Gamma/H^4)_N \lesssim 10^{-3}$ for $N \simeq 30 - 40$, then implies that

$$\exp[-4N/(\omega + 1/2)] \lesssim 10^{-3} \qquad \text{or} \qquad \omega \lesssim 4N/7 \simeq \mathcal{O}(20). \qquad (6.8)$$

This is essentially the constraint obtained previously by Weinberg [15]. Since solar-system tests of gravitational theory require that $\omega \gtrsim 500$ [24], the simplest model of first-order inflation cannot both be consistent with the "big-bubble" constraint and the solar-system tests. It may however be possible to resurrect the theory, e.g., by giving the Brans-Dicke field a mass [11], which would anchor the Brans-Dicke field, thereby making the solar-system constraint irrelevant.

Finally, recall that during inflation the effective value of the gravitational constant varies as t^{-2}; after inflation, the value of the gravitational constant evolves very little. Therefore, in extended inflation the current value of the gravitational constant is determined by the time when the phase transition completes! Through the analogue of the Friedmann equation for the Brans-Dicke-Jordan theory,

$$H^2 = \frac{8\pi}{3m_{\text{eff}}^2} \rho_{\text{VAC}}, \qquad (6.9)$$

we can relate the effective value of the Planck mass squared to the parameters of the phase transition. Since the value of H^4 at the end of the phase transition is about equal to Γ, it follows that

$$m_{\text{eff}}^2 = \frac{8\pi \rho_{\text{VAC}}}{\sqrt{\Gamma}} \simeq \frac{8\pi}{3} e^{A/2} m_U^2; \qquad (6.10)$$

where the final expression follows by writing $\rho_{VAC} = m_U^4$ and $\Gamma = m_U^4 e^{-A}$. In the *simplest* model of extended inflation the value of the Planck mass is related to the unification scale and the tunneling action: $m_{Pl} \sim e^{A/4} m_U$. In more complicated models, this may not be the case [11].

(b) Exponential nucleation rate revisited

In our earlier discussion of this model where $\Gamma = C e^{-A(t)}$ we assumed that the phase transition occurred in less than a Hubble time or so, which allowed us to ignore the effects of the expansion. Let us now relax that assumption, in which case we cannot ignore the effects of the expansion: $I(t)$ is now given by Eq. (6.1), dN/dr is by Eq. (6.5), and the comoving radius of a bubble nucleated at time t' at time t is

$$r(t,t') = \int_{t'}^{t} du/R(u) = H^{-1} R(t') \left[1 - e^{-H(t-t')}\right];$$

where $R(t) = e^{Ht}$ (H = constant). Expanding $A(t)$ in a Taylor as before, $A(t) = A_\bullet + \dot{A}_\bullet (t - t_\bullet)$, it then follows that

$$I(t) = \frac{4\pi}{3} \frac{\Gamma(t)}{H^3 |\dot{A}_\bullet|} \left[1 - \frac{3}{1+\alpha} + \frac{3}{1+2\alpha} - \frac{1}{1+3\alpha}\right]; \qquad (6.11)$$

where $\alpha \equiv H/|\dot{A}_\bullet|$. Recall that $|\dot{A}_\bullet|$ sets the timescale for the phase transition: $\delta t \simeq 2|\dot{A}_\bullet|^{-1} \simeq 2\alpha H^{-1}$; in the limit of a fast transition, $|\dot{A}_\bullet| \gg H$ ($\alpha \ll 1$), Eq. (6.11) reduces to our previous expression, Eq. (4.1), as the quantity in square brackets goes to $6\alpha^3$ in this limit. We have already thoroughly discussed fast transitions in Section IVa. Here we will explore the opposite limit, a slow transition: $\alpha \gg 1$; in this case the expression in square brackets goes to 1, and

$$I(t) \simeq \frac{4\pi}{3} \frac{\Gamma(t)}{H^3 |\dot{A}_\bullet|} \qquad \text{(for } H \gg |\dot{A}_\bullet|). \qquad (6.11')$$

Using the above expression for $I(t)$, valid in the slow-transition regime, $\alpha = H/|\dot{A}_\bullet| \gg 1$, it is a simple matter to compute δt and $(\Gamma/H^4)_\bullet$:

$$\delta t = \frac{\ln(M/m)}{|\dot{A}_\bullet|} \sim 2\alpha H^{-1}; \qquad \left(\frac{\Gamma}{H^4}\right)_{t_\bullet} = \frac{3M|\dot{A}_\bullet|}{4\pi H} = \frac{3M}{4\pi\alpha}, \qquad (6.12)$$

where as before $M \simeq 10 - 100$ is a large, unspecified number, and $m \simeq 0.1 - 1$ is a small, unspecified number: $I(t) = M$ defines the end of the phase transition and $I(t) = m$ defines the beginning of the phase transition.

The physical volume of space remaining in the false vacuum begins decreasing when

$$3H - \frac{4\pi\Gamma(t)}{3H^3} \leq 0.$$

[For arbitrary α, $\Gamma(t)$ in this expression is multiplied by the quantity in square brackets in Eq. (6.11).] This condition is satisfied for $t \geq t_e$, where

$$t_e = t_\bullet + \frac{\ln(3H/M|\dot{A}_\bullet|)}{|\dot{A}_\bullet|} = t_\bullet + \ln(3\alpha/M)\alpha H^{-1} = t_\bullet + \ln(3\alpha/M)\frac{\delta t}{2}. \qquad (6.13)$$

The value of Γ/H^4 at $t = t_*$ is

$$\left(\frac{\Gamma}{H^4}\right)_{t_*} = \frac{9}{4\pi} \simeq 1.$$

As before, the duration of the phase transition is controlled by $|\dot{A}_*|^{-1} = \alpha H^{-1}$. As we will see shortly, the big-bubble constraint restricts α to be less than $\mathcal{O}(10)$. In this case, t_e and t_* occur at about the same time, the phase transition lasts a few Hubble times, and $(\Gamma/H^4)_e \simeq (\Gamma/H^4)_* \simeq 1$. Likewise, the typical bubble size at the end of the phase transition is of order a few Hubble lengths.

Before we go on to the distribution of bubble sizes at the end of the phase transition, let us quickly derive the big-bubble constraint: The uniformity of the CMBR and the successful outcome of primordial nucleosynthesis constraint the value of (Γ/H^4) $N \simeq 30 - 40$ e-foldings before the end of inflation to be less than about 10^{-3}:

$$\left(\frac{\Gamma}{H^4}\right)_N \simeq \exp(\dot{A}_* N H^{-1})\left(\frac{\Gamma}{H^4}\right)_{t_*,t_e} \simeq \exp(-N/\alpha) \lesssim 10^{-3}; \qquad (6.14a)$$

this leads to the constraint,

$$\alpha \lesssim N/7 \simeq \mathcal{O}(10), \qquad (6.14b)$$

which is consistent with our previous estimate, cf. Section Vd.

The big-bubble constraint barely allows the transition to be in the "slow regime:" δt is at most $20H^{-1}$ (corresponding to $\alpha = 10$). Thus, our previous analysis of fast transitions in Section IVa covered essentially the entire parameter space for successful inflation. (Of course we should remind the reader that when deriving the big-bubble constraint we assumed that the two-term Taylor expansion for $A(t)$ is adequate to compute Γ N e-foldings before the end of inflation; e.g., if $A(t)$ increased at early times faster than $\exp(-\dot{A}_* t)$, one could have $\alpha \gtrsim 10$ and still satisfy the big-bubble constraint.)

Finally, consider the distribution of bubble sizes at the end of inflation ($t = t_*$). To evaluate Eq. (6.5) for dN/dr we need to obtain $t(r)$. By inverting the expression $r(t_*, t) = H^{-1}[\exp(-Ht) - \exp(-Ht_*)]$ it follows that

$$H(t_* - t) = \ln(1 + r/r_*); \qquad (6.15)$$

where $r_* = \exp(-Ht_*)H^{-1} = H^{-1}/R(t_*)$ is the comoving size of the Hubble radius at the end of inflation. It is then straightforward to use Eq. (6.15) to find that

$$\left(\frac{dN}{dr}\right)_{t_*} \propto (1 + r/r_*)^{-4-1/\alpha} \exp[-M(1 + r/r_*)^{-1/\alpha}]. \qquad (6.16)$$

This distribution is shown in Figs. 7d,e. Note that in the large α limit ($\Gamma \to \text{const}$), the distribution becomes identical to that of old inflation: $dN/dr \propto (1 + r/r_*)^{-4} \to r^{-4}$. Note for the $(1 + r)^{-4}$ distribution it follows that: $\langle r \rangle = r_*/6$ and $\langle (r - \langle r \rangle)^2 \rangle = r_*^2/6$.

The width of this distribution is controlled by α. To see this, as we have before, write $dN/dx \propto \exp[-f(x)]$, where $x = r/r_* = r_{\text{phys}}/H^{-1}$, r_{phys} is the physical size of a bubble

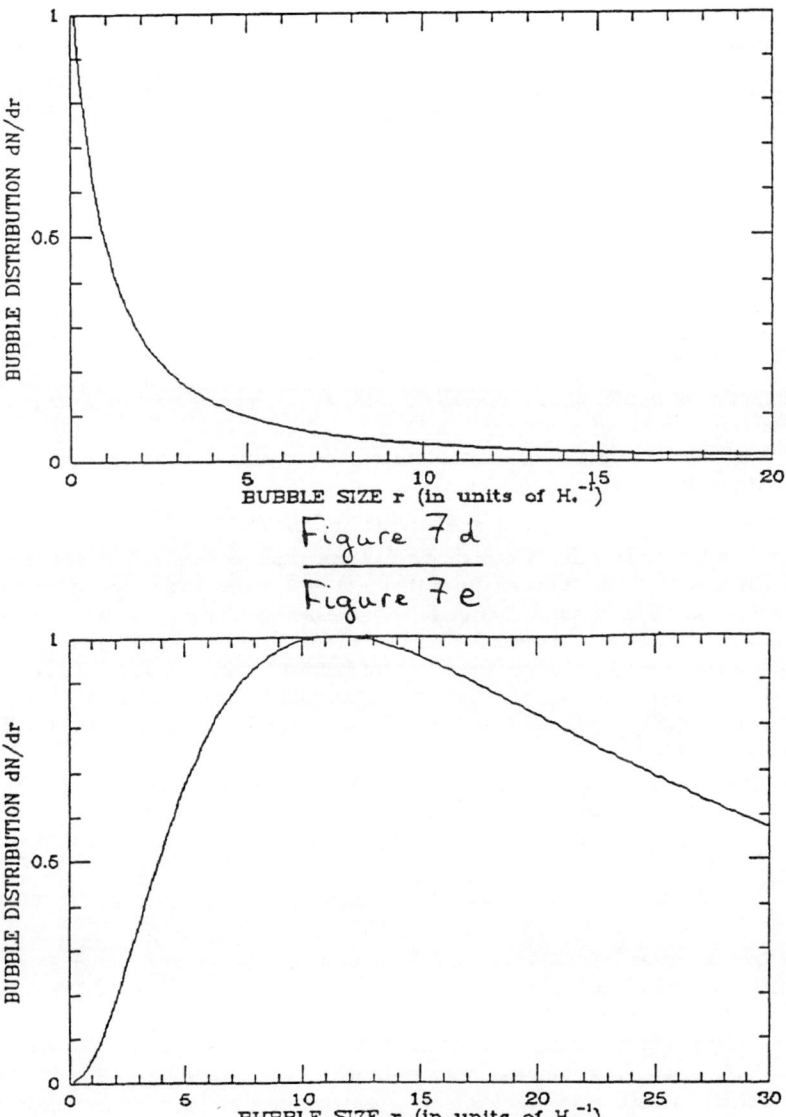

Figure 7d

Figure 7e

at the end of inflation, and $f(x) = (4 + \alpha^{-1}) \ln(1 + x) + M(1 + x)^{-1/\alpha}$. Then expand $f(x)$ in a Taylor series about x_0:

$$f(x) = f(x_0) + \frac{(x - x_0)^2}{2!} f''(x_0) + \cdots ,$$

where $f'(x_0) = 0$, and

$$1 + x_0 = \left(\frac{M}{4\alpha + 1} \right)^\alpha .$$

Using the first three terms in the Taylor expansion for $f(x)$ we find that

$$\frac{dN}{dx} \propto \exp\left[-\frac{(x + x_0)^2}{2\sigma^2} \right]; \qquad \sigma^2 = \frac{\alpha^2}{M} \left(\frac{M}{4\alpha + 1} \right)^{2\alpha + 1} . \qquad (6.17)$$

The distribution of bubble sizes is formally peaked at $x = x_0 \simeq -1$ with a gaussian width σ. For the largest value of α consistent with the big-bubble constraint, $\alpha = 10$, $x_0 \simeq -1$ and $\sigma \simeq 0.4$. This corresponds to a monotonically decreasing distribution of bubble sizes with a width of order a few Hubble lengths.

(c) Almost-eternal inflation

We will now consider a fairly generic class of marginal models. If $\Gamma(t)$ does not drop precipitously at early times, it may happen that $p(t) \to 0$, while the fraction of the physical volume still in the false vacuum continues to increase: $d[R(t)^3 p(t)]/dt > 0$. That this can happen is easy to see: Suppose for purposes of illustration that $\Gamma(t)$ is constant at early times; then $\ln[p(t)] = I(t) \simeq (\Gamma/H^4) \int H dt$ increases with the number of e-foldings of inflation—and must inevitably become large, no matter how small Γ is. However, if $\Gamma \lesssim H^4$, then $V_{\text{phys}}^{-1} dV_{\text{phys}}/dt = 3H - I' \simeq 3H(1 - \Gamma/3H^4)$ will still be positive. This is what occurs in old inflation [15].

As we shall now see, all is not lost provided that at some point in time, at least 60 or so e-foldings after the beginning of inflation, $I(t)$ increases very rapidly. Suppose that this occurs at $t = t_{\text{rapid}}$, after many ($\gg 60$) e-foldings of inflation. By rapid we shall mean that

$$\frac{dI(t_{\text{rapid}})}{dt} \equiv MH, \qquad (6.18)$$

where $M \gg 1$, is say at least 30 or so, and $I' \gg H$ is maintained long enough so that the change in I is also much greater than 1; see Fig. 10. For example, if $I' = MH$ is sustained for a Hubble time, then $\Delta I \simeq M$.

Before the rapid increase in $I(t)$ most of physical space is still in the false vacuum, in spite of the fact that I may be very large, so that most of comoving space is in the true vacuum. Because of this, a *typical* region of physical space is still in the false vacuum. We shall focus on a suitably large region of space still in the false vacuum, one that has inflated the 60 e-foldings required to solve the horizon/flatness problems, and is not scarred by too many big bubbles.

Since most of the physical volume of space is still in the false vacuum, one would expect that such regions are "plentiful." The requirement that the region not be scarred by too many big bubbles is also easy to satisfy, provided that during the last 60 e-foldings Γ/H^4

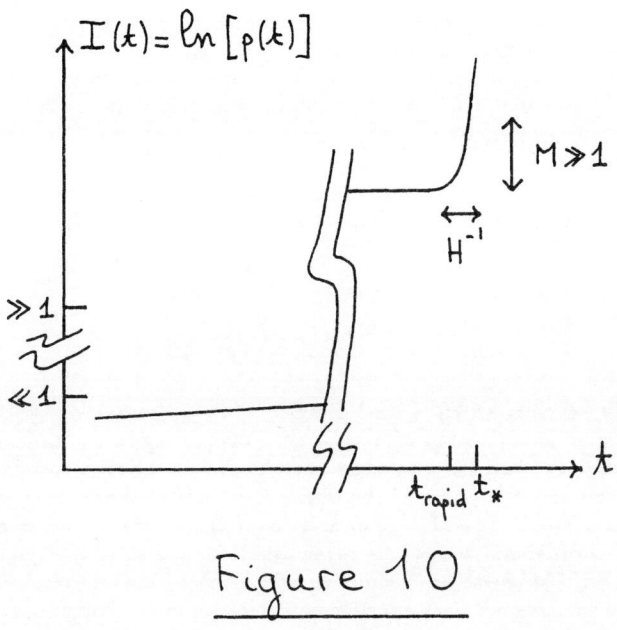

Figure 10

has been less than about 10^{-3}. Recall that the fraction of the volume of the Universe that is expected to be in bubbles of a given size is equal to the value of Γ/H^4 when bubbles of such size were nucleated; see Section Va. The dangerous bubbles are those that were nucleated 30 to 50 e-foldings before the end of inflation. If Γ/H^4 is less than 10^{-3} or so during the 60 e-foldings before $t_{\rm rapid}$, the number of big bubbles will be acceptably small.

What is relevant for the selected region is not,

$$I(t) = \int_{t_0}^{t} \Gamma(t')R^3(t')V(t,t')dt', \qquad (6.19a)$$

but rather,

$$I_{\rm rapid}(t) = \int_{t_{\rm rapid}}^{t} \Gamma(t')R^3(t')V(t,t')dt'. \qquad (6.19b)$$

This is because by choice, we know that in spite of the large value of $I(t)$ at $t = t_{\rm rapid}$, this particular region is *still* in the false vacuum. For this region the clock has been reset so to speak. If the condition for a rapid increase in $I(t)$ is satisfied, i.e., $I' = MH \gg H$, and sustained, then at time $t_{\rm rapid}$: (i) the physical volume in the false vacuum will begin to decrease since $3H - I' \simeq -MH \leq 0$; and (ii) $I_{\rm rapid}(t)$ will, in order a Hubble time increase from 0 to order $M \gg 1$. This means that for the selected region of space the phase transition will complete on a timescale $\mathcal{O}(H^{-1})$ or less, corresponding to a fast transition. While we have not been specific enough to allow the computation of the exact duration of the phase transition or the distribution of bubble sizes, it should be clear that $\delta t \sim (30/M)H^{-1} \lesssim H^{-1}$ and that the distribution of bubble sizes will be sharply peaked and characterized by a size $\mathcal{O}(\delta t)$.

In short, we have shown that a class of models that satisfy the following conditions should successfully undergo a graceful exit from inflation (in most regions of the volume of physical space): (i) Rapid increase in $I(t)$, quantified by $I'(t_{\rm rapid}) = MH \gg H$, maintained long enough so that $\Delta I \gg 1$; (ii) At least 60 e-folds of inflation before $t_{\rm rapid}$; (iii) Volume of physical space in the false vacuum prior to $t_{\rm rapid}$ still increasing; and (iv) Bubble-nucleation "quiet" for the 60 e-foldings of inflation prior to $t_{\rm rapid}$: $(\Gamma/H^4) \lesssim 10^{-3}$.

(d) Two-field inflation

Models have been studied where two fields are involved in inflation: the Higgs field, σ, that is trapped in the false vacuum; and a second field, ϕ, which couples to the Higgs field and is slowly involving toward the minimum of its potential [25]. As the ϕ field evolves the tunneling action for the Higgs field varies: $A = A(\phi)$. The idea then is that the ϕ field "triggers" the phase transition: ϕ slowly rolls and as it does the tunneling action decreases; eventually A is sufficiently small ($A = A_*$ for $\phi = \phi_*$) that bubbles are rapidly nucleated and the phase transition is completed.

Several general remarks can be made: (i) the field ϕ is a "slow roller," and the timescale for its evolution controls the amount of inflation; (ii) if this is to be first-order inflation, rather than slow-rollover inflation, the vacuum energy associated with ϕ must be much smaller than that associated with σ, so that the expansion rate is controlled by V_σ; and (iii) curvature perturbations arise due to quantum fluctuations in the ϕ field, and one must check that they are at most $\mathcal{O}(10^{-4})$.

As we have discussed the quantity determines the success (or nonsuccess) of bubble nucleation is

$$\dot{A}_* \equiv \left(\frac{\partial A}{\partial t}\right)_{t_*}.$$

If we suppose that A depends upon some multiplicative power of ϕ, $A \propto \phi^\beta$, we can write

$$\dot{A}_* = \beta\left(\frac{A_*}{\phi_*}\right)\left(\frac{\partial \phi}{\partial t}\right)_{t_*}. \tag{6.20}$$

Next, recall that the phase transition completes when $\Gamma = Ce^{-A_*} \sim H^4$ (unless $|\dot{A}_*| \gg 1$); if we write $C = m_U^4$ and $H^2 \sim m_U^4/m_{\rm Pl}^2$, it follows that

$$A_* \simeq 4\ln(m_{\rm Pl}/m_U) \sim 3 \times 10^{-2}.$$

Since the "trigger-field" ϕ is a slow roller, its equation of motion is

$$\frac{\partial \phi}{\partial t} \simeq \frac{-V'}{3H},$$

where $V' = \partial V/\partial\phi$.

We will now integrate this equation, assuming that $H =$ const (remember, H depends only weakly upon ϕ). We will consider two cases: first, where ϕ *decreases* as it evolves; e.g., if $V(\phi) = \lambda\phi^4$, or $m^2\phi^2/2$. Let ϕ_i be the initial value of ϕ (i.e., at the beginning of inflation); it then follows that the number of e-foldings of inflation is given by

$$N(\phi_*) \equiv \int H\,dt = 3H^2 \int_{\phi_i}^{\phi_*} d\phi/V' \simeq \frac{3H^2\phi_*}{(p-1)V'(\phi_*)}, \tag{6.21}$$

where the final expression follows by assuming that $V' \propto \phi^p$.* The quantity $N(\phi_*)$ is the total number of e-foldings in the scale factor during the time it takes ϕ to roll from its initial value ϕ_i to the value that it has when $\phi = \phi_*$, and therefore must be at least 60 or so to solve the horizon/flatness problems.

Now consider the case where ϕ *increases* during inflation; e.g., $V(\phi) = \text{const} - \lambda\phi^4 + \cdots$. If we again suppose that $V' \propto \phi^p$, it follows that

$$N(\phi_*) \equiv \int H\,dt = 3H^2 \int_{\phi_*}^{\phi_0} d\phi/V' \simeq \frac{3H^2\phi_*}{(p-1)V'(\phi_*)}, \tag{6.21'};$$

where ϕ_0 is the minimum of $V(\phi)$. In this case $N(\phi_*)$ is not the total number of e-foldings of the scale factor during inflation; instead, it is the number e-foldings that it would take for ϕ to evolve to its minimum (had inflation not ended when $\phi = \phi_*$). While the quantity that determines the amount of inflation is the number of e-foldings that occur during the time it takes ϕ to evolve from its initial value to ϕ_*, in absence of "fine tuning" we might expect that $N(\phi_*)$ is also a large number, say 100 or so. In fact in either case, ϕ increasing

---

* In the case that $p = 1$, the $(p-1)$ factor in Eq. (6.21) becomes $[\ln(\phi_i/\phi_*)]^{-1}$, which is also of order unity.

or decreasing as it evolves, in the absence of "fine tuning" one would expect $N(\phi_*)$ to be a large number, say well in excess of 100.

With Eqs. (6.21) in hand, we can now estimate $\alpha \equiv H/|\dot{A}_*|$ for a generic two-field model. From the previous equations it follows that

$$\alpha = \frac{H}{|\dot{A}_*|} = \frac{\phi_*}{\beta A_*} \frac{H}{\partial \phi/\partial t} \simeq \frac{1}{4\ln(m_{\rm Pl}/m_U)} \frac{(p-1)N(\phi_*)}{\beta}. \tag{6.22}$$

Recall that the big-bubble constraint implies that α can be at the very most $\mathcal{O}(10)$. If $m_U \sim 10^{15}$ GeV, then the first factor is order $1/40$; unless things are arranged so that the action just becomes small enough to end the phase transition when $\phi = \phi_0$, $N(\phi_*)$ must be order 100 or so; for many simple models p is 3. Thus $\alpha \simeq N(\phi_*)/20\beta$ is expected to be of order a few (give or take a factor of 10). That is, a generic two-field inflation model will, at best, be close to the hairy edge when it comes to satisfying the requirements for successful bubble nucleation.

In a two-field model, the Higgs field σ is responsible for the conversion of vacuum energy to radiation (i.e., reheating), while the field ϕ triggers the phase transition and is responsible for the production of density perturbations.† Density perturbations arise due to quantum fluctuations in ϕ. The amplitude of the density perturbations at the post-inflationary horizon crossing is easy to estimate using the fact that while a given mode is outside the horizon the gauge-invariant quantity $\zeta \equiv \delta\rho/(\rho + p)$ remains constant [4]. The Higgs field makes no contribution to $\rho + p$ ($\rho_\sigma + p_\sigma = 0$) or $\delta\rho$ (σ field fluctuations are suppressed because in the false-vacuum state, $m_\sigma^2 \gg H^2$); for the ϕ field: $\rho_\phi + p_\phi = \dot{\phi}^2$ and $\delta\rho = \Delta\phi V' \simeq HV'/2\pi$. From this it follows that

$$\left(\frac{\delta\rho}{\rho}\right)_{\rm HOR} \simeq \frac{3H^3}{V'}; \tag{6.23}$$

where as usual the right-hand side of the equation is to be evaluated the scale of interest crossed outside the horizon during inflation, 40-60 e-foldings before the end of inflation for the astrophysically interesting scales.

This is the familiar formula for the amplitude of inflation-produced density perturbations, though we should remember that H is determined by the σ field and not the ϕ field: $H^2 \simeq 8\pi V_\sigma/3m_{\rm Pl}^2$. Recognizing this fact we can see that $(\delta\rho/\rho)_{\rm HOR}$ takes on the value that it would have in the absence of the σ field times $(V_\sigma/V_\phi)^{3/2}$. Moreover, inflation ends when the σ makes the transition to the true vacuum, and not when the ϕ field reaches its minimum, which implies that the ϕ field is in general further away from its minimum 40-60 e-foldings before the end of inflation then if the σ field was not present. In the case that $V(\phi) = \lambda\phi^4$ or $V(\phi) = m^2\phi^2/2$, the value of ϕ will in general be larger, leading to smaller density perturbations; on the other hand, if $V(\phi) = {\rm const} - \lambda\phi^4 + \ldots$, the value of ϕ will be smaller, leading to larger density perturbations.

---

† Because the ϕ field is very weakly coupled, one has to worry about the energy density left in the coherent oscillations of the ϕ field, which behave as nonrelativistic matter: the "Polonyi problem." Unless the ϕ field is unstable and has a relatively short lifetime, this can be a serious problem.

There is another interesting relation that must hold in two-field models. It evolves the relationship between formulas (6.21) and (6.23). If we suppose again that $V' \propto \phi^p$ and that the value of ϕ is increasing, then the value of ϕ 60 e-foldings before the end of inflation is given by a formula similar to Eq. (6.21'):

$$60 = \frac{3H^2 \phi_{60}}{(p-1)V'(\phi_{60})}.$$

A similar looking relation can be derived by using Eq. (6.23) and demanding that the density perturbations on astrophysically interesting scales be $\mathcal{O}(3 \times 10^{-5})$:

$$3 \times 10^{-5} \simeq \frac{3H^2}{V'(\phi_{60})}.$$

From these two we can solve for the value of ϕ 60 e-folds before the end of inflation

$$\frac{\phi_{60}}{H} \simeq 2(p-1) \times 10^6.$$

This relationship is dictated by the density-perturbation constraint and the requirement for sufficient inflation, and must hold for two-field models where ϕ increases as it evolves; e.g., $V(\phi) = \text{const} - \lambda \phi^4 + \cdots$.

Most important for our present considerations is the fact that $\alpha \equiv H/|\dot{A}_*|$ can be related to $(\delta\rho/\rho)_{\text{HOR}}$:

$$\alpha = \frac{1}{\beta A_*} \left(\frac{\delta\rho}{\rho} \right)_{\text{HOR}} \frac{\phi_*}{H} \sim \frac{10^{-6}}{\beta} \frac{\phi_*}{H} \sim \frac{\phi_*}{10^{17} \text{ GeV } \beta (m_U/10^{15} \text{ GeV})^2}; \qquad (6.24)$$

where in the final expressions we have taken $(\delta\rho/\rho)_{\text{HOR}} \simeq 3 \times 10^{-5}$ and $A_* \simeq 3 \times 10^{-2}$. Provided that ϕ_* is less than about 10^{18} GeV, density perturbations of the proper size and a value of α consistent with the big-bubble constraint can both be achieved. For potentials of the form $V(\phi) = \lambda \phi^4$ or $m^2 \phi^2/2$ the value of ϕ_* is several times the Planck mass, so that both conditions cannot be met.

References

[1] For comprehensive reviews of the "inflationary paradigm" see e.g., E.W. Kolb and M.S. Turner, *The Early Universe* (Addision-Wesley, Redwood City, CA, 1990), Chapter 8; K.A. Olive, *Phys. Repts* 190, 307 (1990); A.D. Linde, *Particle Physics and Inflationary Cosmology*.

[2] Slow-rollover includes the original models of "new inflation," see A. Albrecht and P.J. Steinhardt, *Phys. Rev. Lett.* 48, 1220 (1981) and A.D. Linde, *Phys. Lett. B* 108, 389 (1982), as well as so-called chaotic inflation, see A.D. Linde, *Phys. Lett. B* 129, 177 (1983). For exhaustive discussions of slow-rollover inflation we refer the reader to Refs. [1].

[3] A.H. Guth, *Phys. Rev. D* 23, 347 (1981).

74

[4] J.M. Bardeen, P.J. Steinhardt, and M.S. Turner, *Phys. Rev. D* **28**, 679 (1983); A.H. Guth and S.-Y. Pi, *Phys. Rev. Lett.* **49**, 1110 (1982); S.W. Hawking, *Phys. Lett. B* **115**, 295 (1982); A.A. Starobinskii, *Phys. Lett. B* **117**, 175 (1982).

[5] R. Holman, P. Ramond, and G.G. Ross, *Phys. Lett. B* **137**, 343 (1984).

[6] S. Coleman, *Phys. Rev. D* **15**, 2929 (1977); C.G. Callan and S. Coleman, *ibid* **16**, 1762 (1977); S. Coleman and F. De Luccia, *ibid* **21**, 3305 (1980).

[7] A.H. Guth and E.J. Weinberg, *Nucl. Phys.* **B212**, 321 (1983); E.J. Weinberg, *Phys. Repts.*, in preparation (1991).

[8] R. Watkins and L.M. Widrow, *Phys. Rev. D*, in press (1991).

[9] D. La and P.J. Steinhardt, *Phys. Rev. Lett.* **62**, 376 (1989).

[10] See e.g., E.W. Kolb, *Physica Scripta* **T36**, 199 (1991).

[11] E.W. Kolb, D. Salopek, and M.S. Turner, *Phys. Rev. D* **42**, 3925 (1990); also see the recent paper by, A.H. Guth and B. Jain, MIT preprint CTP-1964, submitted to *Phys. Rev. D* (1991).

[12] M.S. Turner and F. Wilczek, *Phys. Rev. Lett.* **65**, 3080 (1990).

[13] A. Kowsowsky, M.S. Turner, and R. Watkins, in preparation (to be submitted to *Phys. Rev. D*) (1991).

[14] M.S. Turner, E.J. Weinberg, and L.M. Widrow, to be submitted to *Phys. Rev. D* (1991).

[15] E.J. Weinberg, *Phys. Rev. D* **40**, 3950 (1989).

[16] A.H. Guth and E.J. Weinberg, *Phys. Rev. D* **23**, 876 (1981).

[17] A.H. Guth and E.J. Weinberg, *Nucl. Phys.* **B212**, 321 (1983); E.J. Weinberg, *Phys. Repts.*, in preparation (1991).

[18] C.J. Hogan, *Phys. Lett. B* **133**, 172 (1983).

[19] R. Watkins and L.M. Widrow, Fermilab preprint FERMILAB-Pub-91/164-A, submitted to *Phys. Rev. D* (1991).

[20] See e.g., the recent review by T.P. Walker et al., *Astrophys. J.* **376**, 51 (1991).

[21] See e.g., the recent review by R. Malaney, *Phys. Repts.*, in press (1991).

[22] H. Gush et al., *Phys. Rev. Lett.* **65**, 537 (1990); J. Mather et al., *Astrophys. J.* **354**, L37 (1990); *ibid*, in press (Oct. 1991).

[23] R. Holman et al., *Phys. Lett. B* **237**, 37 (1990).

[24] See e.g., C.M. Will, *Theory and Experiment in Gravitational Physics* (Cambridge Univ. Press, Cambridge, 1981).

[25] See e.g., F. Adams and K. Freese, *Phys. Rev. D*, (1991).

INFLATION, GREAT ATTRACTOR AND ANISOTROPIES OF THE RELIC RADIATION

V.N. LUKASH
Astro Space Centre
Academy of Sciences
Profsouznaya 84/32, 117810 Moscow
USSR

ABSTRACT : We consider two lines of predictions for $\Delta T/T$ expected on large scales (i.e. $\theta > 1°$) : theoretical and observational. The first topic, built beginning from the high energies (Big Bang), is basically backed by Inflation today. The second stems from the large scale streaming motions, and does not assume any primordial perturbation spectrum.

1. Introduction

What can COBE do for the modern cosmology in view of $\Delta T/T$ observations ? A detailed answer is given at this school. In my lecture, I will dwell upon two possibilities coming from the gravitational instability mechanism:

(i) testing Inflation,

(ii) testing large-scale structure in the Universe.

For the first case the quadrupole anisotropy is a key point. For the second, it is the $\Delta T/T$ on the COBE smallest scale that matters. In both cases, the expected optimistic prediction is at the level $\Delta T/T \approx 10^{-5}$, which is hopefully reachable within a year of observations [1]. There seems to be the following possible ways out of the problem. If $\Delta T/T$ fluctuations are not to be discovered at the level $\approx 10^{-5}$, the Gaussian perturbation theories may be in trouble. Also, it may well be that some fluctuations would be found by COBE at a low confidence level ($\approx 2\sigma$). Even in this case it will be a great challenge for cosmology, especially in view of the RELIC II mission which will have all chances to confirm or disconfirm our ideas about how structure has appeared in the Universe.

My number $\Delta T/T \approx 10^{-5}$ is a factor of two larger than that predicted by other speakers. The reasons are as follows :

(i) If we really want to discuss Inflation seriously, then let us remember than most models available predict some deviations from the Harrison-Zeldovich post-inflationary spectrum (HZPS) – namely, the perturbation amplitude grows (at least, logarithmically) relative to HZPS with scale growing. For us it means that making the normalization of the primordial spectrum at Great Attractor scale (≈ 40 h^{-1} Mpc), we have some gain in the amplitude on the horizon scale (≈ 3000 h^{-1} Mpc) which is resposible for the quadrupole anisotropy.

(ii) To relate $\Delta T/T$ fluctuations with bulk velocities, we must assume that the Great Attractor is a typical object in the Universe (the Kopernikus Principle). To make GA statistically

---

[1] All estimates for $\Delta T/T$ are given for h = $H_0/(100$ km s^{-1} Mpc$^{-1}) = 0.5$

M. Signore and C. Dupraz (et al.), The Infrared and Submillimetre Sky after COBE, 75–85.
© 1992 *Kluwer Academic Publishers.*

meaningful, not to say that half of its volume is still to be measured, we need modelling – which is more than just the current observations. I will show here that the modelling gives a gain factor of about two.

Below, both topics are briefly analysed.

2. Chaotic Inflation

Since Mike Turner made a good introduction to Inflation Paradigm, I will speak mainly on topics which were not properly covered and are prior to my understanding of the problem.

Let me first stress that, if we want to speak about the *well fixed things*, I would rather give some preference to the theories of the Very Early Universe which have the cosmological standards rather than the particle physics ones, since physics is not yet fixed well at high energies. This point needs some explanation.

In the situation which is ours today, there are two ways to make inflation. The first is to probe inflation for N particle theories including modified gravities as well, where N is a large number. As this mode is coupled to fundamental physics, it can give results only in view of the physics progress, not to say that the major problem here seems to be the Λ-term one rather than the inflation itself.

Another way is to try a cosmological approach to inflation before the true physics is to be confructed. This is justified by the following arguments. The first is testing : amongst different inflationary predictions, there are few of vital cosmological interest, e.g., the flat Friedmann model, post-inflationary primordial perturbation spectrum etc. The next reason comes from the purely cosmological standards of modelling the Universe, which proved to be a powerful tool of the Universe exploration. They are very simple : introduce some basic hypotheses (postulates), then develop a theory from them and test it observationally. One exemple is the Friedmann-Robertson-Walker (FRW) standard models, based on the spatial homogeneity and isotropy (cosmological postulate), which described successfully our Universe on a large scale.

The idea is to find basics for the inflationary models in General Relativity which could create the Friedmannian region (patch) we live in now. Below, I argue for cosmologically standard inflation theory existing today.

First, inflationary theories coupled to particular physics revealed that there exists the broad class of models with the common feature : their evolution is driven by a scalar field ϕ (inflaton) with the potential energy predominant. Linde (1983) first noticed that just initial domination of the potential energy ($\approx \lambda\phi^4$) is quite enough to create inflation. He called his scenario "chaotic" inflation for the reasons of the initial conditions. Recently, the idea has developed to the general theory. Here, we present this theory fundamentals and also calculate the spectra of the primordial perturbations generated.

3. Start Inflation and the Perturbation Spectra

Let us consider some particle physics with the standard gravity [2] :

$$W = \int (L - 1/2\,R)\,(-g)^{1/2}\,d^4x \tag{1}$$

where $R = R(g^{ik})$ is the Ricci tensor in the metric g^{ik}. The basic postulate of the chaotic inflation suggests that the field Lagrangian L is dominated by an inflaton ϕ, weakly coupled

---

[2] Our units imply : $c = h = 8\pi G = 1$

with all of the other fields :

$$L = L(\phi,w) + o \text{ (all other fields)}$$

where $w^2 = \phi_{,i}\phi^{,i}$ is the kinetic term. The weak coupling is the self-consistent condition of the theory (see end of the Section), allowing for metric to be considered as that induced by the inflaton :

$$W = W[\phi,g^{ik}]$$

There is a set of generalizations for the ϕ-Lagrangian :

$$L = -V(\phi) + w^2/2 \qquad (2a)$$
$$L = -V(\phi) + p(w) \qquad (2b)$$
$$L = L(\phi,w) \qquad (2c)$$

where $V = V(\phi)$, $p = p(w)$, $L(\phi,w)$ are the arbitrary functions of their arguments. Theory can be elaborated in the most general form of Eqs. (2b,c) (see Lukash & Novikov 1991). However, since we are interested in the case when the potential energy is predominant, the structure of the kinetic term is not of principal importance any more. For this reason, it is enough for us to consider here the trivial case of Lagrangian (see Eq. (2a)) with inequality providing for accelerated expansion of the local volume :

$$V(\phi) > w^2/2 \qquad (3)$$

Under Eq. (3), the cosmological inflation is caused by the gravity (better to say, "anti-gravity") of ϕ-field slowly rolling down the potential. Obviously, that can happen in regions of a monotonic dependence of $V(\phi)$ on ϕ. Without loss of generality, we may assume :

$$V(\phi) > 0 \quad , \quad dV/d\phi > 0 \quad \text{for } \phi > 0 \qquad (4)$$

with a stable minimum of $V(\phi)$ at $\phi = 0$, and $V(0) = 0$ (which is necessary to hold $\Lambda = 0$).

An example of potential (4) is given on Fig. 1. There are three dynamical ϕ-regimes driving the Universe expansion :

I. *Oscillations* near the equilibrium point $\phi = 0$, $|\phi| < \phi_0 \approx 1$ (decelerated expansion) ;
II. *Slow-rolled* evolution, $\phi_0 < \phi < \phi_F$ (chaotic inflation) ;
III. *Stochastic* evolution, $\phi_F < \phi < \phi_1$ (stochastic inflation).

Quantum fluctuations of the inflaton predominate in the ϕ evolution at the third region (and they are responsible for the highly non-linear global Universe today). During this random stochastic process, ϕ-field can happen to evolve below the critical value ϕ_F at some space-time point [3]. The further ϕ-evolution proceeds as a classical aperiodic motion from ϕ_F to ϕ_0, during which the horizon is boosted by inflation presenting just inside the Friedmannian patch with $\Omega_{tot} \approx 1$. By ϕ_0-time, the patch scale l_F inflates far away (more than 60 folds) beyond causal horizon. The space-time separation in the Friedmann patch is fixed by the long-wave modes developed in the stochastic region by the beginning of the second period at the given ϕ_F-point.

---

[3] By the "point" dimension in regions II and III, we imply the Hubble radius H^{-1}, where H-function is formally introduced as $H \approx (V/3)^{1/2}$. Very often, the Hubble radius is called "horizon" though it is technically incorrect.

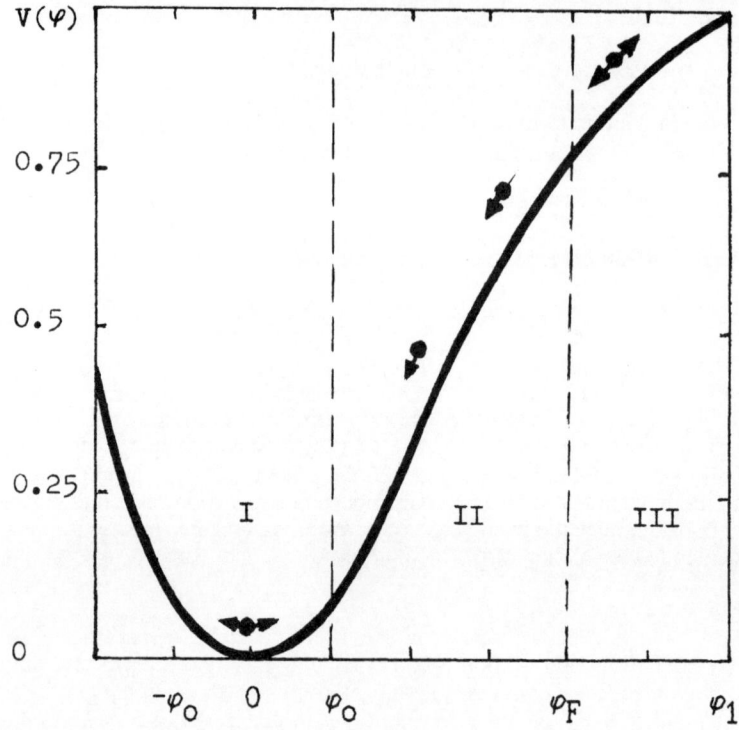

Fig. 1   Potential energy of φ-field at chaotic inflation

The quantum fluctuations developed at the second period have a negligible amplitude : they create the primordial cosmological perturbations inside the Friedmannian patch. Thus as you can see, a very generic feature of the theory is that the perturbation spectrum grows with scale growing just reaching non-linear values on the Friedmann boundary.

The process of inflation stops at Planckian field, $\phi \approx \phi_0 \approx 1$ (Eq. (3) breaks), transferring smoothly to periodic regime where ϕ-field decays into particles (the Reheating).

There are a few important things to be emphasized :

(i) In inflationary regions (II and III), dynamics of the larger-than-horizon part of ϕ-field is governed by the Langevin equation, and the corresponding metric is locally a quasi-isotropic one (see Landau & Lifshitz 1967) with the Hubble parameter $H = u_k \cdot^k/3 \approx (V/3)^{1/2}$, where $u^k = \phi_{,}^k/w$ is the inflaton 4-velocity. The large-scale fields are classical objects, while quantum perturbations play the role of a stochastic force that drives the former. The inflationary solution is the trap separatrix of the dynamical Eq-system so that the stochastic "drift" of classical inflaton proceeds along this separatrix.

(ii) In region II, the classical slow-roll evolution can be given in terms of an asymptotic expansion over the inverse squares of $\phi \gg 1$:

$$H = (V/3)^{1/2} [1 + 1/3\, c^2\phi^{-2} + 0(\phi^{-4})] \tag{5a}$$

$$w = -(2cH/\phi) [1 + 2/3\,(e-1)\,c\phi^{-2} + O(\phi^{-4})] \tag{5b}$$

where $c = 1/2\, d(\text{Ln } V)/d(\text{Ln } \phi)$ and $e = d(\text{Ln } c)/d(\text{Ln } \phi)$. The corresponding metric inflates according to the Friedmann laws on scales $1 < l_F$, $\phi < \phi_F$:

$$ds^2 = dt^2 - a^2 dx^2 \quad , \quad \ddot{a} > 0 \tag{5c}$$

where $a = \exp [\int H\, dt]$, H, $w' = \phi$ and ϕ are functions of time t. Eqs. (5) are true for a very broad class of potentials $V(\phi)$ determined by the slow-rolled conditions (smallness of the ϕ^{-2} terms in Eqs. (5a,b)) :

$$d(\text{Ln } V)/d\phi < 3^{1/2} \quad , \quad |d^2(\text{Ln } V)/d\phi^2| \le 1 \tag{6}$$

(the first inequality provides for the acceleration case).

(iii) In region II, the quantum perturbations of the inflaton $(1 < l_F)$ can be treated in linear approximation against the classical background (5). Therefore, the problem of creation of potential perturbations is reduced of the old one considered in the late 70's, namely to the problem of particle creation in a non-stationary Friedmann Universe (the parametric amplification effect, PAE). In the next section, we remind basics of the PAE theory for potential perturbations (Lukash 1980). The result can be expressed in terms of the q-field (the inflaton + metric coupled perturbations) which, on large scales, presents just the gravitational potential of the "growing" adiabatic mode :

$$ds^2 = dt^2 - a^2 [1 + (2/3)^{1/2}\, q(x)]\, dx^2 \quad , \quad |\nabla q| < Hq \tag{7}$$

Eq. (7) holds for any expansion law, in particular at the post-inflation epoch. The random Gaussian function $q(x)$ has the following spectrum :

$$q_k = (2\pi^2)^{-1/2} [V^{3/2}/(dV/d\phi)]_k \tag{8a}$$

Here, $\phi = \phi_k$ at the horizon crossing at inflation is given directly in terms of the wave number :

$$k = \dot{a} = a(\phi)\, H(\phi) \tag{8b}$$

Because the scale function in Eq. (8b) changes with ϕ much faster than the H-function, we see that q_k grows with k decreasing (i.e., ϕ and scale increasing). So, the critical ϕ_F and l_F can be obtained from Eq. $q_k \approx 1$.

(iv) In principle, Eqs. (8a,b) specify in parametric form any spectrum, depending on the potential and its first derivative shapes (see Eq. (4)). However, for smooth ones :

$$|e| \ll 1 \tag{9}$$

Eqs. (8a,b) resolve explicitly in the Harrison-Zeldovich spectra with only the logarithmic growth to higher scales :

$$V(\phi) \approx \lambda \phi^{2c}/2 \quad , \quad \phi \ge c_0 \approx 1$$

$$\phi_k{}^2 \approx c\,[c_0 + 4\,\mathrm{Log}\,[(k_0/k)\,(\phi/c_0)^{c-1/3}]] \tag{10}$$

$$q_k \approx (\lambda^{1/2}/4\pi c)\,\phi_k{}^{c+1} \approx (\lambda^{1/2}/\pi)\,(4c)^{(c-1)/2}\,(\mathrm{Log}\,k_0/k)^{(c+1)/2}$$

$$\phi_F{}^2 \approx \lambda^{-1/(1+c)} \quad , \quad l_F \approx \exp\,[\lambda^{-1/(1+c)}]\,(\mathrm{cm})$$

Here, λ and c are slowly varying ϕ-functions, $k_0 \approx \mathrm{mm}^{-1}$, $\phi_0 \approx c_0$.

(v) The slow-rolled conditions that led to Eqs. (8a,b) can easily be transformed into the potential and spectrum index restrictions :

$$V^{1/2} \le 7 q_k \quad , \quad |d(\mathrm{Log}\,q_k)/d(\mathrm{Log}\,k)| \le 2 \tag{11}$$

Since the potential at the horizon crossing goes to larger wavelength, we see that it must be rather flat, $V \approx 10^{-8}$ within the structure scale range $k^{-1} \approx (10–10^4)\,h^{-1}$ Mpc, which implies a very strict restriction for the coupling parameter ($\lambda \le 10^{-10}$). On the other hand, the spectrum in the scale range considered can deviate very significantly from the Harrison Zeldovich one so that it appears to be very sensitive to the potential. All this makes chaotic inflation a very testful and predictive theory for both Particle Physics and the large-scale structure of the Universe.

(vi) To start inflation, one should provide for a spatially smooth initial distribution of the inflaton on some finite scale l_{SI} to be larger than the horizon (see Eq. (3)) [4] :

$$\phi/|\nabla\phi| > l_{SI} \equiv \phi H^{-1} > H^{-1} \tag{12}$$

For a massive ϕ-field ($c = 1$, $\lambda = m^2$), $l_{SI} \approx m^{-1}$ is just the Compton wavelength, which prompts the clear physical interpretation of Eq. (12). Note in this connection that the weak coupling of the inflaton (see point (v)) is not at all a "disadvantage" but rather the self-consistency of the theory : both requirements, that l_F is larger than contemporary horizon and $q_k \ll 1$ (at galactic scales), can be achieved simultaneously for small coupling constants $\lambda \ll 1$ only (Eqs. (10)). It is Eq. (12) – the only principal condition of the theory – that reduces the cosmological postulate (homogeneity of the model) to the real and solvable problem.

---

[4] Because of the trap-separatrix property of inflation, the time derivative term $\dot{\phi}$ in Eq. (3) is not important.

4. Parametric Amplification of Potential Perturbations

The main reason which makes it possible to employ the PAE theory in chaotic inflation is that the large-scale modes there inflate outside horizon, and, therefore, become classical before the small-scale quantum fluctuations develop. So, each fluctuation evolves against local background already prepared by the large-scale modes [5], and then converts to the latter and changes the background at next step. More of that, in the second region the classical background is a stable Friedmannian one since the perturbations are small.

All this justifies the validity of the linear potential perturbation theory in Friedmann Universe for scales $l \ll l_P$. Below, we briefly remind its conclusions.

A basic object of the theory is gauge-invariant q-scalar, that is the linear superposition of perturbations of the inflaton $\delta\phi \equiv \Phi$ and one of the gravitational potentials $\delta g_{ik} = h_{ik}$:

$$\phi = \phi^{(F)} + \Phi \quad , \quad g^{ik} = g^{ik(F)} - h^{ik}$$

$$\Phi/w = X + 1/2 \, (C + D_{,i}u^i)$$

$$h_{ik} = Y \, (2 \, u_i u_k - g_{ik}) + (Cu_{(i)};_k) + D_{;ik}$$

$$q = -\xi^{-1} \, (HX + Y/2) \tag{13}$$

Here, $\phi^{(F)}$, $g^{(F)}_{ik}$, u^i, H, and w are Friedmannian functions (see Eqs. (5), "F" is omitted whenever possible) ; X, Y, C, and D are coefficients of the general potential decompositions : X and Y are gauge-invariant 4-scalars, whereas C and D are arbitrary functions (the gauge freedom). The constant factor ξ can be chosen regarding the Lagrangian normalization ($\xi = 6^{-1/2}$ for Eq. (2a)). Inverse transformations relate the original X and Y functions to the q-scalar :

$$X = P/2a - \xi q/H \quad , \quad Y = -HP/a \tag{14}$$

where $P = \int a\gamma q \, dt$, $\gamma H^2 = \xi \, \partial L/\partial(\text{Log } w)$. The Lagrangian for q-field is found by expanding the integrand of Eq. (1) up to the second order in Φ and h^{ik}, with the total divergent terms excluded :

$$W^{(2)} \, [\Phi, h_i{}^k] = W \, [q] = \int L(q) \, (-g^{(F)})^{1/2} \, d^4x$$

$$L(q) = 1/2 \, D^{ik} q_{,i} q_{,k} \quad , \quad D_{ik} = 1/2 \, \alpha^2 \, (u_i u_k + \beta^2 P_{ik}) \tag{15}$$

where $\alpha^2 = \xi\gamma/\beta^2$, $\beta^{-2} = w \, (\partial^2 L/\partial^2 w)/(\partial L/\partial w)$, $P_{ik} = g_{ik} - u_i u_k$.

We see that q is a massless field although the original Lagrangian depends explicitly on the ϕ-field. Also, an important thing is that the theory of gravitating potential perturbations proved to be equivalent to the standard theory of a test minimally coupled q-field in unperturbed Friedmann Universe, with all the known consequences [6]. The Equation of motion of the q-field,

$$(D^{ik} q_{,i});_k = 0 \tag{16}$$

is just the second-order hyperbolic-type Equation describing an oscillator with the friction term ($\approx H\dot{q}$) generally responsible for PAE (i.e. for the creation of phonons, the q-quanta, in the process of cosmological expansion).

---

[5] The characteristic lapse time here is just horizon H^{-1}

[6] Similar property have gravitons, see Grishchuk (1974)

The rest is to solve Eq. (16) in the case of the inflationary separatrix (5). To do it, let us first rewrite Eq. (16) for the conformal field $\bar{q} = \alpha a q$ in the Minkowski coordinates ($\eta \equiv \int dt/a$, x) :

$$\bar{q}_k'' + [(\beta k)^2 - U]\,\bar{q}_k = 0 \quad , \quad U \equiv (\alpha a)''/\alpha a \tag{17}$$

Then the substitution to Eqs. (5) yields :

$$\beta = 1, \qquad U = 2\,(aH)^2\,[1 - c_2\phi^{-2} + O(\phi^{-4})]$$

$$= 2\eta^{-2}\,[1 + 3\,c_3\phi^{-2} + O(\phi^{-4})] \tag{18}$$

$$\eta = -\,(aH)^{-1}\,[1 + 2\,c^2\phi^{-2} + O(\phi^{-4})]$$

where : $c_2/c = c + 3\,(1{-}e)$, $c_3/c = c + (e - 1)$. The conformal time $\eta < 0$ and the initial conditions are :

$$\bar{q}_k = (2k)^{-1/2}\exp(-ik\eta) \quad , \quad \text{for } k|\eta| \gg 1 \tag{19}$$

Eqs. (17–19) can be solved in a general form by matching two explicit solutions. The first one is obtained under the condition $k|\eta| > O(\phi^{-4})$, allowing for the U-approximation at $k|\eta| \approx 1$ as $U \approx \text{const}/\eta^2$:

$$\bar{q}_k = (2k)^{-1/2}\,(1 - i/k\eta)\,(e^{-ik\eta} + O(\phi^{-2}))$$

The second solution is generally true for $k|\eta| < 1$:

$$\bar{q}_k \approx i\pi 2^{1/2}\,k^{-3/2}\,\alpha a\,q_k$$

Fitting them into the overlapping region $\phi^{-4} < k|\eta| < 1$, we get the spectrum q_k (Eq. (8)). As we see, the spectrum grows to larger scales. Say, for $\lambda\phi^4$ theory (c = 2) we have :

$$q_k \approx 500\,\lambda^{1/2}\,[1 + 1/40\,\text{Log}\,(\lambda/\lambda_g)]$$

where λ_g is a galactic scale. So, the gain factor on the horizon reaches $\approx 20\%$ in comparison with galactic scales where the spectrum is normalized.

Notice, in conclusion, that the q-state taken to obtain the q-spectrum is identical with the |in>-vacuum field state ($k|\eta| \gg 1$, see Eq. (19)) with *zero* temperature. The PAE for q-field just briefed before, has nothing to do with the Gibbons-Hawking state (with *non-zero* temperature) in the de Sitter Universe.

5. The Great Attractor Normalization

The conservative approach to the GA data is mainly based on two points (see, e.g., Bertschinger *et al.* 1990). :

(i) Let us measure the bulk velocity of the matter within some sphere around us, say : $V\,(R_{TH} = 60\,h^{-1}\,\text{Mpc}) \approx 327\,\text{km s}^{-1}$, and compare it with the r.m.s. velocity for the given model – for CDM : $V\,(R_{TH} = 60\,h^{-1}\,\text{Mpc}) = 224\,b^{-1}\,\text{km s}^{-1}$. Then the comparison will be OK for CDM : $\approx 1.5\,b$ standard deviations. (Here, h and b are the Hubble and biasing parameters respectively).

(ii) Let us compare the overdensity in GA centre smoothed by some Gaussian filter, say : $\delta\,(R_f = 14\,h^{-1}\,\text{Mpc}) \approx 1.2$, which corresponds to $\delta \equiv \delta\rho/\rho \approx 0.7$ extrapolated to z = 0 according to the linear theory, with the r.m.s. perturbations, e.g., for CDM : σ_0 ($R_f = 14$ h^{-1} Mpc) $\approx 0.34 \, b^{-1}$. Then, it is OK for CDM again : $\nu = \delta/\sigma_0 \approx 2b$.

However, both points are confused by the following conter-arguments :

(i) *Position effect*. The Local Group (LG) is as far from GA as the GA size itself. Thus, if we center the top hat (TH) sphere on LG, we underestimate the effect.

(ii) *Scale effect*. If we center the TH-sphere on GA, then parameter ν (versus R_{TH}) will be negligible for both small and large R_{TH}. So, it is the position of the ν-maximum that indicates the GA peak scale, while the ν-parameter for smaller or larger scales underestimates the effect as well. According to our calculations, the GA peak scale is found within 25 to $35 \, h^{-1}$ Mpc.

The quantitative description requires modelling. We will assume the following basics for it :

(*) *GA definition*. GA is a large-scale coherent flow-field of galaxies, with a characteristic correlation length larger than the averaging filter scale used to get this velocity field : $l_c \geq 40 \, h^{-1}$ Mpc $> R_f \approx 10 \, h^{-1}$ Mpc.

(**) *Statistical hypothesis*. A real density peak to have developed into the GA is close to the mean density peak that can be derived in the theory. As we know, in case of high peak ($\nu_{max} \gg 1$), the latter has a spherically symmetric profile $\delta(r)$ proportional to the correlation function.

Under these hypotheses, Hnatyk et al. (1991) considered the GA formation in the two component post-recombination medium – dark WIMPs (Weakly Interacting Massive Particle) and primordial baryonic matter – with $\Omega_{tot} = 1$ and the different primordial spectra. The results belong to three groups :

(i) Some parameters of linear (initial) GA peaks are practically independent of the model. E.g., the central overdensity in linear peaks is : $\delta(0) = 1$–$2 \, (1+z)^{-1}$, while their non-linear evolution results in : $\delta_{nl}(0) > 3$. (If we do average the actual overdensities $\delta_{nl}(0)$ with the Gaussian filder $R_f = 14 \, h^{-1}$ Mpc, they reduce to 1–1.6 h^{-1} Mpc, in accordance with the POTENT data). The maximum of the ν-parameter, which is the ratio of GA peak density averaged within TH-sphere to the r.m.s. density in this sphere, lies between $\nu_{max} = 3.5$–4.5 b for the TH-radius $R_{TH} = 25$–$35 \, h^{-1}$ Mpc. It gives the gain factor ≈ 2 when normalizing to the GA.

(ii) The realization probability for GA peaks is very different for different models depending on many parameters most important, among which is GA symmetry. Say, as a single positive peak GA can form in CDM, HDM, and hybrid models, at the levels $\delta(0) \approx 4$–5 σ_0, 1–2 σ_0, and 2–4 σ_0 respectively, depending on h, b, etc. Note that middle peaks can significantly deviate from sphericity.

(iii) The central region of the peaks with $\delta(0) > (1+z)^{-1}$ collapses before $z = 0$, resulting in the galactic contraflows, shork waves and other observable phenomena.

In summary, we see that the standard CDM model can now be found in a position similar to that of HDM some years ago : CDM (HDM) is good for galaxy (GA) origin, but helpless for GA (galaxy) formation. The way out is to have the spectrum with large power on both scales.

One of the possible solutions is to develop the hybrid (cold+hot) dark matter models, leaving Harrison-Zeldovich post-inflationary spectrum. The calculation for $\Omega_{tot} = \Omega_c + \Omega_h = 1$ shows that having only about 30% of the total density in the form of neutrino-type particles adding to CDM, can ensure the necessary power enhancement on GA scales (see Fig. 2, Lukash 1990). There are other possibilities open.

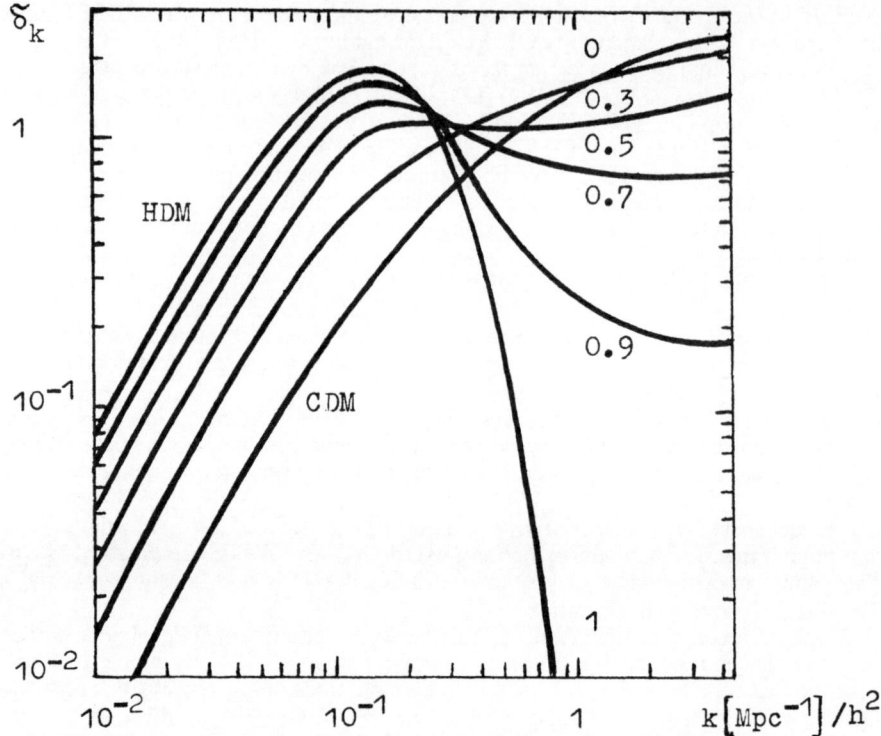

Fig. 2 Primordial perturbation spectra $\delta_k = \sqrt{k^3 P(k)/2\pi^2}$ for hybrid models with hot matter density parameter Ω_h (P(k) is the power spectrum, $\sigma_o(R_f = 0.35h^{-1} \text{Mpc}) = 1$).

6. Conclusions

The chaotic inflation gives a certain advantage to design any post-inflationary spectrum of primordial perturbations. The result depends on the analytical properties of the potential energy term $V(\phi)$ of the inflaton ϕ-field, thus, providing us with the very sensitive test of the theory. For smooth potentials, the spectrum grows moderately to larger scales in comparison with the Harrison-Zeldovich spectrum.

The GA modelling, understood as the coherence large-scale distortion of the galactic Hubble flow, provides such a normalization of post-recombination perturbation amplitude that appears a factor of two higher than that due to the standard normalization on the bulk velocities.

The main assumption necessary to relate the GA phenomenon with $\Delta T/T$ predictions is that there exist many GA's within the contemporary horizon. Here, the symmetry of the local GA is an important test in itself : if this deviates from sphericity, we deal with a typical object in the Universe. In the latter case, we easily derive a model-independent prediction $\Delta T/T$ $(\Theta \geq 1°) \approx 10^{-5}$, which, in the case of the Harrison-Zeldovich post-inflationary spectrum, scales up to quadrupole $(\Theta \approx 90°)$.

CDM standard model cannot account for GA as a statistically representative phenomenon. For this we need the spectrum enhancement of $\approx 30\%$ on GA scales. Hybrid models with hot and cold particles can realize this requirement.

The absence of the galactic contraflows and of X-ray gas in the central region of GA gives evidence for small initial overdensities, $\delta(0) < 1.5 \, (1+z)^{-1}$, and thus for a broad distribution of the GA mass. Nevertheless, the non-linear evolution leads to $\delta_{nl}(0) > 3$.

The confrontation of GA and $\Delta T/T$ observations may result into the direct detection of the post-recombination spectrum of primordial density perturbations.

REFERENCES

Bertschinger, E., Dekel, A., Faber, S.M., Dressler, A., and Burstein, D. :
    Ap.J. **364**, 370 (1990)

Grishchuk, L.P. : JETP **67**, 825 (1974)

Hnatyk, B.I., Lukash, V.N., and Novosyadly, B.S. : Astron. Zh. Lett. **17**, 3991 (1991)

Landau, L.D., and Lifshitz, E.M. : *The Field Theory*, Nauka, Moscow (1967)

Linde, A.D. : Phys. Lett. **B129**, 177 (1983)

Lukash, V.N. : JETP **79**, 1601 (1980)

Lukash, V.N. : *Great Attractor – Challenge to Theory*, Proc. Texas-ESO-CERN
    Symposium held at Brighton, Dec. 16–21 (1990)

Lukash, V.N., and Novikov, I.D. : Lectures on the Very Early Universe,
    preprint DEMIRM 91012, Observatory Paris-Meudon (1991)

STRING THEORY IN COSMOLOGY

N. SANCHEZ
Observatoire de Paris
Section de Meudon, DEMIRM
F–92195 Meudon Principal Cedex
FRANCE

ABSTRACT : In this lecture, I will describe those aspects of string theory which are relevant for Cosmology, with emphasis, for the purposes of this meeting, on the problem of connecting string theory to observational reality. I will talk on :
    o fundamental strings,
    o cosmic strings,
    o the possible connection among them.

1. Fundamental Strings

There exists only three fundamental dimensional magnitudes (length, time and energy) and so three fundamental dimensional constants :

$$(c, \hbar, G).$$

All other physical parameters being dimensionless, they must be calculable in an unified quantum theory of all interactions including gravity ("theory of everything"). The present interest in the theory of fundamental strings comes largely from the hope that it will provide such a theory. (There is no hope to construct a consistent quantum theory of gravity in the context of point particle field theory.) As it is known, fundamental strings cannot exist in arbitrary space time, and in particular they are consistent only in well determined (critical) dimensions $D_c > 4$. The unification of all interactions described by these theories takes place at the critical dimensions $D = D_c > 4$ where the characteristic unification scale is the Planck scale :

$$l_{Pl} = (2G\hbar/c^3)^{1/2} = 10^{-33} \text{ cm} \quad , \quad m_{Pl} = (\hbar c/2G)^{1/2} = 10^{-5} \text{ g}$$

corresponding to energies of order : $E_{Pl} = 10^{19}$ GeV. Here, $\mu = 1/\alpha'$ (α' is the characteristic string slope parameter with dimension of $(length)^2$), and $\mu^{1/2}$ characterizes the energy of string excitations. The problem of connecting string theory to reality is that of relating the theory in $D > 4$ dimensions where :

$$(G\mu)_D \approx O(1),$$

to the real world where $D = 4$ and :

M. Signore and C. Dupraz (et al.), The Infrared and Submillimetre Sky after COBE, 87–99.
© 1992 *Kluwer Academic Publishers.*

$$(G\mu)_{D=4} \ll 1,$$

problem currently handled within the so-called compactification schemes, or alternatively by the "four dimensional models" (see for example Refs. [1] and [2]). The effective low energy point particle field theory (containing GUTs and classical General Relativity) obtained by compactification is governed by physical couplings expressed in terms of μ via dimensionless numbers. (These numbers are vacuum expectation values of the different fields of the theory). The gauge (g) and gravitational couplings are at the tree level related by the Kaluza-Klein (heterotic type) relation :

$$g^2 = G\mu \quad \text{in} \quad D = 4. \tag{1}$$

The four and D dimensional couplings are related by :

$$g_D{}^2 = g_4{}^2\, V_{D-4} = (G\mu)_4\, V_{D-4} \tag{2}$$

where V_{D-4} is the volume of the (D–4) dimensional compact manifold K_{D-4}, of size l_{Pl} (assuming the ground state of the metric to be $R^4 \times K_{D-4}$). It is interesting here to plot the dimensionless parameter $G\mu$ against $M_{GUT}{}^2/\mu$, as required by string unification constraints [3] onto the renormalization group equations. Any perturbative compactification scheme leads to the non-shaded triangle delimited in Fig.1 (double logarithmic plot). It is natural to require : $M_{GUT} < \mu^{1/2}$, which excludes the upper shaded region. It is also natural to require that loop string corrections be under control, that is the ratio (loops/trees) $\ll 1$, which excludes the lower shaded region. The left shaded vertical region is also excluded in order to reproduce the value of the fine structure constant, $g_{em}{}^2 = 1/137$. This constraint is relevant here. The renormalization group equations :

$$M\,(dg^2/dM) = 2/3\pi\,(g^2)^2$$

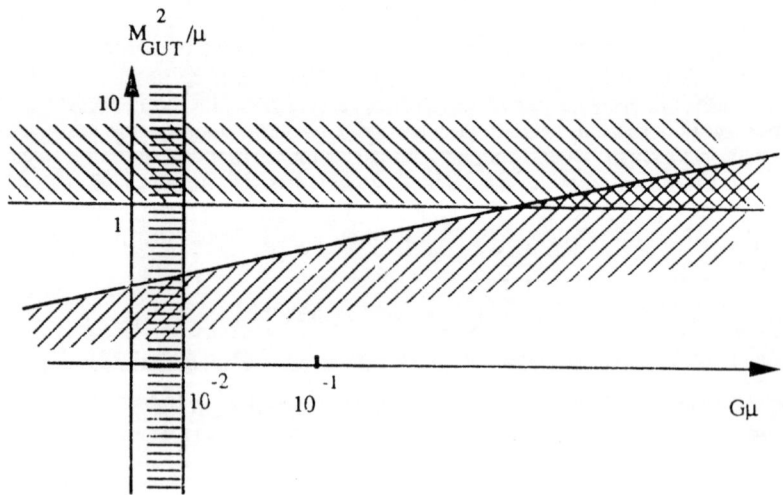

Fig.1. *The elementary unshaded triangle restricting the values of $G\mu$ according to current compactification schemes (log-log plot as in Ref. [3]).*

yield $\quad g^2(M_0) = \dfrac{g^2(M)}{1 + (2/3\pi)\, g^2(M)\, \ln(M/M_0)}$ $\qquad\qquad$ (3)

For : $M_F < M < M_{GUT}$, the coupling constant $g(M)$ increases with M as it should be (QED is infrared stable), and with : $M_0 = M_F = 200$ GeV, $M = M_{GUT} = 10^{16}$ GeV, we see :

$$1/g^2(M_{GUT}) = 1/g^2(M_F) - 6.35,$$

i.e. : $\qquad\qquad g^2(M_{GUT}) > g^2(M_F) = 1/137.$

Since the string unification relation Eq. (1), one must have :

$$G\mu > 1/137 \approx 10^{-2}. \qquad\qquad (4)$$

Let us recall now that cosmic strings are macroscopic objects, finite energy solutions (vortices) of gauge theories coupled to Higgs fields. They have a mass per unit length μ :

$$G\mu = (<\Phi>/m_{Pl})^2,$$

and a radius $\delta \approx 1/M_{UNIF}$, where $<\Phi>$ is the vacuum expectation value of the Higgs field. For grand-unified theories :

$$<\Phi>_{GUT} = 10^{-3}\, m_{Pl},$$

and thus : $\qquad\qquad G\mu = 10^{-6} \qquad\qquad (5)$

These thin strings are described by the Nambu equations.

On the other hand, cosmic strings should be derived from fundamental strings. The thermodynamical model of strings (ideal gas of free strings at finite temperature) predicts the existence of a phase transition at the Hagedorn [4] critical temperature $\approx \mu^{-1/2}$ to a finite energy state in which all the energy is concentrated in a very long string [5]. When applied to the primordial universe (at $t \approx 10^{-32}$ s), this transition can be interpreted as the condensation of fundamental strings into macroscopic cosmic strings and this stringy phase could be the ancestor of the adiabatic era [6,7]. But if cosmic strings are created in this way, they must satisfy the constraint Eq. (4).

Timing of millesecond pulsars as well as cosmic microwave radiation data on $\Delta T/T$ (see Sect. 2) sets up an upper limit to the string parameter $G\mu$ such that : $G\mu < 10^{-7}$–10^{-6}.

At present, we have two ways of connecting string theory to observational reality and thus of limiting the string parameter $G\mu$ [8] :
(i) radioastronomy observations (millisecond pulsar timing ; also microwave background data), which puts :
$$G\mu < 10^{-7}\text{–}10^{-6} ;$$

(ii) elementary particle phenomenology (compactification schemes ; also "four dimensional models"), which requires :

$$G\mu > 10^{-2}.$$

Cosmic strings derived from grand-unified theories agree with (i). For cosmic strings derived from fundamental strings or for fundamental strings themselves there is contradiction between the possibilities (i) and (ii). One of these scenarios connecting string theories to reality must be revised or the connection between fundamental and cosmic strings rejected. Meanwhile, FIRAS/COBE data can help to solve the problem, select one scenario or reject both of them.

1.1. FUNDAMENTAL STRINGS IN COSMOLOGICAL BACKGROUNDS

We have first studied quantum string propagation in de Sitter space [9]. We have found the mass spectrum and vertex operator and found an string instability in de Sitter space. The lower mass states are the same as in flat space-time, but heavy states deviate significantly from the linear Regge trajectories. We found that there exists a maximum (very large) value of order $1/g^2$ for the quantum number N and spin J of particles. There exists real mass solutions only for :

$$N < N_{max} = \pi/2g^2 + O(g^{-2/3}) \quad , \quad g = 10^{-61}.$$

Moreover, for states in the leading Regge trajectory the mass monotonously increases with J up to the value :

$$J_{max} = 1/g^2 + O(1),$$

corresponding to the maximal mass : $m_{max}^2 = 0.76 + O(g^2)$. Beyond J_{max}, the mass becomes complex. These complex solutions correspond to unstable states, already present here at the tree (zero handle) level.

From the analysis of the mass spectrum, we find that the critical dimension for bosonic strings in de Sitter space-time is $D = 25$, instead of $D = 26$ in Minkowski space-time. This result is confirmed by an independent calculation of the critical dimension from the path intergral Polyakov's formulation, using heat-kernel techniques : we find that the dilaton ß-function in D-dimensional de Sitter space-time must be :

$$ß^\phi = (D + 1 - 26)/(48\pi^2 \alpha') + O(1).$$

It is a general feature of de Sitter space-time to lower the critical dimensions in one unit. For fermionic strings, we find $D = 9$, instead of the flat value $D = 10$.

We have found that for the first-order amplitude, $\eta^i(\sigma,\tau)$ ($i = 1,...,D-1$ refers to the spatial components), the oscillation frequency is :

$$\omega_n = [n^2 - (\alpha'mH)^2]^{1/2},$$

instead of n, where H is the Hubble constant. For high modes, $n \gg \alpha'mH$, the frequencies $\omega_n \approx n$ are real. The string shrinks as the universe expands. This shrinking of the string cancels precisely the expanding exponential factor of the metric, and the invariant spatial distance does not blow up. Quantum mechanically, these are states with real masses (i.e., $m^2H^2 < 1$). This corresponds to an expansion time H^{-1} very much bigger than the string period $2\pi/n$, i.e., many string oscillations take place in an expansion period H^{-1} (in only one oscillation the string does not see the expansion).

For low modes, $n < \alpha'mH$, the frequencies become imaginary. This corresponds to an expansion time very short with respect to the oscillation time $2\pi/n$ ("sudden" expansion i.e., the string "does not have time" to oscillate in one time H^{-1}). These *unstable* modes are analyzed as follows. The $n = 0$ mode describes just small deformations of the center of mass motion, and it is therefore a physically irrelevant solution. When $\alpha'mH > 1$, relevant unstable modes appear. Therefore, the $n = 1$ mode dominates $\eta^i(\sigma,\tau)$ for large τ. Hence, if

$\alpha'mH > 2^{1/2}$, η^i diverges for large τ, i.e., fluctuations become larger than the zero order and the expansion breaks down. However, the presence of the above unstability is a true feature as it has been confirmed later by further analysis [10].

The physical meaning of this instability is that the string grows driven by the inflationary expansion of the universe : the string modes couple with the universe expansion in such a way that the string inflates together with the universe itself. This happens for inflationary (i.e., accelerated expanding) backgrounds. In Ref. [10] we have studied the string propagation in Friedman-Robertson-Walker (FRW) backgrounds (in radiation as well as matter dominated regimes), and interpreted the instability above discussed as Jeans-like *instabilities*. We have also determined under which conditions the universe expands,when distances are measured by stringy rods. It is convenient to introduce the *proper amplitude* $\chi^i = C\eta^i$, where C is the expansion factor of the metric. Then, χ^i satisfies the equation :

$$\ddot{\chi}^i + [n^2 - \ddot{C}/C]\,\chi^i = 0.$$

Here, a dot means a τ-derivative. Obviously, any particular (non-zero) mode oscillates in time as long as $\ddot{C}/C$ remains < 1, and in particular when $\ddot{C}/C < 0$. A time-independent amplitude for χ is obviously equivalent to a fixed proper (invariant) size of the string. In this case, the behaviour of strings is stable and the amplitudes η shrink (like $1/C$).

It must be noticed that the time component, χ^0 or η^0, is always well behaved and no possibility of instability arises for it : the string time is well defined in these backgrounds. (i) for non-accelerated expansions (e.g., for radiation or matter dominated FRW cosmologies) or for the high modes $n \gg \alpha'mH$ in de Sitter cosmology, string instabilities do not develop (the frequencies $\omega_n \approx n$ are real). Strings behave very much like point particles : the centre of mass of the string follows a geodesic path, the harmonic-oscillator amplitudes η shrink as the universe expands, in such a way as to keep the string's proper size constant. As expected, the distance between two strings increase with time, relative to its own size, just like the metric scale factor C.
(ii) for inflationary metrics (e.g., de Sitter with large enough Hubble constant), the proper size of the strings grows like the scale factor C, while the co-moving amplitude η remains fixed ("frozen"), i.e., $\eta \approx \eta(\sigma)$.

Although the methods of Refs. [9] and [10] allow to detect the onset of instabilities, they are not adequate for a quantitative description of the high instable (and non-linear) regime. In Ref. [11] we have developed a new quantitative and systematic description of the high instable regime. We were able to construct a solution to both the non-linear equations of motion and the constraints in the form of a systematic asymptotic expansion in the large C limit, and to classify the (spatially flat) Friedman-Robertson-Walker (FRW) geometries according to their compatibility with stable and/or unstable string behavior. An interesting feature of our solution is that it implies an asymptotic proportionality between the world sheet time τ and the *conformal time* T of the background manifold. This is to be contrasted with the stable point-like regime which is characterized by a proportionality between τ and the *cosmic time*. Indeed, the conformal time (or τ) will be the small expansion parameter of the solution : the asymptotic regime (small τ limit) thus corresponds to the large C limit only if the background geometry is of the inflationary type.The non-linear, high instable regime is characterized by string configurations such that :

$$|X'^0| \ll |\dot{X}^0| \quad , \quad |\dot{X}^i| \ll |X'^i|$$

with : $\quad X^0(\sigma,\tau) = C\,L(\sigma) \quad , \quad L(\sigma) = (\delta_{ij}\,X'^i X'^j)^{1/2}$

$$X^i(\sigma,\tau) = A^i(\sigma) + \tau^2\,D^i(\sigma)/2 + \tau^{1+2\alpha}\,F^i(\sigma)$$

where A^i, D^i and F^i are functions determined completely by the constraints, and α is the time exponent of the scale factor of the metric : $C = \tau\, L^{-\alpha}$. For power-law inflation, with $1 < \alpha < \infty$, $X^0 = \tau\, L^{1-\alpha}$. For de Sitter inflation, with $\alpha = 1$, $X^0 = \ln(-\tau HL)$. For Super-inflation, with $0 < \alpha < 1$, $X^0 = \tau\, L^{1-\alpha} + $ const. Asymptotically, for large radius $C \to \infty$, this solution describes string configurations with expanding proper amplitude.

These highly unstable strings contribute with a term of negative pressure to the energy momentum tensor of the strings. The energy-momentum tensor of these highly unstable strings (in a perfect fluid approximation) yields to the state equation $\rho = -P\,(D-1)$, ρ being the energy density and P the pressure ($P < 0$). This description corresponds to *large radius* $C \to \infty$ of the universe.

For *small radius of the universe*, highly unstable string configurations are characterized by the properties :

$$|\dot{X}^0| \gg |X'^0| \quad , \quad |X'^i| \ll |\dot{X}^i|$$

The solution for X^i admits an expansion in τ similar to that of the large radius regime. The solution for X^0 is given by L/C, which corresponds to *small radius* $C \to 0$, and thus to small τ. This solution describes, in this limit, string configurations with shrinking proper amplitude, for which CX'^i behaves asymptotically like C, while CX^i behaves like C^{-1}. Moreover, for an ideal gas of these string configurations, we found :

$$\rho = P\,(D-1),$$

with *positive pressure* which is just the equation of state for a gas of massless particles.

More recently [12], these solutions have been applied to the problem in which strings became a dominant source of gravity. In other words, we have searched for solutions of the Einstein plus string equations. We have shown that an ideal gas of fundamental strings is not able to sustain, alone, a phase of isotropic inflation. Fundamental strings can sustain instead a phase of anisotropic inflation, in which four dimensions inflate and simultaneously, the remaining extra (internal) dimensions contract. Thus, fundamental strings can sustain simultaneously inflation and dimensional reduction. In Ref. [12], we derived the conditions to be met for the existence of such a solution to the Einstein and string equations, and discussed the possibility of a successful resolution of the standard cosmological problems in the context of this model.

2. Cosmic Strings : Observational Tests

Recently (January 1990), a preliminary spectrum of the cosmic microwave background radiation (MBR) measured by FIRAS/COBE between 500 μm and 1 cm from regions near the North Galactic pole, has been reported [13]. It is a "genuine" spectrum of a black body with a temperature of 2.735 ± 0.06 K : the deviation from a black body is less than 1% of the peak intensity over the range 500 μm to 1 cm ; this COBE spectrum is not consistent with the Berkeley-Nagoya excess reported by Matsumoto *et al.* (1988) [14].

A stochastic background of gravitational waves can be detected by means of pulsar timing observations [15]. The stable rotation and sharp radio pulses of PSR 1937+21 make this millisecond pulsar a clock whose frequency stability can exceed that of the best atomic clocks. Up to now, the data yield a firm upper limit $\rho_g < 3.5\ 10^{-36}$ g cm^{-3} for the energy density of a cosmic background of gravitational radiation at frequencies $\nu \approx 10^{-8}$ Hz [16]. This limit corresponds to approximately $2\ 10^{-7}$ times the density required to close the Universe. On the other hand, accurate time of arrival measurements of pulses from the binary pulsar PSR 1913+16 over the last 14 years have led in particular to an accurate observed orbital decay rate $\dot{P}b$. The difference between predicted and observed orbital decay rates $\dot{P}b$ of PSR 1913+16 [17] can be used to place the best available limit on the total energy density of a cosmic gravitational wave background : $\Omega_{total} < 4\ 10^{-2}\ h^{-2}$ ($h \approx 0.7$), at the ultra low frequencies of 10^{-9} to 10^{-13} Hz.

Cosmic superconducting strings (SCS) raise current considerable interest in astroparticle physics [18,19]. If SCS exist, loops formed as a consequence of string interactions, decay during the expansion of the universe by emitting gravitational and electromagnetic waves. The radiation emitted by loops at different epochs, from their formation at $t \sim 10^{-32}$ seconds till now, adds up into a stochastic gravitational wave background characterized by the (dimensionless energy density) spectrum $\Omega_g(\omega)$ known on a wide range of frequencies in three different regions [20] (as it is shown in Fig. 2) :

$$\text{(I) } \omega < \omega_1 \approx 10^{-15}\text{ Hz} \quad , \quad \text{(II) } \omega_1 < \omega < \omega_2 \approx 10^{-13}\text{ Hz} \quad , \quad \text{(III) } \omega_2 < \omega$$

In Ref. [21], we have studied the constraints which can be placed on SCS through the results of five different measurements :

(i) $\Omega_{gl}(\omega) = (\omega/\rho_{crit})\,d\rho_{gl}/d\omega \quad , \quad \rho_{crit} \approx 4\ 10^{-29}\text{ g cm}^{-3}$;

(ii) $\Omega_{total}(\omega) = \int_{10^{-13}\text{ Hz}}^{10^{-8}\text{ Hz}} d\omega\ \Omega_g(\omega)/\omega \quad$, through the binary pulsar residual R_b (in region II);

(iii) MBR temperature fluctuation $\Delta T/T$ for angular separation of about $1°$ (region III);

(iv) chemical potential μ_0 (region I), and

(v) Comptonization parameter y (region II), characterizing the MBR spectral distortions.

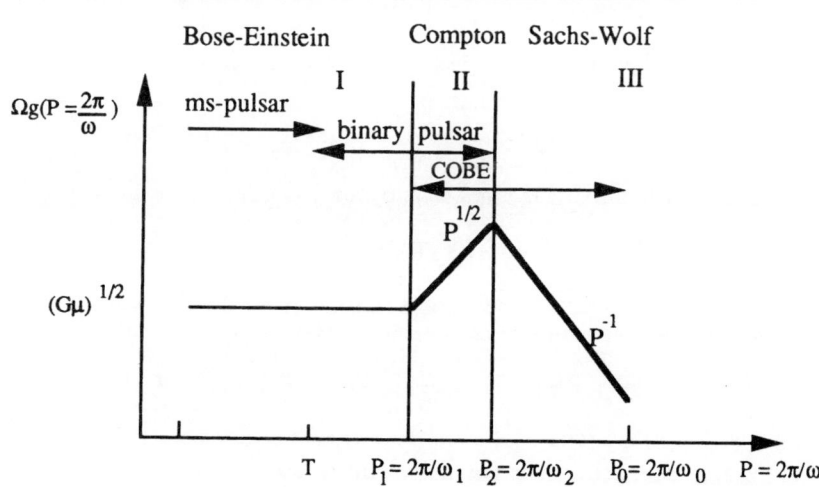

Fig.2. *Schematic logarithmic spectrum $\Omega_g(P)$, as a function of the present period, $P = 2\pi/\omega$ of the gravitational waves generated by cosmic strings ; $P_1 = 10^4$ yr, $P_2 = 10^6$ yr, $P_0 = 10^{10}$ yr. The observational time span is $T \approx 5$ yr for millisecond pulsars, $T \approx 10–10^6$ yr for the binary pulsar. The main effects of distortions in the cosmic microwave background due to electromagnetic and/or gravitational waves in each of the three regions are shown.*

From the compatibility among these measurements, including the absence of distortion in the MBR spectrum recently reported by the FIRAS/COBE experiment, we have obtained : (a) the coefficient f relating the electromagnetic and gravitational radiation rates released by SCS, (b) the chemical potential μ_0(SCS) produced by SCS, (c) the string loop evolution parameters (α, β, γ), and (d) the string parameter $G\mu$. Here α measures the initial loop size relative to the horizon size, β accounts for the loop formation rate and γ for the total power energy emitted by each loop.

Recently, the measurable quantities (i)–(v) above have been expressed explicitly in terms of $G\mu$, the set (α, β, γ) and f covering the whole spectrum and SCS evolution [22] :

$$\Omega_{gI} = 1.8 \; 10^{-6} \, [G\mu/10^{-6}]^{1/2} \, (\alpha^{3/2}\beta) \, \gamma^{-1/2} \, (1+f)^{-3/2}$$

$$\Omega_{total} = 4.6 \; 10^{-4} \, [G\mu/10^{-6}]^{1/2} \, (\alpha^{3/2}\beta) \, \gamma^{-1/2} \, (1+f)^{-3/2}$$

$$\Delta T/T \leq 2.5 \; 10^{-6} \, [G\mu/10^{-6}] \, (\alpha\beta\gamma)^{1/2} \, (1+f)^{-3/4}$$

$$\mu_0 = 0.2 \, [G\mu/10^{-6}]^{1/2} \, (\alpha^{3/2}\beta) \, \gamma^{-1/2} \, f \, (1+f)^{-3/2}$$

$$y = 0.4 \, [G\mu/10^{-6}]^{1/2} \, (\alpha^{3/2}\beta) \, \gamma^{-1/2} \, f \, (1+f)^{-3/2}$$

At the present time, observations place the following constraints:
(i) millisecond pulsar timing measurements lead to :

$$\Omega_g(\omega) < 2 \; 10^{-7} \quad \text{for} : \omega/2\pi \approx 0.2 \text{ cy/yr} \tag{6}$$

(Taylor 1989 [16]), but more generally :

$$\Omega_g(\omega) < 4 \; 10^{-7} \, (2\pi/T)^4 \, R_\mu^2 \quad \text{for} : \omega > 2\pi/T \tag{7}$$

where T is the observation time-span in years and R_μ the residual in ms.
(ii) binary pulsar timing measurements lead to :

$$\Omega_{total} < 4 \; 10^{-2} h^{-2} \quad \text{for} : 10^{-12} \text{ Hz} < \omega < 10^{-9} \text{ Hz} \tag{8}$$

(Taylor and Weisberg 1989 [17]), and more generally :

$$\Omega_{total} < (2H^2)^{-2} \, (R_b/T_b)^2 \tag{9}$$

where $H = h \times 100 \text{ km s}^{-1} \text{ Mpc}^{-1}$ is the Hubble constant ($h \approx 0.7$).
(iii) angular temperature anisotropy measurements yield :

$$\Delta T/T < 6 \; 10^{-5} \quad \text{for angular separations } \theta \approx 1° \tag{10}$$

(Wilkinson 1986, Partridge 1988)[23].
(iv) latest Rayleigh-Jeans MBR spectrum measurements yield :

$$\mu_0 < 10^{-2} \quad \text{with} : T_{RJ} \approx 2.74 \pm 0.02 \text{ K} \tag{11}$$

(Smoot et al. 1987)[24].

The recent results from COBE [13] have infirmed the Nagoya-Berkeley spectral distortion, and show a pure black body spectrum at 2.735 ± 0.006 K within 1% of the peak intensity over the range 500 μm to 1 cm. Using their conservative 1% error bands, Mather et al. [13] set a 3σ upper limit on the Comptonization y-parameter of 0.001, and then :

$$y < 10^{-3} \quad \text{for} : T \approx 2.735 \pm 0.06 \text{ K} \tag{12}$$

From a fit to a pure Bose-Einstein spectrum with a chemical potential μ_0 independent of frequency, Mather et al. gave also the 3σ upper limit :

$$\mu_0 < 0.9 \; 10^{-2} \tag{13}$$

Note that this preliminary value of μ_0 is about an order of magnitude greater than the value (12) on y .

Now, we can express the string parameter Gμ in terms of these measurable quantities with their present upper limits :

$$G\mu \approx 1.2 \; 10^{-8} \; (\Omega_{gl}/2 \; 10^{-7})^2 \; (\alpha^{3/2}\beta)^{-2} \; \gamma (1+f)^3 \tag{14}$$

$$G\mu \approx 0.4 \; 10^{-7} \; (\Omega_{total}/10^{-4})^2 \; (\alpha^{3/2}\beta)^{-2} \; \gamma (1+f)^3 \tag{15}$$

$$G\mu \approx 2.4 \; 10^{-5} \; ([\Delta T/T]/6 \; 10^{-5}) \; (\alpha\beta\gamma)^{-1/2} \; (1+f)^{3/4} \tag{16}$$

$$G\mu \approx 2 \; 10^{-9} \; (\mu_0/0.9 \; 10^{-2})^2 \; (\alpha^{3/2}\beta)^{-2} \; \gamma (1+f)^3 \; f^{-2} \tag{17}$$

$$G\mu \approx 6.2 \; 10^{-12} \; (y/10^{-3})^2 \; (\alpha^{3/2}\beta)^{-2} \; \gamma (1+f)^3 \; f^{-2} \tag{18}$$

Note that in Eq. (14), the quoted upper limit 10^{-4} for Ω_{total} is not yet reached. From Eqs. (14) and (16) on the one hand, and on the other hand from Eqs. (18) and (16), we get the following relations for the string evolution parameters and the ratio between electromagnetic and gravitational radiation rates :

$$\alpha^5 \; (\beta/\gamma)^3 \approx 6.8 \; 10^{-14} \; (1+f)^{9/2} \; f^{-4} \; (y/10^{-3})^4 \; ([\Delta T/T]/6 \; 10^{-5})^{-2} \tag{19}$$

$$\alpha^5 \; (\beta/\gamma)^3 \approx 0.25 \; 10^{-6} \; (1+f)^{9/2} \; (\Omega_{gl}/2 \; 10^{-7})^4 \; ([\Delta T/T]/6 \; 10^{-5})^{-2} \tag{20}$$

$$f \approx 2.3 \; 10^{-2} \; (y/10^{-3}) \; (\Omega_{gl}/2 \; 10^{-7})^{-1} \tag{21}$$

Finally, with the latest upper limits on $\Omega_{gl}(\omega)$, $\Delta T/T$ at 1°, y and the value (24) for f, from Eqs. (14), (16), (18) we have for the string parameter :

$$G\mu < 1.2 \; 10^{-8} \; (\alpha^{3/2}\beta)^{-2} \gamma \tag{22}$$

$$G\mu < 2.4 \; 10^{-5} \; (\alpha\beta\gamma)^{-1/2} \tag{23}$$

$$G\mu < 1.2 \; 10^{-8} \; (\alpha^{3/2}\beta)^{-2} \gamma \tag{24}$$

From the latest millisecond pulsar timing and from the preliminary COBE data, we obtain the following upper limits on $G\mu$:

	generic values	latest numerical simulation values	kinky strings
	$\alpha \approx \beta \approx 1$	$\alpha \approx 2.8 \ 10^{-2}$	$\alpha \approx 10^{-2}$
	$\gamma \approx 100$	$\beta \approx 850, \gamma \approx 100$	$\beta \approx 750, \gamma \approx 100$
PSR 1937+21	$G\mu < 1.2 \ 10^{-6}$	$0.7 \ 10^{-7}$	$0.2 \ 10^{-7}$
COBE	$G\mu < 1.2 \ 10^{-6}$	$0.7 \ 10^{-7}$	$0.2 \ 10^{-7}$

The first set of values for (α,β,γ) corresponds to generic values [25], the second set to the latest numerical simulation results [26,27] and the third set to "kinky" string values [28]. We see that the upper limits put by COBE on $G\mu$ *converge* to those placed by the latest PSR 1937+21 timing measurements, and that generic values of (α,β,γ) yield $G\mu \approx 10^{-6}$ while numerical values yield $G\mu \approx 10^{-7}$.

We have placed new constraints on SCS and on the parameters governing their evolution by requiring that electromagnetic and gravitational energies released by loops be compatible with the preliminary data on the MBR spectrum from FIRAS/COBE experiment, with the already performed MBR angular $\Delta T/T$ measurements (Sachs-Wolf effect) for $\theta \approx 1°$, and with the latest millisecond pulsar timing measurements. These limits continue to descend but the breaking point is not yet reached. The absence of distortion lies within 1% of the peak intensity, but the FIRAS sensitivity for each spectral element is expected to be 1/1000 of the peak of the 2.735 K spectrum. In addition, new constraints are expected to be placed from limits on MBR anisotropies : from the anisotropy experiment [29] DMR on COBE satellite, the sensitivity on $\Delta T/T$ in a field of view of about $7°$ is expected to be increased by an order of magnitude with respect to previous experiments. This will set up stringent limits on $G\mu$ for large straight moving SCS which produce $\Delta T/T$ fluctuations, through lensing-type effects.

3. Connection between Fundamental Strings and Cosmic Strings. Macroscopic Behaviour of Fundamental Strings

Fundamental strings exhibit a macroscopic behaviour which appears quite similar to those of topological cosmic strings. These include :
(i) the reconnection or self-intersection processes,
(ii) the decay rate of fundamental strings into gravitational radiation,
(iii) the thermodynamical behaviour of strings at high temperatures.
It is important to know whether highly excited fundamental strings behave in a classical way. A quantity which is interesting and which can be derived analytically is the decay rate of highly excited (open or closed) strings. One can compare then the gravity wave emission computed from the decay rate of a quantized closed bosonic string with the classical calculations for gravitational radiation already done for cosmic strings. Recently [30], the vertex operator for the absorption and emission of states from a closed string has been computed to obtain the one-loop scattering amplitude whose imaginary part gives the total decay rate Γ (total means summing over all the decay products). The computation is more easily performed for the states lying on the leading Regge trajectory corresponding to spin J massive particles with squared mass $m^2 = 4(J-2)$. Classically, these states correspond to classical trajectories which are rotating pairs of straight strings joined at the ends (with the

ends moving at the speed of light c) ; the quantum state (which is an occupation number eigenstate) is the superposition of such classical trajectories. The decays in which one of the products is massless form the largest class. The amplitude Γ can be converted into a rate of radiation which gives :

$$P = \Gamma E \approx 360 \, G\mu^2,$$

E being the energy of the radiated massless particles.

Let us recall the gravitational radiation rate from cosmic strings, evaluated classically for some particular string trajectories, using the quadrupole formula :

$$P = \gamma\mu^2 \quad \text{with } \gamma \approx 100.$$

We see that a highly excited fundamental string exhibits a classical behavior by emitting as it would be a *classical oscillating quadrupole.* It is very interesting that the purely quantum mechanical calculation of decaying fundamental strings into gravitons agrees with the classical results of cosmic strings. For fundamental strings, the value of γ is larger than that obtained from cosmic strings, but this is due to the particular chosen trajectory (the leading Regge trajectory state corresponding classically to a rigid rotating rod for which the mass is proportional to the length).

The value of γ depends on the particular trajectory. It would be interesting to compute the rate Γ for other states (different from the leading Regge trajectory) and to compare with the values obtained for cosmic strings.

It must be noticed that beside the graviton, the rate Γ contains decays into the massless dilaton and antisymmetric tensor fields, and that, in general, the expression of Γ in terms of the string parameter $G\mu$ is highly dependent on the compactification scheme. The results in D = 4 and D = 26 are rather different. (For instance, in D = 26, Γ is inversely proportional to the string length, and decays into massive states are not suppressed as it is the case in D = 4).

But in spite of these connections, important questions remain to be solved or to be revised, and we are not yet in a position to make contact with the precise results of observations, as has been done for cosmic strings :

a) we would like to know not only the parameter γ, but also the analogues of the α and β evolution parameters of cosmic string interactions, from a quantum fundamental string computation.

b) it would also be interesting to compute the rate Γ of fundamental strings for other states (different from the leading Regge trajectory), and to compare with the values obtained for cosmic strings.

c) and more important, to determine the value of $G\mu$ in D = 4 of fundamental strings, for which compactification and perturbative renormalization group techniques are in conflict with the radioastronomical constraints on cosmic strings.

REFERENCES

[1] M. Dine and N. Seiberg, Phys. Rev. Lett. 55, 366 (1985)
     V. Klapunovsky, ibid., 103
     P. Candelas, G.T. Horowitz, A. Strominger and E. Witten, Nucl. Phys. B285, 46
     (1985)
     D.J. Gross, J.A. Harvey, E. Martinec and R. Rohm, Nucl. Phys. B267, 75 (1986)

[2] K.S. Narain, Phys. Lett. B169, 41 (1986)
     K.S. Narain, M.H.Sarmadi and E. Witten, Nucl. Phys. B279, 369 (1987)

[3] R. Petronzio, G. Veneziano, Mod. Phys. Lett. A2, 707 (1987)

[4] R.Hagedorn, Nuovo Cimento Suppl. 3, 147 (1965)
     S. Fubini and G.Veneziano, Nuovo Cimento A64, 1640 (1969)

[5] S. Frautschi, Phys. Rev. D3, 2821 (1971)
     R.D. Carlitz, Phys. Rev. D5, 3231 (1972)
     B. Sundborg, Nucl. Phys. B254, 583 (1985)
     M.J. Bowick and L.C.R. Wijewadhana, Phys. Rev. Lett. 54, 2485 (1985)

[6] Y. Aharanov, F. Englert and J. Orloff, Phys. Lett. B199, 366 (1987)

[7] F. Englert, J. Orloff and T. Piran, Phys. Lett. B212, 423 (1988)

[8] N. Sanchez and M. Signore, Phys. Lett. B214, 14 (1988)

[9] H.J. de Vega and N.Sánchez, Phys. Lett. B197, 320 (1987)

[10] N. Sánchez and G. Veneziano, Nucl. Phys. B333, 253 (1990)

[11] M. Gasperini, N. Sánchez and G. Veneziano, CERN-TH 5893/90, DFTT-30/90
     and Meudon-DEMIRM 90091 preprint (to appear in IJMPA)

[12] M. Gasperini, N. Sánchez and G. Veneziano, CERN-TH 6010/91, DFTT-06/91
     and Meudon-DEMIRM 91004 preprint

[13] J.C. Mather et al., Ap. J. Letters 354, L37 (1990)
     J.C. Mather et al., COBE preprint n° 90-03 (1990)
     J.C. Mather et al., COBE preprint n° 90-05, to be published in the Proceedings of
     "After the First Three Minutes", Workshop October 15-17 1990, University
     of Maryland, USA

[14] T. Matsumoto et al., Ap. J. 329, 567 (1988)

[15] B. Bertotti, B.J. Carr and M.J. Rees, Monthly Notices Roy. Astron. Soc. 203, 945
     (1983)
     L.A. Rawley, J.H. Taylor, M.M. Davis and D.W. Allan, Science 238, 761 (1987)

[16] J.H. Taylor, "Timing binary and millisecond pulsars",talk given at Observatoire de
     Paris-Meudon (September 1988)

[17] J.H. Taylor and J.M. Weisberg, Ap. J. 345, 434 (1989)

[18] E. Witten, Nucl. Phys. B249, 557 (1985)
      A. Vilenkin, TUTP 89-1 preprint, to be published in the Proceedings of the 14th
          Texas Symposium on Relativistic Astrophysics
      J.P. Ostriker, C. Thompson and E. Witten, Phys. Lett. B180, 221 (1986)

[19] J.P. Ostriker and C. Thompson, Ap. J. Letters 323, L97 (1987)
      B. Rudàk and M. Panek , Phys. Lett. B199, 343 (1987)

[20] R. Brandenberger, A. Albrecht and N. Turok, Nucl. Phys. B277, 605 (1986)

[21] N.Sánchez and M. Signore, Phys. Lett. B214, 14 (1988)
      M. Signore and N. Sánchez, Mod. Phys. Lett. A4, 799 (1989)
      N. Sánchez and M. Signore, Phys. Lett. B219, 413 (1989)

[22] N. Sánchez and M. Signore, Phys. Lett. B241, 332 (1990)
      M. Signore and N. Sánchez, I.J.M.P. A6, 1591 (1991)
      M. Signore and N. Sánchez, to be published in the Proceedings of "After the First
          Three Minutes", Workshop October 15-17 1990, Univ. of Maryland, USA

[23] D.T. Wilkinson, Phil. Trans. R. Soc. London A320, 595 (1986)
      R.B. Partridge, Rep. Prog. Phys. 51, 647 (1988)

[24] G.F. Smoot et al., Ap. J. Letters 317, L45 (1987)

[25] T. Vachaspati and A. Vilenkin, Phys. Rev. D31, 3052 (1985)

[26] A. Albrecht and N. Turok, Phys. Rev. D40, 973 (1989)
      A. Albrecht and N. Turok, Fermilab-Pub 89/140A, PUPT-89-1133 (1989)
      F.S. Accetta and L.M. Krauss, Phys. Lett. B233, 93 (1989)

[27] D. Bennett and F.R. Bouchet, Princeton University preprint PUPT-89-1126 (1989)

[28] B. Allen and R.R. Cadwell, "Small Scale Structure on a Cosmic String Network"
      submitted to Phys. Rev. D. (1990)

[29] C.L. Bennett and G.F. Smoot, preprint LASP 88-14.
      G.F. Smoot et al., "COBE : The Differential Microwave Radiometers", talk given at
          "The 175th Meeting of the American Astronomical Society", Washington DC,
          January 9-13 1990

COSMIC STRINGS: AN INTRODUCTION
TO THEIR FORMATION, EVOLUTION, AND
THEIR MICROWAVE BACKGROUND SIGNATURE

FRANÇOIS R. BOUCHET

Institut d'Astrophysique de Paris, CNRS

75014, Paris, France

ABSTRACT: The goal of these lectures is to present estimates of the cosmic microwave background anisotropies that are produced by a network of cosmic strings. In order to provide some physical insights into the processes that give rise to these anisotropies, I give an introduction to their nature and their generic effects, followed by an overview of the time evolution of a string network. I then describe how to proceed to make actual predictions of the anisotropies of the microwave sky, and how to use these predictions to constrain the one parameter describing strings, their mass per unit length.

1. Introduction

Cosmic strings are topologically stable linear defects that form in many grand unified theories (GUT) during a symmetry breaking phase transition in the early Universe (Zel'dovich, 1980; Vilenkin, 1981). They might also be fundamental string remnants of an earlier phase. Contrary to monopoles and domain walls (the zero- and two-dimensional defects), they are not obviously a disaster for cosmology. In fact, the idea that they might account for the formation of galaxies and large scale structure has generated a lot of interest since Zel'dovich (1980) proposition. If the string tension μ is at the GUT scale (*i.e.* $\mu \sim (10^{16} GeV)^2$, see below), they could provide appropriate seeds for the matter accretion in the matter era, or for the Ostriker – Thomson – Witten explosions (Witten, 1985; Ostriker, Thompson, and Witten, 1986), if they are superconducting. Furthermore they have interesting observable signatures, like a non-zero residual of millisecond pulsar timing measurements (Hogan and Rees, 1984; Taylor, 1989; Stinebring *et al.*, 1990; Bouchet and Bennett, 1990a), or their gravitational lensing effects (Vilenkin, 1984; Gott, 1985, or the expected step-like discontinuities in the microwave background (Kaiser and Stebbins, 1984; Bouchet, Bennett, and Stebbins, 1988). For the value of μ aforementioned, these might soon be detectable.

The following lecture notes are not intented to be a fair review on the rather large and sprawling string litterature. Instead, I have tried to give a pedagogical introduction to the basic features of strings which are of cosmological relevance, with special emphasis on their effect on the Cosmic Microwave Background (hereafter CMB). I did include though some short paragraphs which are somewhat more technical, but which may be skipped without impeding the understanding of the remaining of the notes. The current understanding of cosmic strings physics stems from the work of numerous authors, many of whom I did

M. Signore and C. Dupraz (et al.), The Infrared and Submillimetre Sky after COBE, 101–127.

not do justice to in these short lecture notes of reduced scope. The reader interested in a more in depth introduction to strings should see the review by Vilenkin (1985) and references therein, the book "Cosmic Strings, The current status" of the proceedings of the Yale workshop (1988), and the most recent book "The Formation and Evolution of Cosmic Strings" of the proceedings of the Cambridge workshop (1989). Finally, in some of the more technical sections, I have drawn liberally from already published works I did with my long time collaborators David Bennett and Albert Stebbins. I would like to take this opportunity to thank them for letting me use some of our unpublished material, and for years of quite enjoyable and productive collaboration.

2. What are Cosmic Strings?

The idea of a symmetry breaking phase transition in the early universe is a natural cross-product of current conceptions in particle physics and cosmology. Indeed, in order to reach a unified description of the forces of Nature, particle physicists usually assume that the equations of physics at high energy are invariant under certain symmetry transformations. Such symmetries are nevertheless not observed at low energy; they must thus be "broken" at some point(s) when going down the energy scale. On the other hand, the standard framework for cosmology is the Hot Big Bang model; it is observationnly well supported, in particular by the existence and properties of the microwave background itself. In this model, the temperature generally decreases with time. As a result, very early on, forces are undifferentiated. When the expanding Universe cools down to a symmetry breaking energy scale, a phase transition occurs. Since arbitrarily remote regions cannot be causally connected, different regions may end up in different low-energy state, thereby rising the possibility of creation of defects of various dimensionality, depending on the specific broken symmetry.

2.1. FRIEDMAN-LEMAITRE MODELS IN A NUTSHELL

Although much of cosmic strings physics may be understood with little cosmology knowledge, some basic notions will be used throughout these notes, like the idea of a causal Horizon. For an ordered and pedagogical introduction to the current "Standard cosmological model", one may read "Physical Cosmology" by P.J.E. Peebles (1980), or "Gravitation and Cosmology" of S. Weinberg (1972), as well as "The early Universe" by E. Kolb and M. Turner (1988). Still, in order to make these notes reasonnably self-contained, I recall a few basic concepts which may help in the following.

The standard Hot Big Bang model assumes that the universe is homogeneous and isotropic, at least when averaged on sufficiently large scales. This large scale homogeneity and isotropy imply that the metric $ds^2 = g_{\lambda\nu} dx^\lambda dx^\nu$ (we use the notation of implicit summation on repeated indices) be Robertson-Walker

$$ds^2 = dt^2 - a^2(t) \left[\frac{dr^2}{1 - kr^2} + r^2 d\theta^2 + r^2 \sin^2 \theta d\psi^2 \right] ,$$

where a is the metric expansion factor, k is a constant which by redefinition of r can be limited to take only the values $k = 0$, $+1$, -1 (we also use units such the speed of light c is unity), and the spatial (spherical) coordinates are called comoving coordinates. Einstein equations relate the geometrical properties of space time, as embodied by the Ricci tensor $R_{\lambda\nu}$ (which derives from the metric tensor $g_{\lambda\nu}$), to its energy content distribution described by the energy-momentum stress tensor $T_{\lambda\nu}$. In the absence of a cosmological constant, these equations may be written as

$$R_{\lambda\nu} - g_{\lambda\nu}R = 8\pi G T_{\lambda\nu},$$

where G is Newton's constant. If we describe the content of the model universe by a perfect, homogeneous fluid of density ρ, velocity U_λ ($U^t = 1$, $U^i = 0$), and pressure p, whose $T_{\lambda\nu}$ is given by $T_{\lambda\nu} = p(t)g_{\lambda\nu} + [p(t) + \rho(t)]U_\lambda U_\nu$, the time evolution of the metric expansion factor (if dots denotes time derivatives) is governed by Friedman equation[1]

$$\left(\frac{\dot{a}}{a}\right)^2 = 8\pi G\rho/3 - k/a^2 \ ,$$

while the translationnal invariance of $T_{\lambda\nu}$ ($T^\nu_{\lambda,\nu} = 0$) yields the the energy conservation law

$$\frac{d}{da}(\rho a^3) = -3pa^2 \ .$$

The expansion rate then follows once an equation of state $p = f(\rho)$ has been specified.

The universe today (as indexed by the subscript 0) is observed to be expanding ($H_0 = \dot{a}/a = 100\,h$ km/s/Mpc, with $0.5 \lesssim h \lesssim 1$) and nearly flat ($k = 0$, for the critical, or closure, density $\rho_c = 3H^2/8\pi G$, and we know that $\Omega_0 = \rho/\rho_c$ is in the range $0.05 \lesssim \Omega_0 \lesssim 2$). As we go back in time, a decreases, ρ increases faster than a^{-2} (the simple volume effect already implies $\rho \propto a^{-3}$), and the curvature term becomes ever more negligible. For our purposes, the flat Einstein de Sitter case $k = 0$ will always be appropriate. For non-relativistic matter (pressure $p_m \ll \rho_m$), $\rho_m \propto a^{-3}$ and thus

$$a_m \propto t^{2/3}.$$

For radiation, $\rho_r = p_r/3$, and thus[2] $\rho_r \propto a^{-4}$, which implies

$$a_r \propto t^{1/2}.$$

For a two-component universe of matter and radiation, at late times, the model content is dominated by matter, and the first case applies ($a \propto t^{2/3}$). As one goes back in time, the radiation becomes dominant[3], and the expansion rate is slower.

[1] The time (0,0) component of Einstein equations gives $3\ddot{a} = -4\pi G(\rho + 3p)a$, while the spatial components yield $a\ddot{a} + 2\dot{a}^2 + 2k = 4\pi G(\rho - p)a^2$. Friedman equation follows by eliminating the acceleration $\ddot{a}$.

[2] The wavelength of a radiation will increase proportionnaly to the expansion so that, in addition to the volume effect ($\propto a^3$), there is an extra factor of a in the variation of the energy density with a.

[3] For a temperature of the microwave background today $T_0 \sim 2.7$K, $\rho_{r0} \sim 4.4\,10^{-34}$g/cm^3, while $\rho_{m0} = 2\,10^{-29}\,\Omega_0\,h^2$ g/cm^3, so that $\rho_r = \rho_m$ occurs when $z_{eq} \equiv a_0/a_{eq} \sim 4\,10^4\,\Omega_0\,h^2$, $T_{eq} \sim 10^5\,\Omega_0\,h^2$ K, $t_{eq} \sim 3\,10^{10}\,(\Omega_0\,h^2)^{-2}$ s $\sim 1000\,(\Omega_0\,h^2)^{-2}$ years.

The obervations tell us that there are of the order of 10^8 photons for any baryon (protons, neutrons, etc...), so that entropy comes essentially from radiation, and the entropy density s is $\propto T^3$. It is conserved as long as the expansion of the Universe is adiabatic, *i.e.* $sa^3 =$ constant, so that the temperature decreases proportionally to the inverse of the expansion factor

$$T \propto 1/a.$$

Indeed, this universe is very early on in a hot dense state and progressively cools as expansion proceeds. Finally let us compute the maximal distance that a particle may travel. Photons follow null geodesics $ds^2 = 0$, and the distance covered by a light ray along the ($\theta =$ constant, $\psi =$ constant) line is given by

$$a\,dr = dt.$$

As a result, no information may be propagated between time t_i and t_f on a comoving scale greater than

$$R = \int_{t_i}^{t_f} dr = \int_{t_i}^{t_f} \frac{dt}{a}\,.$$

The corresponding physical distance $\Delta = a(t_f)R$ is $\Delta_m = 2(t_f - t_i)$ in the radiation-dominated era, and $\Delta_m = 3(t_f - t_i)$ in the matter-dominated era. In any case, for $t = t_f \gg t_i$, there is both in the Matter and Radiation dominated era a limit $H \propto t$ to the distance over which causal effects may occur. In the following, we will call H the causal horizon[4].

2.2. A TOY MODEL

In order to introduce the basic idea of a cosmic string, let us consider a toy model of "global cosmic strings", objects which are particle physics analogs of vortex tubes in superfluids. Let ϕ be a Higgs field, to be added to the collection of fundamental fields which make up matter and radiation, and whose purpose is precisely to provide a symmetry breaking mechanism. In this model, we consider only the Higgs field itself, which is a complex scalar field, *i.e.* a function, $\phi(\mathbf{x}, t)$ which assigns a complex number to each point $\mathbf{x}$ in space and time. The complex field ϕ may be rewritten $\phi = \Phi\, e^{i\alpha}$ where Φ is the modulus of the field which takes values between zero and infinity and α is the phase of the field which may take values between 0 and 2π. The energy density of such a field is $\frac{1}{2}(\dot{\phi}^2 + \nabla\phi^2) + V(\phi)$, where $\nabla\phi$ is the spatial gradient of ϕ, $\dot{\phi}$ is the rate of change of ϕ with time, and $V(\phi)$ is the potential which only depends on the modulus of ϕ. Clearly the lowest energy state of the field would have ϕ be constant in space and time and $V(\phi)$ take its minimum value. This field allows topological defects if the minimum value of the potential is not at $\Phi = 0$.

---

[4] As defined, H is different from the particle Horizon which is the maximal distance a particle may have propagated since the Big Bang at $t_i = 0$, *i.e.* the maximum over which events may ever have been in causal contact. If there happens to be another era at very early time, *e.g.* vacuum dominated, where the expansion factor increases exponentially, this particle horizon can be much larger than H. But in the following, we will be interested by the scale over which causal processes may have influenced the strings since their formation at $t_i \ll t$, and H will then be the relevant scale.

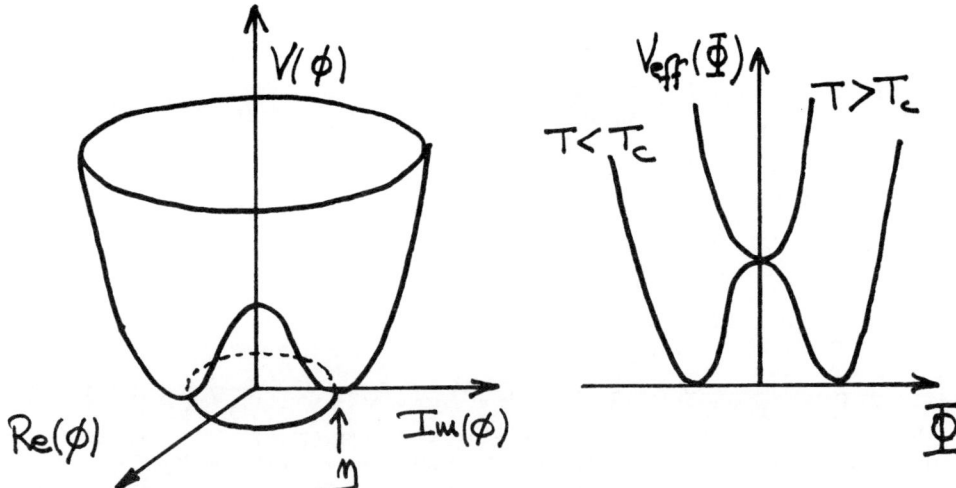

Figure 1: On the left is the mexican hat potential $V(\phi)$, while the effective potential V_{eff} is shown on the right as a function of the field modulus Φ, for two different temperatures.

Suppose the potential for ϕ is (see figure 1) a mexican hat which may be written as $V(\phi) = \frac{\lambda}{2}(\phi^*\phi - \eta^2)^2$. The minimum of the potential is at $\Phi = \eta$. If ϕ_0 is a ground state, so is $\phi_0\, e^{i\delta}$: the ground state is degenerate, and may be parametrised by α. Note that even though the rotational symmetry is broken for any given state with a specific value of α, the ensemble of possible solutions is symmetrical as a whole. This is the situation at zero temperature in the classical limit. At a finite temperature T, thermal excitations (corresponding to variations $\delta\phi$ around the field mean value) may simply be described by an effective potential $V_{eff}(\phi) = V(\phi) + AT^2\,\phi^*\phi$. This implies the existence of a critical temperature $T_c \sim \eta$ (as in the most of the following, we use units where Boltzman constant, k, Planck's constant, h, and the speed of light, c, are all unity). When $T > T_c$, the thermal excitations are energetic enough for the field to go over the potential barrier, so that the symmetry is restored. The ground state expectation value $\langle\Phi\rangle$ is then zero while, at $T < T_c$, $\langle\Phi\rangle$ is still η. This is clearly a phase transition à la Landau, with a two-dimensional order parameter ϕ.

If we now examine this theory in a cosmological context (Kibble, 1976; 1980), we see first of all that the ground state is initially unique and symmetrical $\langle\Phi\rangle = 0$. When the temperature decreases to T_c, a symmetry breaking phase transition occurs. The crucial additional ingredient is that in Cosmology there is an intrinsic limit to the extention ξ of regions where the field can take the same orientation to minimize spatial gradient. This limit is the distance traveled by light over the duration of the phase transition (with $\xi \lesssim H \propto t$, where t is the age of the universe). This amounts to a "quenching" situation and it naturally raises the possibility of forming defects. The kind of defects that arise here is most easily seen by following close paths in space on a scale $\gtrsim \xi$. The orientation of the field changes along the path, which we monitor by following the associated path in the manifold of ground states as is shown in figure 2.

Since the vacuum expectation value $\langle\Phi\rangle$ is single valued, the phase along the path can only change by multiples of 2π, i.e. $\Delta\alpha = 2\pi n$. If the winding number $n = 0$ (dotted line of fig. 2), we can shrink the path in the abstract space to a point at no energetical cost, so the field will rearrange later to have a single orientation on scale $\sim H$ and minimize the gradient

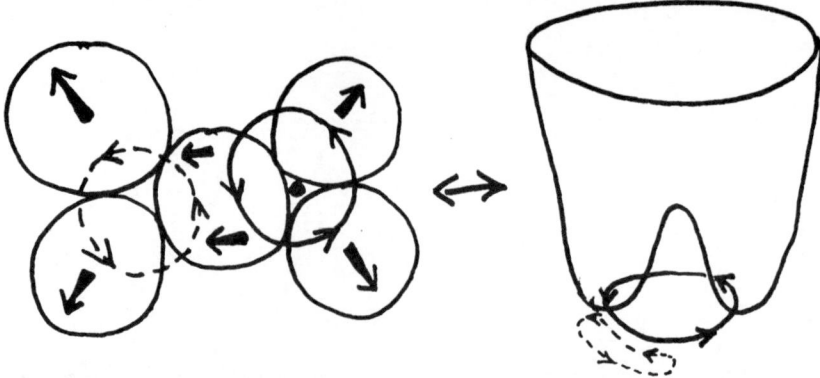

Figure 2: On the left are sketched a few regions of space of size ξ where the Higgs field orientation is denoted by an arrow. On the right is the mexican hat potential of fig. 1 whose minima define the manifold of degenerate ground states. The doted and solid closed path in real space correspond respectively to the dotted and solid closed path in this manifold.

energy. On the other hand[5], if $n = 1$, (solid line of fig. 2), we cannot continuously shrink the path without going through a singular point where α is undefined which correspond to $\langle \Phi \rangle = 0$. To untie this "knot", one would need to slide the path in the abstract space over the maximum of the potential, which is not any more possible at $T < T_c$. Furthermore, if we move the path elsewhere in real space, without going through a singular point, we must still encounter a singular point when we shrink it since otherwise it would be possible to contract to a point the path in abstract space without crossing the $\Phi = 0$ point. In other words, there must an unending line of singular points. Cosmic strings are such linear defects *with no ends* where the universe is trapped in a high energy state of unbroken symmetry.

2.3. MORE TOY MODELS

As we mentionned earlier, the type of defects formed depends on the broken symmetry. If the Higgs field (the order parameter) was to have the same rotational invariance, but was one-dimensional, the defects formed would be planar; these are generically called "walls". If it was three-dimensional, the defects would be point-like, monopoles being of this type. The currently fashionable texture is another product of this type of mechanism, which can occur if the Higgs field is four-dimensional. Although not stable[6], this type of defects can

---

[5] As was shown by Laguna and Matzner (1989), strings with large values of n are unlikely to be stable and would be "peeled down" to low n strings.

[6] A straight infinite string is the simplest field configuration. Since it is invariant under translations along the axis of the string, the field configuration $\Phi e^{i\alpha}$ may be described by specifying the mapping $\alpha(\theta)$, where θ is the azimuth angle around the string (and $\Phi = \eta$ everywhere but very close to the singular point). Clearly, the energetically favored configuration will be $\alpha \equiv \theta$ to minimize the gradient energy. For texture, the minimum of $V(\phi)$ will be a three-sphere hypersurface S_3, instead of a one-sphere (circle) as for strings, or a two-sphere as for monopoles. Let us use polar coordinates (r, θ, ψ) in real space, and $(\chi, \theta\prime, \psi\prime)$ in S_3. A ground state configuration then writes $\phi = \eta(\cos\chi, \sin\chi \sin\theta\prime \cos\psi\prime, \sin\chi \sin\theta\prime \sin\psi\prime, \sin\chi \cos\theta\prime)$. It is illuminating to consider the spherically symmetric knot corresponding to the mapping $\chi \equiv \chi(r), \theta\prime \equiv \theta, \psi\prime \equiv \psi$, where

also have interesting cosmological consequences. In fact, it has recently been argued that these defects may provide a satisfactory large scale structure formation scenario, although it is unclear that the resulting CMB anisotropies will be compatible with the COBE limits.

Up to now we have considered the spontaneous symmetry breaking of global symmetries, *i.e.* invariance under operations which are applied everywhere in the same fashion. For instance, the Lagrangian of the previous toy model

$$L(\mathbf{x}) = \partial_\lambda \phi^\star \partial_\lambda \phi - V(\phi)$$

was globally invariant under the phase $(U(1))$ transformation $\phi(\mathbf{x}) \to e^{-i\delta}\phi(\mathbf{x})$, where δ is an arbitrary real constant, independent of $\mathbf{x}$. Much more powerful is the requirement that the theory be invariant under a local (gauge) transformation, where δ is now an arbitrary function of $\mathbf{x}$. Of course, if we do that, the kinetic part $\partial_\lambda \phi^\star \partial_\lambda \phi$ of the Lagrangian is not invariant anymore. In order to restore the Lagrangian symmetry, one *needs* to introduce a coupled gauge field A which must transform in such a way as to offset the variations arising from the $\partial_\lambda \delta$ terms $(A_\lambda \to A_\lambda + \partial_\lambda \delta)$. This leads to the Abelian Higgs model

$$L(\mathbf{x}) = D_\lambda \phi^\star D_\lambda \phi - \frac{1}{4} F_{\lambda\nu} F^{\lambda\nu} - V(\phi),$$

where the partial derivatives have been replaced by covariant derivatives $D_\lambda \equiv \partial_\lambda - igA_\lambda$. Of course the full Lagrangian now also contains the Lagrangian of the uncoupled gauge field $L_\gamma = -\frac{1}{4} F_{\lambda\nu} F^{\lambda\nu}$, with $F_{\lambda\nu} = \partial_\lambda A_\nu - \partial_\nu A_\lambda$. This theory of an electromagnetic and a Higgs field is now locally (gauge) invariant. It is the simplest toy model leading to the formation of *local* strings, which are then oriented objects. One can for instance define the string direction as the direction of the magnetic gauge field $(B = \nabla \times A)$ trapped inside the string. These local strings are the ones we shall consider in all the following. Finally, let me conclude by mentionning that breaking a rotational symmetry is not the only way to get strings. As a matter of fact, there will be strings formed everytime the manifold of degenerate ground states contains unshrinkable loops, *i.e.* everytime the manifold will not be simply connected (*e.g.* Z_2).

2.4. STRINGS CHARACTERISTICS

From a macroscopic point of view the only important parameter which characterizes a cosmic string is the mass per unit length, which is conventionally denoted by μ. For a fundamental quantum field theory with only one characteristic energy, T_c, all properties of the field theory will depend only on the parameters, T_c, and dimensionnal analysis

---

the latitude angle $\chi(r)$ on S_3 goes from 0 at the origin to π at infinity. Variations of $\chi(r)$ are energetically costly at large r due to the large volume involved $(d\rho \propto (\nabla\phi)^2 \, r^2 dr)$. As a result, when a knot becomes causally connected, the time evolution will tend to concentrate the χ-reversal of the direction of the field close to the origin to minimize the overall gradient energy. This collapse proceeds at the speed of light up to the point when the gradient energy is concentrated in a region of the order of the inverse GUT scale, and is of the order of $V(0)$; the field configuration then unwinds and the knot disappears while releasing its energy as an outgoing spherical shell of massless particles. In the string case, the unwinding energy is unavailable since the lower dimensionality favors an even spread of the gradient energy.

yields $\mu \sim T_c^2$. In the previous models[7] we considered, $T_c \sim \eta$, and thus $\mu \sim \eta^2$. While it is possible to construct models where μ takes practically any value you might like, there is really only a small range for μ which would be cosmologically interesting. In particular, values[8] near 10^{22} g/cm are preferred. Anything much larger is already excluded by astronomical observations while anything much smaller would produce no effects which we are ever likely to see. While 10^{22} g/cm (~ 1000 tons per fermi $\sim 10^7$ Solar mass per parsec) may seem like a very large number, the curvature it produces in space is actually very small. Suppose we consider a piece of string of length ℓ in a volume ℓ^3, the density perturbation will be $\delta\rho = \mu\ell/\ell^3$. In a Friedman-Robertson-Walker universe $G\bar{\rho}t^2 \sim 1$ (G being Newton's constant), and thus $\delta\rho/\bar{\rho} \sim G\mu(t/\ell)^2$. Since the previous value of μ corresponds to $G\mu \sim 10^{-6}$, the perturbations on cosmological scales, close to the Horizon are indeed quite small! It turns out that such value for μ correspond to a symmetry breaking transition when $T_c \approx 10^{16\pm1}$ GeV (when $t \sim 10^{-38\pm2}$), i.e. at the natural GUT scale, which has encouraged many to look more closely at the effects of cosmic strings. For such high energies, the thickness of the string is extremely small, typically $T_c^{-1} \sim 10^{3\pm1}$ Planck length or $10^{-30\pm1}$ cm.

Let me finish this section by a few consideration on the expected statistical characteristics at the outcome of a symmetry breaking phase transition. As was noted earlier, the Higgs field direction cannot be correlated[9] on scales greater than $\xi < H$. As a result, one does not expect the string direction itself to be correlated[10] on a scale greater than ξ, so it should essentially look like a brownian walk of step size ξ. Furthermore, one expects to find of the order of one string length ξ per correlation volume ξ^3. Detailed numerical simulations have shown (Vachaspati and Vilenkin, 1984; Albrecht and Turok, 1985; Scherrer and Frieman, 1986; Aryal et al., 1986) that most of the string pertain to infinite strings whose large scale behaviour is indeed brownian. In fact, about 80% of the string length is in infinite strings, the remaining 20% being in a scale invariant distribution of closed loops. As was shown by Everett (1981), the motions of the strings which are initially heavyly damped by friction with the surrounding matter becomes essentially free as soon as the universe cools down $T \sim 10^{12}$GeV.

3. What do strings do?

Much of the interesting string physics does not depend on the string internal structure and may be understood at a classical level. Indeed, particle physics fixes the transverse dimension of strings ($\sim 10^{-30}$cm for GUT scale strings), while the length of strings of

---

[7]The thickness of a global string is of the order of the inverse Higgs boson mass $\delta_\phi \sim m_\phi^{-1} \sim (\sqrt{\lambda}\eta)^{-1}$. The string mass per unit length is of the order of the vacuum energy density times δ_ϕ^2, which indeed yields $\mu_\phi \sim \eta^2$. For a local string, the size of the region where A is sizable is $\delta_A \sim m_A^{-1} \sim (\sqrt{g}\eta)^{-1}$, so again $\mu_A \sim \eta^2$.

[8]It is easily checked that the only combination of Planck's constant, Boltzmann's constant, and of the speed of light which gives a quantity with dimensions of mass per unit length is $(kT_c)^2/(c^3h)$. Introducing Newton's constant, we get c^2/G.

[9]I do not consider here the case of an earlier inflationnary phase.

[10]Although the absence of ϕ correlations does not imply the absence of correlations on the strings.

interest in cosmology is rather of the order of the Horizon (an extremely large number today since it is about ten billion light-years!). As a result, a useful approximation is to first consider the strings as infinitely thin.

3.1. STRAIGHT INFINITE STRINGS

The idealised case of a *straight* infinite string (Nielsen and Olesen, 1973; Vilenkin, 1985) is particularly illuminating. For definiteness, let us choose the z–axis along the string. This string configuration, and the associated "energy momentum stress tensor" $T_{\lambda\nu}$ is invariant under 2 types of rotations. It is for one thing invariant under usual rotations about the z axis, which implies that the string tension in the transverse (x and y) directions must be equal. And since the string has no structure in the z direction, it must be invariant under Lorentz-boosts in that direction, *i.e.* relativistic rotations in the (z, t) plane, so that the longitudinal tension must be equal to the string's mass per unit length ($\times c^2$). One then has, if $\tilde{T}_{\lambda\nu}$ stands for $T_{\lambda\nu}$ averaged over the string cross section,

$$\tilde{T}_{\lambda\nu} = \delta(x)\delta(y) \int T_{\lambda\nu} \, dx \, dy = \mu \, \mathrm{diag}(1, 0, 0, -1)\delta(x)\delta(y).$$

This means that, if stretched, the mass of such a string will simply increase proportionally to the amount of stretching. This has important consequences on the coupling of strings with gravity.

Indeed, let us recall that for matter which is static and of stress tensor of the form $T_{\lambda\nu} = \mathrm{diag}\,(\rho, -P_1, -P_2, -P_3)$, the Einstein equations yield in the Newtonian limit the Poisson equation

$$\nabla^2 \Phi = 4\pi G(\rho + P_1 + P_2 + P_3);$$

we discover that straight infinite strings have no Newtonian gravity! If one now works out the metrics resulting from the Einstein equations[11], one finds (*e.g.* Gott, 1985)

$$ds^2 = dt^2 - dz^2 - dr^2 - (1 - 4G\mu)r^2 d\theta^2.$$

The geometrical structure of this metric is made obvious by following circular paths. If the path does not go around the string, the ratio Θ of the circle's circumference to its radius will be equal to 2π; we recover the Newtonian conclusion that the surrounding space is everywhere (locally) flat. But it the path encircles the string, the angle Θ will be reduced to $2\pi(1 - 4G\mu)$: the structure of the two dimensional space perpendicular to the string axis is thus that of a cone. The effect of a string on the otherwise plane Minkowskian geometry is thus simply obtained by removing a wedge of angle $\epsilon = 8\pi G\mu$ whose tip is at the intersection of the string with the plane, and by identifying the edges (see fig. 3).

This simple geometrical construction tells us the effect of a string on particles paths. First, suppose we draw 2 parallel lines on a piece of paper, and then cut out a wedge whose tip is in between the lines (see figure 4); when we bring the edges together the lines intersect,

---

[11]The following metric is not in a cosmological context; it only gives the metric perturbation induced by a string, in an otherwise empty universe. The problem would be quite more complex if one were to consider an expanding background, with a cosmological horizon. But this simple case is quite enough to see the metric perturbation locally induced by a string.

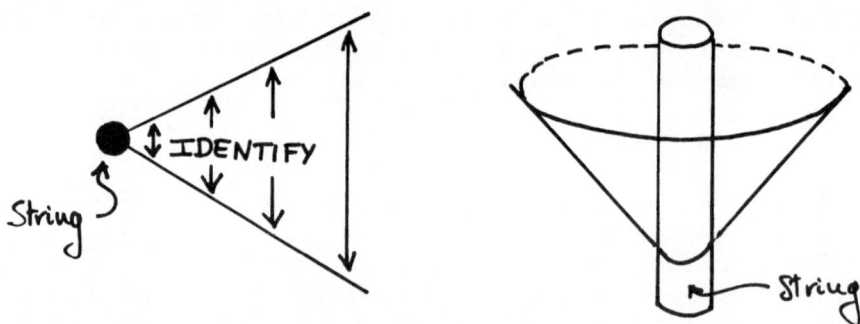

Figure 3: Curvature in a two-dimensionnal space. If it is flat, as for a plane or a cylinder, one recovers the usual Euclidean geometry. But if it is positively curved, like the surface of a sphere for instance, the measure of the circle circumference will be larger for a same diameter and the ratio Θ will be greater than 2π. Perpendicular to the string, space is conical.

with an angle equal to the deficit angle. If a point source is located behind the string, the string focusing of the light rays (provided they pass on different sides of the string) will give rise to two images, whose separation will depend on the location of the string relative to the oberver, but which will be in any case smaller than ϵ. Thus a background object will be lensed by the string (if its angular separation to the plane defined by the string and the observing point is less than ϵ), and the two images will be separated by ϵ.

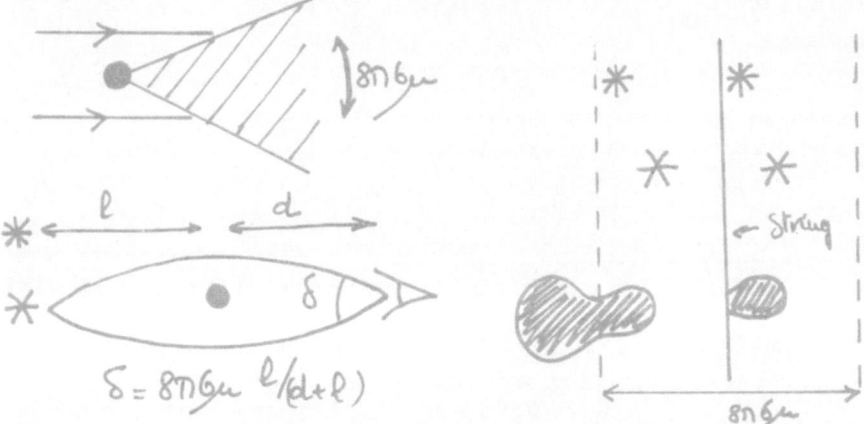

Figure 4: One the left are graphical illustrations of the focusing effect of a metrics conical geometry, while the figure at right is a sketch of the sky replication effect.

If we now look at the sky in the direction of a string, which we suppose for simplicity to be perpendicular to the line of sight, i.e. in the plane of the sky, all sources lying less than ϵ away from the string will give rise to 2 images, so that a narrow band of the sky on each side of the string will be replicated. If an extended source happens to straddle the boundary of the replicated region, we will then see a complete image as well as a partial image of the source, with a very sharp edge to it. This unique effect should be detectable with a high resolution telescope, like the Hubble space telescope, since for a GUT string, the deficit angle ϵ is a few seconds of arc (Paczynski). The feasability of such a search of course depends on how frequently this might occur, which depends on the number of strings expected in a given solid angle. This clearly requires knowledge of the statistical properties in a recent past of a stringy universe.

Up to now, we have implicitly assumed that the string was not moving. This is generally not the case, and in fact a number of interesting effects arise for rapidly moving strings. First, let us consider matter at rest and a string moving with velocity v_s; in the reference frame of the string there is a laminar flow ahead of the string, with equal but opposite velocity. By using the same geometrical construction than before, we see that after the flow passes the string, the two flows on either side of the string start to converge on each other. Thus it is as if the string communicates a constant velocity to the flow which is independent of the distance to the string and is perpendicular to the direction of the flow. The size of this transverse velocity is just $\frac{1}{2}\epsilon v_s/\sqrt{1-v_s^2}$ assuming ϵ is significantly smaller than unity. If the flow is made up of weakly interacting particles, like neutrinos, then the two streams of particles can simply pass through each other. If, unlike neutrinos, the particles have also a small random velocity, the region where the two streams overlap is also a wedge with opening angle ϵ, and the density in this region is just twice the density elsewhere. However, if the velocity dispersion of the particles is greater than ϵ times the speed of the string then the opening angle will be larger and the density in the wedge will be smaller. The same phenemona will occur for a fluid of strongly interacting particles (*i.e.* like a normal fluid made of protons, neutrons, etc...).

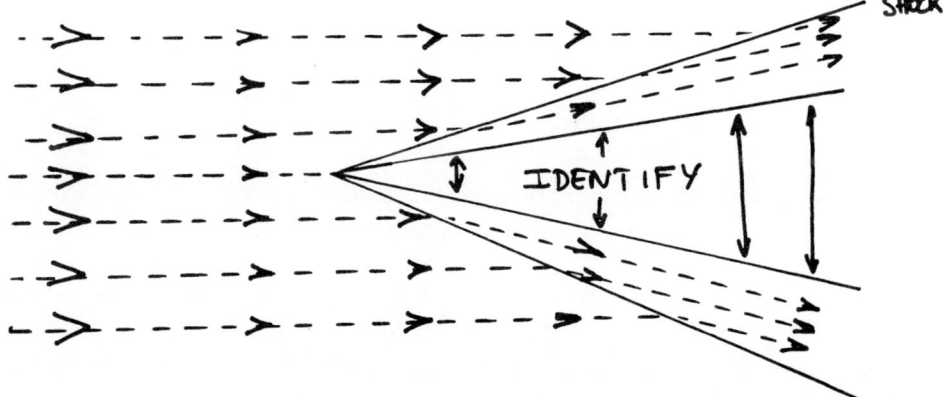

Figure 5: Pictorial presentation of the formation of a wake in the back of a moving straight infinite string. At this point, the reader is encouraged to personnalize his copy of the book by actually cutting the wedge out with scissors, and bring the edges together. This is not recommended for library books, though, and I will not assume responsability for it!

In any case, a moving string in a medium will leave behind an overdense wake. Such wakes are quite interesting since they might provide appropriate initial conditions for gravitational instability to generate the large scale sheets (Vachaspati, 1986) observed in the distribution of galaxies. Of course, the viability of this idea depends on wether the statistical properties of the superposition in time of many wakes are adequate (*e.g.* is there a typical wake separation consistent with the observed scale of voids in large galaxy samples?). Again, the answer to that question depends on the evolution of string networks, and it has not yet been fully worked out, although analytical estimates are promising (Stebbins *et al.*, 1987; their conclusions were not optimistic, but these authors used numerical values which are now considered inacurate which changes these conclusions).

Let us now suppose that we are looking at a uniform light source and that a string perpendicular to the line of sight moves from left to right and is at this time just between us and this source. As before, for an observer moving with the string (see figure 6), both the terrestrial observer, and the source of light are receding to the left with a velocity equal, but in opposite direction to the string motion. If we cut a wedge whose tip is on the string and identify the edges, the part of the light source viewed by the terrestrial observer in the back of the string will seem to move toward him as compared to those seen in the front of the string. This differential Doppler shift is $\epsilon v_s / \sqrt{1 - v_s^2}$, identical of course to the relative velocity of the two flows converging in the back of a moving string.

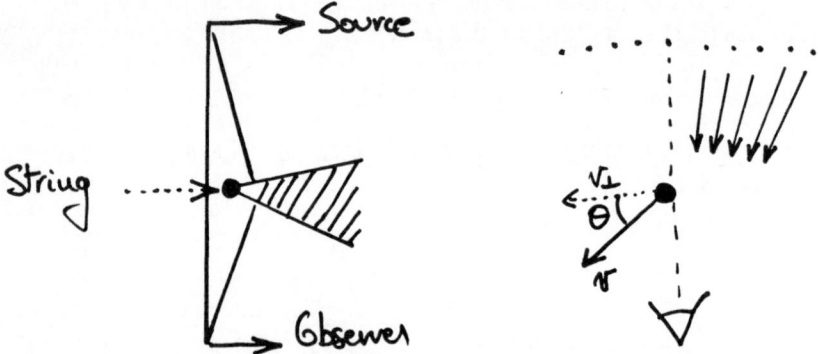

Figure 6: Pictorial effect of a moving string on a background source.

Big Bang Cosmology provides us with a uniform background source of light, the microwave background radiation, and this source is known to have essentially no relative motions. A string signature will thus be a step-like discontinuity (along the string) of the "temperature" of this background, with a magnitude

$$\Delta T / T \sim 8\pi G \mu\, v_s \cos\theta,$$

where $v_s \cos\theta$ is the projected string velocity perpendicular to the line of sight. Of course a more detailed prediction of the expected pattern of microwave background anisotropies requires knowledge of the strings dynamics.

3.2 CURVATURE

Of course, all that was said above strictly applies only to the case of a straight infinite string. But it remains a good approximation as long as one considers distances which are small as compared to the local curvature radius of the string. We shall see that for many strings this curvature radius is very large, comparable in fact to the horizon. On the other hand, one can show, by integrating over infinitesimal straight string pieces, that in the limit of great distances as compared to this curvature radius, a string loop will have the same effect as a point-like mass (of extremely large mass), a reasonable result indeed. This loops have in fact been considered by many authors as possible seeds for starting the accretion process ultimately responsible for the formation of galaxies and clusters of galaxies.

The precise description of the intermediate case of scales comparable to the string curvature radius is rather involved; but the only thing we need consider here is that the conical nature of space around a string remains a valid description up to distances of the order of this curvature radius. At greater distances, it gets progressively modified until it blends into an homogeneous unperturbed space. This has the effect of removing the rather unphysical artificialities of the infinite straight string model. In particular, the doppler shift given to passing photons will not extend to arbitrarily large scales, nor will be the momemtum communicated to the particles of a medium in the back of a moving string.

Cosmic strings, as mentioned above, have an internal tension. So another effect of a string curvature will be to create a restoring force. Thus apart from the very special and irrealistic case of a straight infinite string, a string will have quite complicate motion. In Minkowski space, one can show (see the footnote in section 4.2) that the string equation of motion is simply

$$\mu\ddot{\mathbf{x}} = \mu\mathbf{x}'' ,$$

where overdots denotes partial derivatives with respect to time, and primes partial derivative with respect to the length parameter along the string. This shows that, since the string tension is equal to its mass per unit length (still in units such that $c = 1$), the characteristic velocity μ/μ is the speed of light. In the case of a loop, $i.e.$, when a string comes back onto itself, it will be oscillating periodically, with a period given by its length ($/c$). Of course, the periodic motion of the string will induce some periodic variation of the curvature of the space in which it is embedded. These ripples of the space curvature which are propagating outward are nothing else than gravitational waves, and the combined effect of many loops must produce a stochastic background of such waves. Although we will return to this after later, it is interesting to note at this point that strings, if they exist, should be by far the largest sources of gravitationnal waves known and could be detected fairly soon.

Still, gravity waves are very difficult to detect. The usual terrestrial experiments to detect gravity waves (either using large bars of aluminum or lasers) are not sensitive to the very low frequency gravitational radiation expected from strings. Fortunately we may also use astronomical objects to put limits on the amount of gravity radiation. In particular pulsars, which are rapidly rotating very compact stars made of neutrons, may be used. We detect these stars by the radio pulses they emit every period of rotation. The fastest rotating pulsars are expected to be very good clocks, $i.e.$ their period is extremely stable (and in the millisecond range). In fact, over long time periods they may prove more accurate than the best atomic clock. If gravitational waves are present along the path light between the pulsar and the earth, then the effective time of light propagation should vary with the same frequency than the gravity wave.

Various other expected sources of modulation of the period (due to, for instance, to the earth rotation around the sun, the pulsar binary orbital motion, etc...) can be substracted with great precision by accumulating timing data over several years. This allows an acurate determination of the pulsar parameters, but the match between the observed and computed phases is not perfect. These residual differences correspond to the sources of noise which have not been not accounted for. One can thus derive derive an upper bound on the energy density of a possible stochastic gravitational wave background. Stinebring *et al.* (1990) derive the remarkable limit on the contribution to the closure energy density (per logarithmic interval near a 7 years period) $\Omega_G < 4 \times 10^{-7}$ at the 95% confidence limit. We shall see below the implied limit on $G\mu$.

3.3 INTERACTIONS

Up to know, we have used pure geometrical means to study the effect of strings on their environment. On the other hand, one has to come back to the underlying field theory to know what happens when two pieces of string hit each other. There are essentially two possibilities: either the two pieces of string simply pass through each other and remain unaffected, or they intercommute, that is to say that they break at their crossing point and reconnect the other way. Deciding what happens is a difficult problem, and there is no general analytical answer, although general arguments favor intercommutation. This is why one has for the time being to resort to numerical simulations[12]. It turns out, though, that the answer is rather simple (*e.g.* Shellard, 1987; Matzner, 1988). For nearly all relative angles and approach velocities, the strings intercommute. An exemple of intercommutation is shown in figure 7. Note that the discontinuity formed at the time of intercommutation t_I separates into two discontinuities traveling in opposing direction.

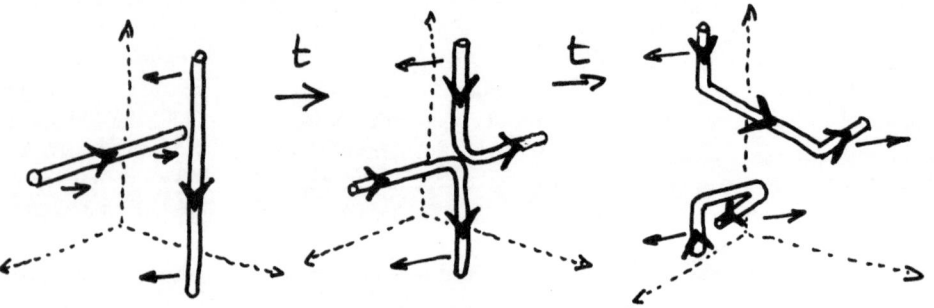

Figure 7: Schematic presentation of strings pieces interaction. A numerical simulation of the process would be a blow up of the interaction region, down to the point when the thickness of the string is resolved.

The strings intercommutation leads to two very interesting possibilities. First of all, if one considers the interaction of a loop and a long string, one sees that the loop can reconnect to

---

[12]In these simulations, one specifies the configuration of the fundamental (Higgs) fields initially so that it corresponds, for instance, to two pieces of string at right angle approaching each other. One then follow the time evolution of the field configuration, which tells how the specific patterns called strings evolve.

the long string, thereby becoming part of it. Conversely, a curved long string may give rise to a loop by intercommutation of two its sections (see figure 8). As a result, a long string may straigthen out by "emitting" loops, which can then slowly decay into gravitational radiation, if they do not hit another long string to which they might reconnect. Of course, a loop can also split into smaller loops making it less likely for the pieces to be hit by a long string. In the course of time, strings will have complicate motions, and intercommutations will further complicate the evolution by changing the topology of the network of strings.

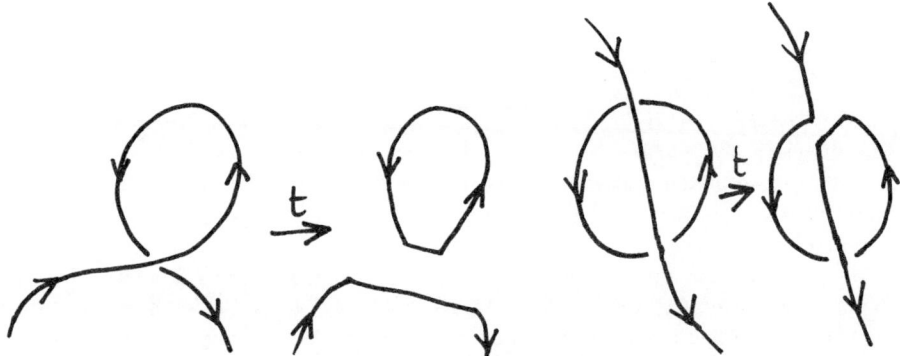

Figure 8: pictorial description of the process of loop creation/destruction.

4. Strings Dynamics

As described in section 2, we expect to start very early on from a situation where there is of the order of a length of string of horizon size per horizon volume. This is only a small perturbation ($\sim 10^{-6}$) to the total energy content of this horizon volume, which at this epoch is dominated by the radiation energy. It is thus legitimate to consider the Universe as a framework, a host to these objects whose behaviour is determined irrespectively of its string content; a string is then simply a local variation of no relevance in determining the evolution of the host. An interesting question then arises: when the Universe evolves, and its energy density decreases like radiation, does the string contribution remain a constant fraction of the energy content so that the overall picture remains unaltered? Or does it vary with time? If the string contribution decreases with time, this means that the strings become an ever more negligible component, and they dilute to oblivion, in effect leaving no remains for us to witness. In effect, they can do nothing, and no astronomical observation will ever be made to decide if they had a fleeting existence or not. On the contrary, if the string contribution increases, at some point it must be taken into account self-consistently in determining the time evolution of the Universe, and one is likely to end up with a string dominated Universe. For GUT strings, the universe has already expanded by a factor of a $10^{19\pm1}$ between the symmetry breaking epoch and, say, the primordial nucleosynthesis, which would give ample time for a transition to a string dominated regime to occur. This would completely change the delicate balance of events which leads to the so successful nuclear elements synthesis, and rule out the string existence.

4.1. SINGLE STRING

So now we must answer the question of the time evolution of the strings contribution to the energy density of the Universe. Before we do that, we must first of all ask ourselves how does the energy of a single string evolve. If we have an isolated loop of size much smaller than the horizon, then we may neglect the effect of the expansion on the small volume of space in which the string loop is oscillating. In this case, neglecting the energy lost in radiating gravitational wave, the energy (*i.e.* the length) of the string remains a constant. A loop of string which is small as compared to the horizon can thus be considered as a simple but extremely massive particle, in keeping with its effect on the surrounding matter. Of course, even though it takes a very long time for a string to radiate a substantial amount of energy, ultimately, the loop will disappear after having converted all its energy into gravitational waves.

But all strings are not small loops. Let us now consider again our toy model of a straight infinite string whose extremities can be viewed as anchored on both sides of the horizon. In this other extreme case, the string can only be stretched by the expansion of the Universe ($\ell \propto a$, a being the metrics scale factor, see below). Since it does not oscillate, there is no radiation emitted, and its energy increases proportionally to the amount of stretching, *i.e.* to the expansion. This works the same way than a spring stores energy when someones pulls on its extremities. Here, "someone" is the Universe itself, and it does so at the expense of its own energy. Even if the original energy is small, sooner or later the string energy will become dominant ($\rho_s \propto \ell/a^3 \propto a^{-2}$) and change the expansion rate itself. In the intermediate cases, the evolution of the energy of a string will be somewhat increasing, although at a slower rythm than the expansion. In any case, we are now faced with a serious problem, since the string energy density will decay somewhat slower than for non relativistic matter (Turok and Bhattacharjee, 1984), while the rest of the Universe early on behaves as radiation, whose energy density decreases quite faster ($\rho_r \propto a^{-4}$). So sooner or later the Universe should become string dominated.

The interesting aspect of strings is thus that many strings are infinite, which insures that they do not all disappear in a finite time in gravitational radiation as loops do. But this is also the source of a seemingly fatal problem, namely that the contribution to the energy density of the Universe of such strings grows with time, and would lead to an unacceptable string dominated Universe. There has been something missing in this analysis though, and that is that in a system of strings in motion, pieces of different strings, or of the same string can interact when they cross.

4.2. SCALING SOLUTION

In order not to conflict with observations on one hand, and not disappear on the other, the string contribution to the energy density of the Universe must be constant. In the radiation era, this amounts to require that the total string energy density evolves like radiation, *i.e.*, decreases like a^{-4}. Since the metric scale factor in this era is growing like $t^{1/2}$, the string energy density must decrease like t^{-2}. In other words, the total string energy, and thus length, per horizon volume ($\propto t^3$) must be proportionnal to the horizon length $\propto t$, so that the density is proportionnal to t^{-2}. In the matter era, the energy density of the Universe goes as a^{-3}, and the expansion factor then grows as $t^{2/3}$ so that, again, the string energy density must decrease like t^{-2}. Thus, irrespective of the era, one needs to have a constant number of strings of horizon size (energy) per horizon volume, althought the proportionality constant may depend on the era (radiation or matter dominated). This means that, the universe should look (statistically) identical, provided one uses as a meter stick the only *a priori* available scale, the horizon size H. If this is the case, one has reached a "scaling solution", which is the central concept of string dynamics (Vilenkin, 1981; Turok and Bhattacharjee, 1984; Kibble, 1985).

The existence of a scaling solution is a difficult question, since the answer depends on the delicate balance between string stretching (which increases a string energy), and the energy transfer to small loops which can turn into radiation after a while, provided they do not return by reconnection their energy (or part of it) to the long strings. Despite many efforts (Kibble, 1985; Bennett, 1986a,b; Mitchell and Turok, 1987a,b), analytical approaches were unable to establish the existence of a scaling solution. This question has recently been answered by numerical means. How does one proceed? The general line (Albrecht and Turok, 1985) is that one first generates "reasonnable" initial conditions in a big box whose dimension contains many horizon volumes, then one evolves the string configuration forward in time by applying the strings equations of motion[13], detecting along the way the intercommutations and implementing them. One can then look at the configuration at various times, for instance by extracting at random a volume of the size of the horizon to see if the length in long strings seem to settle to a constant value (see figure 9). More quantitatively, one may compute the string contribution to the energy density of the Universe as a function of time, and see if it tends toward a constant. If it does, then one performs another experiment from different initial conditions, with more, or less, strings per initial horizon volume, or with different initial velocities, etc...

David Bennett and myself have shown (Bennett and Bouchet, 1988; 1989; 1990) that there is a stable scaling solution, *i.e.* whatever the initial conditions might be, the string system always relaxes to the same state (when expressed in horizon units). Such a result confirmed earlier findings by Albrecht and Turok (1985), although with different numerical values. The newer simulations by Allen and Shellard (1990) appear to confirm our numerical values. The new results of Albrecht and Turok (1989) are different but might be compatible since they aknowledge large possible systematic errors (by a factor of 4, which is precisely

---

[13]They are obtained by minimizing the Nambu action which describes strings of zero thickness, and which is just proportionnal to the surface of the space-time sheet swept by a string (for a point it would be the length of its universe line which would be minimized). By using comoving coordinates $\mathbf{r}$ and the conformal time τ (such that $dt = a\,d\tau$), the Friedmann-Robertson-Walker metric takes the form $ds^2 = a^2(-d\tau^2 + d\mathbf{r}^2)$, and we can locate the string by its comoving position $\mathbf{x}(\tau, \sigma)$ as a function of time τ and of the length parameter along the string σ. If we choose a gauge in which the unphysical components of the velocity parallel to the string vanish, *i.e.* $\dot{\mathbf{x}} \cdot \mathbf{x}' = 0$ (dots are derivatives versus τ, and primes are spatial derivatives, *i.e.* versus σ), one obtains $\ddot{\mathbf{x}} + 2(\dot{a}/a)\dot{\mathbf{x}}(1 - \dot{\mathbf{x}}^2) = (1/\epsilon)(\mathbf{x}'/\epsilon)'$, where ϵ stands for the string energy per unit (comoving) σ ($\epsilon \equiv \sqrt{\mathbf{x}'^2/(1 - \dot{\mathbf{x}}^2)}$). In the flat space time limit, the velocity redshifting due to the expansion disappears ($\dot{a} = 0$), ϵ is constant, and one recovers the equation given before in the text.

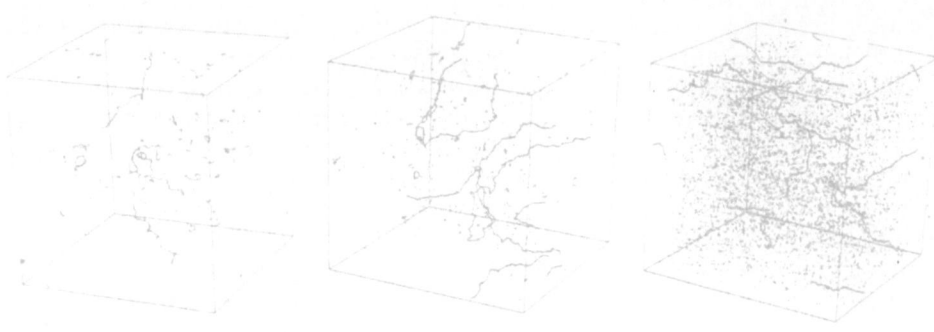

Figure 9: Snapshots of a volume of size t^3 extracted from simulations in the radiation era $(H = 2t)$ by Bennett and Bouchet, at different times in the simulation.

the difference between our value and theirs). We find that the long string system may be viewed as a collection of brownian walks with persistence length $\xi \sim H/3$, and the long string density is of the order of 50 Horizon length H of long string per horizon volume H^3 in the radiation era, and about $30H/H^3$ in the matter era. This shows that indeed the long string density at the horizon scale is a fixed and small fraction of the total energy density $(\delta\rho_{LongStrings}/\rho_{tot} \sim 50G\mu$ in the radiation era, $G\mu \sim 10^{-6}$ for GUT strings). The reason for the stability of the relaxed state is that if, in a particular region of space, the string density becomes smaller than its equilibrium value, there are less chances for intercommutations, strings are straighter (and more likely to be stretched by the expansion), resulting in an energy density decrease which is less rapid than for the rest of the energy content of the universe. On the other hand, if the string density becomes too large, it means that the string network appears more convoluted, and chances of splitting off loops are greatly enhanced, which tends to reduce its contribution again.

4.3. LOOPS

One should be aware that although the long strings dynamics is fairly well understood, the loop sizes distribution is not well known (Bennett and Bouchet, 1989). From our result, we can only set an upper limit on the typical loop sizes chopped off the long string network, which turns out to be tiny ($R_{Loop} < 10^{-4}H$). An interesting question then arises: what sets the scale for the size of the emitted loops? Naïvely one would expect that, since there seem to be only one scale in the problem, namely the horizon scale, strings should be curved on that scale, be separated by distances of the order of that scale, and give rise to loops of about that size. We argued that an ingredient had been overlooked in this analysis, which is the structure left behind every time two pieces of strings intercommute. Reconnection amounts to link together two pieces which had initially different direction (as defined by the tangent vector) and different velocities. This discontinuity in the time and space derivatives is called a kink. It is in fact a temporary superposition of two kinks, one left-moving and one right-moving kink, which then separate by going in opposite directions (see figure 7).

These four kinks (two on each pieces) travel along the strings at the speed of light.

The existence of these kinks happens to alter radically the naïve evolution of a string system. As computer simulations showed (Bouchet and Bennett, 1990b), even if one starts from perfectly smooth long strings, the number of kinks on the strings quickly increases with time, superposing to the large scale string pattern a "saw-tooth" like small scale structure. In some range of scale, the string may be represented as a fractal curve of dimension slightly greater than one. This happens because smoothing mechanisms are rather inefficient. Indeed, kinks get progressively smoothed by the expansion of the Universe that "pulls" on the string extremities; but we showed that the Universe needs to expand hundreds of time for this to be noticeable (Bennett and Bouchet, 1990). Another possibility would be for the loops to be preferentially emitted in regions were the kink density is high, bringing down the kink density. This might happen at some level, but we showed that it was not enough to prevent the kink density to increase. We argued that the only remaining smoothing mechanism is the emission of gravitational radiation, and it was recently shown (Quasnok and Spergel, 1990; Hindmarsh, 1990) to be controlled by the kink density on the emitter. As a result, the small scale pattern on strings cannot be a real fractal extending down to arbitrarily small scales, since the emission of gravitational radiation imposes a natural lower cutoff (which cannot be attained in current numerical simulations).

Apart from the fact that a string length (and thus its energy) is longer than its end-to-end distance, this small scale structure imposes the size of the non-self-intersecting loops, although the details of the mechanism are still a matter of debate. The general outline is that, when an intercommutation of the long strings creates a highly curved pieces of string, this small scale structure promotes many more intercommutations, and the highly curved piece of string ends up in a "cloud" of tiny loops whose size is governed by the smallest scale where there is power, *i.e.*, the lower cut-off scale. This has also an indirect consequence: the loops formed are so small that the probability that they be hit by a long string is very small, and the reconnection process becomes rather unimportant. Finally, even though the exact distribution of loops is not known yet, the tight upper limit on the typical loop sizes changes much of the naïvely expected astrophysical consequences.

4.4. ASTROPHYSICAL CONSEQUENCES OF LOOPS' SMALL SIZES

The first important consequence of the grinding of strings into tiny loops is that the expected amount of gravitational radiation due to the decay of these loops is substancially lowered compared to what one would expect if there size was comparable to the horizon (Bouchet and Bennett, 1990a; Bennett and Bouchet, 1991). As a matter of fact, the period of oscillation of a loop is proportional to its length, so small loops oscillate faster, and disappear more rapidly, in a time precisely proportional to their length. Thus for a given amount of energy to dissipate (which is fixed by the scaling requirement, *i.e.*, that the Universe does not become string dominated), the conversion occurs faster if smaller loops are formed. This means that the corresponding gravitational wave energy is emitted earlier, so that it gets more redshifted by the expansion, and the corresponding energy density today is lowered. Still, there is one unknown in predicting the exact expected value of the energy density of this background of randomly superposed gravitational waves, and this unknown is the mass per unit lengh of the strings μ to which all expected observational effects are proportional. By using the simulations, we can use the analysis of measurements of pulsar timing (Stinebring *et al.*, 1990) to limit the value of $G\mu$ to be $\lesssim 4 \times 10^{-5}$ (Bouchet and Bennett, 1990a).

The lack of large stable loops has another important astrophysical consequences. Let us suppose that the loop production were to stabilize at the largest value consistent with our numerical results, which is probably a drastic overestimate. Then one can show[14] that, all told, the total contribution of the loops to $\delta\rho/\rho$ on scales of order $1 \, h^{-1}\mathrm{Mpc}$ is more than a factor of 6 smaller than the contribution from the long strings, and on scales larger than this loops are even less important (Bouchet and Bennett, 1990b). This implies that the long string wakes (Stebbins *et al.*, 1987) will give the dominant density perturbations and that loops will play only a minor role in string induced galaxy formation. As mentionned earlier, there is some hope that strings will be able to account for the apparent sheet-like nature of the galaxy distribution (Geller and Huchra 1989, Peebles 1989).

David Bennett and myself are currently using our numerical code to do these calculations in collaboration with Albert Stebbins. The biggest problem when comparing a model with observations of the large scale structure is to predict where the bright galaxies (which is all we observe) are going to end up. Still, our preliminary results suggest that a scenario based on string wakes might work in a universe with hot dark matter because the small sheets formed at early times may be prevented from collapsing by the neutrinos' random velocities leaving only the very large sheets on which galaxies lie. On the other hand, in a universe with cold dark matter, there seems to be too many small sheets to correspond with the sheets of galaxies we observe today. However it is not clear whether the gravitational merging of the sheets might not lead to an acceptable distribution of galaxies. In any case, a successfull scenario for the formation of large scale structures will require a minimal value of $G\mu$ since it controls the overall amplitude of the initial density perturbations. As we shall see, this will imply a minimal amplitude for the microwave background anisotropies.

5. Strings and the Microwave Background

We saw that the basic effect of a moving strings on the appearance of an homogeneous

---

[14]Let us denote by α the mean value of the loop length at production. (Loops are weighted by their length in order to determine the mean.) The largest loops around at the transition epoch of Radiation-Matter domination will be those that have just formed with that size ($\alpha = 0.005t$). If we ignore the long strings for a moment, these loops will give the dominant density fluctuations on the scale of their mean separation (density fluctuations are frozen in the radiation era, and start growing at the transition time to matter domination). The mean separation of loops of size between αt and $\alpha t/2$ will be $H_{eq}/23$, which is $\sim 1.5 h^{-1}$ Mpc (comoving), and the total energy density in these loops is only about half of the energy density in the long strings. One might expect that the density fluctuations of these loops might be enhanced by the correlations between the positions of the loops, but this is not the case because the loops are formed with initial velocities $\sim 0.7c$ which quickly wash out the initial correlations of the loops (Bennett and Bouchet 1989a). These large velocities also cause the loops to move more than 10 times their mean separation before they stop, so that a typical cubic Mpc will see about 10 of these loops passing through it. This reduces the contribution from loops to $\delta\rho$ on these scales by about $\sqrt{10}$. The mean separation of the long strings is about $5 h^{-1}$ Mpc at this time, so there the entire density of the long string network contributes to $\delta\rho$ on scales smaller than this. This yields the numbers quoted in the text.

photon background could be fully understood by using the simple straight infinite strings model. On the other hand, the CMB anisotropies generated by "real" (*i.e.* not straight) strings are quite difficult to compute, in particular when one considers scales comparable to the horizon. Nevertheless, for strings embedded in Minkowsky space, Stebbins (1985) has showed that, in the limit of an infinitely remote observer

$$\Delta T/T \sim -4\frac{G\mu}{c^2} \oint \frac{\mathbf{u} \cdot (\mathbf{x}_\perp - \mathbf{r}^{proj})}{|\mathbf{x}_\perp - \mathbf{r}^{proj}|^2} \, d\sigma,$$

where σ is the string length parameter. $\mathbf{x}_\perp$, $\mathbf{r}^{proj}$, and $\mathbf{u}$ all live in the plane orthogonal to the photon direction (unit wavevector $\mathbf{k}$), and $\mathbf{x}_\perp$ marks the photon direction, $\mathbf{r}^{proj}$ marks the string element (so that $|\mathbf{x}_\perp - \mathbf{r}^{proj}|$ is the distance in the plane of the string element to the photon), and

$$\mathbf{u} = \left(1 - \frac{(r\prime \cdot \mathbf{k})^2}{(1 - \dot{\mathbf{r}} \cdot \mathbf{k})^2)}\right)(\dot{\mathbf{r}} - (\dot{\mathbf{r}} \cdot \mathbf{k})\mathbf{k}).$$

The meaning of the above formula is made clearer if one notes that it can be cast in the form of a (two-dimensional) Laplace's equation

$$\nabla^2(\frac{\Delta T}{T}) = -8\pi\frac{G\mu}{c^2} \oint \mathbf{u} \cdot \nabla_{x_\perp} \delta^{(2)}(\mathbf{x}_\perp - \mathbf{r}^{proj}) \, d\sigma.$$

In this limit, the temperature anisotropy field is given by a sum over a collection of infinitesimal dipoles (gradient of δ) along the string, the dipole moments being $\mathbf{u}$. It is thus easy to compute in Fourier space, provided one knows the source term, *i.e.* the time evolution of the interacting system.

Of course, the photon background has no reason to be completely homogeneous. As usual, there could be temperature, velocity, potential, or baryonic perturbations present on the last scattering surface, as well extra perturbations due to the gravitationnal wave background emitted by strings before recombination. In addition, there is also the possibility, as for any standard scenario, of secondary anisotropies, and/or contamination by foreground sources. All of those would add linearly, and can be computed as usual. So let us ignore them for a moment, and concentrate on the previous string specific anisotropies. In principle, the simulations provide all the required information to apply Stebbins' formalism. There is nevertheless a potential problem, since we have seen that the loops in the simulations have not relaxed to their scaling distribution. Contrary to the long string case, the loops distribution cannot be trusted. We thus need to know the dominant contribution to the anisotropies at various angular scales.

In the matter era, the (total) projected angular length of string in the redshift interval $[z_i, z_f]$ scales on average as $\theta_{string} \propto \sqrt{z_i} - \sqrt{z_f}$, *i.e.* most of the string length we see in that interval is concentrated at high redshift. This means that most of the temperature anisotropies will be imprinted near the last scattering surface. If there is no reionization episode in the history of the universe, then the last photon scatterings occur at a redshift $z_{ls} \sim 1000$, and the angle subtended by the horizon at that time is only $\theta_H \sim 1.8°$ (10° for $z_{ls} \sim 30$), which is not much larger than the resolution of small angular scale CMB anisotropy experiments (typically $\sim 1°$). Thus one expects that all the CMB anisotropies of present day experiment will be imprinted by the long strings only. As we shall see, this is confirmed by the topology of the maps we generated and by computing their power spectrum which is peaked at a scale $\sim 3/\theta_H$. Thus, although the loops's size distribution is poorly known, we can safely predict the expected CMB anisotropies, even at scales as small as a degree .

5.1. A PRACTICAL EXEMPLE

I now describe some work I did in collaboration with David Bennett and Albert Stebbins (Bouchet, Bennett, and Stebbins, 1988, herafter BBS) to make actual predictions. In order to generate some CMB maps, we propagate photon planes in a simulation box along different axes during matter era simulations. At each time step, we record the positions and the transverse velocity (*i.e.* in the plane) of the strings poking through or in the photon plane. At the end of the simulation the recorded position in the plane perpendicular to the photon direction of propagation simply correspond to the projected string configuration on the backward observer's light cone. We take the last scattering surface to be the face of the cube from which the photon plane is launched. We are thus missing no visible strings from behind the simulation cube, but we are missing strings from between the front face of the cube and us. By how much? Since in such simulations the horizon increases from $H = L/4$ to $H = L$ (where L is the box size), $z_f = z_i/16$, *i.e.* we miss 25% of the total string length, if $z_i = z_f \sim 1000$ (remember, $\theta_{string} \propto \sqrt{z_i} - \sqrt{z_f}$). We compensate for this by superposing to this result the result of another simulation scaled by a factor four (1/16 of its resulting map), *i.e.* we place two simulation end-to-end, in effect covering 256 expansion factors. If the last scattering surface occurred at $z_{ls} \sim 1000$, we now miss only 6% of the projected string length, and we have 18% too many strings if the last photon scattering occurs at $z_{ls} \sim 30$, which seems a reasonable compromise. An example of such a superposition of projected string configuration on the backward observer's light cone is shown in figure 10–a. Since the horizon is initially a quarter of the box size, the angle subtended by the box is four times the horizon angle at the last scattering epoch, which thus sets the angular scale of our maps $\theta_L \sim 7°$ if $z_{ls} \sim 1000$, and $\theta_L \sim 30°$ if $z_{ls} \sim 30$.

The resulting $\delta T/T$ maps can be appropriately visualized by using color graphics as in figure 10–b. Clearly, the line of discontinuity correspond to the strings position, as may be seen from figure 10–c. Of course, the amplitude of the discontinuity is determined by the strings' velocity component lying in the plane perpendicular to the line of sight, which can vary, and even change sign. The non-gaussian character of the string induced anisotropies is made obvious by Fourier transforming the CMB image of figure 10–b, randomizing the phases while keeping the amplitudes of the various Fourier components, and transforming back. The resulting image (figure 10–d) has the same power spectrum than the original one, and is thus in some sense the closest gaussian analog of the initial field. In the gaussian case, different wavelength have no phase relations, while short and long wavelength conspire in the string case to form sharp discontinuities. On the other hand, away from the discontinuities, there are very little anisotropies.

In order to compare with the results of real observations, it is best to emulate as closely as possible the experimental procedure. This is fairly easy in this case since we have available the "source" of the signal, and we just need to convolve our maps with a beam pattern appropriate for a given experiment. It is fortunate that the current experimental set-ups are

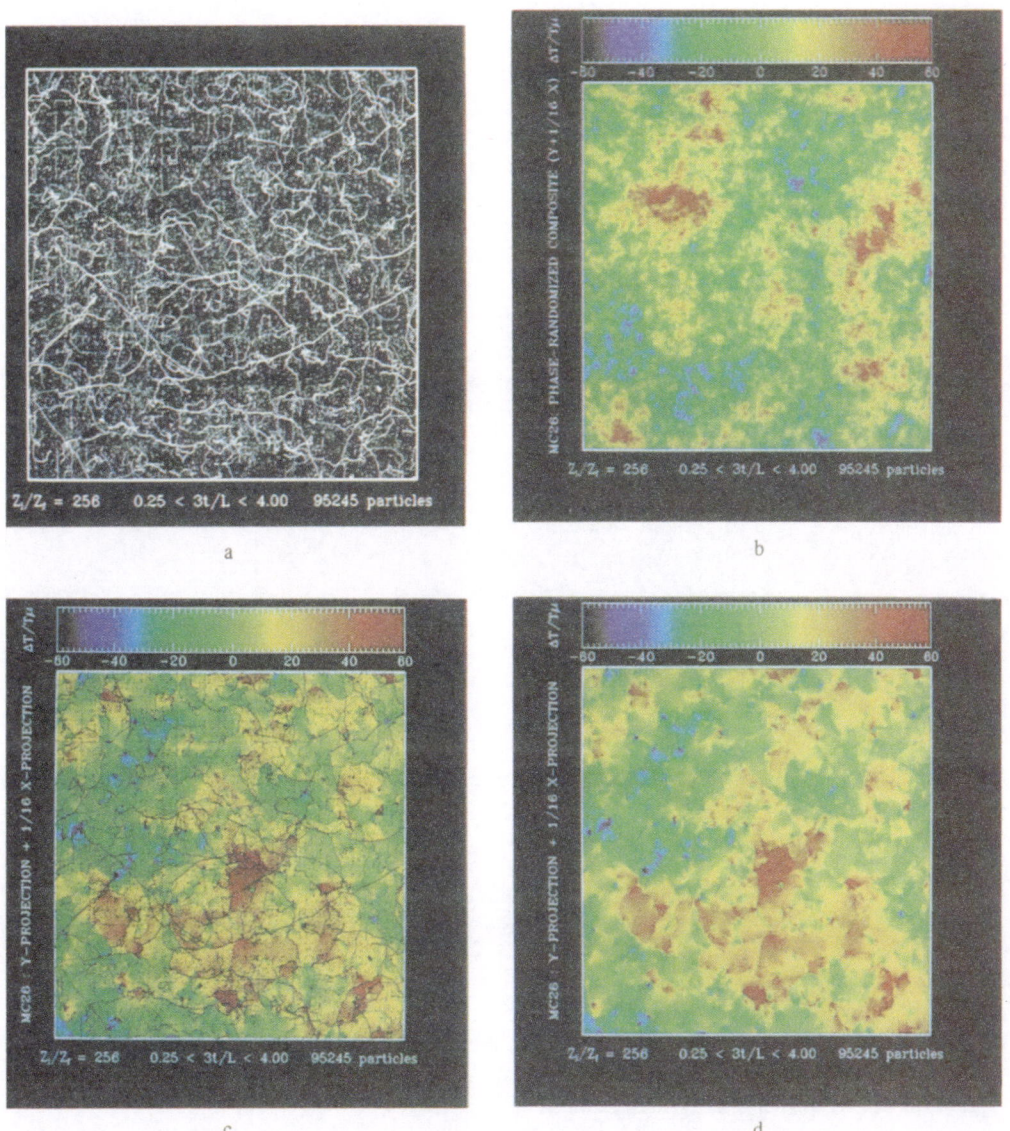

Figure 10: Microwave Background sky from strings. a) Projected String configuration. Points along a string correspond to different times, when the photon plane intersect that string at those points. b) Induced anisotropies (the color coding is arbitrary; in particular, positive $\Delta T/T$ are denoted in red, although they correspond to a blueshift). c) = a) + b). d) Same than b), once the phases of the fourier transformed image have been randomized.

rather appropriate to detect string induced anisotropies. Indeed the anisotropy patterns are lines of temperature discontinuity and the recorded signal is the difference between what is received from one patch of the sky and what is received from another patch (which might be the average of the signal received in 2 different directions as is the case in the three beam switching experiments). The experiments thus record a differential map which is in principle fairly optimal (Gott *et al.*, 1990) for the detection of line discontinuities.

On the other hand a truly optimal set–up does depend on the characteristic (angular) spacing between the discontinuities. It is easy to see in the simple case of an infinitely fine beam that for a very large beam throw the 2 patches of the sky will be many discontinuities away from each other and one thus expects a quasi- gaussian signal. On the other hand, for a tiny beam throw, in most cases, the 2 patches will not be separated by a temperature discontinuity at all, thereby yielding a null signal (in the absence of other sources of signal). In a few instances though, the patches will correspond to regions on each side of a discontinuity yielding a strong signal which would be extremely improbable in the gaussian case.

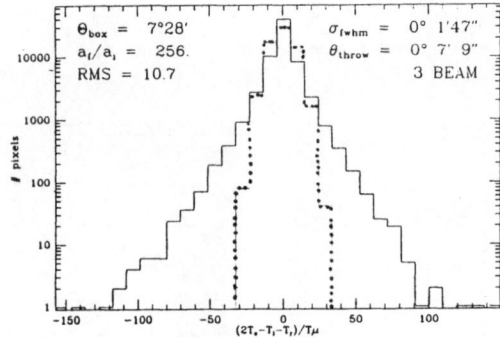

Figure 11: Number of pixels (out of 65 536) of a map convolved with a three beam pattern of beam throw of about 8 arcs minute, each gaussian beam having a FWHM of 1 arc minute 47 arc second. The "gaussian" case (dotted line) was generated from a string induced one by Fourier transforming the map, randomizing the phases, and transforming back.

Figures 11 and 12 present the result of two convolutions corresponding to two three-beam-switching experiment of different geometrical configuration as a number of pixels as a function of the corresponding $\Delta T/T$.

One can obviously now use the convolved maps in a more quantitative fashion in order to extract a limit on $G\mu$ (which is just an overall multiplicative factor of the maps) that a particular experiment places. But for rough estimates, one cannot use directly the quoted numbers by the observers since their analysis procedure generally assumes gaussian statistics. It is thus necessary to apply the same analysis procedure (based on maximum likelihood techniques) to extract a reliable number. It turns out that the tightest constraint at present comes form the Owens valley experiment (provided $z_{ls} \sim 1000$) since it yields $G\mu \lesssim 5\,10^{-6}$. This appears to be the best limit available so far.

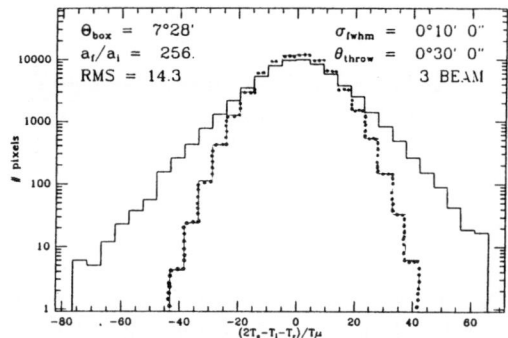

Figure 12: Same as Figure 11, but with a beam throw of 1/2 degree and a beam FWHM of 10 arcs minutes.

5.2. FURTHER CONSIDERATIONS

One should stress at this point that the previous upper limit is fairly conservative since the perturbations induced by density and velocity perturbation on the last scattering surface have not been accounted for. A full calculation of these perturbations will require to propagate in time the metrics perturbations induced by the strings up to the last-scattering surface, which involves a numerical integration of the (linearized) Einstein equations whose source term may be obtained at each time step of the simulations from the stress-energy tensor of the perturbing strings. Such a calculation does not pose any problem of principles and should be undertaken in a near future. Still, these perturbations are expected to be in the same angular range and of the same amplitude than those already computed (but not of the same geometry). They will also add incoherently, thereby tightening the limit by a factor that should be of the order of $\sqrt{2}$.

The full sky COBE map have a resolution of 7 degrees, which is about the total size of the previous maps. Due to the difference in angular scales involved, one needs to generate a new series of full sky maps for strings. The steps involved are in principle identical, but it is quite more involved from a pratical point of view. Indeed, one is now interested in computing large scale properties, on scales much larger than the Horizon at the time of an anisotropy imprint. This is clearly out of the realm of the Minkowskian formalism we used. Fortunately, the general problem of the calculation of metrics perturbations causally generated by strings embeded in a Friedmann-Robertson-Walker background has recently been solved by Shoba Veeraraghavan and Albert Stebbins (1990). Still the numerical problem is rather demanding since, among other things, the problem becomes really three-dimensionnal instead of being in effect two dimensionnal. In any case, we hope to produce such maps and analyse them in the not too distant future...

REFERENCES

Albrecht, A., and Turok, N. 1985, *Phys. Rev. Lett.* , **54**, 1868.

Albrecht, A., and Turok, N. 1989, *Phys. Rev.*, **D40**, 973.

Allen, B., and Shellard, E. P. S. 1990, *Phys. Rev. Lett.*, **64**, 119.

Aryal, M., Everett, A.E., Vilenkin, A., and Vachaspati, T. 1986, *Phys. Rev.*, **D34**, 434.

Bennett, D. P.1986a, *Phys. Rev.*, **D33**, 872.

Bennett, D. P.1986b, *Phys. Rev.*, **D34**, 3592.

Bennett, D. P., and Bouchet, F. R. 1988, *Phys. Rev. Lett.*, **60**, 257.

Bennett, D. P., and Bouchet, F. R. 1989, *Phys. Rev. Lett.*, **63**, 2776.

Bennett, D. P., and Bouchet, F. R. 1990, *Phys. Rev.*, **D41**, 2408.

Bennett, D. P., and Bouchet, F. R. 1991, *Phys. Rev. Brief Rep.*, **D43**, 2733.

Bouchet, F.R., and Bennett, D. P. 1990a, *Phys. Rev. Rapid Comm.*, **D41**, 720.

Bouchet, F.R., and Bennett, D. P. 1990b, *Astrophys. J. Lett.*, **354**, L41.

Bouchet, F.R., Bennett, D. P., and Stebbins, A. 1988, *Nature*, **335**, 410.

Cambridge Cosmic String Workshop 1989, *The Formation and Evolution of Cosmic Strings*, Edited by Gibbons, G.W., Hawking, S.W., and Vachaspati, T., Cambridge University Press, The Pitt Building, Trumpington Street, Cambridge CB2 1RP, England.

Everett, A.E. 1981, *Phys. Rev.*, **D24**, 858.

Gott, J. R. 1985, *Astrophys. J.*, **288**, 422.

Gott, J.R., Park, R., Juszkiewicz, R., Bies, W. E., Bennett, D. P., Bouchet, F. R., and Stebbins, A. 1990, *Astrophys. J.*, **352**, 1.

Hogan, C., and Rees, M. 1984, *Nature*, **311**, 109.

Hindmarsh, M. 1990, *New Castle preprint NCL-90 TP11*.

Kaiser, N., and Stebbins, A. 1984, *Nature*, **310**, 391.

Kibble, T.W.B. 1976, *J. Phys.*, **A9**, 1387.

Kibble, T.W.B. 1980, *Phys. Rep.*, **67**, 183.

Kibble, T.W.B. 1985, *Nucl. Phys.*, **B252**, 227.

Kolb, E.W., and Turner, M.S. 198, *The early Universe*, Frontiers in Physics.

Matzner, R. 1988, *J. Comput. Phys.*, **2**, 51.

Mitchell, D., and Turok, N. 1987a, *Phys. Rev. Lett.*, **58**, 1577.

Mitchell, D., and Turok, N. 1987b, *Nucl. Phys.*, **B294**, 1138.

Nielsen, H.B, and Olesen, P. 1973, *Nucl. Phys.*, **B61**, 45.

Ostriker, J. P., Thompson, C., and Witten, E. 1986, *Phys. Lett.*, **B180**, 231.

Peebles, P.J.E. 1980, *Physical Cosmology*, Princeton University Press.

Quashnock, J.M. and Spergel, D.N. 1990, *Phys. Rev.*, **D42**, 2505.

Scherrer, R.J., and Frieman, J.A. 1986, *Phys. Rev.*, **D33**, 3556.

Stebbins, A. 1985, *Astrophys. J.*, **288**, 422.

Shellard, E.P.S. 1987, *Nucl. Phys.*, **B283**, 624.

Stebbins, A., Veeraraghavan, S., Brandenberger, R., Silk, J., and Turok, N. 1987, *Astrophys. J.*, **322**, 1.

Stinebring, D.R., Ryba, M.F., Taylor, J.H., and Romani, R.W. 1990, *Phys. Rev. Lett.*, **65**, 285.

Taylor, J. H. 1989, *Princeton University preprint to be published in the proceedings of the US-USSR Workshop on High Energy Astrophysics, Tbilisi and Moscow, June, 1989.*

Turok, N., and Bhattacharjee, P. 1984, *Phys. Rev.*, **D29**, 1557.

Vachaspati, T. 1986, *Phys. Rev. Letters*, **57**, 1655.

Vachaspati, T., and Vilenkin, A. 1984, *Phys. Rev.*, **D30**, 2036.

Veeraraghavan, S., and Stebbins, A. 1990, *Astrophys. J.*, **365**, 37.

Vilenkin, A. 1981, *Phys. Rev. Lett.*, **46**, 1169,.
    1496

Vilenkin, A. 1984, *Astrophys. J.*, **282**, L51.

Vilenkin, A. 1985, *Physics Reports*, **121**, 263.

Weinberg, S. 1972, *Gravitation and Cosmology*, Wiley and Sons.

Witten, E. 1985, *Nucl. Phys.*, **B249**, 557.

Yale Cosmic String Workshop 1988, *Cosmic Strings, The Current Status*, Edited by Acceta,
    F.S., and Krauss, L.M., World Scientific, P.O. Box 128, Farrer Road, Singapore 9128.

Zel'dovich, Ya.B. 1980, *Mon. Not. R. Astron. Soc.*, **192**, 663.

THE COSMIC MICROWAVE BACKGROUND RADIATION

Joseph Silk
Departments of Astronomy and Physics, and Center for Particle Astrophysics,
University of California, Berkeley

ABSTRACT.
I review the implications of the spectrum and anisotropy of the cosmic microwave background for cosmology. Thermalization and processes generating spectral distortions are discussed. Anisotropy predictions are described and compared with observational constraints. If the evidence for large–scale power in the galaxy distribution in excess of that predicted by the cold dark matter model is vindicated, and the observed structure originated via gravitational instabilities of primordial density fluctuations, the predicted amplitude of microwave background anisotropies on angular scales of a degree and larger must be at least several parts in 10^6.

1. INTRODUCTION

Discovered in 1964, the cosmic microwave background (CMB) was actually predicted to exist as a consequence of helium synthesis in the hot big bang. Not only is it a relic of this early hot phase, but its ideal blackbody spectrum and extreme degree of uniformity provide persuasive arguments that favor a cosmological origin in the very early universe. A firm prediction of the Big Bang theory is that the relic radiation field from the first instants of the expansion should be blackbody radiation. There is a well defined epoch, prior to which thermalisation processes were rapid compared to the expansion time scale. Injection of photons at a later epoch would remain as a distortion to the primeval blackbody spectrum. No distortions have been found, but COBE has provided a remarkable confirmation of the blackbody spectral prediction. The gravitational instability theory for the origin of large-scale structure, galaxies, galaxy clusters and superclusters, from infinitesimal fluctuations in the very early universe leads to the prediction that the microwave background should be highly isotropic. After two decades of intensive searching for temperature fluctuations. limits on $\Delta T/T$ are approaching 1 part in 10^5 over angular scales from 5 arc-min to 90 degrees, with no relic fluctuations yet having been confirmed, apart from the dipole anisotropy, of amplitude $\sim 10^{-3}$. These lectures present a pedagogical review of the theory of the CMB spectral and angular distortions.

2. THERMALISATION

The important cosmological processes that absorb and create photons are bremsstrahlung $e + p \rightarrow e + p + \gamma$ and double Compton scattering $e + \gamma \rightarrow e + 2\gamma$. At high Ω_b, the former process dominates; at lower Ω_b, the latter process is more important. One compares these two processes by separately evaluating the latest epoch at which they are effective in creating blackbody photons[1]. The timescale associated with the bremsstrahlung process is

$$t_{ff} = \frac{n_\gamma}{4\pi \int j_{ff} d\nu (h\nu)^{-1}} \ ,$$

M. Signore and C. Dupraz (et al.), The Infrared and Submillimetre Sky after COBE, 129–141.
© 1992 Kluwer Academic Publishers.

where the free-free emissivity is

$$j_{ff} = 5 \times 10^{-39} \, n_e n_i Z^2 T^{-1/2} \, exp(\frac{-h\nu}{kT}) \text{ erg cm}^{-3} \text{ s}^{-1} \text{ st}^{-1} \text{ Hz}^{-1}$$

and the number density of blackbody photons is

$$n_\gamma = 60.2(\frac{kT}{hc})^3 \text{ cm}^{-3}.$$

The result of the integration incorporates a logarithmic term $ln(\frac{kT}{h\nu})$ for which the low frequency divergence is suppressed by a cut-off determined by absorption and Compton heating of the low frequency photons:

$$\tau_{es}(4kT/m_e c^2)\kappa_{es} = \kappa_{ff}\nu_{min},$$

where

$$\kappa_{ff} = j_{ff}(\frac{c^3}{2kT\nu^2}), \tau_{es} \approx c\sigma_T n_e,$$

and

$$\kappa_{es} = \sigma_T/m_p;$$

here σ_T is the Thomson scattering cross-section. Consequently, the thermalisation time scale via free-free photon production and absorption is

$$t_{ff} \approx 1.4 \times 10^{24} \, (\Omega_b h^2)T^{-5/2} \text{ sec},$$

Thermalisation is therefore effective at $t_{ff} < t_{exp} \approx 10^{20}T^{-2}$ sec, provided that

$$T > T_{ff} \equiv 2 \times 10^8 \text{ K}.$$

The cross-section for photon production by double Compton scattering is approximately proportional to $\sigma_T(\frac{kT}{m_e c^2})^2$, so that the thermalisation time-scale by this process is

$$t_{2comp} = (n_e \sigma_{2comp} c)^{-1} \approx 4 \times 10^{40}T^{-5}(\Omega_b h^2)^{-1} \text{ sec},$$

whence one infers that double Comptonisation is effective at

$$T > T_{2comp} \equiv 7 \times 10^6(\Omega_b h^2)^{-1/3}.$$

I conclude that the dominant thermalisation process is via double Compton scattering, and is effective at $t < 2 \times 10^6(\Omega_b h^2)^{2/3}$ sec or $T \gtrsim 1$keV. Note that last scattering of the cosmic background radiation occurred much later, at approximately 10^{13} sec or 0.1eV.

3. RELIC SPECTRAL DISTORTIONS

Once the temperature drops below 10^7 K, any further heat input to the radiation background can no longer be thermalised. Bremsstrahlung drives photons to lower frequencies, Compton scattering heats photons. This competition is described symbolically by the following equation for the redistribution of photon energies:

$$\frac{dI_\nu}{dt} = -3HI_\nu + \frac{\partial I_\nu}{\partial t}\Big|_{brem} + \frac{\partial I_\nu}{\partial t}\Big|_{comp},$$

in terms of the specific intensity I_ν. The bremsstrahlung term equals $c(j_{ff} - \kappa_{ff}I_\nu)$, and the Comptonisation term is is obtained from following the evolution of the photon energy distribution subject to repeated Compton scattering by non-relativistic electrons, studied via the Boltzmann equation for the evolution of the photon phase space density $n(\nu)$ due to scattering by free electrons in an isotropic thermal medium. Expansion of the Boltzmann equation in the non-relativistic electron limit, where the fractional energy transfer per scattering is small, leads to the Kompaneets equation[2]:

$$\frac{\partial n}{\partial t}\Big|_{comp} = \frac{kT_e}{m_e c^2}\frac{n_e\sigma_T}{c}x^{-2}\frac{\partial}{\partial x}\left(x^4\left(\frac{\partial n}{\partial x} + x + x^2\right)\right),$$

where the photon occupation number n is defined by

$$n(\nu) = \frac{c^3}{8\pi h\nu^3}I_\nu,$$

$x = h\nu/kT_e$, and T_e is the electron temperature: in general $T_e \neq T_\gamma$.

The Comptonisation equation reduces in the equilibrium limit $\frac{\partial n}{\partial t} = 0$ to the simple solution

$$n = \frac{1}{e^{(h\nu - \mu)/kT_e} - 1},$$

where the chemical potential μ is an integration constant. Only if $\mu = 0$ does this describe black-body radiation; if $\mu < 0$, this solution corresponds to a Bose distribution in which the number of photons is conserved: such a distribution describes photons which are in thermal equilibrium with the electrons but for which photons are not created or destroyed despite the energy transfer. Compton scattering conserves the number of photons, and hence cannot produce blackbody radiation. There are too few photons present at any given electron temperature for a blackbody if $\mu < 0$, and the low frequency (limit $h\nu < |\mu|$) of the photon distribution is $I_\nu \propto \nu^3$ instead of ν^2 for a blackbody. In the high frequency limit, a Wien distribution is recovered: $I_\nu \propto \exp(-h\nu/hT)$. One obtains this time independent limit in the case of early energy injecton to the universe, when the rate of photon diffusion towards low frequencies of photons near the blackbody peak via bremsstrahlung exceeds that of Comptonisation heating of the photons:

$$\frac{h\nu_{min}}{kT_e} \approx 1300(\Omega_b h^2)^{1/2}T_e^{-3/4} < 1.$$

In this case, the μ–distortion describes the relic spectral distortion, with μ/kT measuring, for a near blackbody distortion, the fractional energy distortion : if $\mu < 0$, $|\mu| < kT$, $\rho_\gamma \approx exp(\mu/kT)aT^4$.

At $T < 10^4$ K, Comptonisation dominates late energy input. Assume that $T_e \gg T_\gamma$, whence the Kompaneets equation reduces to ($x \ll 1$))

$$\frac{\partial n}{\partial t} = \frac{kT_e}{m_e c^2}\frac{n_e\sigma_T}{c}x^{-2}\frac{\partial}{\partial x}\left(x^4\frac{\partial n}{\partial x}\right).$$

Define the Comptonisation y–parameter which is the product of the average fractional photon energy change per scattering and the mean number of scatterings.

$$y = \int_t^{t_o} \frac{k}{m_e c^2}(T_e - T_\gamma)n_e\sigma_T c\, dt,$$

A factor of 4 has been omitted in this definition of y. The factor of 4 arises because for a Bose-Einstein photon distribution (applicable in the ideal limit of pure Compton scattering which neither creates nor destroys photons) $\langle h\nu \rangle = 3kT$ and $\langle (h\nu)^2 \rangle = 12kT$. The Kompaneets equation simplifies to

$$\frac{\partial n}{\partial y} = x^{-2} \frac{\partial}{\partial x} \left(x^4 \frac{\partial n}{\partial x} \right),$$

$$y = \tau_0 \int_0^z \frac{kT_e}{m_e c^2} \frac{1+z}{\sqrt{1+\Omega z}} \, dz,$$

and

$$\tau_0 = \frac{\Omega_b n_0 \sigma_T c}{H_0}.$$

If deviations from a Planck function are small, the following solutions may be obtained[3]:

$$\frac{\Delta I_\nu}{I_\nu} = \frac{\Delta n}{n_0} = xy \frac{e^x}{e^x - 1} \left(x \frac{e^x + 1}{e^x - 1} - 4 \right)$$

where $n_0 = (e^x - 1)^{-1}$, whence

$$\frac{\Delta T}{T} = y \left(x \frac{e^x + 1}{e^x - 1} - 4 \right).$$

The function $\Delta I_\nu / I_\nu$ reaches a minimum at $x = 1.78$ or $\lambda = 2.7$ mm, and changes sign at $x = 3.83$, $\lambda = 1.25$ mm, being negative at larger wavelengths. In the Rayleigh-Jeans limit ($x \ll 1$),

$$\frac{\Delta T_{RJ}}{T} \approx -2y, \quad \frac{\Delta T_{RJ}}{T_{RJ}} \approx \frac{-16 x^2 y}{3}.$$

Comptonisation depletes photons in the Rayleigh-Jeans region, and augments photons in the Wien tail ($x \ll 1$):

$$\frac{\Delta T_{Wien}}{T} \approx yx$$

If deviations from a Planck function are large, but characterised by a grey body with effective temperature T, in the Rayleigh-Jeans region, the photon "cooling" is described by ($T_e \ll T_\gamma$)

$$\frac{\dot{T}}{T} = -2\sigma_T n_e c k \frac{T_e}{mc^2},$$

whence

$$T = T_{blackbody} \; e^{-2y},$$

while the radiation density satisfies

$$\rho_\gamma = \rho_{\gamma,blackbody} \; e^{12y}.$$

The current observational limits on μ and y-distortions of the cosmic blackbody radiation are $\mu/kT < 7 \times 10^{-3}$ and $y < 4 \times 10^{-4}$. The y limit in particular constrains the temperature of the uniform intergalactic medium to satisfy

$$T < 4 \times 10^7 \; (1+z)^{-3/2} \; (\Omega_b h^2)^{-1/2}.$$

This eliminates the possibility of an intergalactic medium that is hot enough ($kT \approx 40$ keV) to account for the $2 - 100$ keV diffuse X-ray background radiation.

4. RECOMBINATION SPECTRAL DISTORTIONS

Once there are too few photons with energy above the hydrogen ionization threshold of 13.6 eV, the hydrogen recombines. To make a first estimate of the epoch of recombination, note that

$$\frac{n_\gamma}{n_b}(h\nu > 13.6eV) = \frac{16\pi}{n_b}(\frac{kT}{\hbar c})^3 e^{-I_H/kT}$$

$$\approx \frac{3 \times 10^7}{\Omega_b h^2} \exp(-160000/T)$$

suffices to maintain a fully ionized universe to $T \sim 10^4 K$. Recombination photons provide a negligible source of spectral distortions, amounting to

$$\delta\rho_\gamma/\rho_\gamma \sim (10.2eV/2.7kT_{rec})(n_b/n_\gamma) \sim 10^{-8}$$

if each proton recombines only once, although this estimate is boosted because the first excited level of hydrogen has a supra-thermal population during the recombination epoch.[4,5] The ionization level decreases to a level of order t_{rec}/t, once photoionizations are ineffective. With $t_{rec} \sim 2 \times 10^9(n_H/n_e)(\Omega_b h^2)^{-1}sec$ and $t \sim 7 \times 10^{12}(\Omega h^2)^{-1/2}sec$, the relic level inferred for n_e/n_H is $\sim 3 \times 10^{-4}(\Omega h^2)^{1/2}(\Omega_b h^2)^{-1}$.

5. DECAYING PARTICLES

Consider the out-of-equilibrium decay of a massive particle $\chi \to \chi' + \gamma$ that was thermally produced in the early universe. The decay must occur at $kT \leq 1keV$, once thermalization is ineffective to produce a signature in the CMB spectrum. Hence the particle mass, for a distortion near the peak, must be $\leq 1keV$. The resulting distortion $\Delta\rho_\gamma/\rho_\gamma \propto m_x/T_\gamma$ for particles whose radiative decays are a factor η in number density that of the CMB photons, so that $\Delta\rho_\gamma/\rho_\gamma \propto \eta m_x \tau_{decay}^{1/2}$, where τ_{decay} is the radiative decay lifetime. The observed limit on distortions near the peak is $\Delta\rho_\gamma/\rho_\gamma \leq 0.01$, whence the restriction

$$m_x \leq 1(\frac{\Delta\rho_\gamma/\rho_\gamma}{0.01})(\frac{10^5 \sec}{\tau})^{1/2}(\frac{0.1}{\eta})eV$$

may be derived. However such particles must also be thermally produced in the advanced stages of stellar evolution, when temperature of $\geq 10keV$ are attained. In fact, horizontal branch morphologies would be adversely affected by decays of the form $e^+ + e^- \to \overline{\chi} + \chi'$, an inevitable consequence of $\chi \to \chi' + \gamma$, if the particle mass and lifetime were in the range required to show up as a distortion near the CMB peak. Red giant evolution excludes particle masses and lifetime that satisfy[6]

$$1eV \leq m_x(\tau/10^5 yr)^{1/3} \leq 1000eV.$$

6. PREGALACTIC STARS AND SPECTRAL DISTORTIONS

The radiation density associated with pregalactic stars may be appreciable. For a mass fraction Ω_* in such objects at redshift z_* and a nuclear burning efficiency ϵ (~ 0.004 for massive stars), one infers that

$$\Omega_{rad} \sim 4 \times 10^{-7}(\Omega_*/0.01)(100/(1+z_*))(\epsilon/0.004),$$

as compared to the CMB contribution to the critical density $2.6 \times 10^{-5}h^{-2}$. Pregalactic stars are likely to copiously produce heavy elements and be shrouded by dust. Thus the stellar radiation would be reradiated in the far infrared by dust. The detailed spectral distortions depend on the wavelength dependence of the dust emissivity, but the integrated distortion will be

$$\Delta\rho_\gamma/\rho_\gamma = \Omega_{rad}/\Omega_{CMB} = 0.02h^2.$$

If a fraction f of this energy were directed into heating the intergalactic medium at $z \gg 10$, Compton scattering would distort the CMB by an amount as large as

$$4y \sim f\Omega_{rad}/\Omega_{CMB}.$$

7. RECOMBINATION

A more precise computation of recombination must take account of several additional considerations. These include the fact that excited hydrogen atoms, whose abundance is augmented by build-up of Lyman alpha photons, can be ionised by photons of energy less than 13.6 eV, and that these excited levels can also decay into 2-photon decays, thereby providing a further source of ionizing photons. The additional contributions to the ionization result in an ionization level that exceeds that inferred from the Saha equation.

The following expression approximates the residual ionization over $800 \leq z \leq 1500$:

$$\frac{n_e}{n_H} \approx \frac{206}{z^2}\frac{(\Omega h^2)^{1/2}}{f\Omega_b h^2}[1 + 2.26 \times 10^4 z \exp(-1.462 \times 10^4/z)]$$

where $f = (1 + 1.45z/z_{eq})^{-1/2}$, $z_{eq} = 4 \times 10^4\Omega h^2$. The numerical results show that only at $z \leq 500$ does the ionization fraction freeze out, to values between 4.3×10^{-3} ($\Omega = 1$) and 1.6×10^{-4} ($\Omega = 0.1$); in both cases $\Omega_b = 0.1$ has been assumed.

Last scattering of the cosmic microwave background photons, if no subsequent reionisation occurred, may be inferred by evaluating $e^{-\tau}d\tau/dz$, the probability density for last scattering at redshift z, where the optical depth to Thomson scattering

$$\tau \approx \tau_0(\Omega h^2)^{-1/2}(\Omega_b h^2)\int_0^z \frac{n_e}{n_H}z^{1/2}dz$$

and $\tau_0 = n_0\sigma_T c/H_0 = 0.06h^{-1}$. One obtains[7]

$$e^{-\tau}\frac{d\tau}{dz} = 5.26 \times 10^{-3}z_{1000}^{13.25}e^{-0.37z_{1000}} \quad ; z_{1000} = z/1000.$$

The function is approximated by a Gaussian centered at $z = 1070$ and of width $\Delta z = 80$: these parameters define the *surface of last scattering*, the effective photosphere of the universe as viewed at microwave frequencies.

8. COSMIC MICROWAVE BACKGROUND ANISOTROPY: PREDICTIONS

Some of the assumptions underlying cold dark matter, such as adiabaticity, scale-invariance, or random phases, may have to be relaxed to reconcile theory with observations of large-scale structure. Rival theories are emerging, which drop one or more of the underlying assumptions of the canonical model for primordial fluctuations, and the various implications for the CMB need to be critically assessed in the light of these other theoretical options. In what follows, I will review the various possibilities that have emerged for estimates of $\frac{\delta T}{T}$.

8.1 Primeval curvature fluctuations

The simplest inflation models result in adiabatic, or curvature, fluctuations, baryon synthesis resulting in a universal value of the specific entropy. There are several different contributions to the amplitude of the resulting $\frac{\delta T}{T}$. While a numerical solution of the collisionless Boltzmann equation is necessary to give precise results, the various effects may be considered separately as source terms in this equation, and their dependence on cosmological parameters discussed analytically.

On large angular scales, the predominant effect is due to the gravitational redshifting of photons that emerge from local minima in the gravitational potential on the surface of last scattering (the Sachs-Wolfe effect).[8] In order of magnitude, one finds that, since $|\delta_k|^2 \propto k$ and therefore $\frac{\delta\rho}{\rho} \propto L^{-2}$ over physical scales $L \gtrsim L_{eq}$,

$$\frac{\delta T}{T} \sim \frac{1}{3}\frac{|\delta\phi|}{c^2} \sim \frac{1}{3}|\frac{\delta\rho}{\rho}|_{ls}|\frac{L}{ct}|_{ls}^2$$

for scales $L \gtrsim L_{ls}$. The causal horizon at last scattering L_{ls} subtends an angle $\sim 2\Omega^{1/2}$deg in the standard model (no reionization). This angular scale increases if there are post-decoupling ionization sources, but cannot exceed $\sim 10\Omega^{1/2}(z_{ls}/60)^{1/2}$deg.

On somewhat smaller angular scales down to $\sim (\Omega/z_{eq})^{1/2} \sim 0.4\Omega^{1/2}h^{-1}$deg, corresponding to the range subtended by scales between L_{ls} and L_{eq}, the horizon size at the epoch of equality of matter and radiation densities, the leading contribution to $\frac{\delta T}{T}$ arises from the Doppler shifts across fluctuations at last scattering, namely

$$\frac{\delta T}{T} \sim |\frac{v}{c}|_{ls} \sim |\frac{\delta\rho}{\rho}|_{ls}|\frac{L}{ct}|_{ls}$$

over scales $L_{eq} \lesssim L \sim L_{eq}$. The corresponding comoving scales are $L_{eq} \sim 12(\Omega h^2)^{-1}$ Mpc and $L_{ls} \sim 200(z_{ls}/10^3)^{-1/2}h^{-1}$ Mpc.

Over still smaller angular scales, corresponding to physical scales $\lesssim \frac{1}{3}L_{eq}$, the radiation era suppression of fluctuation growth results in a residual density fluctuation spectrum of approximately constant amplitude (power spectrum asymptoting to $|\delta_k|^2 \propto k^{-3}$). In this limit, the primordial adiabat $\frac{\delta T}{T} \sim \frac{1}{3}|\frac{\delta\rho}{\rho}|_{ls}$ determines the resulting amplitude of the primordial temperature anisotropies, subject to the fuzziness arising from the finite thickness of the surface of last scattering. This typically has a spread of order $\Delta z/z \sim 0.1$ due to the non-instantaneous nature of the hydrogen recombination process, and this translates proportionately into an angular scale ~ 10 arc-min below which the fluctuations are progressively suppressed roughly as θ^2.

In a baryon-dominated universe, primordial curvature fluctuations have been eliminated as an option for seeding large-scale structure. Temperature anisotropies $\frac{\delta T}{T} \gtrsim 10^{-4}$ are produced, and are enhanced by a factor $\sim \Omega^{-1}$ in open models. Introduction of a dominant component of non-baryonic matter transforms the situation, however. Sub-horizon fluctuation growth now can occur prior to matter-radiation decoupling at $z \sim 1000$ once the universe is dominated by non-relativistic matter.

This has three dramatic implications for the amplitude of the resulting anisotropies. On small scales, essentially scales smaller than the horizon at z_{eq}, $\frac{\delta T}{T}$ is reduced by the additional growth factor, namely $\sim \frac{1+z_{eq}}{1+z_{dec}}$. On larger scales, the power in the fluctuation spectrum is reduced because there is no longer the feature at L_{dec} that arose in the baryon-dominated case because effective fluctuation growth only occurred at $z \lesssim z_{dec}$. Finally, on all scales, there is a further reduction that arises from the renormalization to large-scale structure necessitated by the new shape to the power spectrum of the associated density fluctuations, along with the requirement for a bias factor if $\Omega = 1$. Normalization is usually effected at the scale where the variance in the galaxy counts, averaged over spherical cells, has unit amplitude, $8h^{-1}$ Mpc. In practise, one utilizes an integral over the two-point galaxy correlation function, $J_3 \equiv \int_0^r \xi(r) r^2 dr$, which directly measures luminous mass fluctuations on large scales. The resulting mass variance is simply calculated in linear theory for the residual matter density fluctuation spectrum provided one specifies the ratio of total mass to luminous mass densities over the relevant scales. A convenient parametrization for this ratio is the bias factor, defined to be the ratio of galaxy (i.e. luminous matter) to mass fluctuations over an arbitrarily located sphere of radius $8h^{-1}$ Mpc.

Most of the power for the predicted temperature anisotropies in the standard CDM model is distributed between $l = 100$ and $l = 1000$ for a spherical harmonic decomposition of the sky. If the universe always remained ionized, fluctuations are suppressed on small angular scales, and the distribution is shifted into the range $l = 10$ to $l = 100$. The adopted baryon density modifies the predicted amplitude in two ways: the thickness of the last scattering surface is increased as Ω_b is raised, boosting the Doppler contribution to $\frac{\delta T}{T}$, and the fluctuation growth rate is reduced between z_{eq} and z_{dec}, this latter effect being important if $\Omega_b \gtrsim 0.2$.

8.2. Primeval entropy fluctuations

The isocurvature fluctuation mode requires the existence of primeval entropy fluctuations. These are defined as follows: no net curvature implies

$$\rho_m \delta_m + \rho_r \delta_r = 0,$$

and a finite entropy perturbation is

$$\delta s \equiv \frac{3}{4}\delta_r - \delta_m,$$

where δ_m and δ_r are the fractional matter and radiation density perturbations. The factor of $3/4$ arises because specific entropy is just the number of photons per baryon. Once the universe becomes matter-dominated, one ends up with a residual CMB fluctuation

$$\delta_r = \frac{4}{3}\delta s$$

that is potentially observable on large angular scales. Indeed, for a scale-invariant spectrum, $\delta s = constant$, the resulting anisotropy is unacceptably large.[9] On scales below the horizon scale at matter-radiation equality, this corresponds to $n = -3$, since $|\delta s| = |\delta_m| \propto k^{\frac{n+3}{2}}$ at large redshift, where the spectral index of the primordial fluctuation spectrum is defined by $|\delta_k|^2 \propto k^n$. Because isocurvature fluctuations undergo no growth in the radiation era, the sub-L_{eq} spectrum is flat, unlike the CDM spectrum which diverges logarithmically towards decreasing scale. This further enhances the predicted CMB anisotropy, when the spectrum is normalized to the galaxy distribution, relative to CDM.

The residual matter fluctuations at $z \ll z_{eq}$ are generated at horizon crossing by pressure gradients and grow as $\delta_m \sim \delta s(1 + z_{eq})/(1 + z)$, fall off on large scales as $|\delta_k|^m \propto k^4$ at $k^{-1} \gg k_{eq}^{-1}$, and have a peak in power at $k^{-1} \sim k_{rec}^{-1} \sim 50(\Omega h^2)^{-1}$. An empirical fit to the galaxy correlations $< \frac{\delta \rho}{\rho} >^2 \sim r^{-2}$ is $n = -1$ over scales of order $(0.1 - 3)L_{eq}$. This results in reduced large-scale power in the CMB anisotropy that should however be testable in forthcoming experiments. One natural implication of the isocurvature fluctuation model is that structure formation can occur very early, immediately after recombination has occurred. This is unavoidable if $n \gtrsim -1$, since the density fluctuations are normalized to unit variance on scale $\sim 8h^{-1}$Mpc and therefore have amplitude $\sim 10^{-3}[(M/10^{14}\,\mathrm{M}_\odot\Omega h^{-1}]^{-a}$ with $1/2 \gtrsim a \gtrsim 1/3$ for $0 \gtrsim n \gtrsim -1$. This is greater than the Jeans mass immediately after decoupling, and hence the first structures will develop by gravitational instability on these scales. Fragmentation and star formation are likely to be the inevitable outcome.

The ensuing ionizing photon production by massive stars should be capable of reionizing the universe. The last scattering surface is thereby shifted to a much lower redshift,

$$1 + z \gtrsim 60\frac{\Omega^{1/3}}{(0.06/\Omega_b)^{2/3}}.$$

Primary anisotropies are erased on angular scales below a few degrees, but secondary temperature fluctuations are generated due to moving ionized gas, of order $\frac{\delta T}{T} \sim \delta \mathbf{n} \cdot \mathbf{v}$. These fluctuations vanish in first order, but correlated fluctuations in δn and v survive in higher order. Their magnitude is as large as $\sim 10^{-5}$ on arc-second scales, but falls off steeply towards larger angular scales.[10] Secondary ionization also affects galaxy formation via Compton drag, which is effective at $z \gtrsim 100$ in braking any motions of diffuse, uncondensed ionized gas.

8.3. Open cosmological models and broken scale-invariance

Mounting evidence for large-scale power in excess of that allowed by the scale-invariant Zeldovich-Harrison density fluctuation spectrum meshes well with the continuing preponderance of evidence for low Ω. Scale invariance must be broken if $\Omega < 1$, and this leads to a large-scale signature in the CMB anisotropy. This arises at the curvature radius, defined by $cH^{-1}(1-\Omega)^{-1/2}$, which becomes less than the horizon scale $2c/H\Omega$ if $\Omega < 0.85$. A unique multipole signature, essentially due to curvature focussing, arises in the low order CMB multipoles[11,12]. The predicted amplitude should be 10^{-6} or larger, even in isocurvature fluctuation models[13].

8.4. Topological defects and non-gaussian fluctuations

Difficulties in reconciling $\Omega = 1$ and primordial scale-invariant fluctuations with the observed large-scale power have provoked alternative schemes for galaxy formation. The early phase transition associated with GUT scale symmetry braking may not only drive inflation but also be a source of topological defects that act as seeds for large-scale structure. While some of these defects are a potential disaster (monopoles, walls) because they soon overdominate the universe, others (strings, textures) annihilate or decay to provide a self-similar distribution of non-linear objects that provide a small contribution to the energy density. However the resulting response of the ambient matter in the matter-dominated era allows galaxy formation to commence at a much earlier epoch than in the curvature-dominated fluctuation models, yielding qualitatively different large-scale phenomena.

Strings are intrinsically one-dimensional non-Gaussian objects with a mass density and width corresponding to the energy scale and associated Compton wavelength of the $T \sim 10^{16}$GeV phase transition when they formed. String-seeded models of large-scale structure impose a distinct strategy on CMB experiments that requires coverage of many ($\gtrsim 100$) independent patches of sky in order to extract their distinctive signature, which is sharp only at arc-second angular resolution but is always characteristically non-Gaussian[14]. Strings, as do textures, induce isocurvature mass fluctuations.

Global texture is a three-dimensional topological "knot" in otherwise empty space consisting of a Higgs field.[15] Once the horizon scale becomes comparable to the radius of a texture, the texture collapses due to the tension associated with the deformation of the Higgs field. As the texture collapses, it drags in with it all of the associated energy density. When it reaches a size of $\sim 10^{-30}$ cm, the knot unwinds itself, emitting a burst of weakly interacting massless particles known as goldstone bosons. The energy density associated with the texture, and also with the expanding spherical shell of bosons, pushes ordinary matter around, thereby forming the large-scale structure we see today.[16] The energy density of the texture and its associated potential well produce a CMB distortion for a global texture, in the form of a small frequency shift, which we would interpret as a temperature fluctuation in the CMB towards the direction of the texture. Any CMB light ray passing a texture after collapse follows the infalling texture folds into the potential well and is blueshifted. If it passes the center before the collapse, the light ray travels through the incoming texture folds as it climbs out of the potential well, thereby being redshifted. The spherical symmetry of the collapsing texture produces a distribution of *disk*-shaped patches of lower and higher CMB temperature on the sky of degree angular scale. The maximum amplitude of the temperature fluctuations is calculated to be $\delta T/T \sim 10^{-5}$, and the predicted distribution of amplitudes is non-gaussian[17].

8.5. Minimal models

Arbitrary fluctuation spectra can be designed by appropriately fine-tuning the inflationary potential, the gradients in this potential being responsible for the final emergent fluctuation spectrum. Hence it is useful to ask, given the observed requirements on large-scale power in the universe, what are the minimal CMB fluctuations?

Two cases may be distinguished . First, suppose as our earlier prejudices require, that the observed structure arises from primordial density fluctuations. If gravitational instability is indeed responsible for the observed fluctuation amplitude today, one can infer this amplitude from the large-scale peculiar velocities of galaxies , where the power spectrum is still linear, by solving the perturbed continuity equation

$$\nabla \cdot v = -\partial \delta_m / \partial t.$$

These same fluctuations can be extrapolated back to last scattering where they induce a gravitational potential fluctuation in the CMB regardless of whether the primordial fluctuation mode is curvature or isocurvature. Provided the peculiar velocity fields measured out to ~ 5000 km s^{-1} are "typical", they can be used to estimate these Sachs–Wolfe fluctuations[18]. The predicted CMB fluctuations amount to at least $\frac{\delta T}{T} \sim 5.10^{-6}$ over a degree. An alternative approach uses a much deeper galaxy sample to provide a more representative probe of the large–scale galaxy distribution[19] in terms of the angular correlation function $w(\theta)$ for the APM survey. The minimal fluctuations consistent with the data are

$$\frac{\delta T}{T} = 10^{-5} g \frac{\Omega^{0.3}}{b} [\theta \ J(\theta)/5 \times 10^{-5}]^{1/2}_{\theta=10\mathrm{deg}} \quad ,$$

where $J(\theta) = \int w(\theta)\theta d\theta$ and g is a factor that depends on the experimental configuration and degree of beam smoothing: with $\sim 1.5°$ FWHM, $g \gtrsim 1$ in models with non-zero Λ. Up to a factor 10 reduction is possible over these intermediate angular scales in the minimal fluctuations predicted for a cosmological constant–dominated cosmology because of the shift of the observed length scale inferred from the galaxy angular correlations at low redshift to the larger angular scales subtended at high z on the surface of last scattering.

A second possibility allows even lower fluctuations in the CMB. This is the situation that arises when the density fluctuations are not primordial, but are generated at a late, post-last scattering phase transition[20]. This situation is motivated by a neutrino mass of about 0.01 eV, as perhaps may be inferred if the deficiency in solar neutrino fluxes reported in a series of three independent experiments is interpreted in terms of neutrino oscillations. Such a low mass can arise in a symmetry braking phase transition that occurs at the corresponding energy scale, $T \sim 0.01\mathrm{eV}$.

Provided that the phase transition is first order, one may then spontaneously generate, at a redshift of 10 or so, a domain structure on the appropriate horizon scale, ~ 100Mpc. The domain walls retain a frozen-in energy density of the prior phase, with thickness of order the Compton wavelength of the neutrino, ~ 1Mpc, and drive structure formation for as long as they survive. The walls have near light velocities, and their subsequent evolution is controversial: they need to survive long enough to fulfill their role as structure seeds before conveniently disappearing as a consequence, for example, of dissipation via scalar boson radiation. The predicted CMB fluctuations are due to the potential differences along photon geodesics that traverse these time-dependent structures, $\frac{\delta T}{T} \sim (v/c)^3 \sim 10^{-6}$. Larger contributions may also arise because of wall curvature on horizon scales at formation, which produces gravitational focussing effects, and ionized gas motions, following the top-down development of structure expected in such a scheme, that induce Doppler contributions to $\frac{\delta T}{T}$.

9. CONFRONTATION WITH OBSERVATIONS

Extraction of precise upper limits from the data requires detailed comparison of the experimental configuration and systematic sources of noise in an individual experiment with the predictions of the theoretical model being considered. Generally, the models yield temperature fluctuations that are correlated on the sky, and may, in some instances, even be non-gaussian. Each experiment, and each telescope, is likely to have its own idiosyncracies that, in an ideal situation, should be carefully simulated on an artificial data set that is generated for each theoretical model.

A typical double beam experiment measures temperature correlations over some switching angle θ_s, smoothed over an effective beam size, θ_b. To optimize the signal, one requires $\theta_s >> \theta_b$. In this limit, the predicted amplitude of the rms temperature fluctuations has the following dependence on parameters, approximately valid for CDM with $0.3 \gtrsim \Omega_b \gtrsim 0.03$:

$$\frac{\delta T}{T} \propto \Omega_b{}^{1/3} b^{-1} \Omega^{-1} h^{-4/3}.$$

A well-matched experiment with respect to the standard inflationary model, with decoupling on schedule, was one performed at the South Pole with $\theta_b = 13$ arc-min, $\theta_s = 30$ arc-min for which the CDM prediction is[21] $\frac{\delta T}{T} \approx 1.3 \times 10^{-5}$ for a bias factor of 2, a Hubble constant of 50 km/s/Mpc, and $\Omega_b = 0.06$. In other sensitive experiments, the CDM predictions are $\frac{\delta T}{T} \approx 8 \times 10^{-6}, 4 \times 10^{-6}$, for $\theta_b = 0.8$ arc-min (OVRO) and $\theta_b = 7°$ (COBE), respectively.

Ideally, one should compute the predicted correlation function for temperature fluctuations, $C(\theta_b, \theta_s)$, make many realizations of the predicted data set, and evaluate the likelihood function for a detection of intrinsic fluctuations on the sky. Whenever a positive result is obtained (hopefully soon!), one can begin to worry about galactic and extragalactic sources of anisotropy. Once galactic contributions can be subtracted by their frequency dependence, the extragalactic signal is likely to include contributions from Sunyaev-Zeldovich fluctuations in remote galaxy clusters and groups[22,23] as well as distant radio galaxies.[24] In the meantime, null detections already are leading to a restrictive series of upper limits on angular scales that begin to challenge the standard CDM model:[25,26,27]

$\delta T/T(1'.27, 4'.5) < 3.0 \cdot 10^{-5}$,

$\delta T/T(0'.76, 7'.15) < 2.1 \cdot 10^{-5}$,

$\delta T/T(13', 1°) < 2.8 \cdot 10^{-5}$.

One concludes[28] that little parameter space remains for open adiabatic fluctuation models. In particular, if the Hubble constant is 50 km s^{-1} Mpc^{-1}, then $\Omega \gtrsim 0.6b^{-1}$. Large-scale velocity field constraints, along with direct mass determinations of galaxies, groups, and clusters, require $1 \lesssim b \lesssim 2$. Hot dark matter (HDM) inflationary models with $m_\nu \sim 20$eV produce similar angular correlations to the CDM models, but the lack of small-scale power in the residual matter fluctuation distribution forces the normalization to give an uncomfortably recent epoch for galaxy formation if light traces mass on large scales. *Anti*biasing, $b \lesssim 1$, helps circumvent this problem, but only at the expense of generating an unacceptably large amplitude for the temperature anisotropies. HDM adiabatic models are effectively ruled out unless galaxy formation occurs very recently ($z < 1$) and the Hubble constant is close to the maximum allowable value. However reports of a 17 keV mass for the τ neutrino may give some credence to HDM. While cosmology necessarily requires a neutrino of such a mass to be unstable, confirmation of this result would make it plausible from the particle physics point of view that one of the lighter neutrinos has a cosmologically interesting mass. Seeded HDM theories, invoking non-linear seeds such as strings or textures, provide a possible means of resurrecting HDM by allowing structure formation to occur at an acceptably early epoch.

10. FUTURE PROSPECTS

Spectral distortions to the CMB generated by galaxy formation processes are inevitable at a level of order $y \sim 10^{-4}$, due to the hot electrons that must inevitably have been produced in galaxy halos and in the intergalactic medium. The COBE experiment is expected to reach this level of sensitivity provide that galactic dust contributions can be sufficiently well modelled. The current limit of $y \lesssim 4 \times 10^{-4}$ does not compel the universe ever to have undergone a neutral phase, given the baryon density inferred from primordial nucleosynthesis constraints. Complementary observations of fine-scale structure at sub-millimeter wavelengths will be necessary to extract galaxy and galaxy cluster–scale signatures. This may not be achievable until large infrared telescopes are launched, such as ISO and SIRTF.

The hypothesis of primordial density fluctuations implies that, at least in the context of inflationary models, the last scattering surface of the CMB should be exposed by experiments that are either ongoing or in preparation. The CMB on large angular scales probes the universe back to the inflationary epoch, and the CDM template predicts that $\frac{\delta T}{T} \sim 10^{-5}$. COBE, with an anticipated limiting sensitivity of $\sim 4 \times 10^{-6}$ over large angular scales ($\gtrsim 10$deg) should be capable of definitively testing this model. Other experiments, on balloon-borne gondolas and at the South Pole, are probing angular scales between 0.5 deg and 10 deg with comparable sensitivity. The galaxy is likely to prove a major obstacle at the $\frac{\delta T}{T} \sim 10^{-5}$ level. Synchrotron emission from cosmic ray electrons in old supernova remnants, thermal emission from diffuse HII regions, and dust emission from interstellar cirrus, are all contributing inhomogeneously at essentially unknown amounts, to the general galactic background at this level. However gravitational instability models for structure formation predict that $\frac{\delta T}{T} \gtrsim 10^{-6}$, and this is within the capability of experiments presently under consideration for the next decade.

ACKNOWLEDGEMENTS

I am grateful to my collaborators J. Bartlett, K. Gorski and N. Vittorio for their indispensable advice during the course of preparing this review. The Director of the Mount Stromlo and Siding Spring Observatories, Professor A. Rodgers, generously provided hospitality and support during

1990, when part of this review was written. I also acknowledge partial support from NASA grant NGR 05-003-578 at the University of California, Berkeley.

REFERENCES

1. Danese, L., and De Zotti, G. 1982, *Astron. Astrophys.*, **107**, 39.

2. Kompaneets, A. 1957, *Sov. Phys. JETP*, **4**, 730.

3. Zel'dovich, Ya. B., and Sunyaev, R.A. 1969, *Astrophys. Space Sci.*, **4**, 301.

4. Zel'dovich, Ya. B., Kurt, V. G. and Sunyaev, R. A. 1968, *Soviet Physics JETP*, **55**, 278.

5. Dubrovich, V. K. 1975, *Soviet Astonomy Letters: Pisma Astron. Zh.*, **1**, 3.

6. Dearborn, D. *et al.* 1990; *Ap. J.*, **336**, 61.

7. Jones, B. J. T. and Wyse, R. F. G. 1981, *Astr. Ap.*, **149**, 144.

8. Sachs, R.K., and Wolfe, A.M. 1967, *Ap.J.*, **147**, 73.

9. Efstathiou, G. and Bond, J.R. 1986, *Mon. Not. R. astr. Soc.*, **218**, 103.

10. Vishniac, E. 1987, *Ap. J.*, **322**, 597.

11. Wilson, M.L. 1983, *Ap. J.*, **273**, 2.

12. Abbott, L.F., and Schaeffer, R.K. 1986, *Ap.J.*, **308**, 546.

13. Górski, K.M., and Silk, J. 1989, *Ap.J. Letters*, **346**, L1.

14. Bouchet, F.R., Bennett, P.P., and Stebbins, A. 1988, *Nature*, **335**, 410.

15. Davis, R. L. 1987, *Phys. Rev.* **D**, **35**, 3705.

16. Turok, N. 1990, *Phys. Rev. Lett.*, **64**, 2736.

17. Turok, N. and Spergel, D. 1990, *Phys. Rev. Lett.*, **64**, 2736.

18. Juszkiewicz, R., Gorski, K. and Silk, J. 1987, *Ap. J. Letters*, **323**, L1.

19. Kashlinsky, A. 1991, *Ap. J. Letters*, (submitted).

20. Hill, C.T., Schramm, D.N., and Fry, J.N. 1989, *Comm. Nucl. Part. Phys.*, **19**, 25.

21. Vittorio, N. and Silk, J. 1991, submitted.

22. Schaeffer, R., and Silk, J. 1988, *Ap. J.*, **333**, 509.

23. Cole, S. and Kaiser, N. 1988, *Mon. Not. R. astr. Soc.*, **233**, 637.

24. Franceschini, A., Toffolatti, L., Danese, L., and De Zotti, G. 1989, *Ap. J.*, **344**, 35 .

25. Uson, J. and Wilkinson, D.T. 1984, *Ap.J.*, **277**, L1.

26. Readhead, A.C.S., Lawrence, C., R., Meyers, S.,T., Sargent, W.,L.,W., Hardebeck, H.E., Moffet, A.T. 1989, *Ap.J.*, **346**, 566..

27. Meinhold, P. and Lubin, P. 1991, *Ap. J. Letters*, **370**, L1.

28. Vittorio, N., Meinhold, P., Muciaccia, P. F., Lubin, P., and Silk, J. 1991, *Ap. J. Letters*, **372**, L1.

The Spectral Distortions of the Cosmic Microwave Background Radiation, LEP and Heavy Neutrinos

P. SALATI

Theory Division, CERN

CH-1211, Geneva 23

Switzerland

ABSTRACT : In the early Universe, energy may have been released into the primordial plasma, generating spectral distortions of the cosmic background radiation (CBR) which, incidentally, would be today the precious signatures of the primeval epochs. In the first three sections, the delicate interplay between radiation and plasma is investigated. When thermal contact is established between photons and electrons, the radiation spectrum relaxes towards a Bose-Einstein distribution as a consequence of inverse Compton scatterings. The spectrum latterly achieves a Planck distribution as bremsstrahlung and double Compton emissions proceed. The μ and y spectral distortions are discussed. Then, the primordial behaviour of particles, such as the neutrino, is generically presented in order to analyse the implications of the recent LEP and COBE results as regards the cosmological rôle of heavy neutrinos. The latter are no longer viable candidates for cosmological dark matter. Finally, in the last section, the hypothetical 17 keV neutrino recently reported in some β-decay experiments faces the COBE constraints on the CBR spectrum.

1 - The Kompaneets equation

As a result of the linearity of Maxwell's equations, electromagnetic waves do not interact with each other. Photons achieve thermal equilibrium because they merely collide with free electrons which, therefore, act as the seeds of the radiation thermalization. The Thomson-Compton cross section may be expressed as a function of the classical radius of the electron $r_0 = e^2/4\pi m_e c^2 = 2.818 \times 10^{-13}$ cm

$$\sigma_T = \frac{8\pi}{3} r_o^2 \simeq 6.67 \times 10^{-25} \, \text{cm}^2 \ . \tag{1}$$

Note that when an incident photon with energy $h\nu << m_e c^2$ collides with an electron at rest, the angular distribution of the scattered photon favours the forward and backward directions

$$\frac{1}{\sigma_T} \frac{d\sigma_T}{d\Omega} = \frac{3}{16\pi} \left(1 + cos^2\theta \right) \ , \tag{2}$$

143

M. Signore and C. Dupraz (et al.), The Infrared and Submillimetre Sky after COBE, 143–173.
© 1992 *Kluwer Academic Publishers.*

where θ denotes the angle between the initial and final photon momenta. Energy is transferred from the photon to the electron

$$\nu - \nu' \simeq \frac{h\nu^2}{m_e c^2} \left(1 - cos\theta\right) . \tag{3}$$

As soon as recombination occurs and electrons neutralize with protons into hydrogen atoms, the cosmic background radiation (CBR) drops out of equilibrium and its features get quenched.

This section is devoted to the intricate interplay between the radiation and the plasma of free electrons. Our aim is to derive the Kompaneets equation [1] which accounts for the relaxation of the photon distribution $\eta(\nu)$ towards a Bose-Einstein spectrum in thermal equilibrium at temperature $T_\gamma = T_e$. We focus on non-relativistic electrons and consider photons with energy $h\nu << m_e c^2$.

1.1 - A heuristic approach

To commence, we assume a hot electron gas. Photon energies $h\nu$ are small with respect to the electron temperature T_e. This condition is met for instance in the Rayleigh-Jeans region of the spectrum. We specially focus on an electron which, in the cosmological frame $\mathcal{R}_C$, has velocity $\vec{\beta} = \beta \vec{U}$, with $\beta << 1$. When a photon with initial momentum $(h\nu_i/c) \, \vec{U}$ collides with that electron, its frequency undergoes, on average, a Doppler shift.

In the rest frame $\mathcal{R}_e$ of the electron, the frequency of the incident photon is

$$\nu_i' = \sqrt{\frac{1 - \beta}{1 + \beta}} \, \nu_i . \tag{4}$$

Since $h\nu_i' << m_e c^2$, relation (3) implies that the frequency ν' is not changed by the scattering in the rest frame $\mathcal{R}_e$. In addition, the final momentum may be crudely assumed, with equal probabilities, either aligned with, or opposite to the incident direction $\vec{U}$. In the first case, the scattered photon points towards the same direction as the electron, and its frequency, in $\mathcal{R}_C$, is not modified. In the opposite situation, the scattered photon and the electron are back to back so that, in $\mathcal{R}_C$, the frequency ν_f is red-shifted with respect to the initial value ν_i. In the cosmological frame $\mathcal{R}_C$, the situation may therefore be summarized by

$$\nu_f(\rightarrow) = \nu_i(\rightarrow) \quad \text{while,} \quad \nu_f(\leftarrow) = \left(\frac{1 - \beta}{1 + \beta}\right) \nu_i(\rightarrow) . \tag{5}$$

If the photon scatters from the forward direction, the same reasoning applies and we readily infer a blue-shift of the frequency

$$\nu_f(\leftarrow) = \nu_i(\leftarrow) \quad \text{while,} \quad \nu_f(\rightarrow) = \left(\frac{1 + \beta}{1 - \beta}\right) \nu_i(\leftarrow) . \tag{6}$$

Photons with incident momenta along the same direction as the electron undergo an average frequency shift

$$< \nu_f > (\text{after collision}) \simeq \left(2 + \frac{1 - \beta}{1 + \beta} + \frac{1 + \beta}{1 - \beta} \right) \left(\frac{\nu_i}{4} \right) , \tag{7}$$

which, once expanded up to second order in the electron velocity β, leads to

$$\frac{\Delta \nu}{\nu} \simeq \beta^2 . \tag{8}$$

Photons with incident momenta transverse to the electron velocity are also, on average, scattered perpendicularly so that their frequencies are not shifted. That possibility has two chances out of three of occurring in each collision.

Since every photon undergoes $\sigma_T c n_e$ collisions each second, a third of which only are not sterile and lead to the frequency shift (8), we infer the average increase

$$< \Delta \nu > \simeq \frac{1}{3} (\sigma_T c n_e) < \beta^2 > \nu \Delta t , \tag{9}$$

during the time interval Δt. Since electrons, whose density is denoted by n_e, behave as a perfect gas, the mean value $< \beta^2 >$ may be expressed as a function of T_e and m_e. The parameter y is defined as

$$dy = \frac{d\nu}{\nu} = (\sigma_T c n_e) \left(\frac{kT_e}{m_e c^2} \right) dt = \frac{dt}{t_C} , \tag{10}$$

where t_C is the typical time scale for the inverse Comptonization of the photons. As is clear from expressions (9) and (10), thermal contact between a hot electron gas and the radiation leads to the so-called y spectral distortion [2]

$$\nu(t) = \nu(0) e^y , \tag{11}$$

where the parameter y evolves according to

$$y = \int \frac{dt}{t_C} . \tag{12}$$

Since the frequency of the distorted spectrum is rescaled upwards by the factor e^y, the number of photons per logarithmic interval $dLog\nu$ is constant as is clear on the upper panel of figure 1. The solid line stands for a pure black-body spectrum at temperature T_0

$$\frac{dn}{dLog\nu} = \frac{\nu^3}{e^{h\nu/kT_0} - 1} , \tag{13}$$

while the short dash curve exhibits a $y = 0.5$ distortion and is shifted towards the high energy region of the diagram.

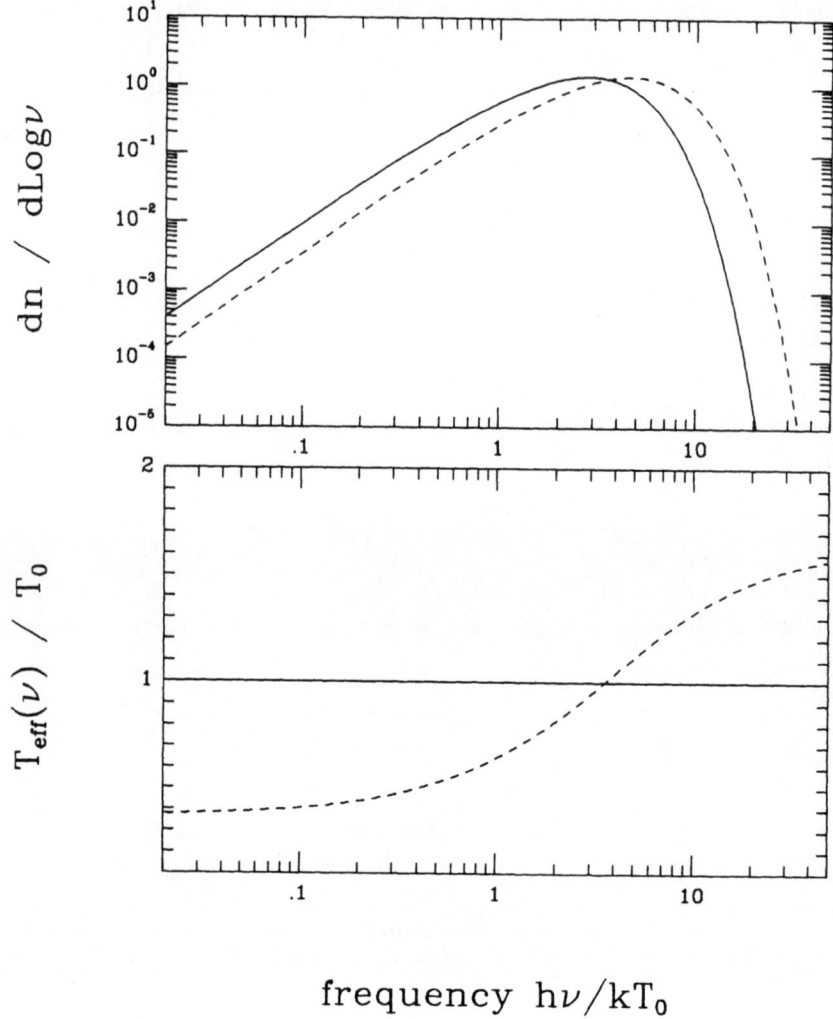

Figure 1

Upper panel : A Planck distribution (solid curve) experiences a $y = 0.5$ spectral distortion which results into a shift of the spectrum (dashed curve) towards higher frequencies.

Lower panel : In the Rayleigh-Jeans region, the effective black-body temperature $T_{eff}(\nu)$ of the distorted spectrum (dashed curve) is shifted downwards by an e-fold with respect to the initial temperature T_0 (solid line).

The lower panel displays the effective black-body temperature T_{eff}. For the pure Planck spectrum (solid line), the temperature has constant value $T_{eff} = T_0$. Since photons collide with hot electrons and gain energy, high frequencies are populated at the expense of low energies. If on the one hand, the distorted spectrum (short dash curve) exhibits an excess in the Wien region, on the other hand, the Rayleigh-Jeans part is deficient. Note that T_{eff} is indeed shifted downwards with respect to the pure black-body curve. For low frequencies, the spectrum merely reduces to $\propto \nu^2 T$, so that the effective Rayleigh-Jeans temperature decreases by a factor e^{-2y}

$$T_{\mathcal{RJ}}(\nu) = e^{-2y} T_0 \ . \tag{14}$$

Since in our example $y = 0.5$, the Rayleigh-Jeans temperature is shifted downwards by an e-fold, as is clear in figure 1.

We have analysed the spectral distortion generated by the thermal contact between hot electrons and low energy photons with $h\nu << kT_e$. We have neglected the frequency shift which photons undergo in the electron rest frame in which, moreover, the radiation is *not* isotropic. We have also disregarded stimulation effects which may be important during Thomson collisions. Notice, however, that the y-distortion effect is typical of the Rayleigh-Jeans region of the spectrum for which the condition $h\nu << kT_e$ is always fulfilled.

1.2 - The energy transfer between electrons and photons

We first assume a pure black-body radiation with temperature T_0. In the cosmological frame $\mathcal{R}_C$, the radiation is isotropic. However, in the rest frame $\mathcal{R}_e$ of an electron moving with velocity $\vec{\beta}$, photons from the forward direction are blue-shifted, while those from the backward direction are red-shifted. In $\mathcal{R}_C$, the photon distribution is

$$\eta(\nu) = \frac{1}{e^{h\nu/kT_0} - 1} \ . \tag{15}$$

Since that distribution is a Lorentz invariant, it may be expressed as

$$\eta(\nu', \theta) = \frac{1}{e^{h\nu'/kT_\theta} - 1} \tag{16}$$

in the electron rest frame $\mathcal{R}_e$. The angle between the electron and photon momenta is denoted by θ, and ν' stands for the frequency. We readily infer an anisotropy of the temperature distribution in $\mathcal{R}_e$

$$T_\theta \simeq (1 - \beta\cos\theta) T_0 \ , \tag{17}$$

for non-relativistic $\beta << 1$ electrons. In $\mathcal{R}_e$, photons with initial direction θ transfer to the electron an average momentum $(h\nu'/c)\cos\theta$ per collision. Since there are many more photons coming from the forward ($\theta = \pi$) than from the backward ($\theta = 0$) directions, the net transfer of momentum results into the drag force

$$F_{drag} = \int \left(\frac{h\nu'}{c} \cos\theta \right) (\sigma_T c) \left(\eta(\nu', \theta) \frac{2\nu'^2 d\nu' d\Omega}{c^3} \right) . \tag{18}$$

Once frequencies have been integrated out, that expression simplifies

$$F_{drag} = \left(\frac{8\pi^5}{15} \frac{k^4}{h^3 c^3} \right) T_0^4 \sigma_T \int \left(\frac{T_\theta}{T_0} \right)^4 \cos\theta \left(\frac{d\Omega}{4\pi} \right) , \tag{19}$$

and the photon energy density ρ_γ may be factored out. The drag force is therefore a mere function of the electron velocity β and of the energy density ρ_γ

$$\vec{F_{drag}} = -\frac{4}{3} \sigma_T \rho_\gamma \vec{\beta} . \tag{20}$$

Radiation acts as a viscous medium with respect to the electrons which experience a damping of their propagation. That friction results into the heating of the photon gas at the expense of the electrons, whose kinetic energy is transferred to the radiation with the rate

$$\epsilon_{e \to \gamma} = < F_{drag} \beta c > n_e , \tag{21}$$

per unit volume. For non-relativistic free electrons, that energy transfer simplifies finally into

$$\epsilon_{e \to \gamma} = 4 \rho_\gamma (\sigma_T c n_e) \left(\frac{kT_e}{m_e c^2} \right) = 4 \rho_\gamma \frac{dy}{dt} . \tag{22}$$

The careful reader will have noticed that stimulation is absent from our discussion. However, we have not neglected that important effect. In $\mathcal{R}_e$, if a photon with momentum $\vec{p_1}$ scatters into the state $\vec{p_2}$, it gives to the electron a momentum proportional to $\eta_1 (1 + \eta_2)(\vec{p_1} - \vec{p_2})$. The reverse reaction is possible and results in the momentum transfer $\eta_2 (1 + \eta_1)(\vec{p_2} - \vec{p_1})$. The stimulation factors vectorially cancel because $d\sigma_T/d\Omega$ depends only on the angle $(\vec{p_1}, \vec{p_2})$ between the photon momenta. In addition, since $h\nu << m_e c^2$, frequencies in the initial and final states are identical and $\nu_1 = \nu_2$.

For a hot electron gas, i.e., $T_\gamma << T_e$, the energy flow from radiation to electrons is negligible. The energy density ρ_γ evolves therefore as

$$d\rho_\gamma \simeq \epsilon_{e \to \gamma} dt = 4 \rho_\gamma dy , \tag{23}$$

and exponentially increases with the y parameter

$$\rho_\gamma(t) = \rho_\gamma(0) e^{4y} . \tag{24}$$

As a matter of fact, we could have readily guessed that result. Since $T_\gamma << T_e$, a y-distortion obtains and frequencies are shifted according to the exponential growth (11).

Remember that ρ_γ is an integral over $\nu^3 \, d\nu$, and is proportional to the fourth power of the average frequency which increases as e^y.

Before COBE, the X-ray background was understood as the thermal emission of very hot electrons with $T_e \sim 300$ keV, heated up by, say, quasars at a red-shift $z \sim 5$. An exciting signature of this model is the resulting y spectral distortion of the CBR which was predicted to be

$$y = \frac{\sigma_T \, c \, n_e{}^0}{H^0} \; \Omega_{IGM} h \; \frac{kT_e}{m_e \, c^2} \int_0^5 \sqrt{1+z} \, dz \; , \qquad (25)$$

for a flat Universe. The parameter h is the Hubble constant expressed in units of $H^0 = 100$ km/s/Mpc while Ω_{IGM} stands for the density of the inter-galactic medium. For $\Omega_{IGM} \sim 0.2$ and $h = 1/2$, the distortion was inferred to be $y \simeq 0.07 \, (kT_e/m_e \, c^2)$. COBE [3] has placed the stringent constraint $y \leq 10^{-3}$, so that the inter-galactic medium temperature cannot exceed ~ 10 keV. Hot electrons are therefore ruled out as a plausible source of the X-ray background.

In order to derive the energy transfer $\epsilon_{\gamma \to e}$ from the radiation to the electrons, we will assume the latter at rest in the cosmological frame $\mathcal{R}_C$. When a photon with energy $h\nu$ collides with an electron, it loses an average energy

$$h\Delta\nu \simeq \left(\frac{h\nu}{m_e \, c^2} \right) h\nu \; , \qquad (26)$$

where the factor $cos\theta$ of relation (3) has been integrated out. That energy transfer is second order in β since, if we grossly assume $T_\gamma \sim T_e$, the fraction $(h\nu/m_e \, c^2) \sim (kT_\gamma/m_e \, c^2)$ and is of order $(kT_e/m_e \, c^2) \sim \beta^2$. Stimulation effects do not cancel now and, after integration over the photon distribution $\eta(\nu)$, the rate $\epsilon_{\gamma \to e}$ may be expressed as

$$\epsilon_{\gamma \to e} = \sigma_T \, c \, n_e \int \frac{8\pi\nu^2}{c^3} \frac{(h\nu)^2}{m_e \, c^2} \, \eta(\nu) \, [1 + \eta(\nu)] \, d\nu \; . \qquad (27)$$

1.3 - The Kompaneets equation

A proper derivation of the Kompaneets equation [1] may be found, for instance, in [4] and is fairly lengthy. We will obtain that equation quite crudely, still rapidly. Frequencies may be conveniently expressed in units of the electron temperature so that we define

$$x = \frac{h\nu}{kT_e} \; . \qquad (28)$$

The energy density of the photon bath is

$$\rho_\gamma = 8\pi \frac{(kT_e)^4}{(hc)^3} \int x^3 \, \eta(x) \, dx \; , \qquad (29)$$

and varies according to the energy balance between radiation and electrons

$$\frac{d\rho_\gamma}{dt} = \epsilon_{e\to\gamma} - \epsilon_{\gamma\to e} = 8\pi \frac{(kT_e)^4}{(hc)^3} \int x^3 \, dx \, (4 - x - x\eta) \frac{\eta}{t_C} \; . \tag{30}$$

The variation of ρ_γ is directly related to the time derivative of the photon distribution so that, after integration by part of the various integrals, the Kompaneets equation obtains

$$t_C \left. \frac{\partial \eta}{\partial t} \right|_K = \frac{1}{x^2} \frac{\partial}{\partial x} \left\{ x^4 \left(\frac{\partial \eta}{\partial x} + \eta + \eta^2 \right) \right\} \; , \tag{31}$$

where $\partial\eta/\partial t$ is expressed as a function of the first and second derivatives of the photon distribution η with respect to the frequency x.

As an illustration, we derive directly from the Kompaneets equation the time evolution of the Rayleigh-Jeans temperature, *i.e.*, relation (14). The distribution of low-frequency photons at temperature T_γ simplifies into

$$\eta(x) \simeq \frac{kT_\gamma}{h\nu} = \frac{\alpha(t)}{x} \; , \tag{32}$$

where the time-dependent ratio T_γ/T_e is denoted by α, and is assumed to be very small with respect to 1. In the Rayleigh-Jeans region of the spectrum, the ordering $x \ll \alpha(t) \ll 1$ obtains, and we may neglect η and η^2 with respect to $\partial\eta/\partial x$ in equation (31). The evolution of α is readily inferred to be

$$t_C \frac{d\alpha}{dt} = -2\,\alpha(t) \; , \tag{33}$$

and leads to relation (14) with an exponential decrease of the Rayleigh-Jeans temperature.

Note finally that, under the action of Thomson-Compton collisions, a photon spectrum evolves until it reaches the equilibrium configuration for which

$$\frac{\partial \eta}{\partial x} + \eta + \eta^2 = 0 \; . \tag{34}$$

The previous relation is satisfied for a Bose-Einstein spectrum

$$\eta_{BE}(x) = \frac{1}{e^{(x+\mu)} - 1} \; , \tag{35}$$

where the constant of integration μ of the differential equation (34) may be translated into the chemical potential $-(\mu \times kT_e)$ of the Bose-Einstein distribution.

2 - The relaxation of the chemical potential $\mu \to 0$

When heat is injected into radiation, the average energy per photon increases and a spectral distortion develops. After a time $\sim t_C$, Thomson collisions redistribute the additional energy, and the spectrum relaxes towards a Bose-Einstein distribution with chemical potential $\mu > 0$ and temperature $T_\gamma = T_e$. Thomson collisions conserve the number of photons. If they operate alone, the spectrum (35) does not evolve towards the pure black-body shape

$$\eta_{Pl}(x) = \frac{1}{e^x - 1} \ . \tag{36}$$

In fact, as a result of the combined action of bremsstrahlung and double Compton reactions, for which the number of photons may increase, μ relaxes towards 0.

2.1 - The bremsstrahlung process

The bremsstrahlung reaction $e\,p \rightarrow e\,p\,\gamma$ replenishes the radiation with additional photons. If an electron is accelerated by an ion of charge Z_i, the corresponding electric dipole moment $\vec{d}$ of the electron-ion pair radiates a power

$$P = \frac{2}{3c^3} \frac{\partial^2 d}{\partial t^2} \ , \tag{37}$$

so that the total energy emitted at frequency $\nu = \omega/2\pi$ during the passage of the electron near the ion is

$$\frac{dW}{d\omega} = \frac{8\pi}{3c^3} \omega^4 \hat{d}^2(\omega) \ . \tag{38}$$

Note that $\hat{d}$ stands for the Fourier transform of the dipole moment

$$\omega^2 \hat{d}(\omega) = \frac{Z_i e^3}{\pi m_e v b} \ , \tag{39}$$

where an undeflected electron trajectory with velocity v and parameter of impact b, has been assumed. Each electron radiates an energy dW per second

$$\frac{dW}{d\omega} = \frac{16}{3c^3} \frac{Z_i^2 e^6}{m_e^2} \frac{n_i}{v} \int_{b_{min}}^{b_{max}} \frac{db}{b} \ , \tag{40}$$

where n_i is the density of ions. Above $b_{max} \sim v/\omega$, the period associated to the frequency ω is so small with respect to the duration of the passage of the electron in the vicinity of the ion, that the Fourier transform of $\vec{d}$ is completely chopped out and vanishes. On the other hand, if b is too small, deflection may be important, and quantum effects come into play as soon as b gets smaller than the de Broglie wavelength $h/m_e v$ of the electron. Integration over those electrons whose kinetic energy exceeds $\hbar\omega$ leads to the bremsstrahlung evolution equation

$$t_{\gamma e}\, x^3 \left.\frac{\partial \eta}{\partial t}\right|_B = Q\, e^{-x}\, g(x)\, [1 - \eta(x,t)(e^x - 1)] \ , \tag{41}$$

with the Q factor of the form

$$Q = \frac{\alpha}{(2\pi)^{7/2}} \left(\frac{hc}{kT_e}\right)^3 \left(\frac{m_e c^2}{kT_e}\right)^{1/2} \sum n_i Z_i^2 \ . \tag{42}$$

The Gaunt factor $g(x)$ may be expressed as

$$g(x) = Log\left(\frac{2.2}{x}\right) \qquad x \leq 1$$

while, \hfill (43)

$$g(x) = \frac{Log(2.2)}{\sqrt{x}} \qquad x \geq 1 .$$

The typical collision time of photons with electrons is denoted by $t_{\gamma e} = (n_e c \sigma_T)^{-1}$. Note that the bremsstrahlung production and absorption processes are in balance as soon as a Planck distribution obtains.

2.2 - The Double Compton reaction

The differential cross section of the *double* Compton reaction $e\,\gamma(\nu') \rightarrow e\,\gamma(\nu_1)\,\gamma(\nu_2)$ is of order α with respect to the Thomson-Compton cross section

$$\frac{d\sigma_{DC}}{d\nu_2} = \frac{4\alpha}{3\pi} \left(\frac{h\nu'}{m_e c^2}\right)^2 (1 - cos\theta_1) \frac{1}{\nu_2} d\sigma_T . \tag{44}$$

One of the final photons has approximately the same energy as its progenitor, *i.e.*, $\nu_1 \simeq \nu'$. When integrated over the photon and electron distributions, the evolution of the radiation spectrum owing to double Compton emission and absorption takes the form [5]

$$t_{\gamma e} x^3 \left.\frac{\partial\eta}{\partial t}\right|_{DC} = \frac{4\alpha}{3\pi} \left(\frac{kT_e}{m_e c^2}\right)^2 \mathcal{I}(t) \left[1 - \eta(x,t)(e^x - 1)\right] . \tag{45}$$

The integral $\mathcal{I}(t)$ over the field distribution of photons is the driving term for the double Compton process

$$\mathcal{I}(t) = \int x'^4 \eta(x',t) \left[1 + \eta(x',t)\right] dx' , \tag{46}$$

where $x' = h\nu'/kT_e \simeq h\nu_1/kT_e$. Should not there be a single photon, double Compton would be completely ineffective.

2.3 - The relaxation of a Bose-Einstein distribution towards a Planck spectrum

Suppose that at some initial time $t = 0$, a pure black-body radiation is thermally coupled to a hot plasma of ions and electrons at temperature $T_e > T_\gamma$. The photon distribution $\eta(x,t)$ evolves under the combined action of Thomson collisions, and bremsstrahlung and double Compton reactions

$$\left.\frac{\partial\eta}{\partial t}\right|_{tot} = \left.\frac{\partial\eta}{\partial t}\right|_K + \left.\frac{\partial\eta}{\partial t}\right|_B + \left.\frac{\partial\eta}{\partial t}\right|_{DC} , \tag{47}$$

where $x = h\nu/kT_e$. This equation must be integrated numerically. The main features of the evolution of $\eta(x,t)$ may nevertheless be sketched. Comptonization stops as soon as a Bose-Einstein spectrum is established, while bremsstrahlung and double Compton drive the photon distribution towards a Planck spectrum.

If on the one hand, Comptonization occurs on the time scale t_C, on the other hand, bremsstrahlung and double Compton absorptions are respectively associated with the time scales

$$t_B^{abs} \;=\; \frac{x^2}{Q\,g(x)}\,t_{\gamma e} \tag{48}$$

and

$$t_{DC}^{abs} \;=\; \frac{3\pi}{4\alpha}\left(\frac{m_e\,c^2}{kT_e}\right)^2 \mathcal{I}^{-1}(t)\,x^2\,t_{\gamma e} \;, \tag{49}$$

in the low energy part of the spectrum. Provided its frequency is lower than the critical value

$$x_B \;=\; \left\{g(x_B)\,\frac{Q}{8}\,\left(\frac{m_e\,c^2}{kT_e}\right)\right\}^{1/2} \;, \tag{50}$$

for which $t_B^{abs}(x_B) = t_C/8$, a low energy photon is more likely to be absorbed by inverse bremsstrahlung than to migrate towards higher frequencies. Below x_B, bremsstrahlung is therefore more efficient than electron scattering, and a Planck spectrum readily obtains. For the double Compton process, that critical frequency is

$$x_{DC} \;=\; \left\{\frac{\alpha}{6\pi}\left(\frac{kT_e}{m_e\,c^2}\right)\mathcal{I}(t)\right\}^{1/2} \;. \tag{51}$$

As shown in [5], double Compton dominates bremsstrahlung whenever

$$\frac{n_\gamma}{n_e} \geq 0.1 \left(\frac{m_e\,c^2}{kT_e}\right)^{5/2} \;, \tag{52}$$

i.e., for a hot and diluted gas. For a colder and denser medium, bremsstrahlung takes over.

Lightman [5] has numerically investigated the evolution of an initial radiation spectrum embedded inside a homogeneous plasma at temperature $kT_e = m_ec^2/10$ and density $n_e = n_i = 10^{23}\,\mathrm{cm}^{-3}$, under the combined action of electron scattering and bremsstrahlung (figure 2) or double Compton emissions (figure 3).

In figure 2, no photons are present initially. After $t \sim 0.4\,t_{\gamma e}$, the low energy part of the spectrum has already relaxed towards a pure black-body distribution. Below x_B, photons are created and, due to electron scatterings, they migrate from the Rayleigh-Jeans region to higher frequencies. The spectrum eventually achieves a Planck form (dashed line) at $t \sim 1.5 \times 10^6\,t_{\gamma e}$. Note that x_B is essentially constant during the evolution.

154

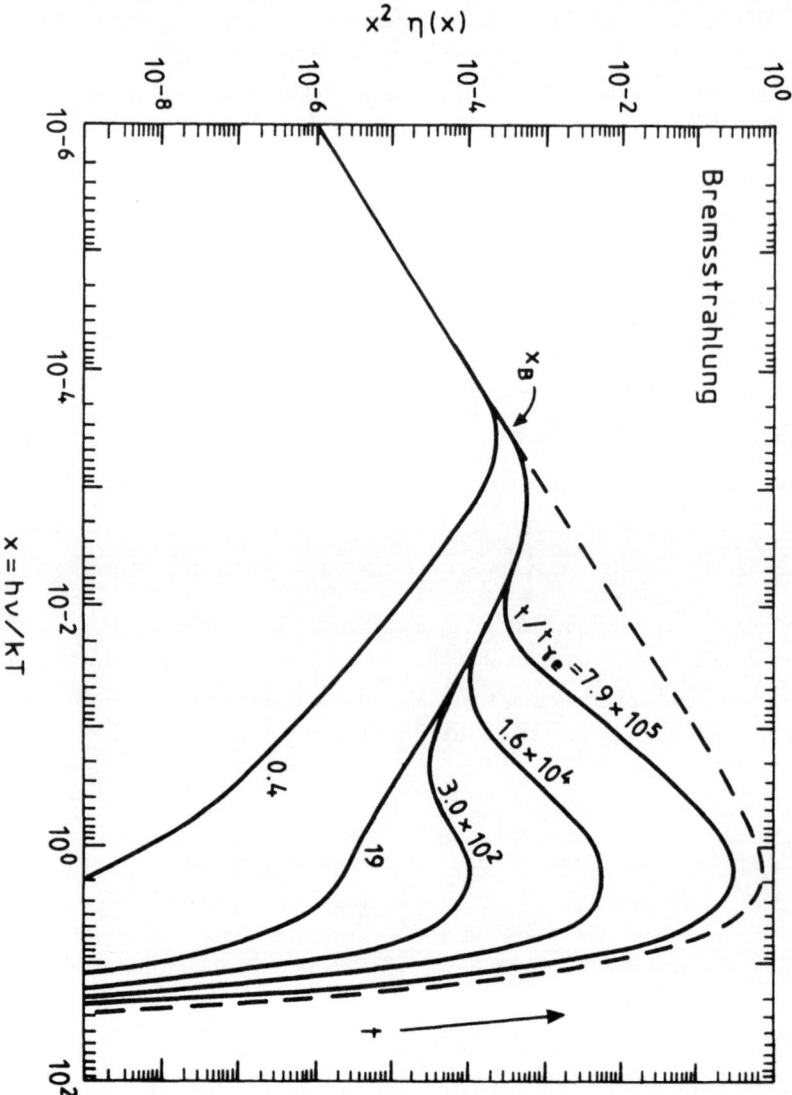

Figure 2

Time evolution of the radiation spectrum in a homogeneous plasma under the processes of bremsstrahlung emission and absorption and of electron scattering. No photons are present initially. Note that x_B is nearly constant. The photon distribution eventually achieves a Planck spectrum (dashed curve) at $t \sim 1.5 \times 10^6 \, t_{\gamma e}$.

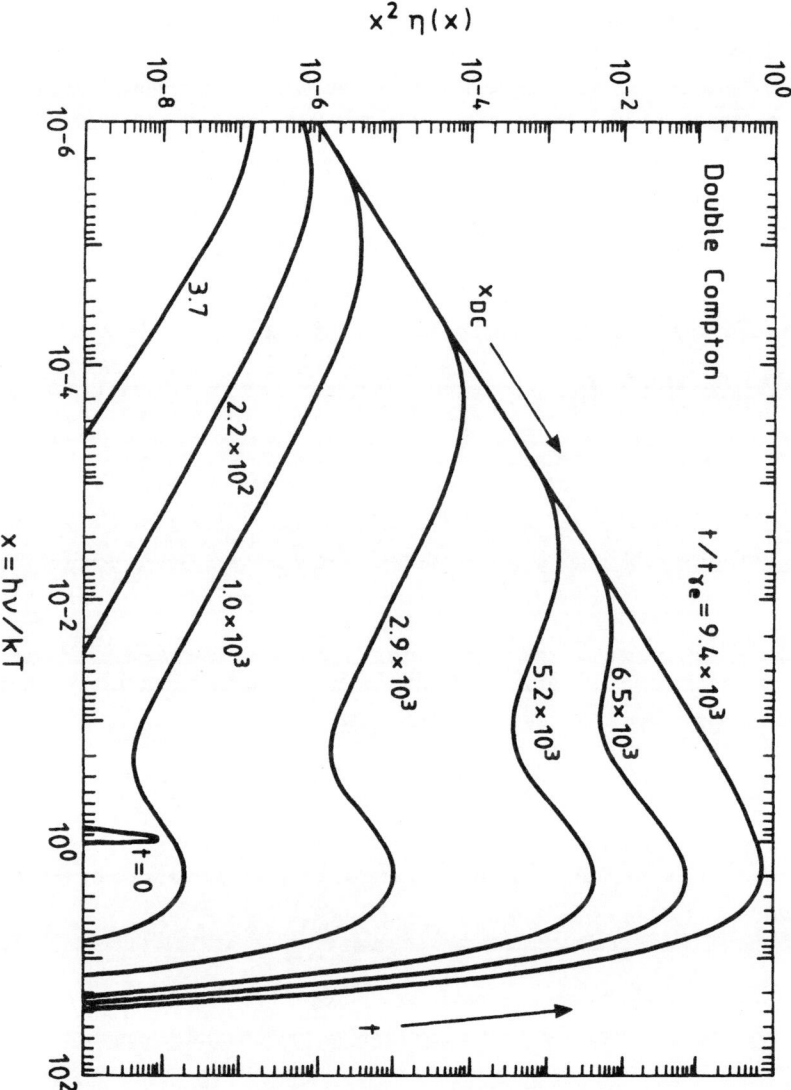

Figure 3

Time evolution of the radiation spectrum in a homogeneous plasma under the processes of double Compton emission and absorption and of electron scattering. A small number of seed photons is assumed initially at x = 1. Note the increase of x_{DC} and the self-similar evolution of the photon distribution which rapidly relaxes towards a Planck spectrum.

Under the action of double Compton processes, the radiation spectrum of figure 3 evolves self-similarly. The initial condition is a small number of seed photons at $x = 1$. Since field photons are scarce at early times, the evolution starts slowly. The larger $\mathcal{I}(t)$, the faster the relaxation of the photon distribution. The Planck spectrum obtains a hundred times faster than for pure bremsstrahlung emission. As may be guessed from relation (51), the critical value x_{DC} increases dramatically with $\mathcal{I}(t)$, in contrast with the evolution of figure 2 where x_B is constant.

3 - The cosmological distortions of the microwave background spectrum

3.1 - The various time scales

As discussed above, the typical collision time of photons with free electrons is $t_{\gamma e} = (n_e c \sigma_T)^{-1}$. When recombination proceeds, the electron density n_e drops down while $t_{\gamma e}$ increases.

The redistribution of energy among photons, described by the Kompaneets equation (31), is due to the inverse Compton scatterings of low frequency photons by energetic electrons, and is associated with the typical time scale

$$t_C \simeq \left(\frac{m_e c^2}{kT_e}\right) t_{\gamma e} \; . \tag{53}$$

The evolution of the electron temperature is due to the energy balance between matter and radiation

$$3n_e k \frac{dT_e}{dt} = \epsilon_{\gamma \to e} - \epsilon_{e \to \gamma} \; . \tag{54}$$

Note that protons contribute an additional factor $3kn_e/2$ to the heat capacity of the plasma. The electron temperature evolves according to the differential equation

$$\frac{dT_e}{dt} + \frac{4\sigma_T \rho_\gamma}{3m_e c}\left(T_e - T_e^{equi}\right) = 0 \; , \tag{55}$$

and relaxes towards its kinetic equilibrium value

$$T_e^{equi} = \int \frac{8\pi \nu^2}{c^3} \frac{(h\nu)^2}{4k\rho_\gamma} \eta(\nu)\left[1 + \eta(\nu)\right] d\nu \; , \tag{56}$$

with the typical time scale

$$t_{e\gamma} = \frac{3m_e c}{4\sigma_T \rho_\gamma} \; . \tag{57}$$

As shown by Lightman [5], the typical time scale on which bremsstrahlung may replenish a deficient photon population can be expressed as

$$t_B \simeq \frac{2\zeta(3)}{Q} g^{-2}(x_B) t_{\gamma e} , \tag{58}$$

while the restoration of a Planck distribution by double Compton emission occurs on the time scale

$$t_{DC} \simeq \frac{\pi}{8\alpha} \left(\frac{m_e c^2}{kT_e} \right)^2 t_{\gamma e} \tag{59}$$

In figure 4, those various time scales are displayed as a function of the red-shift $1 + z = T_\gamma(t)/2.375$ K, and are compared with the corresponding age $t_U(z)$ of the Universe [6]. The expansion equation has been integrated numerically with two light neutrino species, radiation and a baryon density of $n_B/n_\gamma = 3 \times 10^{-10}$, in agreement with the primordial nucleosynthesis estimates.

3.2 - The y and μ distortions

In figure 4, the solid line stands for the age t_U of the Universe. Note that $t_{e\gamma}$ (dotted curve) is always much smaller than the other relevant time scales, so that the electron temperature T_e relaxes rapidly towards its equilibrium value at any given time and for any given photon distribution.

Photons stop to collide with electrons after recombination. For a red-shift $z \geq 10^3$, $t_{\gamma e}$ (short dashed curve) is smaller than the typical expansion time t_U. However, thermalization of the photon bath is only effective above a red-shift $z_{BE} \sim 2 \times 10^5$ below which the inverse Comptonization time scale t_C (long dashed curve) exceeds the expansion time t_U. The red-shift z_{BE} corresponds to an age of $\sim 10^9$ sec.

For the low baryon abundance which we have adopted, double Compton emission is more efficient than bremsstrahlung. Indeed, the time scale t_B (dotted – short dashed curve) always exceeds t_U. On the other hand, double Compton processes restore the Planck form of the radiation spectrum above the critical red-shift $z_{Pl} \sim 1 - 2 \times 10^7$, at which t_{DC} (short dashed – long dashed curve) crosses t_U. The red-shift z_{Pl} is associated with an age of $\sim 10^5$ sec.

We may therefore sketch the scenario of the spectral distortions of the microwave background radiation. Four possibilities may occur, depending on when the energy is released into the primordial plasma.

158

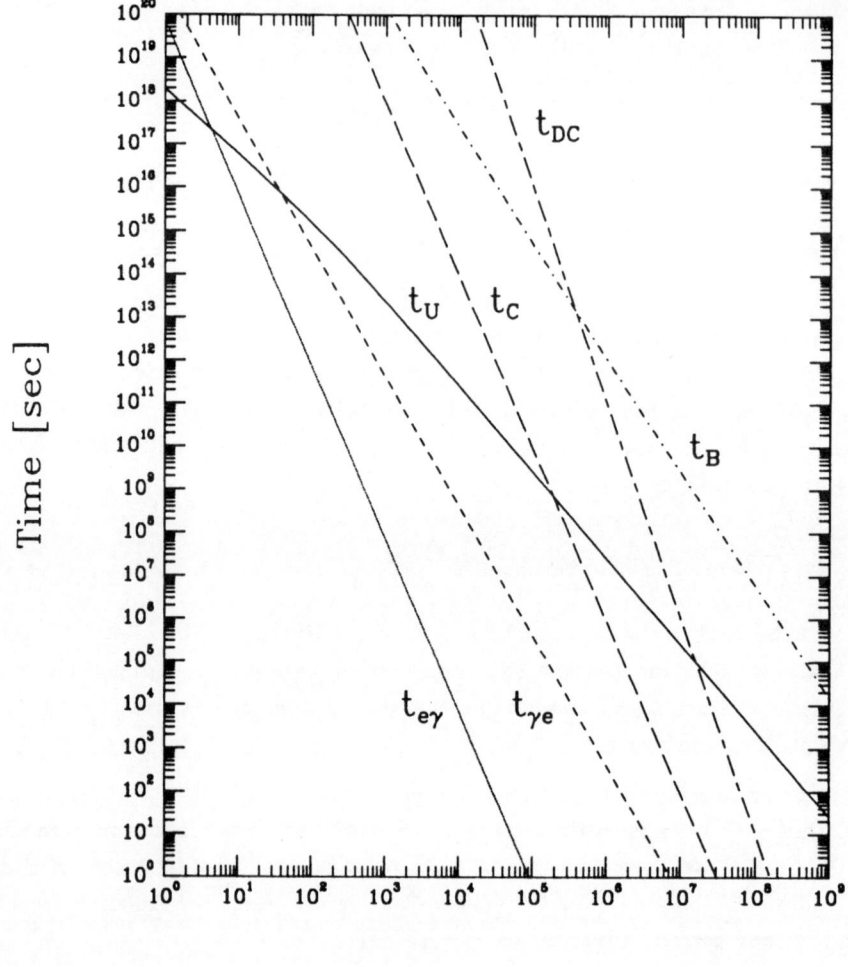

Redshift (1+z)

Figure 4

The age of the Universe t_U (solid) is presented as a function of the red-shift $(1 + z)$ and is compared to the relevant time scales $t_{e\gamma}$ (dotted), $t_{\gamma e}$ (short dashed), t_C (long dashed), t_{DC} (short dashed – long dashed) and finally t_B (dotted – short dashed).

1) $z \geq z_{Pl}$

Double Compton emission acts efficiently and any spectral distortion is erased during an expansion time. No matter how large the energy release is, the microwave background is not affected.

2) $z_{BE} \leq z \leq z_{Pl}$

Any heat release generates a μ-distortion. The radiation spectrum relaxes towards a Bose-Einstein distribution faster than the Universe expands ($t_C \leq t_U$). However, a pure black-body radiation cannot be achieved since double Compton emission is slower than expansion ($t_{DC} \geq t_U$). The photon bath is still in thermal equilibrium, but with an average photon energy larger than for a Planck spectrum, which translates into a non-zero chemical potential μ. In units where $\hbar = k = c = 1$, the energy ρ_{BE}, number n_{BE} and entropy σ_{BE} densities may be expressed as

$$\rho_{BE} = \frac{T_e^4}{\pi^2} I_3(\mu), \quad n_{BE} = \frac{T_e^3}{\pi^2} I_2(\mu) \quad \text{and} \quad \sigma_{BE} = \frac{T_e^3}{\pi^2} \left\{ \frac{4}{3} I_3(\mu) + \mu I_2(\mu) \right\}, \quad (60)$$

where the integral $I_n(\mu)$ is defined as

$$I_n(\mu) = \int_0^\infty x^n \left(e^{x+\mu} - 1 \right)^{-1} dx . \quad (61)$$

If $\Delta\rho_\gamma$ (or Δn_γ) denotes the increase of the energy (or number) density due to the heat release, the plasma temperature T_e and chemical potential μ evolve according to

$$\Delta n_\gamma = \left(\frac{\partial n_{BE}}{\partial T_e} \right) dT_e + \left(\frac{\partial n_{BE}}{\partial \mu} \right) d\mu$$

$$\text{while} \quad\quad\quad\quad\quad\quad\quad\quad\quad\quad\quad\quad\quad (62)$$

$$\frac{\Delta\rho_\gamma}{T_e} = \left(\frac{\partial \sigma_{BE}}{\partial T_e} \right) dT_e + \left(\frac{\partial \sigma_{BE}}{\partial \mu} \right) d\mu .$$

Note that

$$\frac{dI_n}{d\mu} = -n\, I_{n-1}(\mu) . \quad (63)$$

Since in the practical cases of interest μ is small – COBE [3] sets the stringent constraint $\mu \leq 0.009$ – the previous integrals may be evaluated at $\mu = 0$ and simplify into $I_1(0) = \pi^2/6$, $I_2(0) = 2\zeta(3)$ and $I_3(0) = \pi^4/15$.

The electron temperature T_e relaxes towards the radiation temperature T_γ much faster than expansion. In general, high energy photons are injected into the plasma so that $\Delta n_\gamma << n_\gamma$. This leads to a unique relation between T_e and μ, and we may therefore express the variation of the chemical potential parameter μ as a function of the relative increase of the photon energy

$$\frac{\Delta\rho_\gamma}{\rho_\gamma} = \left\{ \left(\frac{2\pi^2}{9\zeta(3)} \right) - \left(\frac{90\zeta(3)}{\pi^4} \right) \right\} d\mu \simeq 0.714\, d\mu . \quad (64)$$

3) $z_{rec} \leq z \leq z_{BE}$

Both t_{DC} and t_C exceed the expansion time t_U. Photons may still collide with electrons so that a y-distortion ensues, which corresponds to the small value $y \sim t_U/t_C$. If energy is released in the primordial plasma, electrons and protons are heated up to the temperature $T_e \gg T_\gamma$. As a result of rare inverse Compton reactions, the Rayleigh-Jeans region is slightly depleted by a factor e^{-2y} with respect to the rest of the spectrum. Remember that COBE [3] sets the bound $y \leq 10^{-3}$.

4) $z \leq z_{rec}$

Below $z_{rec} \sim 10^3$, free electrons combine with protons. Figure 4 has assumed a constant ionization fraction $x_e = 1$. In fact, as soon as recombination drives the population of free electrons to extinction, $t_{\gamma e}$ exceeds t_U and the primordial radiation decouples from the neutralized matter. The microwave spectrum gets quenched and its features are fossilized until now. The cosmic background radiation is therefore of crucial importance as regards the thermal history of the Universe. It provides us with a unique probe of its recent evolution.

4 - The primordial behaviour of particles

As observed by the COBE satellite, the CBR is very close to a pure black body radiation. We would like to investigate now the constraints set by the CBR observations upon the properties of particles such as neutrinos. What we need to understand is the generic behaviour of particles in the early Universe. As we shall see, species decouple from equilibrium and get quenched. The possibility of predicting the present abundance of their remnants is quite exciting. As an example, the CBR is the fossil radiation of the primordial photon gas which decoupled from baryons at recombination.

4.1 - The equilibrium density

A denotes a generic particle with g_A helicity states. Collisions of A particles with other species imply their thermalization at temperature T. Moreover, A may annihilate with antipartners $\overline{A}$ in, say, electron-positron pairs. If the following set of reactions

$$2\gamma \rightleftharpoons e^+e^- \rightleftharpoons A + \overline{A} \rightleftharpoons e^+e^- \rightleftharpoons 3\gamma \tag{65}$$

is in chemical equilibrium, the various chemical potentials are given by

$$\mu_A + \mu_{\overline{A}} = \mu_{e^+} + \mu_{e^-} = 2\mu_\gamma = 3\mu_\gamma . \tag{66}$$

We readily conclude that the equilibrium of reactions (65) implies a vanishing μ_γ and, therefore, a Planck spectrum for the primordial fire-ball. Moreover, the chemical potentials of matter and antimatter are related to each other by

$$\mu_A = -\mu_{\overline{A}} . \tag{67}$$

If we assume furthermore that there is no asymmetry between the species A and $\overline{A}$, which is indeed the case for baryons whose density is small, $n_B/n_\gamma = 3 \times 10^{-10}$, we infer $\mu_A = \mu_{\overline{A}} = 0$. The number density of A's at equilibrium may therefore be expressed as

$$n_A^e = \int \frac{d^3\vec{p}}{8\pi^3} g_A \frac{1}{e^{E/T} + \epsilon} , \tag{68}$$

where $\epsilon = 1$ (fermions) or -1 (bosons). The energy E is related to the momentum p since $E = \sqrt{M^2 + p^2}$, where M denotes the mass of the particle. Depending on the ratio M/T, two regimes may be distinguished.

• The ultra-relativistic limit

When $M/T \to 0$, i.e., for high temperatures and momenta, the A population is ultra-relativistic. Since $E \simeq p$, the previous expression simplifies into

$$n_A^e = \frac{g_A}{\pi^2} \zeta(3) T^3 \left\{ 1 \text{ (B) or } \frac{3}{4} \text{ (F)} \right\} . \tag{69}$$

Since at early times $T \propto R^{-1}$, the codensity of an ultra-relativistic population, i.e., the density of particles per volume which expands with the expanding Universe, is constant

$$n_A^e R^3 \propto \frac{n_A^e}{T^3} = f_A^e = \frac{g_A}{\pi^2} \zeta(3) \left\{ 1 \text{ (B) or } \frac{3}{4} \text{ (F)} \right\} . \tag{70}$$

Since $(k/\hbar c)^3 = 83.22 \text{ cm}^{-3} \text{K}^{-3}$, the codensity is $\simeq 20 \text{ cm}^{-3} \text{K}^{-3}$ for bosons with two helicity states – as photons – and $\simeq 15 \text{ cm}^{-3} \text{K}^{-3}$ for fermions. We readily infer from the value of the CBR temperature $T_\gamma^0 = 2.735$ K measured by COBE [3] that the photon density, today, is $\sim 400 \text{ cm}^{-3}$.

• The non-relativistic regime

For large values of M/T, the codensity is exponentially suppressed by a factor e^{-a}.

$$f_A^e = \frac{g_A}{(2\pi)^{3/2}} a^{3/2} e^{-a} , \tag{71}$$

where $a = M/T$. In the non-relativistic regime, bosons and fermions have the same behaviour.

4.2 - The thermal decoupling

Thermalization is achieved by the numerous collisions of A's with the other species. By numerous, we mean that the collision rate exceeds the expansion rate so that the temperature T_A has plenty of time to relax towards the temperature of the other populations. In illustration of this idea, we present the thermal behaviour of neutrinos. The rate of their collisions with electrons is given by

$$\Gamma_C = c\sigma_C n_e \simeq \left(G_F^2 T^2\right) \times \left(\frac{3}{\pi^2} \zeta(3) T^3\right) \simeq 4.9 \times 10^{-23} \text{ MeV} \left(\frac{T}{1 \text{ MeV}}\right)^5, \quad (72)$$

where the Fermi constant is $G_F = 1.16 \times 10^{-11} \text{ MeV}^{-2}$. At early times, the curvature of space may be neglected so that the Hubble parameter may be expressed as

$$H = \left\{\frac{8\pi}{3} \frac{\rho}{M_{Pl}^2}\right\}^{1/2} = \left\{\frac{8\pi^3}{45} g_{eff}\right\}^{1/2} \frac{T^2}{M_{Pl}} \simeq 4.5 \times 10^{-22} \text{ MeV} \left(\frac{T}{1 \text{ MeV}}\right)^2. \quad (73)$$

The Planck mass is $M_{Pl} = 1.22 \times 10^{22} \text{ MeV}$ while the energy density ρ has been expressed in units of the photon energy density $\rho_\gamma = \pi^2 T^4/15$, hence an effective number of degrees of freedom $g_{eff} = 43/8$. The collision rate can be compared now with the expansion rate

$$\frac{\Gamma_C}{H} \simeq \left(\frac{T}{2 \text{ MeV}}\right)^3, \quad (74)$$

and we readily infer a decoupling temperature of ~ 2 MeV below which collisions are so rare that neutrinos no longer see their surroundings and become a fossil radiation.

4.3 - Annihilation and the chemical freeze-out

If neutrinos have a mass M larger than the decoupling temperature of ~ 2 MeV, they may substantially annihilate before they freeze out. As an illustration, we focus on heavy neutrinos with a mass $M = 2$ GeV.

At high temperature, for $T > 2$ GeV, heavy neutrinos are in chemical equilibrium. They steadily annihilate into $f\bar{f}$ pairs while the reverse process is also very active. The annihilation – production reaction

$$A + \overline{A} \rightleftharpoons f + \overline{f} \quad (75)$$

is in equilibrium and the A's density relaxes towards its equilibrium value (68).

Below 2 GeV, A and $\overline{A}$ annihilate. As long as the chemical reaction (75) is in equilibrium, the A's density is given by relation (68). As the temperature decreases, the A's are significantly depleted by annihilation and their density drops down. The antiparticles $\overline{A}$ with which A's annihilate, become rare also.

At $T \simeq 100$ MeV, the density n_A is so low, particles and antiparticles are so much depleted that reaction (75) ceases to be in equilibrium. The probability for an A to encounter an antipartner becomes less than 1 per typical expansion time. Annihilations stop under the combined action of the dilution due to the expansion of space, and of the severe depletion of the A species which has occurred between $T = 2$ GeV and $T = 100$ MeV.

Below 100 MeV, annihilations are inhibited and the codensity f_A remains constant. Around $T \simeq 2$ MeV, heavy neutrinos stop colliding with the other species and become mere fossils of the early stages of the Universe. If they are stable, they pervade the intergalactic medium until the present epoch, and may even contribute a significant fraction to the closure mass.

The density n_A evolves according to the differential equation

$$\frac{dn_A}{dt} = -3Hn_A - <\sigma_a v> n_A{}^2 + <\sigma_a v> n_A^{e\,2} , \qquad (76)$$

where the first expression on the right-hand side refers to the dilution resulting from expansion. The second term accounts for the A's annihilations while the last expression describes the retro-creation of $A\overline{A}$ pairs from light fermions and assumes detailed balance. In terms of the codensity, the evolution equation simplifies into

$$\frac{df_A}{dt} + (<\sigma_a v> n_A)\, f_A = <\sigma_a v> T^3 f_A^{e\,2} . \qquad (77)$$

In order to solve the differential equation (77), two typical time scales may be defined.

1) When equilibrium is reached, i.e., when $f_A = f_A^e$, the time derivative $df_A/dt \simeq 0$. The characteristic time scale of the relaxation of f_A towards its kinetic equilibrium value f_A^e is merely related to the annihilation rate

$$\tau_{rel}^{-1} = <\sigma_a v> n_A . \qquad (78)$$

2) The time scale of the variations of the equilibrium f_A^e itself may be expressed as

$$\tau_{eq}^{-1} = -\frac{d}{dt} \text{Log} \left\{ f_A^{e\,2} T^3 \right\} \simeq 2aH , \qquad (79)$$

in the non-relativistic regime. As is clear in figure 5, two phases may be distinguished :

• At high temperature, as long as $\tau_{rel} < \tau_{eq}$, f_A has plenty of time to relax towards the equilibrium f_A^e which evolves at a much slower pace. As a consequence, the annihilation reaction (75) is in chemical equilibrium so that $f_A = f_A^e$. On the upper panel of figure 5, the solid line f_A cannot be distinguished from the dashed curve f_A^e. As is clear from the lower panel, as T decreases and $a = M/T$ increases, relaxation becomes progressively less efficient until the ratio τ_{rel}/τ_{eq} eventually reaches 1 where the decoupling from equilibrium occurs.

164

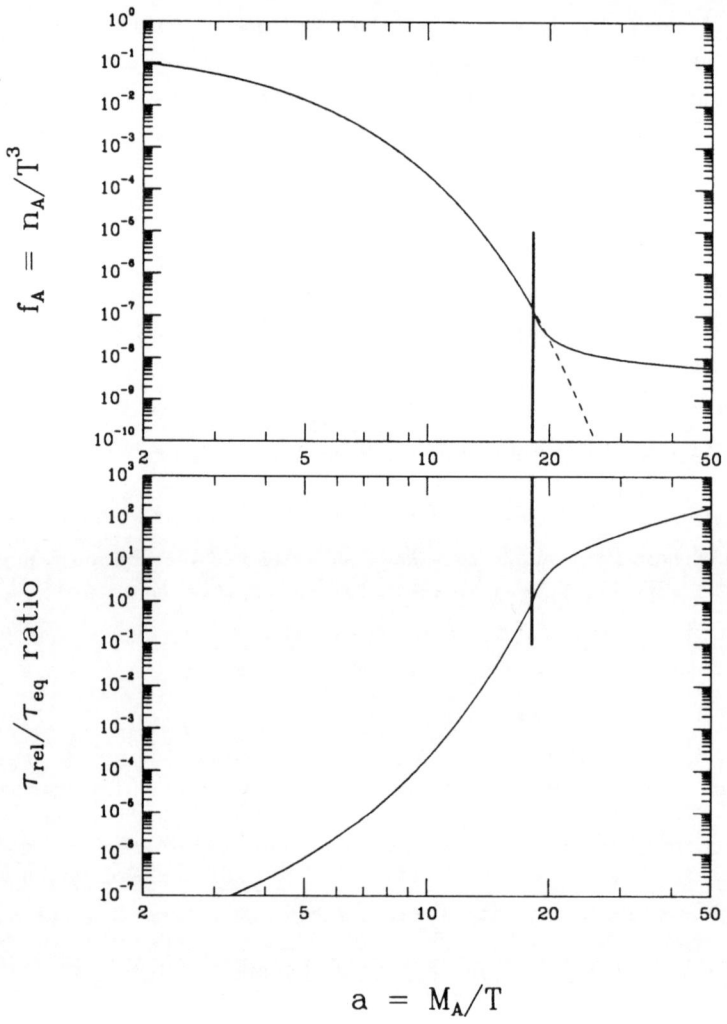

Figure 5

Upper panel : The codensity f_A (solid curve) follows the equilibrium value f_A^e (dashed curve) until the mass-to-temperature ratio a reaches the freeze-out point at $a \simeq 18$. For $a > 18$, f_A decouples from equilibrium and decreases slowly towards its present value.

Lower panel : The ratio of the relaxation time τ_{rel} to the equilibrium time τ_{eq} is plotted as a function of a. Decoupling occurs as soon as τ_{rel} exceeds τ_{eq}.

• Below the freeze-out temperature T_F, the relaxation time τ_{rel} exceeds τ_{eq}. Whilst f_A^e drops down and vanishes, f_A still decreases a little bit. The freeze-out point may be derived from the equality $\tau_{rel} = \tau_{eq}$ which, for heavy neutrinos whose annihilation cross-section is

$$\sigma_a v \left(A\overline{A} \to f\overline{f} \right) = \frac{G_F^2}{2\pi} M^2 N_A \ , \tag{80}$$

translates into

$$\sqrt{a_F}\, e^{a_F} = \frac{3\sqrt{5}}{(2\pi)^4}\, G_F^2\, M^3\, M_{Pl}\, N_A\, g_{eff}^{-1/2} \simeq 0.7 \times 10^7 \left(\frac{M}{1\,\text{GeV}} \right)^3 \frac{N_A}{\sqrt{g_{eff}}} \ . \tag{81}$$

For a 2 GeV neutrino, the effective number of annihilation channels is $N_A \sim 14$, whilst the effective number of degrees of freedom is $g_{eff} = 57/8$ at a temperature $T_F \sim 100$ MeV. We infer from the previous equation a freeze-out point at $a_F \simeq 18.06$. In figure 5, the freeze-out is indicated by a heavy vertical line at $a = a_F$. Note that as soon as the ratio τ_{rel}/τ_{eq} exceeds unity, the codensity f_A (solid line of the upper panel) decouples from the equilibrium value f_A^e (dashed curve). From decoupling until now, f_A has been slightly decreasing according to the relation

$$\frac{df_A}{dt} = - <\sigma_a v> n_A f_A \tag{82}$$

which, in terms of the parameter $x = 1/a$, simplifies into

$$\frac{df_A}{f_A^{\ 2}} = \sqrt{\frac{45}{8\pi^3}}\, g_{eff}^{-1/2} <\sigma_a v> M\, M_{Pl}\, dx \ . \tag{83}$$

Integration of the previous differential equation from $x_F = 1/a_F$ where the codensity decouples from equilibrium, i.e., $f_A(a_F) = f_A^e(a_F) = f_F$, until the present epoch $x = 0$, where f_A reaches its asymptotic value f_A^{asy}, yields

$$f_A^{asy} = \frac{f_F}{1 + 2a_F} \simeq \sqrt{\frac{8\pi^3}{45}}\, g_{eff}^{1/2} (<\sigma_a v> M\, M_{Pl}\, x_F)^{-1} \ . \tag{84}$$

At present, the codensity of a *stable* 2 GeV neutrino is $f_A^{asy} \sim 4 \times 10^{-9}$ which translates into $\sim 3.3 \times 10^{-7}$ cm^{-3} K^{-3}. Note finally the dependence of f_A^{asy} on the annihilation cross section : the larger $<\sigma_a v>$, the lower the relic abundance f_A^{asy}.

5 - Heavy neutrinos confront LEP and COBE

The recent results [7] of the Large Electron Positron collider (LEP) at CERN, Geneva, as well as the COBE measurements [3] of the CBR spectrum have given heavy neutrinos quite a turn as regards their cosmological rôle.

5.1 - The relic density of heavy neutrinos

Slightly after the thermal freeze-out of neutrinos at a temperature ~ 2 MeV, electrons and positrons annihilate. Radiation is reheated and, as a consequence, the photon to neutrino temperature ratio T_γ/T_ν increases by a factor $(11/4)^{1/3} \simeq 1.4$. We infer a neutrino temperature $T_\nu^0 \simeq 2$ K at present. The careful reader may feel uncomfortable with a neutrino temperature which does not make any *thermodynamical* sense below 2 MeV where, indeed, T_ν is more a typical scale factor than a temperature per se, and varies as R^{-1}. Depending on the mass of the neutrino, two regimes may be discussed in order to derive the relic abundance ρ_ν.

• For $M < 2$ MeV, neutrinos decouple before they may annihilate. The resulting fossil population is ultra-relativistic at decoupling so that the present neutrino codensity is merely given by expression (70). If neutrinos are stable, we infer a relic abundance

$$\rho_\nu = \frac{3\zeta(3)}{2\pi^2} T_\nu^0{}^3 M \simeq 110 \,\text{keV}\,\text{cm}^{-3} \left(\frac{M}{1\,\text{keV}}\right) , \tag{85}$$

to be contrasted with the closure density

$$\rho_C \simeq 2 \times 10^{-29} \, h^2 \,\text{g}\,\text{cm}^{-3} \simeq 11.25 \, h^2 \,\text{keV}\,\text{cm}^{-3} , \tag{86}$$

which corresponds to a *flat* Friedmann-Lemaître cosmology. If the mass density exceeds ρ_C, the Universe is spherical and will recontract in the future while, in the opposite case, its geometry is hyperbolic and the expansion will last for ever. The Hubble constant h is expressed in units of 100 km/s/Mpc. The relic density ρ_ν may be conveniently expressed in units of the closure density

$$\Omega_\nu h^2 \simeq \frac{M}{100\,\text{eV}} . \tag{87}$$

The larger M, the larger the relic abundance $\Omega_\nu h^2$ which, as a matter of fact, cannot reasonably exceed ~ 1 under the penalty of overclosing the Universe, whose age should be larger, at least, than the age of our galactic disk, *i.e.*, ~ 9 billion years. We readily infer a cosmological upper bound on the mass of light stable neutrinos of ~ 100 eV.

• If $M > 2$ MeV, neutrinos may significantly annihilate before reaction (75) gets quenched. The freeze-out temperature below which annihilations stop as a result of the severe depletion of the neutrino density may be derived from equation (81) with appropriate values for N_A and g_{eff}.

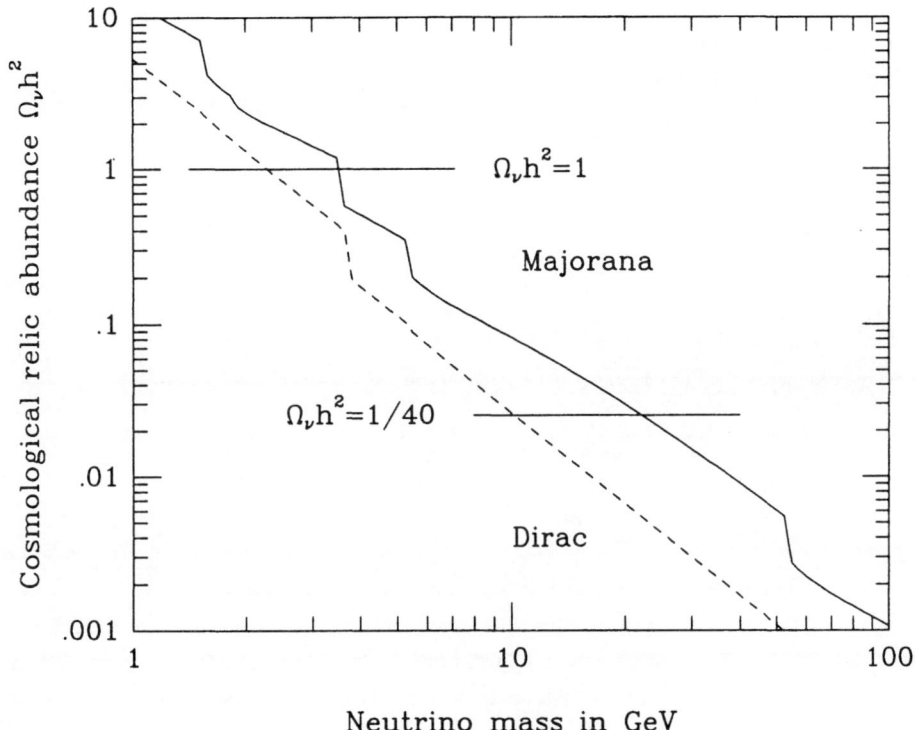

Figure 6

The cosmological relic abundance $\Omega_\nu h^2$ of stable heavy neutrinos is presented as a function of their mass, for Majorana (solid) and Dirac (dashed) species. The heavier the neutrino, the lower its contribution to the present mass density of the Universe.

As may be guessed from (81), the larger M, the larger $a_F = M/T_F$. The relic density of heavy neutrinos may be expressed as

$$f_A^{asy} \simeq 9 \times 10^{-9} \frac{g_{eff}^{1/2}}{N_A} \frac{a_F}{M^3} , \qquad (88)$$

which translates into the ratio

$$\Omega_\nu h^2 \simeq 0.98 \left(\frac{g_{eff}^{1/2} a_F}{N_A} \right) \left(\frac{1\,\mathrm{GeV}}{M} \right)^2 , \qquad (89)$$

to which both neutrinos and antineutrinos contribute. For a neutrino mass $M \sim 2$ GeV, we may take $g_{eff} = 57/8$ and $N_A = 14$ so that we infer a fossil abundance

$$\Omega_\nu h^2 \sim \left(\frac{1.8\,\mathrm{GeV}}{M} \right)^2 , \qquad (90)$$

in fairly good agreement with figure 6 where the annihilation cross sections of heavy *Majorana* (solid curve) and *Dirac* (dashed line) neutrinos have been carefully computed. As is obvious in relation (84), the relic abundance $\Omega_\chi h^2$ of any stable species χ is inversely proportional to the associated annihilation cross section. As a consequence of expression (80), the fossil density $\Omega_\nu h^2$ decreases approximately as M^{-2} as soon as the neutrino mass exceeds 2 MeV (see both curves of figure 6). The neutrino relic abundance $\Omega_\nu h^2$ is peaked around $M \sim 2$ MeV and is fairly well described by expression (87) for $M < 2$ MeV whilst, above 2 MeV, relation (90) is valid.

Since Majorana neutrinos are their own antipartners, their annihilation is suppressed with respect to the Dirac case. As a result, the relic abundance $\Omega_\nu h^2$ of Majorana species is slightly larger than for Dirac particles. Provided a *stable* heavy Dirac neutrino has a mass in the range $\sim 2 - 10$ GeV, it may contribute a significant fraction to the closure mass and may be a plausible candidate to the cosmological dark matter. For a Majorana neutrino, the relevant mass range is between 3 and 20 GeV.

5.2 - LEP, heavy neutrinos and dark matter

Recent experiments at LEP [7] have observed the reaction

$$e^+ e^- \to Z^0 \to f\overline{f} , \qquad (91)$$

where an electron and a positron annihilate into a pair of light fermions. As the center of mass energy $\sqrt{s}$ of the electron-positron pair varies, the cross section of the previous process exhibits a characteristic Breit-Wigner resonance, with a peak at $\sqrt{s} = M_{Z^0}$ which corresponds to the production of the Z^0 boson. The width of this lineshape is merely the total decay rate of the Z^0 boson. From the direct observation of the Z^0 decays into quarks (the constituents of nucleons) and charged leptons (there are three charged leptons : the electron, the muon and the tau), and from the determination of the total decay rate, the invisible width, *i.e.*, the decay rate into invisible neutrinos, has been for the first time measured to be ~ 0.49 GeV. Note that the Z^0 decay rate into a neutrino-antineutrino pair is given by

$$\Gamma_{Z^0} = \frac{G_F}{12\sqrt{2}\pi} M_{Z^0}^3 \, \kappa(\beta) \simeq 165\,\text{MeV}\ \kappa(\beta) \ , \tag{92}$$

where β denotes the velocity of the final neutrinos in the rest frame of the decaying Z^0 boson. The kinematic function $\kappa(\beta) = \beta^3$ for Majorana neutrinos whilst, for Dirac species, $\kappa(\beta) = \beta(\beta^2 + 3)/4$. For massless particles, $\beta = 1$ and $\kappa = 1$, so that the decay width is a mere 165 MeV. The invisible width corresponds to three light neutrino species which were already known to exist in association with the above mentioned charged leptons. A fourth family heavy neutrino may exist but should not contribute to the Z^0 decay width, *i.e.*, it should not be produced in Z^0 decays. We readily infer the lower bound $M > M_{Z^0}/2 \sim 45$ GeV, so that should such a species exist and be stable, its remnants would not contribute significantly to the closure mass density of the Universe as may be readily inferred from figure 6. A heavy neutrino cannot be a good dark matter candidate as was incidentally advocated before the LEP results. LEP has shown that the conventionnal model of the electro-weak interactions is distressingly valid.

5.3 - The 17 keV neutrino

There is new evidence, in β-decay experiments [8], that the electron neutrino oscillates in one percent of the cases into another neutrino of mass 17 keV. According to our present knowledge, the mass of the muon neutrino cannot exceed 10 keV whilst the number of light neutrino generations, N_ν, is three as measured by the LEP experiments [7]. Therefore, the 17 keV neutrino may be either the tau neutrino, or an additional species with vanishing hypercharge. The properties of this neutrino are severely constrained by astrophysics and cosmology.

1) If the 17 keV neutrino is stable, its mass may overclose the Universe by a factor $\Omega_\nu h^2 \sim 170$ as derived from expression (87). A value of the Hubble constant H^0 as low as 50 km/s/Mpc associated with a large closure density $\Omega_\nu \sim 700$ implies a Universe not older than ~ 1 billion years. Such a low value is in conflict with current cosmological observations. In particular, the age of the Universe is known to be at least larger than the age of the earth, *i.e.*, 5 billion years. From the luminosity function of white dwarves and its peculiar cut-off at low luminosities, the age of the galactic disk has been inferred [9] to be ~ 9 billion years.

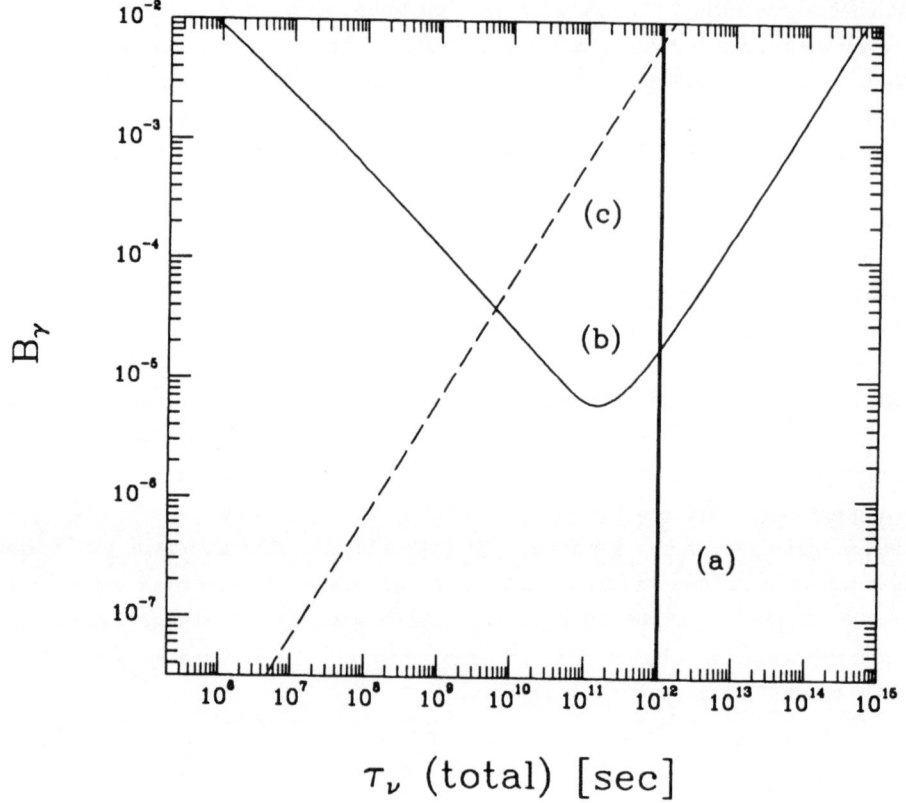

Figure 7

Astrophysical and cosmological arguments constrain the 17 keV neutrino to lie in specific domains of this (τ_ν, B_γ) diagram. The region on the right-hand side of the closure bound (a) (heavy line) is excluded. The domain above curves (b) and (c) is also forbidden. The quasi-Planckian spectrum of the CBR implies the constraint (b). Finally, the dashed curve (c) is derived from the solar maximum mission data.

Finally, the magnitude shift between the main-sequence turn-off and the horizontal branch of the Hertzsprung-Russell diagram of globular clusters is compatible [10] with an age of population II stars greater than 15 billion years. We readily conclude that the 17 keV neutrino must be unstable.

Should the decay products be non-relativistic, they would merely dominate the expansion of the Universe just as if the neutrino were stable. That species must therefore decay into relativistic products whose energy density may be red-shifted. If decays occur at temperature T_γ^D, the decay products contribute today a factor

$$\Omega_{dp} = 170\,h^{-2}\left(\frac{2.735\,\text{K}}{T_\gamma^D}\right) , \qquad (93)$$

to the closure density, so that the earlier the decay, the larger T_γ^D and the lower Ω_{dp}. Provided that the 17 keV neutrino decays before the temperature reaches down ~ 500 to 2000 K, its decay products do not overclose the Universe. We infer an upper bound on the neutrino lifetime τ_ν of $\simeq 10^{12}$ seconds, which corresponds to the heavy vertical line (a) of figure 7.

2) Spectral distortions of the CBR may be generated by the radiative decays of the 17 keV neutrino ν_H. As a result of the process $\nu_H \to \gamma\nu_e$, whose branching ratio is B_γ, the electromagnetic energy density increases with time

$$d\rho_\gamma = n_\nu(t)\frac{m_\nu}{2}\,B_\gamma\,\frac{dt}{\tau_\nu} . \qquad (94)$$

In what follows, we focus only on massless final states. Since the temperature of interest is much less than 17 keV, massive neutrinos may be considered at rest so that the photon energy is $m_\nu/2$ on average. The density of the massive species ν_H exponentially decays in time while the resulting μ-distortion evolves as

$$d\mu = 7.07 \times 10^{-2}\left(\frac{m_\nu}{T_\gamma}\right) B_\gamma\, e^{-t/\tau_\nu}\frac{dt}{\tau_\nu} . \qquad (95)$$

We have grossly assumed that a μ-distortion was the only possible spectral alteration of the CBR. As mentioned in section 3.2, a μ-distortion is only possible between 10^5 and 10^9 seconds. After 10^9 seconds and until recombination, which occurs around 10^{11} seconds, a y-distortion ensues. In order to estimate the implications of the 17 keV neutrino on the CBR spectrum, we have numerically integrated equation (95) and set the constraint [3] $\mu \leq 9 \times 10^{-3}$. The resulting upper bound on B_γ is presented in figure 7 as a function of the neutrino lifetime τ_ν and results into the solid line (b).

3) Finally, data from the Gamma-Ray Spectrometer on the Solar Maximum Mission (SMM) satellite [11] show no excess γ-ray emission coincident with the ~ 10 second neutrino burst from supernova 1987A. From the absence of a detectable signal, an upper

bound on the flux of MeV photons from the supernova has been found to be ~ 1 photon cm^{-2} during a period of observation ~ 10 seconds encompassing the neutrino burst. If neutrinos decay radiatively in flight between the supernova and the solar system, a lower bound on their radiative lifetime may be placed [12] from the SMM observations, which translates into the limit

$$\tau_\nu \geq (1.49 \times 10^{14}\,\text{sec})\,B_\gamma\ , \tag{96}$$

and the dashed curve (c) of figure 7 above which the excluded region lies.

The constraints of figure 7 may be summarized [13] by

$$B_\gamma \leq 4 \times 10^{-5} \quad \text{and} \quad \tau_\nu \leq 10^{12}\,\text{sec}\ . \tag{97}$$

Should the 17 keV neutrino decay into three light neutrinos, the corresponding lifetime τ_ν would exceed $\sim 10^{16-17}$ seconds. Moreover, the radiative branching ratio in such a model would merely be $B_\gamma \sim \alpha \sim 10^{-2}$, well above the constraint (97). The 17 keV neutrino should therefore decay into scalar species such as the axion or the majoron. We hope to have convinced the reader of the importance of COBE as regards the properties of particles. The discovery of a 17 keV neutrino, if confirmed, would be an exciting clue for the existence of a new scalar species, and would revive the interest in axion, majoron or familon theories. Thanks to the COBE results, such a discovery would definitely point towards unconventional physics.

References

[1] Kompaneets, A. S. (1957) *Soviet Phys. – JETP* **4**, 730.

[2] Sunyaev, R. A., Zel'dovich, Ya. B. (1970) *Astrophys. Space Sci.* **7**, 3;
Sunyaev, R. A., Zel'dovich, Ya. B. (1970) *Astrophys. Space Sci.* **9**, 368;
Sunyaev, R. A., Zel'dovich, Ya. B. (1972) *Astron. Astrophys.* **20**, 189;
Sunyaev, R. A., Zel'dovich, Ya. B. (1980) 'Microwave background radiation as a probe of the contemporary structure and history of the universe', *Ann. Rev. Astron. Astrophys.* **18**, 537-560 and references therein.

[3] Mather, J. C. *et al.* (1990) *Astrophys. J. Lett.* **354**, L37.

[4] Peebles, P. J. E. (1971) 'Physical Cosmology'.

[5] Lightman, A. P. (1981) 'Double Compton emission in radiation dominated thermal plasmas', *Astrophys. J.* **244**, 392-405.

[6] Sarkar, S. (1990) private communication.

[7] L3 Collaboration, Adeva, B. *et al.* (1989) *Phys. Lett.* **B231**, 509;
ALEPH Collaboration, Decamp, D. *et al.* (1989) *Phys. Lett.* **B231**, 519;
OPAL Collaboration, Akrawy, M. Z. *et al.* (1989) *Phys. Lett.* **B231**, 530;
DELPHI Collaboration, Aarnio, P. *et al.* (1989) *Phys. Lett.* **B231**, 539;
see also Fernandez, E. (1990) 'Proceedings of the Neutrino 90 Conference', 11-15 June 1990, CERN, Geneva, Switzerland and references therein.

[8] Simpson, J. J. (1985) *Phys. Rev. Lett.* **54**, 1891;
Simpson, J. J. and Hime, A. (1989) *Phys. Rev.* **D39**, 1825;
Hime, A. and Simpson, J. J. (1989) *Phys. Rev.* **D39**, 1837;
Sur, B. *et al.* (1990) LBL preprint;
Norman, E. B. (1990) proceedings of the 14th 'Europhysics Conference on Nuclear Physics', Bratislava, Czecho-Slovakia;
Zliman, I. *et al.* (1990) proceedings of the 14th 'Europhysics Conference on Nuclear Physics', Bratislava, Czecho-Slovakia;
Hime, A. and Jelley, N. A. (1991) Oxford University preprint OUNP-91-01.

[9] Winget, D. E., Hansen, C. J., Liebert, J., Van Horn, H. M., Fontaine, G., Nather, R. E., Kepler S. O. and Lamb, D. Q. (1987) *Astrophys. J. Lett.* **315**, L77.

[10] Iben Jr., I. and Renzini, A. (1984) *Phys. Rep.* **105**, 329.

[11] Chupp, E. L., Vestrand, W. T. and Reppin, C. (1989) *Phys. Rev. Lett.* **62**, 505.

[12] von Feilitzsch, F. and Oberauer, L. (1988) *Phys. Lett.* **B200**, 580;
Kolb, E. W. and Turner, M. S. (1989) *Phys. Rev. Lett.* **62**, 509.

[13] Altherr, T., Chardonnet, P. and Salati, P. (1991) 'The 17 keV neutrino in the light of astrophysics and cosmology', preprint CERN-TH-6110/91.

DARK MATTER CANDIDATES AND METHODS
FOR DETECTING THEM

G. G. RAFFELT
Max-Planck-Institut für Physik
Postfach 401212
W-8000 München 40
Germany

ABSTRACT. We review possible solutions of the cosmological dark matter problem, emphasizing those baryonic or non-baryonic candidates for which realistic detection schemes have been devised.

1. Introduction

The problem of cosmological dark matter was first discussed in Zwicky's famous paper of 1933 on the redshifts of extragalactic nebulae [1]. From his redshift measurements of galaxies in the Coma cluster he derived velocity dispersions of 1500 to 2000 km/s, much too large if this cluster was assumed to be a gravitationally bound system in virial equilibrium so that the average kinetic energy of the galaxies would equal $-1/2$ of the average gravitational potential energy. Zwicky concluded that "the average density in the Coma system would have to be at least 400 times larger than that derived from the observation of luminous matter. If that turned out to be true it would lead to the surprising result that dark matter exists at much larger densities than luminous matter." Subsequent observations confirmed the discrepancy between the amount of directly observed, luminous matter and the dynamically inferred mass density of the universe [2]. Perhaps the most compelling evidence for DM comes from the observed rotation curves of spiral galaxies, *i.e.*, from the rotational velocities measured as a function of galactic radius [3]. These rotation curves stay flat at velocities of $200 - 300$ km/s without ever decreasing as they should if the luminous matter alone were responsible for the gravitational force holding the observed material in orbit. Hence galaxies appear to consist of several times as much DM than visible stuff. For our own Milky Way one infers a halo DM density in the solar neighborhood of

$$\rho_{\odot}^{\mathrm{DM}} = 0.3\,\mathrm{GeV/cm}^3 = 5.35{\times}10^{-25}\,\mathrm{g/cm}^3 \tag{1}$$

with an uncertainty of probably as much as a factor of 2 in either direction [4].

We may express the mass-energy density, ρ, of the universe in dimensionless units as $\Omega = (8\pi/3)\,G_{\mathrm{N}}\,H^{-2}\,\rho$ where G_{N} is Newton's constant and H is the instantaneous Hubble expansion parameter. In an isotropic and homogenous Friedman model of the universe,

M. Signore and C. Dupraz (et al.), The Infrared and Submillimetre Sky after COBE, 175–188.
© 1992 *Kluwer Academic Publishers.*

$\Omega = 1$ is the "critical value" corresponding to vanishing spatial curvature and $\Omega < 1$ corresponds to negative curvature, both with infinite future expansion, while $\Omega > 1$ is a "closed universe" with positive curvature and recollapse in the future. For the present-day density parameter we may write, in term of the present-day density, ρ_0,

$$\Omega_0\, h^2 = \rho_0/1.88{\times}10^{-29}\,\text{g\,cm}^{-3}\,, \tag{2}$$

where $h \equiv H_0/100\,\text{km\,s}^{-1}\,\text{Mpc}^{-1}$ and observationally $0.5 \lesssim h \lesssim 1.0$.

The luminous matter of the universe today amounts to $\Omega_{\text{lum}} \sim 0.01$ or less while the DM in galactic haloes is around ten times as much. On the scale of galactic clusters one infers mass densities even larger, i.e., $\Omega_{\text{cluster}} =$ several 0.1, but there is no compelling observational evidence that the density of the universe would be critical. Still, important theoretical arguments favor a value $\Omega = 1$ so that there would have to be a smooth background component to the matter density which has no effect on the dynamics of structures on the cluster scale or smaller. At any rate, the average DM density of the universe is a few tens to one hundred times larger than the density of the luminous matter [5, 6].

The evidence for the existence of DM is based on the validity of Newton's theory of gravity and Einstein's relativistic generalization, which can not be tested directly at galactic or even cosmic distance scales. Therefore it is prudent to remain open to the possibility that the "dark matter problem" may be resolved by a suitable modification of the established law of gravity [7]. Currently, however, there does not seem to exist a compelling alternative to general relativity that would be able to explain the observations without the need for DM.

It is natural to consider first the possibility of *baryonic dark matter* (BDM) because baryonic material is known to exist while it is uncertain if any of the particle candidates discussed below exist or exist with the right properties. Moreover, the standard theory of primordial nucleosynthesis together with the observed abundances of light elements which are believed to have been produced in the big bang requires an amount of baryonic material today in the range $0.0097 < \Omega_B h^2 < 0.016$ [8]. Because $\Omega_{\text{lum}} \lesssim 0.01$ it is likely that there is more baryonic material in our universe than can be accounted for by luminous matter, especially if $h \sim 0.5$, so that probably there exist at least some "dark baryons". The BDM option is virtually identical with faint or non-luminous stars such as such as old neutron stars, black holes, or brown dwarfs (stars which are too small to ignite hydrogen burning), although baryonic matter may exist in more exotic forms such as quark nuggets [9], or may occur as dust or gaseous distributions of hydrogen, perhaps in molecular form. Moreover, while black holes do not carry baryon number even if they originated from a stellar collapse they are usually lumped together with baryonic candidates.

The second DM option is that of *particle dark matter* (PDM) with its quintessential candidate, a massive neutrino species. The discovery of the cosmic microwave background radiation gave enormous credibility to the hot big bang cosmology (as opposed to a steady state theory), and Gershtein and Zel'dovich almost immediately showed in 1966 that in this picture one would expect an analogous cosmic neutrino background [10]. If neutrinos had a non-vanishing mass, this cosmic background would contribute to the energy density of the universe with an amount proportional to the expected number density of neutrinos times the assumed neutrino mass so that today

$$\Omega_\nu h^2 = (m_{\nu_e} + m_{\nu_\mu} + m_{\nu_\tau})/92\,\text{eV}\,. \tag{3}$$

Gershtein and Zel'dovich were thus able to derive an upper bound to neutrino masses of a few 100 eV, while Cowsik and McClelland [11] seem to be the first authors who, in 1973, explicitly discussed the neutrinos' role as DM candidates.

Since the late seventies a large number of new particles were postulated on the basis of theoretical considerations of elementary particle physics, none of which has so far been detected but many of which could conceivably constitute the DM of our universe. The frontrunners among these non-neutrino PDM candidates are supersymmetric particles and axions. The lightest supersymmetric particle (LSP) that would have to be the DM particle may well be a spin 1/2 state called "neutralino" with properties very similar to the usual neutrinos. Axions, on the other side, are bosons that would have to have masses around $1\,\mu$eV $\equiv 10^{-6}$ eV if they were the DM.

The experimental and observational search for galactic DM has begun in earnest since the mid-eighties. In this lecture I will give a brief overview over these current efforts to identify the physical nature of DM. Methods to search for PDM have been reviewed by Primack, Seckel, and Sadoulet [12], and by Smith and Lewin [13]. The search for both PDM and BDM was reviewed by Spiro [14], and the general issue of DM in the broad context of cosmology is well covered in the recent textbooks by Börner [5] and by Kolb and Turner [6].

2. Attempts to Detect Particle Dark Matter

2.1. NEUTRINOS

Among the *known* particles neutrinos are the only PDM candidates. It is almost universally believed that there should be a "background sea" of neutrinos in the universe analogous to the cosmic microwave background, but it is not known whether any of the neutrinos have a mass of the necessary magnitude given in equation (3). Interestingly, there is a second mass range where they could be the DM. The present-day neutrino density was fixed at a cosmic epoch when the temperature fell below $T_{\text{freeze}} \sim 1$ MeV so that the early universe became too cold and dilute for them to interact effectively ("neutrino freeze-out"). Therefore, the equilibrium number density of a neutrino species with $m_\nu \gg T_{\text{freeze}}$ would be suppressed by a Boltzmann factor, $\exp(-m_\nu/T_{\text{freeze}})$, leading to a density of [6]

$$\Omega_\nu h^2 \sim 3\,(\text{GeV}/m_\nu)^2 \left[1 + \tfrac{1}{5}\ln(m_\nu/\text{GeV})\right]. \tag{4}$$

Therefore a neutrino species with a mass around 1 GeV could provide the DM just as well as one with a mass around 30 eV.

The masses for the known species are experimentally bounded by $m_{\nu_e} \lesssim 10$ eV, $m_{\nu_\mu} \lesssim 250$ keV, and $m_{\nu_\tau} \lesssim 35$ MeV so that GeV masses require the existence of a fourth family. However, any neutrino with $m_\nu < m_Z/2 = 46$ GeV would contribute to the decay width of the neutral gauge boson, Z°, by virtue of the process $Z^\circ \to \bar{\nu}\nu$. About a year ago it became possible to measure this decay width with great precision using the LEP machine at CERN, and the result indicates that there exist exactly 3 neutrino families with $m_\nu < m_Z/2$, excluding the GeV range for DM neutrinos.

The argument leading to equation (4) fails if the cosmic neutrinos have a non-vanishing chemical potential so that the conservation of their lepton number prevents their primordial annihilation just as the conservation of baryon number prevented the disappearance of baryons from our universe. Neutrinos can carry a conserved lepton number only if they are not identical with their own anti-particles, *i.e.*, if they are "Dirac fermions" (as opposed to "Majorana fermions" which are identical to their own anti-particles). In this case a fourth neutrino with a mass exceeding $m_Z/2$ could still provide the DM.

This possibility was tested by a number of direct PDM search experiments. The main idea of these searches for heavy neutrinos or similar PDM candidates (often called WIMPs for "weakly interacting massive particles") is to consider their elastic scattering on nuclei. For a nucleus with charge Z, neutron number N, and mass m_n the relevant non-relativistic Dirac-neutrino scattering cross section is

$$\sigma = \frac{G_F^2}{8\pi} \left[N - (1 - 4\sin^2 \Theta_w) Z \right]^2 \left(\frac{m_\nu m_n}{m_\nu + m_n} \right)^2. \tag{5}$$

This expression does not apply to Majorana neutrinos whose cross section would essentially scale with the square of the nuclear spin. The recoil energy of the struck nucleus is [12]

$$\Delta E = \frac{m_n m_\nu^2}{(m_n + m_\nu)^2} v^2 (1 - \cos\theta), \tag{6}$$

where θ is the scattering angle and v the neutrino velocity. For halo DM and an isotropic cross section the average recoil is $\langle \Delta E \rangle \sim 2\,\text{keV}\, m_n m_\nu^2/(m_n + m_\nu)^2$ where now all masses are understood in GeV [14]. The task of WIMP search experiments is to detect such small nuclear recoil energies.

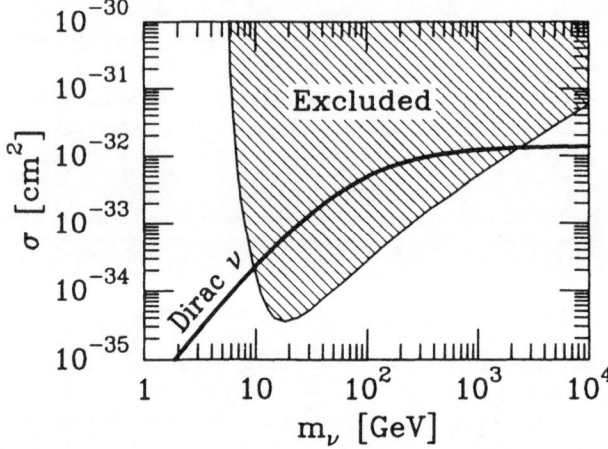

Figure 1. Excluded range (90% CL) of masses and scattering cross section with natural Germanium for galactic halo DM candidates (neutrinos or other particles) according to the Gotthard DM search experiment [15]. The curve for a Dirac neutrino is given by equation (5), using the isotope mixture of natural Ge as a target.

One possibility is to use semiconductor crystals as targets where the nuclear recoil creates free charges that can be detected. Ionization spectrometers made of germanium were originally developed to search for 2β decay, but they could be adapted to a dedicated DM search. Results are now available from the work of a group at the University of South Carolina [16], from a group at UCSB, LBL, and UCB [17, 18], and from the CalTech, Neuchâtel, and Paul Scherrer Institut collaboration [15] shown in Fig. 1. Also, results from the UCSB/LBL/UCB group are now available using Si as a target material [19, 18]. The null results of these experiments exclude certain regions in the plane of WIMP masses and scattering cross sections, and specifically indicate that for halo Dirac neutrinos $m_\nu < 10\,\text{GeV}$. This result, together with the LEP bounds discussed above, entirely excludes the option of neutrino DM in the high mass regime so that only the original option of neutrinos in the 30 eV regime remains viable [20].

Unfortunately, no realistic method is known to detect such low-mass galactic neutrinos because of their absurdly small interaction cross sections [see equation (5)]. However, any positive evidence for non-vanishing neutrino masses would greatly boost the credibility of the light-neutrino option even without their direct detection in the halo. One popular explanation of the solar neutrino problem (MSW solution) involves oscillations between ν_e and ν_μ or ν_τ, a mechanism which requires the existence of neutrino masses [21]. While the existing measurements do not yet prove the actual operation of the MSW effect, the ongoing gallium experiments SAGE and GALLEX will undoubtedly yield unrefutable evidence if the neutrino mass difference and mixing angle are in the regime favored by the Kamioka results [22]. Preliminary results from SAGE [23] already seem to indicate that there may be a substantial deficiency in the low-energy ν_e flux.

2.2. SUPERSYMMETRIC PARTICLES

While neutrinos in the 30 eV range are attractive DM candidates they are afflicted with serious problems related to their small mass. The available phase space in a galaxy can not accomodate enough neutrinos if their mass is too small [24], leading to a *lower* bound on the mass of halo neutrinos which is just barely consistent with the upper bound based on equation (1). Moreover, light neutrinos would remain relativistic until very late in the evolution of the universe, preventing the formation of structures on scales below supercluster dimensions by their "free streaming". Therefore light neutrinos are often referred to as "hot DM" while the heavy mass option is termed "cold DM". These arguments are the reason for much of the attention that has been paid to the now excluded heavy neutrino possibility, and they are the reason for considering alternative candidates for cold DM.

The most attractive option apart from axions are supersymmetric particles. Supersymmetry (SUSY) is a theory of elementary particles in which all fermions have bosonic and all bosons have fermionic partners ("superpartners" or "sparticles"). For example, the photon's partner is the spin-$\frac{1}{2}$ photino and the electron's partner is the spin-0 selectron. SUSY models involve a new quantum number called R-parity which distinguishes between particles and sparticles. If R-parity is conserved, the lightest SUSY particle (LSP) must be stable and would be an ideal DM candidate. Frequently it is assumed that a superposition of photino, zino, and higgsino is the LSP which is then called neutralino. It is a spin-$\frac{1}{2}$ Majorana fermion with unknown mass and an interaction strength similar to that of regular neutrinos. The measurements of the Z° decay width also yield constraints on SUSY particles, none of which have been detected yet. While SUSY theories contain many free parameters so that it is difficult to interpret experimental results unambiguously, it appears that neutralinos must have masses $m_x \gtrsim 15\,\text{GeV}$ [25].

The methods to search for neutralinos are similar to those for heavy Majorana neutrinos, *i.e.*, one must detect nuclear recoils in the keV regime. The relevant cross sections are much smaller than those for Dirac neutrinos because of the p-wave nature of Majorana interactions which do not benefit from the large N^2 coherent enhancement of equation (5). Hence it will be much more difficult to find such particles than it was to exclude heavy Dirac neutrinos. (For neutralino cross sections and detection possibilities see [26, 27]). Currently much effort is going into detector development, especially cryogenic detectors [28], but it will be several years until a realistic experiment can be mounted.

If WIMPs are the halo DM they constantly impinge on the Sun and some of them will scatter on nuclei, lose some of their kinetic energy, and get trapped in the Sun. Subsequently they would build up in the solar center, boosting the probability for their mutual encounter and annihilation. Therefore it is possible that DM will be eventually discovered by the observation of high-energy neutrinos from the Sun that would signal WIMP annihilation there [29, 27]

2.3. AXIONS

We now turn to a PDM candidate, the axion, which is very different from neutrino-like particles both in its physical nature and in the scheme devised for detecting it. The prediction for the existence of axions arose independently from cosmology from an attempt to solve the *CP*-problem of strong interactions, *i.e.*, an attempt to explain the puzzling smallness of a free parameter of QCD, $\overline{\Theta}$, which measures *CP*-violating effects and especially the magnitude of a neutron electric dipole moment. The experimental upper bounds on such a dipole moment indicate that $\overline{\Theta} \lesssim 10^{-9}$ while one should expect $\overline{\Theta} \sim 10^{-3}$. The smallness of $\overline{\Theta}$ and hence the vanishing neutron electric dipole moment can be understood in a scheme devised by Peccei and Quinn [30] which allows one to interpret the "parameter" $\overline{\Theta}$ as new physical field. More precisely, $\overline{\Theta} \rightarrow a(\boldsymbol{r}, t)/f_a$ where $a(\boldsymbol{r}, t)$ is a pseudoscalar particle field, the axion field, and f_a is an unknown energy scale termed "axion decay constant" or "Peccei-Quinn scale". Axions would have a mass $m_a/\text{eV} = 0.60\times10^7\,\text{GeV}/f_a$. The mass term in the Lagrangian represents a potential for the axion field so that its ground state is at $a = 0$, *i.e.*, at $\overline{\Theta} = 0$. In other words, whenever the axion field is different from zero, the potential drives it to its *CP* conserving minimum so that *CP* conservation in strong interactions would be dynamically enforced by the axion mass term.

Axions would interact with gluons and photons, and in some implementations of the Peccei-Quinn scheme also with quarks and leptons. The interaction with photons is described by the Lagrangian density $\mathcal{L}_{a\gamma} = g_{a\gamma}\boldsymbol{E}\cdot\boldsymbol{B}\,a$ where $\boldsymbol{E}$ and $\boldsymbol{B}$ are the electric and magnetic fields, respectively. Hence axions couple to two photons as shown in Fig. 2(a), a vertex which allows for their radiative decay. The coupling constant is

$$g_{a\gamma} = -\frac{\alpha}{2\pi f_a}\xi = -1.94\times10^{-10}\,\text{GeV}^{-1}\,\xi\,\frac{m_a}{\text{eV}} \qquad (7)$$

where $\xi \equiv E/N - 1.92$ and E/N is a free parameter of the specific implementation of the Peccei-Quinn scheme. In the framework of grand unified theories (GUTs) one has $E/N = 8/3$, but it is also possible to construct models with $E/N = 2$, which leads to a near cancellation of the coupling. In any reasonable model, E/N must be a ratio of small integers.

Because the axion mass and the strength of all of its interactions scale as f_a^{-1}, these spin-0 bosons can be arbitrarily light and arbitrarily weakly interacting. Still, for a large range of parameters they would be produced in large numbers in the interior of stars and yet escape almost unscathed, thus carrying away large amounts of energy. This "exotic energy loss" of stars would lead to observable modifications of the standard path of stellar evolution, and a large number of arguments that were raised in this spirit yield a constraint [31]

$$m_a \lesssim 10^{-3}\,\mathrm{eV} \quad \mathrm{or} \quad f_a \gtrsim 10^{10}\,\mathrm{GeV}\,. \tag{8}$$

Of course, the stellar evolution constraints are complemented by laboratory experiments which exclude the "high mass" range where axions would interact too strongly to escape freely from stars.

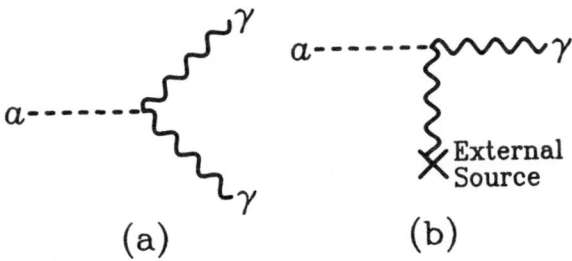

Figure 2. (a) Coupling between axions and two photons which allows for a radiative axion decay mode. (b) Axion to photon conversion in an external magnetic or electric field (Primakoff effect).

Just as neutrinos, axions would have been in thermal equilibrium in the early universe at a certain epoch, but given the constraint equation (8) there would not be enough thermally produced axions to contribute significantly to the DM [32]. However, there is a second mechanism to produce axions in the early universe, namely axions radiated by cosmic strings. Such topological defects would have to form during the Peccei-Quinn phase transition at a cosmic temperature $T \sim f_a$ below which the axion exists as a nearly massless particle together with a very massive scalar state that is of no importance in our low-energy world. At temperatures above the phase transition, both fields appear symmetrically with the same high mass. The expected energy density today is found to be

$$\Omega_a h^2 = (\mathrm{factor}) \times (1\mu\mathrm{eV}/m_a)^{1.175} \tag{9}$$

which *increases* with decreasing axion mass. Unfortunately, the overall factor in this expression is not well known and depends severely on the energy spectrum of the string radiation. Davis and Shellard [33] found a critical axion density, $\Omega_a = 1$, for $m_a \sim 10^{-3}\,\mathrm{eV}$ while Sikivie and collaborators [34] found $m_a \sim 10^{-5}\,\mathrm{eV}$.

The axion production scenario is yet different if cosmic inflation occurred after the Peccei-Quinn phase transition. Then all axion strings would have been diluted away and the axion density depends on the random value of the initial axion field in our part of the universe, the "initial misalignment value" of the axion field. The resulting axion density today happens to be analogous to equation (9) with a factor of order unity or less [35]. Hence the exact prediction for the mass of axions, if they are assumed to be the DM of the universe, depends quite sensitively on details of the evolution of the early universe and on details of the behavior of radiating cosmic strings. With all of these uncertainties a value for m_a anywhere between 10^{-3} eV and 10^{-9} eV or even smaller appears plausible and worth searching for.

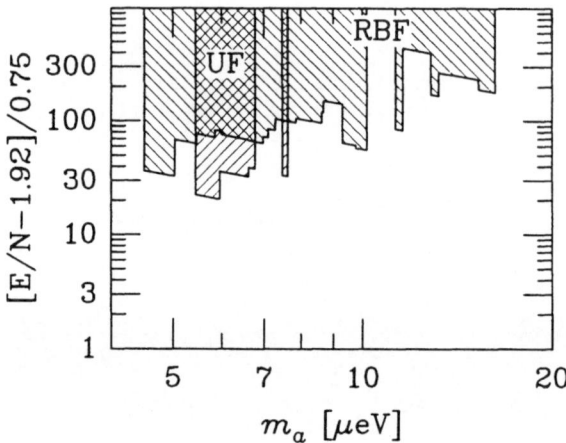

Figure 3. Axion parameters which are excluded at the 95% C.L. by the Rochester-Brookhaven-Fermilab (RBF) Axion Search Experiment [36] and by the University of Florida (UF) Cosmic Axion Search [37], assuming axions are the DM in our galaxy with a local density given in equation (1). The Livermore proposal [38] could cover the entire regime $0.6\,\mu\text{eV} < m_a < 16\,\mu\text{eV}$ down to 1 on the vertical axis, corresponding to $E/N = 8/3$, the preferred value for realistic axion models.

The one known possibility to search for galactic axions, proposed by Sikivie [39] in 1983, is based on the two-photon interaction which allows for the possibility of axion conversion into photons in the presence of an external electric or magnetic field [Fig. 2(b)]. Because galactic DM axions would have to move with velocities less than about $10^{-3}c$ so that they would be bound to the galaxy, the conversion into photons would mean that essentially their rest mass energy, $m_a c^2$, would be converted into electromagnetic radiation of this frequency. A suitable experimental arrangement is to use a microwave cavity placed in a strong magnetic field and to measure its power output. Because microwave frequencies are in the GHz regime, this method is sensitive to axion masses in the μeV range, a plausible regime in the light of the discussion above. If the (n, ℓ) mode of the cavity is tuned to the axion mass one expects a power from axion conversion into this mode of

$$P_{n,\ell} = 1.4 \times 10^{-21} \,\text{Watts}\; C_{n,\ell} \left(\frac{V}{3\,\text{m}^3}\right) \left(\frac{B}{7\,\text{Tesla}}\right)^2 \left[\frac{\text{Min}(Q_l, Q_a)}{10^5}\right] \times$$
$$\times \left(\frac{\rho_a}{5.35 \times 10^{-25}\,\text{g/cm}^3}\right) \left(\frac{m_a}{1\,\mu\text{eV}}\right) \xi^2 ,$$

$$(10)$$

where V is the cavity volume, B the effective static field strength, $C_{n,\ell}$ is a geometric form factor of the cavity mode, ρ_a is the local galactic axion density [see equation (1)], ξ was defined after equation (7), $Q_l \sim 10^5$ is the loaded quality factor of the cavity, and $Q_a \sim 10^6$ is the "quality factor" of the galactic halo axions defined as the ratio of their average energy over their energy spread [38, 39]. However, because the value of the axion mass is not known one has to scan over the largest possible range of cavity frequencies and integrate for a long time at each frequency, a procedure which leads to severe practical limitations for the parameter space that can be covered with this method. In Fig. 3 we show the excluded range of parameters produced by two pilot experiments [36, 37]. By integrating long enough at a fixed frequency of the cavity, i.e., at a fixed axion search mass, it would be possible for either of these experiments to detect or exclude axions at this chosen frequency.

In order to perform a full-scale search over an extended range of masses, however, one needs a larger volume and/or magnetic field strength than were available in the pilot experiments. Recently (July 1990) such an experiment was proposed to be performed at the Lawrence Livermore National Laboratory in California [38] with a cavity volume and magnetic field strength as indicated in equation (10). For comparison, the pilot experiments had typical cavity volumes of $V \sim 10^4 \, \mathrm{cm}^3 = 0.01 \, \mathrm{m}^3$ at $B \sim 6$ Tesla. Because of the extraordinary magnitude of the available volume and field strength it appears possible to cover a range of axion masses $0.6 - 16 \, \mu\mathrm{eV}$ within a search time of approximately 5 years, i.e., in this mass range it should be possible to extend the cross-hatched region in Fig. 3 down to 1 on the vertical axis, which corresponds to $E/N = 8/3$, the preferred value for realistic axion models. To achieve this wide range of frequencies (axion search masses) at this large detection volume it would have to be subdivided into anywhere between 1 and 1024 individual cavities.

3. Attempts to detect faint stars

In a galactic halo, conventional matter can appear in many different forms, especially as a gaseous interstellar medium or as dust, or in condensed forms such as "snowballs", planet-sized objects (Jupiters), normal stars, compact stars (white dwarfs, neutron stars), or as black holes which are usually lumped together with stellar, baryonic DM candidates. It is difficult, however, to hide most of these objects or to hide traces of their existence, leaving only brown dwarfs or black holes in a certain mass range as viable candidates for DM in galactic haloes [40]. Because some of the arguments raised are based on indirect evidence, it is prudent to remain open to a wider class of choices. For example, recently a project was begun to measure the proper motions of a large number of faint stars in order to identify a possible halo component of the local white dwarf population which is normally thought of as belonging to the galactic disk [41]. Such a halo component would give away the existence of at least some halo DM in the form of white dwarfs.

More likely is the option of normal hydrogen stars in the mass range $10^{-6} - 10^{-1} \, M_\odot$. Objects with smaller masses probably would have evaporated within the age of the universe while stars with masses in excess of about $0.08 \, M_\odot$ would ignite hydrogen and thus appear as luminous matter. The one severe argument against these "Jupiters" or "brown dwarfs" is that one would need extremely many of them to account for the halo DM.

It is extremely exciting that Paczyński in 1986 proposed a method to search systematically for these small stars which relies on their gravitational bending of light rays [42]. It was shown a long time ago [43] that a "pointlike" mass (deflector) placed between an observer and a light source creates two distinct images as indicated in Fig. 4, while a non-singular and transparent mass distribution always creates an odd number of images. When the source is exactly aligned behind the deflector (mass M_D) the image would be an annulus instead ("Einstein ring") with a radius ("Einstein radius")

$$R_E = (G_N M_D d)^{1/2} \quad \text{where} \quad d \equiv 4d_{OD}d_{DS}/d_{OS} \tag{11}$$

(see Fig. 4). Because of differential bending of the "rays" which produce the images, the image brightnesses will be different from each other and different form the single image in the absence of gravitational lensing. If the two images can not be resolved because their angular distance α is below the resolving power of the observer's telescope, the only observable effect will be an apparent brightening of the star ("gravitational microlensing") with a magnification ("amplification") factor

$$A = \frac{2 + u^2}{u(4 + u^2)^{1/2}} \quad \text{where} \quad u \equiv R/R_E \tag{12}$$

and R is the distance of the deflector from the line of sight.

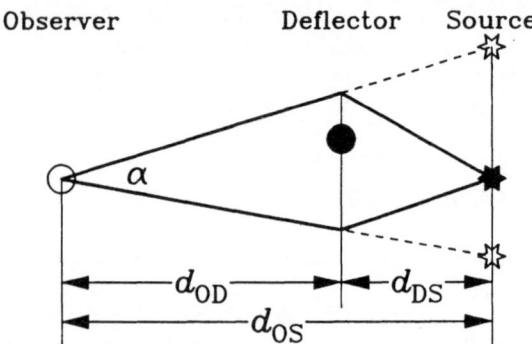

Figure 4. Geometry of light deflection by a pointlike mass, yielding two images of a source viewed by an observer.

If we imagine that a terrestrial observer watches a distant star, we expect that faint DM stars (sometimes called MACHOs for "massive astrophysical compact halo objects") will occasionally pass near the line of sight and thus cause the image of the monitored star to brighten. If the deflector moves with a velocity v transverse to the line of sight, and if its "impact parameter" or minimal distance to the line of sight is R_0, an instant which we identify with time $t = 0$, then one expects a lightcurve as shown in Fig. 5 for several values of R_0/R_E. The time scale for this event is R_E/v.

A convenient sample of stars is provided by the Large Magellanic Cloud (LMC) at a distance of about 50 kpc. Any star in the LMC will be substantially brightened at the time of observation if the line of sight intersects with the circular cross section πR_E^2 around some MACHO. Because the number density of DM stars is inversely proportional to their assumed mass while πR_E^2 is directly proportional to it, the probability for a target star to be lensed at the instance of observation is independent of the mass of the DM objects. For stars in the LMC one finds a probability ("optical depth for microlensing of the galactic halo") of $\sim 10^{-6}$. In other words, if one looked simultaneously at $\sim 10^6$ stars in the LMC one would have a good chance to see at least one of them brightened by a dark halo star.

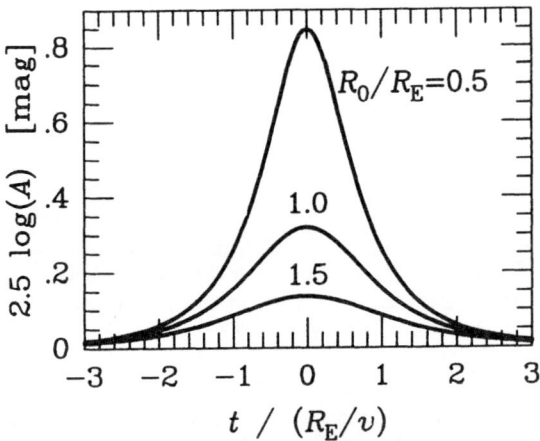

Figure 5. Apparent lightcurve of a source if a pointlike deflector passes the line of sight with a transverse velocity v at an "impact parameter" R_0. A is the magnification ("amplification") factor so that the vertical axis gives the apparent brightness increase in magnitudes.

In order to identify a lensing event one has to monitor this large sample of stars long enough to identify the characteristic light curve shown in Fig. 5. It has the property of being unique, symmetric about $t = 0$, and achromatic, three signatures which should allow one to discriminate against normal variable stars, of which there should be $\sim 10^4$ in an observed sample of 10^6 stars. Because the typical time scale $t_0 \equiv R_E/v$ depends on the deflector mass one must cover a broad range of time-scales in order to be able to detect or exclude MACHOs of a given mass. If the deflector mass is $1\,M_\odot$, the mean microlensing time will be 3 months, for $10^{-2}\,M_\odot$ it is 9 days, for $10^{-4}\,M_\odot$ it is 1 day, and for $10^{-6}\,M_\odot$ it is 2 hours. In order to measure light curves at the longer relevant time scales one can take photographic plates of the LMC on subsequent nights for several weeks or months which then have to be digitized and evaluated with automated methods because of the large number of target stars. The faster variations must be searched for with CCD cameras which feed their data directly into a computer for automatic interpretation.

This search is currently being pursued by a group in France using the facilities of the European Southern Observatory (the LMC is in the southern sky!), and by a Californian-Australian collaboration using the Mt. Stromlo Observatory. While first photographic plates have been taken already to study the feasibility of the project, a full-scale search will undoubtedly take several years. The results of these projects will be of great importance because they will either turn up DM stars or else allow one to exclude the best motivated BDM candidates and thus strengthen the case for PDM. In either event, this search will produce an unprecedented data base of variable stars in the LMC! In [44] we give references to several detailed studies concerning these projects.

4. Summary

A number of dedicated experiments using Ge and Si ionization detectors have excluded a large region in the plane of masses and scattering cross sections for weakly interacting DM candidates. Together with recent precision measurements of the Z° decay width they exclude a fourth generation neutrino in the GeV mass range as a DM candidate, which otherwise would have been a prime candidate. However, ν_μ or ν_τ in the 30 eV mass range remain perhaps the best motivated PDM candidates. Should the ongoing solar neutrino experiments SAGE and GALLEX yield definitive evidence for non-zero neutrino masses, ν_μ or ν_τ would become the undisputed frontrunners among the PDM candidates, although a direct detection of DM neutrinos, if they exist, will not be possible in the foreseeable future. Before a realistic detection experiment for supersymmetric DM candidates can be mounted, much development work for cryogenic or ionization detectors will have to be completed, and although many groups are currently working in this field worldwide, it will be several years before even a pilot experiment can be performed. Two successful pilot experiments have shown the feasibility of axion search experiments using microwave cavities, but they fell short in sensitivity by at least two orders of magnitude in order to cover a meaningful range of axion masses. A realistic full-scale experiment has been proposed but not yet funded. The frontrunner among baryonic DM candidates are faint stars in the $10^{-6} - 10^{-1}$ $M_\odot$ range which can be discovered, in principle, through their gravitational microlensing of stars in the Large Magellanic Cloud. Two independent groups are working on this project, first photographic plates have been taken, and one can hope for results within the next few years. If we are lucky the discovery of the unknown stuff which makes up at least 90% of our universe is just around the corner!

Acknowledgements

I acknowledge helpful conversations on the issues of DM detection with C. Alcock, K. van Bibber, G. Börner, S. Cooper, K. Griest, P. Jetzer, K. Olive, A. de Rújula, P. Sikivie, J. Silk, M. Spiro, G. Steigman, and L. Stodolsky. I am very grateful for the hospitality of the Center for Particle Astrophysics and the Astronomy Department at the University of California in Berkeley were part of this lecture was written, and I acknowledge financial support at Berkeley by grants from NASA, DOE, and NSF.

References

1. F. Zwicky, *Helv. Phys. Acta* **6** (1933) 110.
2. V. Trimble, *Ann. Rev. Astron. Astrophys.* **25** (1987) 425.
3. M. S. Roberts and R. N. Whitehust, *Astrophys. J.* **201** (1975) 327. V. C. Rubin, W. K. Ford Jr., and N. Thonnard, *Astrophys. J.* **238** (1980) 471 and *Astrophys. J.* **261** (1982) 439. D. Burstein, V. C. Rubin, N. Thonnard, and W. K. Ford Jr., *Astrophys. J.* **253** (1982) 70. V. C. Rubin, *Sci. Am.* **248**:6 (June 1983) 88.
4. J. A. R. Caldwell and J. P. Ostriker, *Astrophys. J.* **251** (1981) 61. R. A. Flores, *Phys. Lett.* **215 B** (1988) 73.
5. G. Börner, *The Early Universe—Facts and Fiction* (Springer, Berlin, 1988).
6. E. W. Kolb and M. S. Turner, *The Early Universe* (Addison-Wesley, Redwood City, 1990).
7. M. Milgrom, *Comm. Astrophys.* **13** (1989) 215. R. H. Sanders, *Astron. Astrophys. Rev.* **2** (1990) 1.
8. K. A. Olive, D. N. Schramm, G. Steigman, and T. P. Walker, *Phys. Lett.* **236 B** (1990) 454.
9. E. Witten, *Phys. Rev. D* **30** (1984) 272. C. Alcock and A. Olinto, *Ann. Rev. Nucl. Part. Sci.* **38** (1988) 161.
10. S. S. Gershtein and Ya. B. Zel'dovich, *Pisma Zh. Eksp. Teor. Fiz.* **4** (1966) 174 [*JETP Lett.* **4** (1966) 120]. A similar calculation was performed by R. Cowsik and J. McClelland, *Phys. Rev. Lett.* **29** (1972) 669.
11. R. Cowsik and J. McClelland, *Astrophys. J.* **180** (1973) 7.
12. J. Primack, D. Seckel, and B. Sadoulet, *Ann. Rev. Nucl. Part. Sci.* **38** (1988) 751.
13. P. Smith and J. Lewin, *Phys. Rep.* **187** (1990) 203.
14. M. Spiro, to appear in the Proc. of the Neutrino '90 conferenece, Geneva, June 1990.
15. D. Reusser *et al.*, *Phys. Lett.* **255 B** (1991) 143.
16. S. P. Ahlen *et al.*, *Phys. Lett.* **195 B** (1987) 603.
17. D. Caldwell *et al.*, *Phys. Rev. Lett.* **61** (1988) 510.
18. D. Caldwell, *Mod. Phys. Lett. A* **5** (1990) 203.
19. D. Caldwell *et al.*, *Phys. Rev. Lett.* **65** (1990) 1305.
20. K. Griest and J. Silk, *Nature* **343** (1990) 251.
21. J. Bahcall, *Neutrino Astrophysics* (Cambridge University Press, Cambridge, 1989).
22. K. S. Hirata *et al.*, *Phys. Rev. Lett.* **66** (1991) 9. J. N. Bahcall and H. A. Bethe, *Phys. Rev. Lett.* **65** (1990) 2233.
23. V. N. Gavrin, in the Proc. of Neutrino '90, Geneva 1990.
24. S. Tremaine and J. E. Gunn, *Phys. Rev. Lett.* **42** (1979) 407.
25. J. Ellis *et al.*, *Phys. Lett.* **245 B** (1990) 251.
26. K. Griest *Phys. Rev. Lett.* **61** (1988) 666.
27. G. B. Gelmini, P. Gondolo, and E. Roulet, *Nucl. Phys. B* **351** (1991) 623.
28. A. K. Drukier and L. Stodolsky, *Phys. Rev. D* **30** (1984) 2295. M. Goodman and E. Witten, *Phys. Rev. D* **31** (1985) 3059. I. Wasserman, *Phys. Rev. D* **30** (1986) 2295. K. Pretzl, N. Schmitz, and L. Stodolsky (eds.), *Low Temperature Detectors for Neutrinos and Dark Matter* (Springer, Berlin, 1987).
29. J. Silk, K. Olive, and M. Srednicki, *Phys. Rev. Lett.* **55** (1985) 257. M. Srednicki, K. Olive, and J. Silk, *Nucl. Phys. B* **279** (1987) 804.

30. R. D. Peccei and H. Quinn, *Phys. Rev. Lett.* **38** (1977) 1440. S. Weinberg, *Phys. Rev. Lett.* **40** (1978) 223. F. Wilczek, *Phys. Rev. Lett.* **40** (1978) 279. For reviews see: J. E. Kim, *Phys. Rep.* **150** (1987) 1. H.-Y. Cheng, *Phys. Rep.* **158** (1988) 1.

31. M. S. Turner, *Phys. Rep.* **197** (1990) 67. G. G. Raffelt, *Phys. Rep.* **198** (1990) 1.

32. M. S. Turner, *Phys. Rev. Lett.* **59** (1987) 2489.

33. R. L. Davis, *Phys. Lett.* **180 B** (1986) 225. R. L. Davis and E. P. S. Shellard, *Nucl. Phys.* **B 324** (1989) 167. See also A. Dabholkar and J. Quashnock, *Nucl. Phys.* **B 333** (1990) 815.

34. D. Harari and P. Sikivie, *Phys. Lett.* **195 B** (1987) 361. C. Hagmann and P. Sikivie, Report UFIFT-HEP-90-30 (Univ. of Florida, 1990).

35. J. Preskill, M. B. Wise, and F. Wilczek, *Phys. Lett.* **120 B** (1983) 127. L. F. Abbott and P. Sikivie, *Phys. Lett.* **120 B** (1983) 133. M. Dine and W. Fischler, *Phys. Lett.* **120 B** (1983) 137.

36. DePanfilis *et al.*, *Phys. Rev. Lett.* **59** (1987) 839. W. Wuensch *et al.*, *Phys. Rev. D* **40** (1989) 3153.

37. C. Hagmann *et al.*, *Phys. Rev. D* **42** (1990) 1297.

38. P. Sikivie *et al.*, "Experimental Search for Dark Matter Axions in the $0.6-16\,\mu$eV Mass Range", Proposal to the U.S. Dept. of Energy and the Lawrence Livermore National Laboratory (1990), unpublished.

39. P. Sikivie, *Phys. Rev. Lett.* **51** (1983) 1415, *ibid.* **52** (1984) 695 (E), *Phys. Rev. D* **32** (1985) 2988 and *ibid.* **36** (1987) 974 (E).

40. B. Carr, *Comm. Astrophys.* **14** (1990) 257.

41. C. T. Tamanaha, J. Silk, M. A. Wood, and D. E. Winget, *Astrophys. J.* **358** (1990) 164 and J. Silk, private communication.

42. B. Paczyński, *Astrophys. J.* **304** (1986) 1.

43. O. Chwolson, *Astr. Nachr.* **221** (1924) 329. A. Einstein, *Science* **84** (1936) 506. G. A. Tikhov, *Dokl. Akad. Nauk. S.S.S.R.* **16** (1937) 199. S. Refsdal, *Mon. Not. R. Astron. Soc.* **128** (1964) 295.

44. A. Milsztajn, Report, DPhPE 90-06 (June 1990), to be published in the Proc. of the Moriond Workshop on New and Exotic Phenomena, Les Arcs, January 1990. B. Carr and J. Primack, *Nature* **345** (1990) 478. K. Griest, *Astrophys. J.* **366** (1991) 412. L. Krauss and T. A. Small, Report, YCTP-P10-90 (July 1990). K. Griest *et al.* (MACHO collaboration), Report, CfPA-TH-90-024 (October 1990). A. de Rújula, Ph. Jetzer, and E. Massó, Report, CERN-TH.5812/90 (August 1990) and CERN-TH.5787/91 (January 1991).

FLUCTUATIONS AND ANISOTROPIES IN INTEGRATED BACK-GROUNDS

XAVIER BARCONS
Departamento de Física Moderna
Universidad de Cantabria
39005 Santander, Spain

ABSTRACT. The aim of this lecture is to present a (hopefully) comprehensive overview of the techniques used to gain information on source populations from the angular variations of the corresponding integrated background. Emphasis is made on the angular two–point autocorrelation function and the one–point distribution of fluctuations (the $P(D)$ curve) of the background and their relation to the number–flux relation ($\log N - \log S$) and clustering of the underlying sources.

1. Introduction

The extragalactic sky, at any wavelength, can be generally splitted into two components: a diffuse component and the integrated emission from cosmic sources. By diffuse we mean any component that cannot be resolved with any foreseeable instrumentation. At millimeter wavelengths, the sky (extragalactic will be assumed henceforth) is dominated by the diffuse Cosmic Microwave Radiation, discrete sources producing a negligibly small contribution. At X–ray wavelengths, COBE has ruled out the possibility that a hot intergalactic medium produces a diffuse component (Barcons, Fabian & Rees 1991) and therefore the integrated emission from sources (most of which have not yet been discovered) dominates the X–ray background. The situation is presumably more complicated at submillimetre and infrared wavelengths, where some diffuse emission must be present (especially near the peak of the Planck spectrum) and sources will certainly give a contribution.

Here, we shall be dealing with the integrated component only. If both components are statistically independent, we can study their properties separately and, at the end, we can join them straightforwardly in order to describe the point–to–point variations of the total background. It must also be stressed that although the fractional contribution of sources to the background might be small, they could dominate fluctuations, because of the high isotropy expected in the diffuse component. This possibility has been studied by Franceschini *et al.* (1989) at cm wavelengths.

M. Signore and C. Dupraz (et al.), The Infrared and Submillimetre Sky after COBE, 189–200.

The lecture is organised as follows. In Section 2 a brief discussion on how to describe the source population and how that is linked to the moments of the background intensity is presented. We next discuss with some detail the angular 2–point correlation function of the background and its relation with the 2D correlation function of the sources. Section 4 presents an extensive account on how to relate confusion noise, i.e., the one–point disribution of fluctuations or $P(D)$ curve, to source counts and clustering. For the first time we show how to account properly for source clustering in the $P(D)$ curve. Some models are used to illustrate how source clustering can affect estimates of source counts from $P(D)$ noise. More detailed results will be presented elsewhere (Barcons, Toffolatti & Carrera 1991).

2. Source distribution and background fluctuations

The integrated background intensity, as seen by a certain instrument when pointing towards a direction given by the unit vector $\vec{n}$, can be written as

$$I(\vec{n}) = \int d\Omega_{\vec{n}'} \frac{dS(\vec{n}')}{d\Omega_{\vec{n}'}} G(\vec{n} - \vec{n}') \tag{1}$$

where $\frac{dS}{d\Omega}$ is the flux per unit solid angle (intensity) and $G(\vec{n} - \vec{n}')$ is the beam response function. The shape of this function has important effects on the statistical properties of the 2D random field $I(\vec{n})$. Although for illustrative purposes we are going to use a top hat function, it must be kept in mind that other shapes will produce different results. For example, for a smoothly decreasing beam profile, very off axis strong sources will produce weak signals in $I(\vec{n})$ comparable with those produced by faint on–axis sources.

Let us assume that there are N sources in the sky (N must be really big in order to to give riseto a background and not to a set of resolved sources) with fluxes $S_1, \ldots, S_N$ and a surface brightness profile given by a function $\Sigma(\vec{n} - \vec{n}_i)$, $\vec{n}_i$ being the position of the center of the i-th source. The we have

$$\frac{dS(\vec{n})}{d\Omega_{\vec{n}}} = \sum_{i=1}^{N} S_i \Sigma(\vec{n} - \vec{n}_i) \tag{2}$$

and the eq. (1) can be written as

$$I(\vec{n}) = \sum_{i=1}^{N} S_i \mathcal{G}(\vec{n} - \vec{n}_i) \tag{3}$$

where

$$\mathcal{G}(\vec{n} - \vec{n}_i) = \int d\Omega_{\vec{n}'} \Sigma(\vec{n}' - \vec{n}_i) G(\vec{n} - \vec{n}') \tag{4}$$

is an effective beam function which accounts for the possibility that sources are extended. Then the difference between point sources $(\Sigma(\vec{n} - \vec{n}_i) = \delta(\vec{n} - \vec{n}_i))$

and extended sources is just a correction in the beam function. Our aim is to produce a statistical description of $I(\vec{n})$ as given by eq. (3) where the effective beam function $\mathcal{G}$ is only different from zero on small angles.

In this lecture, we use some definitions and techniques in statistics which may not be too familiar to some readers. In Van Kampen (1981) a good review with abundant examples can be found.

2.1 STATISTICAL DESCRIPTION OF THE SOURCE POPULATION

Here we are only concerned with the 2D distribution of sources. As far as the background is concerned redshift dependence is just integrated and therefore lost. We assume that there is no flux segregation, i.e., the flux and the position of a source are independent random variables. This assumption might be at odds with biasing, but corrections should be relatively easy to introduce.

The probability density function for the flux S is

$$f(S) = \frac{1}{\mu} n(S) \tag{5}$$

where $n(S)$ is the number of sources per unit solid angle and unit flux and $\mu = \int dS n(S)$ is the total number of sources per unit solid angle. The function $n(S)$ is also called the differential source counts function, which is related to the so–called $\log N - \log S$ function by

$$N(> S) = \int_S^\infty dS n(S) \tag{6}$$

A power law parametrization for this function is very often used

$$N(> S) = K S^{-\gamma} \tag{7}$$

where this shape has to flatten to a constant at low fluxes in order to have a finite number of sources in the sky and a finite background intensity. The exponent $\gamma = 1.5$ is referred to as the *euclidean* slope giving rise to the euclidean source counts which are expected at high fluxes (coming mostly from local sources) if there is no evolution. Although this is known to happen at X–ray wavelengths, the source counts function is much steeper than the euclidean one in the radio band (at high fluxes). As we shall see in Section 4, source counts relate directly to background fluctuations at the level of one source per beam, and therefore at flux levels much fainter than the threshold for direct source detection.

The position of the sources in the sky is described here by a 2D random field $\rho(\vec{n})$ representing the surface density of sources in the sky in the direction $\vec{n}$. The statistical properties of a random field (and this applies both to $\rho(\vec{n})$ and $I(\vec{n})$) are fixed when all the k–point correlation functions $\langle \rho(\vec{n}_1) \ldots \rho(\vec{n}_k) \rangle$ are known. With these functions we can build the characteristic functional

$$\begin{aligned}
\Phi_\rho[\omega] &= \sum_{k=0}^\infty \frac{(2\pi i)^k}{k!} \int d\Omega_1 \ldots \int d\Omega_k \omega(\vec{n}_1) \ldots \omega(\vec{n}_k) \langle \rho(\vec{n}_1) \ldots \rho(\vec{n}_k) \rangle \\
&= \left\langle \exp\left(2\pi i \int d\Omega \omega(\vec{n}) \rho(\vec{n}) \right) \right\rangle
\end{aligned} \tag{8}$$

However, a description in terms of *cumulants* is much more appropriate. Cumulants $\langle\langle\rho(\vec{n}_1)\ldots\rho(\vec{n}_k)\rangle\rangle$ are defined in terms of their generating functional $\varphi_\rho[\omega] \equiv \ln(\Phi_\rho[\omega])$ as

$$\varphi_\rho[\omega] = \sum_{k=1}^{\infty} \frac{(2\pi i)^k}{k!} \int d\Omega_1 \ldots d\Omega_k \omega(\vec{n}_1)\ldots\omega(\vec{n}_k)\langle\langle\rho(\vec{n}_1)\ldots\rho(\vec{n}_k)\rangle\rangle \qquad (9)$$

If the sources are uniformly distributed (i.e., they are *not* clustered), all cumulats, except the first order one $\langle\langle\rho(\vec{n})\rangle\rangle = \langle\rho(\vec{n})\rangle = \mu = \frac{N}{4\pi}$ are exactly zero. In this case, the positions of the sources are statistically independent one from each other.

The 2D source correlation function $\xi(\vec{n} - \vec{n}')$ is related to the second order cumulant

$$\langle\langle\rho(\vec{n}_1)\rho(\vec{n}_2)\rangle\rangle = \mu^2\xi(\vec{n}_1 - \vec{n}_2) \qquad (10)$$

Notice that two source distributions having the same 2–point correlation functions can be different (and therefore produce differently fluctuating backgrounds). All cumulants must be given in order to fix the statistical properties of $\rho(\vec{n})$.

2.2 THE k-POINT CORRELATION FUNCTION OF THE BACKGROUND

Given the properties of the source population as discussed in the previous section, we can evaluate the characteristic functional of the 2D random field $I(\vec{n})$:

$$\Phi_I[\omega] = \left\langle \exp\left(2\pi i \int d\Omega\omega(\vec{n})I(\vec{n})\right) \right\rangle_{S,\rho} \qquad (11)$$

where the average has to be carried out over both flux and positions of the sources. The k–point background intensity correlation functions can be obtained as functional derivatives of eq. (11)

$$\langle I(\vec{n}_1)\ldots I(\vec{n}_k)\rangle = \frac{1}{(2\pi i)^k} \frac{\delta^k\Phi_I[\omega]}{\delta\omega(\vec{n}_1)\ldots\delta\omega(\vec{n}_k)}\bigg|_{\omega=0} \qquad (12)$$

Inserting eq. (3) into (11) we find

$$\Phi_I[\omega] = \left\langle \left(\frac{1}{N}\int d\Omega\rho(\vec{n})\int dS f(S) \exp\left[2\pi iS \int d\Omega'\omega(\vec{n}')\mathcal{G}(\vec{n} - \vec{n}')\right]\right)^N \right\rangle_\rho \qquad (13)$$

where the flux average has been explicitly carried out.

Now, taking $N \to \infty$ we have

$$\Phi_I[\omega] = \left\langle \exp\left(\int d\Omega\rho(\vec{n})\int dS f(S)\left[\exp\left[2\pi iS \int d\Omega'\omega(\vec{n}')\mathcal{G}(\vec{n} - \vec{n}')\right] - 1\right]\right) \right\rangle_\rho \qquad (14)$$

where the integrations, in practice, only have to be extended to these domains where the effective beam function is not zero. Now we see that this characteristic functional is the characteristic functional of the 2D field $\rho(\vec{n})$ but with another argument function

$$\Phi_I[\omega] = \Phi_\rho[F[\omega]] \tag{15}$$

where (see equation 8)

$$F(\vec{n}) = \frac{1}{2\pi i} \int dS f(S) \left[\exp\left[2\pi i S \int d\Omega' \omega(\vec{n}') \mathcal{G}(\vec{n} - \vec{n}')\right] - 1 \right] \tag{16}$$

and therefore $\Phi_I[\omega]$ can be obtained as an expansion in terms of the cumulants of the surface density field $\rho(\vec{n})$. Defining $\varphi_I[\omega] = \ln(\Phi_I[\omega])$, we have

$$\varphi_I[\omega] = \sum_{k=1}^{\infty} \frac{(2\pi i)^k}{k!} \int d\Omega_1 \ldots \int d\Omega_k \langle\langle \rho(\vec{n}_1) \ldots \rho(\vec{n}_k)\rangle\rangle F(\vec{n}_1) \ldots F(\vec{n}_k) \tag{17}$$

In the absence of clustering this reduces to just the first term

$$\varphi_I[\omega] = \mu \int d\Omega \int dS f(S) \left[\exp\left[2\pi i S \int d\Omega' \omega(\vec{n}') \mathcal{G}(\vec{n} - \vec{n}')\right] - 1 \right] \tag{18}$$

In principle, if one has enough exposures of the sky in different directions, one can try to measure directly k–point correlation functions and compare them to eq. (17) and consequently to the source distribution functions. However, it is customary (most times due to the unavailability of convenient data sets) to study only one and two point properties of the background intensity.

3. The 2-point background correlation function

The second order cumulant of $I(\vec{n})$ can be easily evaluated either from eq. (3) directly or via eq. (12) using (17), for a completely general source distribution

$$\langle\langle I(\vec{n}_1) I(\vec{n}_2)\rangle\rangle \equiv \langle I(\vec{n}_1) I(\vec{n}_2)\rangle - \langle I\rangle^2$$

$$= \langle S^2\rangle_s \mu \int d\Omega \mathcal{G}(\vec{n}_1 - \vec{n}) \mathcal{G}(\vec{n}_2 - \vec{n}) + \tag{19}$$

$$\langle S\rangle_s^2 \mu^2 \int d\Omega \int d\Omega' \mathcal{G}(\vec{n}_1 - \vec{n}) \mathcal{G}(\vec{n}_2 - \vec{n}') \xi(\vec{n} - \vec{n}')$$

It is, perhaps, more common to quote the autocorrelation function (ACF) defined as $W(\theta) = \langle\langle I(\vec{n}_1) I(\vec{n}_2)\rangle\rangle / \langle I\rangle^2$ where $\cos\theta = \vec{n}_1.\vec{n}_2$. For more details and definitions see, e.g., Dautcourt (1977) or Barcons & Fabian (1988,1989). Sometimes $\Gamma(\theta) = \sqrt{|W(\theta)|}$ is quoted instead (e.g., De Zotti et al. 1990) which gives an insight on what fraction of the intensity is actually fluctuating.

The first term in equation (19) arises from the overlap of the beams, and it is not linked at all to source clustering. Since $\langle I\rangle = \mu\langle S\rangle_s \int d\Omega' \mathcal{G}(\vec{n} - \vec{n}')$, the contribution of this term to $W(\theta)$ goes like the inverse of the number of sources per beam and is called the Poisson term. It goes away when the number of sources increases and when θ is much greater than the beam size. If the beam profile is known, we can use the extrapolation to $\theta \to 0$ of $W(\theta)$ to evaluate $\langle S^2 \rangle_s$ assuming no clustering (see, e.g., Soltan 1991). Notice that for a $\log N - \log S$ function like (7) with $1 < \gamma < 2$, $\langle S^2 \rangle_s$ is dominated by the brightest undetected (i.e., not removed from the background) sources and it is really very insensitive to the faintest sources.

The second term in eq. (19) is obviously due to source clustering and it does not go away when the number of sources per beam increases. In this lecture we assume that all sources have the same spatial distribution, in which case for $1 < \gamma < 2$, this clustering term is dominated by the weakest sources. As already mentioned, this might be a too strong assumption. Biasing models predict that the most luminous (and possibly brightest) sources cluster more strongly than the weaker ones. A flux–dependent correlation function would lead to a weighted average of S in the second term of eq. (19). If, for example, only sources with a flux above some threshold do actually cluster, only those would have to be taken into account when evaluating $\langle S \rangle_s$ in this clustering term.

A few final remarks on ACFs. If the background is the sum of different uncorrelated components (i.e., galaxies, AGNs, diffuse, etc.) then $W(\theta) = \sum f_i^2 W_i(\theta)$, where $W_i(\theta)$ is the ACF of the i-th component and f_i its fractional contribution to the background intensity. If these components are correlated, then $W(\theta)$ calculated as above is only a lower limit. Secondly, ACFs are not always positive. Carrera et al. (1991) have shown that the ACF of the X–ray background might be negative on scales of a few degrees. This is what we expect if the Universe has a void–and–wall structure. And finally, the ACF is most efficient by picking up clustering or anticlustering on scales greater than the beam size. On smaller scales it might be more useful to use confusion noise as explained in next section.

4. The P(D) curve

The one–point probability density function is usually called the $P(D)$ curve, where D stands for the deflection or deviation from the mean ($D = I - \langle I \rangle$). Apart from the mean intensity, the $P(D)$ curve contains all the one–point information of the background as viewed with a given beam. Its second moment $\langle D^2 \rangle = \langle I^2 \rangle - \langle I \rangle^2$, which is often used as a measure of fluctuations in the Cosmic Microwave Background in one beam experiments, has to be handled with caution here. The main reason for this is that $P(D)$ curves are usually non–gaussian and have very extended tails towards positive values of D. Note that $\langle D^2 \rangle = W(0)\langle I \rangle^2$, and from the discussion in last section $\langle D^2 \rangle$ will be dominated by the brightest sources included in the background. Fortunately, the use of the $P(D)$ curve itself avoids most of these difficulties.

$P(D)$ curves have been used to test deep source counts at radio (e.g., Franceschini et al. 1989 and references therein) and X–ray frequencies (e.g., Shafer 1983 and references therein). The basic principle behind this scheme is that a direct detection of a source with a given instrument requires a rather rare event, say 4 or 5 sigma above the mean background, but the shape of the $P(D)$ curve is obviously dominated by the 1 sigma deflections. That means that in a statistical sense, ~ 1 sigma sources can be "seen" without being detected, by fitting the $P(D)$ curve to the observed histogram of deflections.

The way in which source clustering affects the shape of the $P(D)$ curve has been only properly studied as far as the second moment is concerned (Dautcourt 1977, Barcons & Fabian 1988). We show here how to incorporate clustering to the standard $P(D)$ scheme.

4.1 DERIVATION OF THE P(D) CURVE

Since we are only interested in one–point properties of the background, let us take $\vec{n}$ parallel to the z–axis unit vector $\vec{e}_z$, let us call $I \equiv I(\vec{e}_z)$ and define a characteristic function

$$\phi_I(\omega) = \Phi[\omega\delta(\vec{n} - \vec{e}_z)] \equiv \exp(\varphi_I(\omega)) \tag{20}$$

where the cumulant generating function is now

$$\varphi_I(\omega) = \sum_{k=1}^{\infty} \frac{1}{k!} \int_b d^2x_1 \ldots \int_b d^2x_k \mu^{-k} \langle\langle \rho(\vec{x}_1) \ldots \rho(\vec{x}_k) \rangle\rangle \Psi(\omega\mathcal{G}(\vec{x}_1)) \ldots \Psi(\omega\mathcal{G}(\vec{x}_k)) \tag{21}$$

where

$$\Psi(t) = \int dSn(S)\big[\exp(2\pi iSt) - 1\big] \tag{22}$$

Note that since the beam has a small extension (less than a few degrees), $\vec{x} = \vec{n} - \vec{e}_z$ is a cartesian vector and the integrals over d^2x give no contribution outside the beam.

Now, if we know the cumulants of the distribution of sources, the beam function $\mathcal{G}(\vec{x})$ and the source counts function, we can calculate the characteristic function $\phi_I(\omega)$ from which the probability density function for I can be evaluated via a Fourier transform

$$P_I(I) = \int d\omega e^{-2\pi i\omega I} \phi_I(\omega) \tag{23}$$

If the number of counts received per beam and exposure is small, this will have to be convolved with a Poisson counting noise.

The standard way to proceed is to model the cumulants $\langle\langle \rho(\vec{x}_1) \ldots \rho(\vec{x}_k) \rangle\rangle$ and $n(S)$ with some free parameters, then evaluate $P(D) = P_I(D + \langle I \rangle)$ and fit this to the histogram of observed deflections via minimization of χ^2 or some other method. In this way we'll gain information about source properties below the detection threshold.

For illustrative purposes, a circular top hat with angular radius α will be assumed for $\mathcal{G}(\vec{x})$. After rescaling fluxes and intensities to convenient units, eq. (21) will read

$$\varphi_I(\omega) = \sum_{k=1}^{\infty} \frac{\Psi(\omega)^k}{k!} \int_{|\vec{x}_1|<\alpha} d^2x_1 \ldots \int_{|\vec{x}_k|<\alpha} d^2x_k \mu^{-k} \langle\langle \rho(\vec{x}_1) \ldots \rho(\vec{x}_k) \rangle\rangle \tag{24}$$

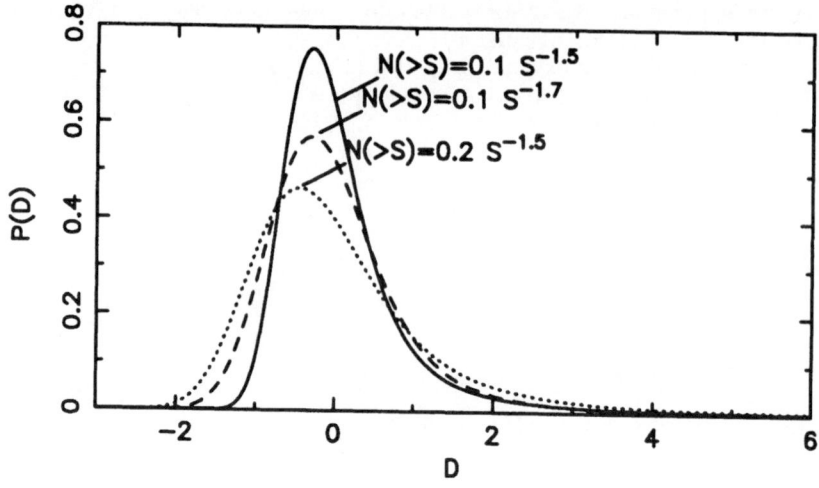

Figure 1. P(D) curves for several flux distributions with no clustering

4.2 UNIFORM DISTRIBUTION OF SOURCES

If sources are not clustered at all, the only non–zero cumulant is the first order one $\langle\langle\rho(\vec{x})\rangle\rangle = \mu^{-1}$ and eq. (24) becomes

$$\varphi_I(\omega) = \pi\alpha^2\Psi(\omega) \equiv \Psi_b(\omega) \tag{25}$$

where

$$\Psi_b(\omega) = \int dS n_b(S)\left[e^{2\pi i\omega S} - 1\right] \tag{26}$$

where $n_b(S)$ is the differential source counts per beam.

In Fig. 1 we show several $P(D)$ curves obtained in this way, for a power law $\log N - \log S$ relation (eq. 7). The width of the $P(D)$ curve grows with K. Deviations from a euclidean form also change the $P(D)$ curve as shown in Fig. 1.

Since this is the classical treatement of confusion noise, we refer to Shafer (1983) for a more complete discussion. We just want to emphasize that the shape of the $P(D)$ curve is dominated by those fluxes for which $N_b(> S) \sim 1$. Fainter (and more numerous) sources produce smaller and smaller gaussian noise and brighter sources only show up in the high D tail. If the beam function is not a top hat, then off–axis bright sources can also influence the central shape of the $P(D)$ curve.

4.3 EXCESS VARIANCE

Source clustering has been usually incorporated to this scheme in the following way. If sources do cluster, it is clear that there will be more fluctuations, which will result in a broadening of the $P(D)$ curve. This broadening is included by convolving the $P(D)$ curve without clustering with a gaussian

$$P_I(I) = \int dI' P_I^{No\,Clus}(I' - I)\eta(I') \tag{27}$$

where $\eta(I)$ is a zero–mean gaussian with standard deviation σ which is called excess variance because it is meant to contain all fluctuations not directly attributable to variations in the number of uniformly distributed sources. The excess variance can be due to several facts: irregularities in the instrument, clustering, etc.. The contribution to σ from clustering is (Barcons & Fabian 1988)

$$\sigma^2 = \langle I \rangle^2 \Xi \tag{28}$$

where

$$\Xi = \frac{1}{(\pi\alpha^2)^2} \int_{|\vec{x}_1|<\alpha} d^2x_1 \int_{|\vec{x}_2|<\alpha} d^2x_2 \xi(\vec{x}_1 - \vec{x}_2) \tag{29}$$

For a gaussian correlation function

$$\xi(\vec{x}) = \xi(0) \exp(-\frac{\vec{x}^2}{2L^2}) \tag{30}$$

we show $\Xi/\xi(0)$ as a function of L in Figure 2.

4.4 GAUSSIAN CLUSTERING

Let us assume that the 2D random field $\rho(\vec{x})$ representing the surface density of sources, is a gaussian one, which means that all cumulants of order greater than 2 are exactly zero. The distribution of galaxies is known to deviate from this model (see, e.g., Peebles 1978) but we can use it as a first approximmation.

In this case, eq. (24) becomes

$$\varphi_I(\omega) = \Psi_b(\omega) + \frac{1}{2}\Xi\Psi_b(\omega)^2 \tag{31}$$

All dependence on clustering is thus contained in the double integral of the source correlation function Ξ. In Figure 3 we show the $P(D)$ curve obtained from this simple gaussian clustering model together with a convolution of the $P(D)$ curve without clustering with the appropriate gaussian. The difference between both $P(D)$ curves depends on the specific parameters of the $\log N - \log S$ relation. The circumstances under which the convolution with a gaussian is a

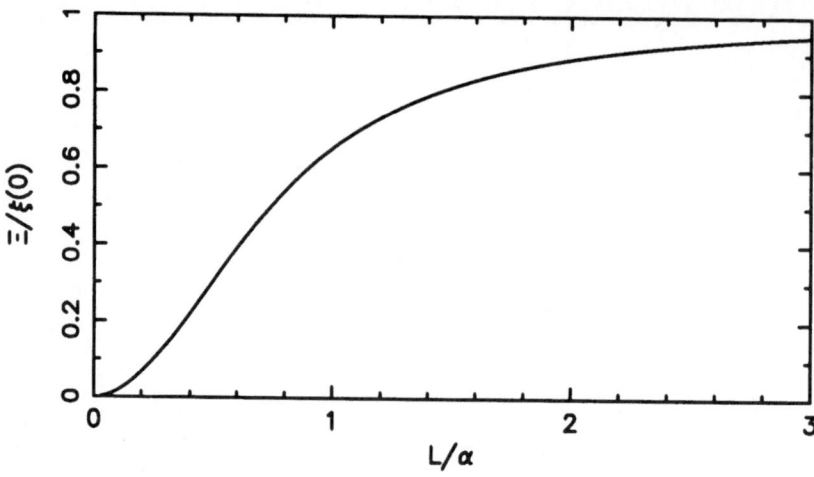

Figure 2. The double integral of the gaussian correlation function given in eq. (30) for different values of the correlation length (see text for details.

good approximmation to gaussian clustering are currently being investigated and will be published elsewhere (Barcons, Toffolatti & Carrera 1991).

4.5 SHOT NOISE CLUSTERING

As an alternative model to source clustering, we assume that there is a uniform distribution of clusters with mean surface density λ clusters per unit solid angle and profile $\eta(\vec{x})$ around the cluster center. This leads to a shot noise (Rice 1945) for the 2D density field $\rho(\vec{x})$. Its cumulants are

$$\langle\langle\rho(\vec{x}_1)\dots\rho(\vec{x}_k)\rangle\rangle = \lambda \int d^2x\,\eta(\vec{x}-\vec{x}_1)\dots\eta(\vec{x}-\vec{x}_k) \tag{32}$$

For the sake of simplicity, we shall take a gaussian cluster profile

$$\eta(\vec{x}) = \frac{\eta_0}{2\pi l^2} \exp(-\frac{\vec{x}^2}{2l^2}) \tag{33}$$

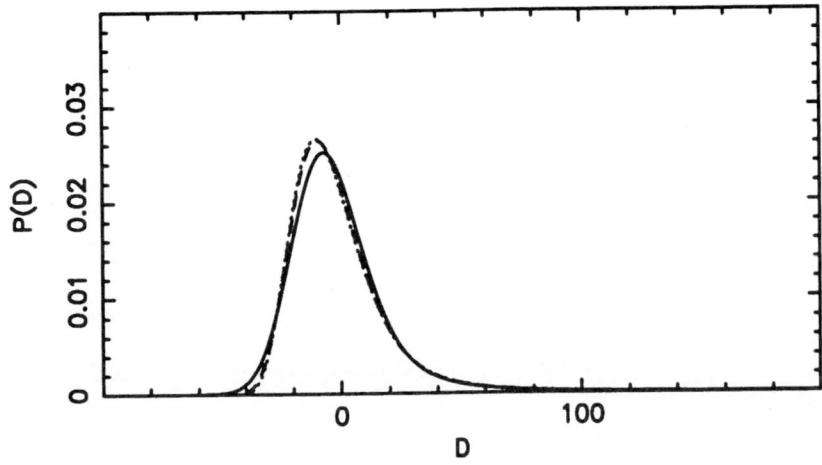

Figure 3. P(D) curves with clustering. The continuous line represents the convolution of the corresponding curve with a gaussian and the dotted one is the gaussian clustering model. Both curves have the same second moment.

which leads to a gaussian two point correlation function (eq. 30) with $L = \sqrt{2}l$ and $\xi(0) = (4\pi\lambda l^2)^{-1}$. Conversely, given the mean surface density of sources and a gaussian correlation function (30) we can take $\lambda = (2\pi L^2 \xi(0))^{-1}$, $l = L/\sqrt{2}$ and $\eta_0 = 2\pi L^2 \xi(0)\mu$ to generate shot noise clustering with the same correlation function.

The function $\varphi_I(\omega)$, as defined in (24), can be written in this case as

$$\varphi_I(\omega) = \lambda \int d^2x \left[e^{2L^2 \xi(0)\Psi_b(\omega)\beta(\vec{x})} - 1 \right] \tag{34}$$

where

$$\beta(\vec{x}) = \frac{1}{2\pi l^2} \int_{|\vec{y}|<\alpha} d^2y \exp\left[-\frac{(\vec{x}-\vec{y})^2}{2l^2} \right] \tag{35}$$

It is actually expected that when the number of sources per beam is < 2, shot noise and gaussian clustering models must converge, since cumulants of order higher than 2 will the be irrelevant.

5. Final comments

I have discussed some recipes on how to use angular variations of integrated backgrounds to probe deep source counts and clustering properties. The theoretically expected ACFs and $P(D)$s can be evaluated with the use of Fast Fourier Transforms and standard integration routines, and therefore require only modest amounts of CPU, to be compared with much more CPU–expensive Monte Carlo techniques.

Clustering on any scale can be taken into account when computing the expected $P(D)$ curve. If this is not done, or it is not done properly, the $\log N - \log S$ curve inferred from fluctuation analyses can be overestimated (i.e., K will be overestimated). This might contribute to source count mismatches between direct detections and fluctuation analyses known to happen in the X–ray (e.g. Boldt 1988) and radio (Franceschini et al. 1989) bands at very deep levels.

Apologies and Acknowledgements

I must really apologize for having used so much equations for so few (if any) results. However, this is meant to be a lecture and the background to which I have been applying these techniques and for which I can present results (the X–ray background) is out of the scope of this conference. I really hope that the techniques presented here can be of some use towards the understanding of the submillimeter and infrared sky. Anyway, I would like to thank my collaborators F.J. Carrera, A.C. Fabian and L. Toffolatti for scientific help and the CICYT for financial support.

References

Barcons, X. & Fabian, A.C., 1988. Mon. Not. R. ast. Soc., **230**, 189
Barcons, X. & Fabian, A.C., 1989. Mon. Not. R. ast. Soc., **237**, 119
Barcons, X., Fabian, A.C. & Rees, M.J., 1991. Nature, in the press
Barcons, X., Toffolatti, L. & Carrera, F.J., 1991. In preparation
Boldt, E.A., 1988. In: Large Scale Structure and Motions in the Universe, eds. Mezzetti, M. et al.. Kluwer, Dordrecht
Carrera, F.J., et al., 1991. Mon. Not. R. ast. Soc., in the press
Dautcourt, G., 1977. Astr. Nachr., **298**, 141
De Zotti, G., Persic, M., Franceschini, A., Danese, L., Palumbo, G.G.C., Boldt, E.A. & Marshall, F.E., 1990. Ap. J., **351**, 22
Franceschini A., Toffolatti, L., Danese, L. & De Zotti, G., 1989. Ap. J., **344**, 35
Peebles, P.J.E., 1980. The Large Scale Structure of the Universe, Priceton University Press.
Rice, S.O., 1945. Bell Syst. Techn., **24**, 46
Shafer, R.A., 1983. Ph. D. Thesis. University of Maryland.
Soltan, A., 1991. Mon. Not. R. ast. Soc., in the press
Van Kampen, N.G., 1981. Stochastic Processes in Physics and Chemistry, North Holland, Amsterdam.

The thermal history of the cosmological gas

A.BLANCHARD
D.A.E.C.
Observatoire de Meudon
Place Janssen
92 190 Meudon
&
Université de Paris VII
75 005 Paris
France

ABSTRACT. The aim of this lecture is to briefly review the thermal history of the baryonic content of the universe after the epoch $Z \approx 1400$, at which the primeval plasma recombines in the standard model. The different processes that might determine its physical state are presented as well as the status of our knowledge of its properties at different epoch and the observational constraints that can be set. The uncertainty on this history is large because of the very poor constraints on possible reheating mechanisms. Nevertheless, such processes should have existed and they might be a fundamental key of galaxy formation.

1   Introduction

After the quark phase transition the baryonic content of the universe is turn in protons and neutrons, the electric neutrality ensures[1] the presence of electrons and protons in exactly the same amount:

$$n_e = n_p$$

Later on, matter and radiation are in a thermal equilibrium and consequently do have the same temperature:

$$T_m = T_\gamma$$

The most important event of the gas history after that period is probably the primordial nucleosynthesis. The hot Big Bang picture has been originally proposed in order to explain

---

[1]Actually the neutrality of the universe is postulated, but the observed rate of expansion of the universe shows that any possible charge asymmetry should be very small.

M. Signore and C. Dupraz (et al.), The Infrared and Submillimetre Sky after COBE, 201–211.
© 1992 *Kluwer Academic Publishers.*

the elements abundances (Alpher et al., 1948). The primeval nucleosynthesis remains one of the very great success of this model. The discovery of the cosmological radiation and the remarquable blackbody shape of its spectrum certainly guarantee that no model will be able to compete with this picture for a long time... The primeval nucleosynthesis determined the composition of the cosmological gas, that is probably not altered until the stars will produce a small amount of heavier elements[2]. As it is well known, around 25% of the gas is turn in Helium, while others elements are produced in much smaller quantity. The observed abundances of these elements, specially the lithium, allow one to determine the density of the baryonic content of the universe, wich is now believed to be known with a 50% accuracy (Olive et al., 1990):

$$\rho_b \approx 3.10^{-31} \text{ g cm}^{-3} \tag{1}$$

corresponding to a number density

$$n = \frac{\rho_b}{m_p} \approx 1 \cdot 8 \times 10^{-7} \text{cm}^{-3}$$

It follows that the main uncertainty on the cosmological density parameter of the baryonic gas lies in the uncertainty in the Hubble constant H_0 ($= 100h\text{km/s/Mpc}$):

$$\Omega_b \approx 0 \cdot 015 h^{-2} \tag{2}$$

This faces us on with an important problem of modern cosmology: the *observed* amount of baryons (in stars and cold gas in galaxies) is appreciably smaller than the above number:

$$\Omega_* = \frac{(M/L)_{*+gas}}{1500h} \approx 0 \cdot 001 - .01 h^{-1} \tag{3}$$

this means that most of the baryons are in a dark form. One could advocate that if the Hubble constant is close to 100 km/s/Mpc, the discrepancy is not significant. However, with such a small value of H_0 , one would have to abandon one or several of the following basic "facts" of cosmology:

- Primeval nucleosynthesis is correct.

- Dynamical measurements of the density of the universe show that $\Omega_0 \geq 0 \cdot 1$.

- The age of the universe t_0 is ≈ 14 Gyr.

- The cosmological constant Λ is zero.

- The dark matter is non-baryonic.

Actually, the most conservative point of vue is to assume that $\Omega_0 = \Omega_b \approx 0 \cdot 1, h \approx 0 \cdot 5, t_0 = 20$ Gyr (but this model is only weakly consistent with observations), the dark

---

[2]Of course this doesn't necessary mean that these heavier elements are not important: for instance, they allow for our existence, and, more directly related to this lecture, they might change the thermal history of the gas, and provide fruitful constraints on it.

matter being then baryonic, as very probably in the form of unseen stars (brown dwarfs or "jupiters" being the most appealing candidates).

However, Inflation has provided a theoretical argument for $\Omega_0 = 1$. Actually, this value is probably the only one that could easily emerge from a theory of the very early universe. In addition there exist now few significant observations that comfort this theoretical prejudice (Bertschinger, 1990, Cowie et al., 1991). In this model a value $h = 0.5$ is almost unavoidable in order to escape a dramatic age problem.

Therefore, in each of the above models, which are able to reproduce reasonably well the observations, the main component of the baryonic component is dark, and this strongly advocates to study the detail of its thermal history. Notice that accordingly to primeval nucleosynthesis the value of Ω_b should be of the order of 0.05 in both models.

After the nucleosynthesis, the temperature of the universe decreases. We will now pay attention to what's happen when the temperature drops down to 4000K. At that time, the electrons begin to recombine with protons to form H atoms, but the ionization is still high enough that photons diffuse on free electrons.

2   Basic elements

2.1   time scale consideration

In the following we will meet several processes that may or may not play in important role in the evolution of the species which are involved. The first consideration to have for any possible process is to examine whether the are efficient or not. We will take as a fist example the case of the diffusion of photons ($Z = 1100$) on the free electrons of the universe.

The fundamental process to consider is the Thomson diffusion. The cross-section of this process for low energy photon ($h\nu \ll mc^2$) is $\sigma_T = 6.6 \times 10^{-25}$ cm $^{-2}$ (the actual cross section should take into account quantum correction, the Klein-Nishina formula should then be taken, but is differs only in the relativistic case). The time scale for the diffusion on free electrons therefore is :

$$t_T \;=\; \frac{1}{c\sigma_T n_e}$$

This diffusion process will be important only if its time scale is shorter than the age of the universe:

$$t \;=\; \frac{2}{3H_0}(1+z)^{-3/2} \;\approx\; 2h^{-1}10^{17}(1+z)^{-3/2}\,\mathrm{s}$$

The electron density is:

$$n_e \;\approx\; \chi n \frac{\rho_b}{m_p} \;=\; \chi 1.8 \times 10^{-7}(1+z)^3\,\mathrm{cm}^{-3}$$

where χ is the ionization fraction (for simplicity we neglected the presence of helium). In the standard picture, the surface of last scattering is located at a redshift close to 1100. The opacity of the universe to Thomson scattering is:

$$\tau_T = \int_0^{+\infty} c\sigma_T n_e \, dt = \int_0^{+\infty} 0 \cdot 01 h^{-1} \chi(z) \frac{(1+z)}{(1+\Omega_0 z)^{1/2}} dz \tag{4}$$

the fractionnal ionization is then found to be $\chi \approx 4 \times 10^{-3}$ by setting $\tau_T = 1$. A comprehensive discussion of the standard recombination could be found in Peebles (1968) pionering work. A more recent analysis, with special care to the last scattering surface is given by Jones and Wyse (1985).

2.2   The neutral period.

After that period matter and radiation are not in contact any more and they evolve independently. The temperature of the photon background evolves as

$$T_\gamma = 2 \cdot 735 \times (1+z)$$

Notice that increasing with redshift means *decreasing* with time. The temperature of the gas can be found from thermodynamic. Let u_m be the internal energy of a given comoving volume, and T_m its temperature:

$$u_m = \frac{3}{2} n V k T_m \tag{5}$$

where n is the total number density of particles. As the universe expands, the pressure works and

$$du_m = -p \, dV = -nk T_m \, dV$$

leading to

$$\frac{dT_m}{T_m} = -\frac{2}{3} \frac{dV}{V}$$

therefore $T_m \propto V^{-2/3} \propto (1+z)^2$. This is the adiabatic cooling. Notice that we have implicitly assumed that there was no other heating or cooling mechanisms. We have also apply the standard laws of thermodynamic. These are relevant only if we are in an equilibrium situation, that is that collisions are efficient enough. It is important therefore to verify this hypothesis. The time scale for collisions between neutral hydrogen atoms is :

$$t_c \approx \frac{1}{n\sigma v}$$

where v is the thermal velocity, $\sigma \approx \pi a_0^2 \approx 10^{-16} \mathrm{cm}^2$. The ratio

$$\frac{t_c}{t} \approx 500 h \, (1+z)^{-5/2}$$

therefore, at small redshift, the gas, if completely neutral, is not in thermal equilibrium any more. However, we can still define its temperature by the relation:

$$\frac{3}{2}kT_m = \frac{1}{2}mv^2$$

the velocity of a freely moving particle in an expanding universe evolves accordingly to

$$v = v_0 \times (1 + z)$$

therefore the temperature decreases in exactly the same way then in the equilibrium case:

$$T_m = T_0 \times (1 + z)^2 \tag{6}$$

2.3 A short comparison with observation

Actually, the picture we have describe above is in complete desagreement with two basic observational facts: firstly the dark matter is clumped (galaxies, clusters,...) and there is no reason why most of the baryons will not be clumped in the dark matter potentials. The second problem comes from the Gunn-Peterson test (Gunn and Peterson, 1967). Observations of distant QSO's reveal no significant absorption even for Lyman α photons. This implies a very ionized medium. The cross section of such photons is $\sigma \approx 7 \cdot 3 \times 10^{-18} \mathrm{cm}^2 (Ry/h\nu)^3$. That is

$$\tau_{GP}(z) \approx 1 \cdot 5 h^{-1} 10^4 (1 - \chi) \frac{(1 + z) \Delta z}{(1 + \Omega_0 z)^{1/2}} dz$$

the measurements imply that $\tau_{GP} \leq 0.05$ at $z \approx 2.6$ (Steidel and Sargent, 1987). Therefore, the neutral fraction is smaller than 4.10^{-6}. Strictly speaking it is possible that the IGM does not exist at all, however such a possibility is very doubtful : as we have seen, the amount of observed baryons is much smaller than predict by primeval nucleosynthesis. Even if the dark matter is baryonic, there should remains some baryons left over of the process of galaxy formation. Further more, a robust argument for the existence of the substantial fraction of gas in the IGM comes from clusters: we see in clusters a gas mass which is several times the mass in stars. Assuming that the $(M/L)_{\mathrm{gas}}$ is universal ($\approx$10–60 from clusters), it does imply the existence of an IGM containing several times the mass seen in stars and and might contain all the baryons predicted by the primeval nucleosynthesis (from the Coma cluster one would conclude that $\Omega_{IGM} \approx 0.04$). Ionization of the IGM need at some level energy injection in the medium (reheating), and the equation 5 has to be modified accordingly:

$$u_m = \frac{3}{2} nVkT_m + \epsilon_h V dt - \epsilon_c V dt \tag{7}$$

where ϵ_h and ϵ_c are respectively the rate of energy input and the total cooling rate of the gas. This leaves us with considerable amount of freedom, because of a variety of possible processes: the interaction with the CBR is able to cool or heat the gas, other background contributions might heat the gas, bulk flows and cosmic rays produce by stars might reinjected energy in the IGM, and at the same time high temperature gas is able to distort the CBR and to produced others backgrounds. In addition the chemical composition modifies the cooling rates, and we have no serious reasons to treat this problem as homogeneous...

3 Cooling and heating mechanisms

Here, the different processes that could be taken into account will be briefly presented, but a full description of the possible scenarios will not be given. Peebles (1971) early discussed these processes in the cosmological context. Basic elements can be found in Rybicki and Lightman 1979. The appendix of Stebbins and Silk (1979) gave the relevant formula to deal with this question, and has been widely used in the preparation of this lecture. No specific heating model will be discussed here. Such heating sources can be astrophysical (quasars, AGN, early galaxies, population III stars), see for instance Couchman (1985) or have a more exotic origin like decaying particle (Stebbins and Silk, 1979, Asselin et al., 1987, Sciama, 1990).
I will now examine the consequences of the existence of cooling or heating mechanism on the evolution of the gas. Hereafter, we assume the medium to be homogeneous.

3.1 time evolution

Eq. 7 can be written in the form or

$$\frac{du_m}{u_m} = -\frac{4}{3}\frac{dt}{t} \pm \frac{dt}{\tau} \tag{8}$$

where

$$\tau = \left|\frac{u_m}{\epsilon}\right|$$

is the cooling time $(-)$ or heating time $(+)$ depending on the process under consideration. Again it is rather clear that only processes that have a characteristic time shorter than the age of the universe are to be taken into account. Generally, the ratio τ/t will depend on t as a power law:

$$\frac{\tau}{t} = \left(\frac{t}{t_e}\right)^\alpha$$

neglecting the adiabatic cooling in the above equation, the energy evolution can be found :

$$u_m \propto exp\left(\pm\frac{1}{\alpha}(t/t_e)^\alpha\right)$$

which means that the temperature will relax very rapidly to an equilibrium value. Therefore if the heating is a short burst of energy which is instantanously injected in the medium, and if an efficient cooling is possible the temperature will go down very fast. In order to maintain the IGM "hot", it is necessary to have source injection which provide energy over more than an expansion time. It is then not necessary to integrate the differential equations (which might be easily unstable). It is rather simpler to evaluate the final condition of the system by looking for the solution of the implicit equation:

$$\epsilon_h\,(n,T) - \epsilon_c\,(n,T) \;=\; 0 \tag{9}$$

which is satisfied when the equilibrium is achieved.

3.2 Inverse Compton cooling

When photons diffuse on electrons by Thomson scattering the energy exchanged is of the order of

$$\Delta E \;\approx\; \frac{h\nu}{m_e c^2}\,(4T_e - h\nu)$$

The rate of energy transfer from the electrons to the photons gas is

$$\epsilon_{cc} \;=\; \left(\frac{4kT}{m_e c^2}\right) c\sigma_T n_e a T_\gamma^4$$

The associated time scale therefore is

$$t_{cc} \;\approx\; \frac{\frac{3}{2}nkT}{\epsilon_{cc}} = \frac{3}{4}\frac{mc}{\sigma_T a T_\gamma^4}$$

leading to

$$\frac{t_{cc}}{t_H} \;=\; \frac{180}{(1+Z)^{5/2}} \quad \text{s}$$

Therefore at redshift greater than 10, the Compton mechanism will efficiently cool the gas. Is easy to verify how efficient this is if the gas is heated at $Z \approx 100$, at a redshift of 10 the temperature is damped by a factor of 10^{-65} ... The energy lost by the gas is transfered to the CBR, and produce distortion of the spectrum (the relaxation toward the planckian shape can not be achieved anymore – see the lecture by P.Salati in this volume). The distortion is characterised by the y_c parameter

$$dy_c \;=\; \frac{4kT}{m_e c^2}\sigma_T n_e\,\frac{c}{H_0}\,\frac{dZ}{(1+Z)^2\,(1+\Omega_0 Z)^{1/2}}$$

Numerically

$$y_c \;=\; 4\times 10^{-8}\,(1+Z_h)^{7/2}\,T_5 \quad \text{for adiabatic cooling}$$
$$y_c \;=\; 10^{-7}\,(1+Z_h)^{3/2}\,T_5 \qquad \text{for T = cste}$$

Z_h being the redshift at which the gas is heated, and T_5 the temperature in unit of 10^5 K. The remarquable blackbody shape of the CBR radiation (Mather et al., 1990) ensures that the temperature of the IGM was not greater than $\approx 10^8$ K at a redshift of 5.

3.3 Bremsstrahlung cooling

Hot electron gas radiates energy via Bremsstrahlung emission

$$\epsilon_B = 1 \cdot 4 \times 10^{-27} T^{1/2} n_e^2 \bar{g} \qquad \text{erg/cm}^3/\text{s}$$

where $\bar{g}$ is the velocity average Gount factor. The associated time scale is

$$t_{cB} \approx 4 \times 10^{20} \frac{T_5^{1/2}}{(1+Z)^3} \qquad \text{s}$$

3.4 Recombinative cooling

Recombinations on exited ground states lead to the emission of a photon, and the corresponding energy is lost for the gas. An accurate formula is given by

$$\epsilon_R = \frac{64}{3} \sqrt{\frac{\pi}{3}} \alpha^4 a_0^2 c \sqrt{\frac{kT}{R_y}} n_e \left[-0 \cdot 0713 + \frac{1}{2} log \left(\frac{R_y}{kT} \right) + 0 \cdot 6 \sqrt[3]{\frac{kT}{R_y}} \right]$$

A good approximation however consists to assume that to each recombination corresponds an energy loss which is equal to one Ry when the mean cinetical energy of a particle of the gas is higher than $13 \cdot 6\text{eV}$, and to kT when the cinetical energy is smaller. The recombination time is equal to

$$t_r = \frac{1}{n_e (\sigma v)_{rec}} = \frac{\sqrt{3\pi}}{n_e 64\pi\alpha^4 a_0^2 c} \left(\frac{kT}{Ry} \right)^{2/3} = 5 \times 10^{19} \frac{T_5^{2/3}}{\chi (1+z)^3} \qquad \text{s}$$

The energy lost is therefore approximated by

$$\frac{dE}{E} \approx \frac{dN_r}{kT} \left(\frac{1}{Ry} + \frac{1}{kT} \right) = \frac{dt}{t_{cr}}$$

leading to

$$t_{cr} \approx \frac{t_r}{1 + \frac{Ry}{kT}}$$

where t_r is the recombination time.

3.5 Collisional exitation cooling

When a free electron undergoes a collision with a neutral atom, the atom can be exited, and then de-exited radiatively to the ground state. The energy lost by this process is

$$\epsilon_{cec} = 7 \cdot 5 \times 10^{-19} n_e^2 \chi (1 - \chi) e^{-Ry/kT}$$

The corresponding cooling time is

$$t_{cec} = 1 \cdot 5 \times 10^{14} \frac{1 + \chi}{\chi (1 - \chi)(1 + z)^3} T_5 e^{+Ry/kT} \qquad \text{s}$$

3.6 The cooling function.

When one is dealing with a cosmological problem in which cooling and heating processes compete, it is possible to use the cooling function to estimate the energy loss of the gas. This function is defined as :

$$\epsilon = n_e^2 \Lambda(T)$$

and the function $\Lambda(T)$, taking all the above radiation mechanisms into account, has been tabulated. One should keep in mind that an equilibrium situation is assumed, which usually needs the density greater than the homogeneous value, and the absence of significant photoionization mechanism. In that case the global cooling time is just

$$t_c \approx \frac{3}{2} \frac{kT}{n_e \Lambda(T)}$$

3.7 Ionization equilibrium

Actually, eq. 9 involves the ionization fraction, so that there are two unknown quantities T and χ. Therefore, a second equation is need. Again we can write down the time evolution for the ionization fraction, but as long as the relevant timescale are shorter then the age of the universe, we can assume an equilibrium situation for which the collisional ionization time t_I is equal to the recombination time t_R From

$$t_I = \frac{1}{n_{HI}\overline{\sigma_c v}} = \frac{1}{n_{HI}7 \cdot 2a_0^2} \left(\frac{m_e}{kT}\right)^{1/2} e^{Ry/kT} \times$$

$$\approx \frac{2 \cdot 2 \times 10^{14}}{(1-\chi)(1+z)^3} T_5^{-1/2} e^{Ry/kT} \qquad s$$

and

$$t_R = \frac{1}{n_e \overline{\sigma_r v}} = \frac{3\sqrt{3\pi}}{n_e c\alpha^4 a_0^2} \left(\frac{kT}{Ry}\right)^{2/3} \times$$

$$\approx \frac{5 \cdot 8 \times 10^{19}}{\chi(1+z)^3} T_5^{2/3} \qquad s$$

Therefore the ionization fraction depends only on the temperature

$$\chi \approx \frac{1}{1 + 2 \cdot 8 \times 10^{-6} T_5^{-7/6} e^{Ry/kT}}$$

therefore to satisfy the Gunn-Peterson test the temperature of the IGM needs to be of the order of $2 \, 10^5$ K. Actually the equilibrium situation is not achieved at redshift $z \leq 5$ as the recombination time is larger than the age of the universe. An other approximation in this calculation is that we neglect photoionization. However photoionization is an efficient way to ionize the medium because of the long recombination time. If any energy heat the cosmological gas, a further equation for the ionization has to be solved

$$d(\chi n) = n_{HI}\frac{dt}{t_I} - n_e\frac{dt}{t_R} + n_X\frac{dt}{t_i}$$

When the gas is photoionized at some redshift, if its temperature reached a value above few 10^3K, it remains ionized because the recombination time is larger than the age of the universe. If the IGM is photoionized, its temperature could be as low as 10^4K(Barcons, X., Fabian, A.C., Rees, M.J., 1991) quasars are the only well-known source of UV photons but they are not numerous enough to achieve the observed level of ionization (Shapiro and Giroux, 1987). The present constrainst do not even prove that the universe ever recombine (Bartlett, J., Stebbins, A., 1991). It is therefore not known whether this level has been achieved by energy injection or by photoionization by unknown sources. An interesting proposition is that the hard X-ray background could heat the gas at a temperature close 10^5 K (Collin-Souffrin, 1991), but this process seems not to be able to explain the Gunn-Peterson test (Shapiro, 1991).

4 The history of the cosmological gas and galaxy formation

Only baryons are observable in the universe. However dynamical measurements show that the main component of the mass is dark. Whether this dark matter is baryonic or not, it should be in a collision-less form. Thus for a long time cosmologists did try to understand galaxy formation by understanding the properties of this collisionless component, evolving under its own gravity. This is the standard gravitational instability picture (Peebles, 1980). Numerical simulations has been widely used to this purpose (White, 1991). Nevertheless, it is rather clear that our understanding of galaxy formation need to a description of the baryonic content of the universe (Binney, 1977, Rees and Ostriker, 1977, Silk, 1977). Taking into account the dissipative processes that determine the cosmological evolution of the gas is a rather despaired tentative, but cosmologists have always been unrealistically optimistic...

The concept of "biasing" has revived the attention on dissipative processes in galaxy formation: galaxies may not be random sub-set of the total content of the universe, and their distribution may not reflect the overall distribution of universe. The recent studies in which cooling processes are taking into account has unambigously showed that star formation from cool gas should have been inhibited in order to prevent an excess quantity of stars by now (Blanchard et al.;1990, 1991; White, Frenk, 1991). However, a general discussion of this problem is out of the scope of this lecture.

Finally, I mentionned the fact that this problem begins to be investigated with the numerical simulations that include a gas component (Calrlberg, Couchman, 1987). These simulations have confirmed that the excess of cooling advocates for the existence of efficient feedback mechanism (Klypin, Kates, 1991). An other consequences is that substantial fraction of gas is turn in hot gas and may lead to observable consequences (Cen et al., 1990).

References

Alpher, R., Bethe, H., Gamov, G.: 1948, *Phys. Rev.*, **73**, 803.

Asselin, X., Girardi, G., Salati, P., Blanchard, A. : 1988, *Nucl. Phys.* **B310**, 669.

Barcons, X., Fabian, A.C., Rees, M.J.: 1991, *Nature*, **350**, 685.

Bartlett, J., Stebbins, A.: 1991, *Ap.J.*, to be published.

Bertschinger, E.: 1990, *Rencontres de Moriond in Astrophysics*, p.411, Ed. Alimi et al., Editions Frontières, Gif-sur-Yvette.

Binney, J.: 1977, *Ap.J.*, **215** , 483.

Blanchard, A., Valls-Gabaud, D., Mamon, G. : 1990, *Rencontres de Moriond in Astrophysics*, p. 403, Ed. Alimi et al., Editions Frontières, Gif-sur-Yvette.

Blanchard, A., Valls-Gabaud, D., Mamon, G. : 1991, *Astron. Astrophys.*, to be published.

Carlberg, R., Couchman, H.: 1989, *Ap.J.*, **340** , 47.

Cen, R.Y., R.Y., Jameson, A., Liu, F. Ostriker, J.P.: 1990, *Ap.J.Lett.*, **362**, L41.

Collin-Souffrin, S.: 1991, *Astron. Astrophys.*, to be published.

Couchman, H.: 1985, *M.N.R.A.S.*, **214**, 137.

Cowie, L.: 1991, preprint.

Gunn, J.E., Peterson, B.A.: 1965, *Ap.J.*, **142** , 1633.

Jones, B.J.T., Wyse, R.F.G.: 1985, *Astron. Astrophys.Lett.*, **149**, 144.

Klypin, A.A., Kates, R.E.: 1991, *préprint* MPA 558.

Mather, J.C., et al.: 1990, *Ap.J.Lett.*, **354** , L37.

Olive, K.A., Schramm, D.N., Steigman, G., Walker, T.P.: 1990, *Physics Letters*, **B236**, 454.

Peebles, P.J.E.: 1968, *Ap.J.*, **153** , 1.

Peebles, P.J.E.: 1971, *Physical Cosmology*, Princeton University Press.

Peebles, P.J.E.: 1980, *The Large-scale Structure of the Universe*, Princeton University Press.

Rees, M.J., Ostriker, J.P.: 1977, *M.N.R.A.S.*, **213**, 75p.

Rybicki, G., Lightman, A.: 1979, *Radiative Processes in Astrophysics*, John Wiley and Sons, NY.

Sciama, D.W.: 1990, *M.N.R.A.S.*, **246**, 191.

Shapiro, P.R.: 1991, *Rencontres de Moriond in Astrophysics*, Ed. B.Rocca et al., Editions Frontières, Gif-sur-Yvette.

Shapiro, P.R., Giroux, M.L.: 1987, *Ap.J.*, **321** , L107.

Sherman, R.D.: 1979, *Ap.J.*, **232**, 1.

Silk, J.: 1977, *Ap.J.*, **211** , 638.

Stebbins, A., Silk, J.: 1986, *Ap.J.*, **300**, 1.

Steidel, S., Sargent, W.L.W. : 1987, *Ap.J.*, **318**, L11.

White: 1991, *Rencontres de Blois: Physical Cosmology*, Ed. A.Blanchard et al., Editions Frontières, Gif-sur-Yvette.

White, S.D.M., Frenk, C.S.: *Ap.J.*, to be published.

White, S.D.M., Rees, M.J.: 1978, *M.N.R.A.S.*, **183**, 341.

SOURCES OF COSMIC INFRARED-SUBMILLIMETRE BACKGROUND RADIATION

B.J.CARR
School of Mathematical Sciences
Queen Mary & Westfield College
Mile End Road
London E1 4NS

ABSTRACT. In this lecture I first review attempts to detect extragalactic background radiation in the infrared and submillimetre bands. Although such observations are exceedingly difficult, and one can currently only place upper limits on the intensity of such a background, the prospects of a detection have greatly improved with COBE. I will then discuss the possible sources of a such background. In the absence of dust, one could expect many astrophysical sources in the period after decoupling to generate a *near*-IR background: in particular, primeval galaxies, pregalactic stars, cosmic explosions and decaying elementary particles. In some circumstances, these backgrounds will have been reprocessed into the *far*-IR and submillimetre bands by cosmological dust, with a spectrum which depends only weakly on the specific source. In this case, an important signature of the scenario will be the anisotropies expected in the background as a result of the dust clumpiness and temperature variations and I discuss the chances of detecting these.

1. OBSERVATIONS OF INFRARED/SUBMILLIMETRE BACKGROUND RADIATION

For the purposes of this lecture, I will regard a background as being cosmological if it derives from unresolved sources at a cosmological redshift ($z>1$). The detection of any background is difficult since one needs an absolute calibration. For a cosmological background, the problem is compounded because there are a large number of *local* backgrounds which have to be subtracted before one can extract the high redshift contribution [1]. In the IR and submillimetre bands, our own atmosphere provides such a background and this necessitates non-ground-based observations. There are also backgrounds due to scattered zodiacal light (ZL) below 3μ, interplanetary dust (IPD) at 3-50μ, and interstellar dust (ISD) at 50-400μ. The cosmic background radiation (CBR) itself hides backgrounds above 400μ, so - despite its immense importance as a probe of the early Universe - it is a nuisance in some respects!

Estimates of these competing backgrounds are shown in Figure (1), though some of them are rather uncertain. One sees that there are minima at around 4μ, 100μ and 400μ, so these are the best "windows" in

M. Signore and C. Dupraz (et al.), The Infrared and Submillimetre Sky after COBE, 213–230.
© 1992 *Kluwer Academic Publishers*.

which to search for an extragalactic background. Although positive detections have been claimed in all of these windows (eg. the Nagoya-Berkeley excess at 700μ [2], the Rowan-Robinson background at 100μ [3], and the Matsumoto et al. background at 2μ [4]), as have distortions near the CBR peak [5], none of these has been subsequently confirmed and currently all one can claim is upper limits on the background radiation density in various bands. Although COBE is much more sensitive than other instruments in the IR and submillimetre bands, its ability to detect a cosmological background is still constrained by the precision with which one can model local emission.

Note that there are, in principle, several ways in which cosmological backgrounds can be differentiated from local ones. Firstly, one can use the Sunyaev-Zeldovich effect [6]: radiation passing through hot gas in any intervening cluster will have its spectrum distorted [7], with a deficit longward and an excess shortward of the spectral peak. If this effect is observed, it implies that the background must have originated at a higher redshift than the cluster. So far, however, it has only been detected for the CBR itself [8]. Secondly, one can use the fact that the peculiar motion of the Sun relative to the cosmological rest frame should induce a dipole anisotropy in any cosmological background. It should also modify the spectral index of the background [9]. Again, this effect has only been detected for the CBR [10] but it has already been used to place upper limits on the far-IR background [11].

The upper limits on the background light density in various wavebands are conveniently expressed in units of the critical density ρ_{crit} by defining a quantity

$$\Omega_R(\lambda) \equiv \frac{4\pi\nu I(\nu)}{c^3 \rho_{crit}} = 7\times10^{-7} h^{-2} \left[\frac{\lambda}{100\mu}\right]^{-1} \left[\frac{I(\nu)}{MJy/sr}\right] \tag{1}$$

where h is the Hubble parameter in units of 100 km/s/Mpc. For comparison, the CBR peaks at $\lambda_{peak}=1400\mu$ with a density $\Omega_R(\lambda_{peak})= 1.8\times10^{-5} h^{-2}$. The limits are summarized in Figure (1). The current FIRAS results [12] imply that any excess background must have an intensity less than 1% of the peak CBR intensity over the range $500-5000\mu$. This implies

$$\Omega_R(\lambda) < 2\times10^{-7} h^{-2}(\lambda/\lambda_{peak})^{-1} \tag{2}$$

although the limit could soon be strengthened by a factor of 10. In particular, this disproves the Nagoya-Berkeley claim [2] of an excess at 700μ. The DIRBE results at the south ecliptic pole [13] give upper limits in the J, K, L, M, 12μ, 25μ, 60μ, 100μ, $120-200\mu$ and $200-300\mu$ bands. However, the limits indicated in Figure (1) are very conservative since they do not include any subtraction for foreground emission.

Stronger but less definite constraints in the far-IR band come from the most recent analysis [14] of the Nagoya-Berkeley data: careful modelling of the interstellar and interplanetary dust contributions gives $\Omega_R(275\mu)<7\times10^{-7} h^{-2}$, $\Omega_R(135\mu)=(1.4\pm0.4)\times10^{-6} h^{-2}$ and $\Omega_R(100\mu)<8\times10^{-7} h^{-2}$. The 135μ result may contain an isotropic component, although it would not necessarily be extragalactic. Some of the IRAS limits are also shown in Figure (1). Although the original claim |3| for the detection of a background at 100μ with $\Omega_R(100\mu)=3\times10^{-6}h^{-2}$ is inconsistent with both the Nagoya-Berkeley and DIRBE results, a combination of the DIRBE and IRAS data may allow better 60μ and 100μ limits [15] than indicated in Figure (1). However, these are not shown because of the uncertainties.

Both the DIRBE and IRAS limits are of order $10^{-4}h^{-2}$ at 12μ and 25μ. They are very weak there because the interplanetary dust emission is so large. Indeed it may never be possible to get good cosmological light limits in this waveband. In the near-IR, the rocket experiment of Matsumoto et al. [4] gave a limit $\Omega_R(1-5\mu)<3\times10^{-5}h^{-2}$, with a possible detection of a "line" at 2.2μ with $\Omega_R(2.2\mu)=3\times10^{-6}h^{-2}$. However, this claim was always controversial and it seems to be disproved by the recent limits of Noda et al. [16], also shown in Figure (1).

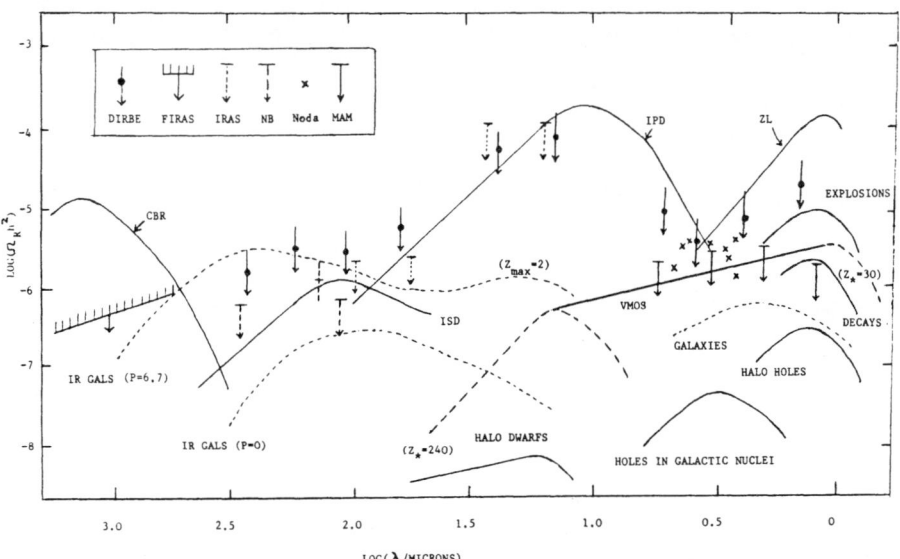

Figure (1). This compares the observational constraints on the cosmological background radiation density from DIRBE, FIRAS, IRAS, Nagoya-Berkeley (NB), Matsumoto et al. (MAM) and Noda et al. with the CBR, local foregrounds and some predicted extragalactic backgrounds.

2. BACKGROUND RADIATION FROM GALAXIES

One inevitable source of background radiation, and the one which has been most studied, is galaxies. Indeed it is possible that galaxies provide the dominant background in both the near-IR and far-IR, so it is important to understand this contribution if one is to discriminate it from other cosmological backgrounds. One can estimate the total energy in this background from metallicity arguments but to obtain detailed spectra more precise models are required. The usual procedure is to extrapolate the luminosity of presently observed galaxies back to high redshifts, although this would not necessarily include the contribution from the burst of star formation associated with primeval galaxies [17].

The *near*-IR background is associated with the redshifted emission of normal (non-dusty) galaxies. In order to calculate this [18], one first needs the evolving spectrum for a *single* galaxy and this requires assumptions about the stellar mass function, the star formation rate and stellar evolution. To get the *total* background, one must then integrate over redshift for the different populations of galaxies and different cosmological models [19-23]. In particular, one must assume values for the deceleration parameter q_0 and the galaxy formation redshift z_G. As an example, predictions for the maximum and minimum evolution cases of Yoshii & Takahara [20] are shown in Figure (2), adapted from Noda et al. [16], along with some of the near-IR observational upper limits and the minimum background associated with the deep galaxy counts of Tyson [24] and Cowie et al. [25]. The minimum evolution prediction is also shown in Figure (1). The important message of Figure (2) is that the theoretical predictions are only slightly below the upper limits.

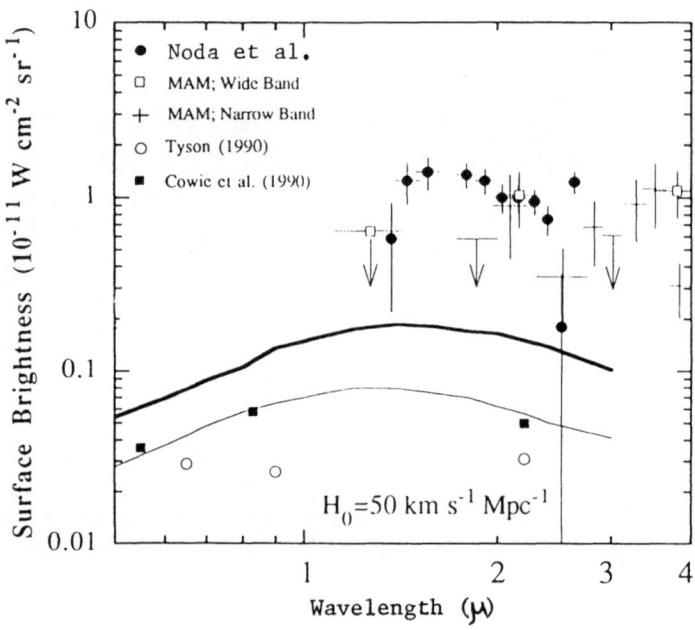

Figure (2). A comparison of the near-IR upper limits from Matsumoto et al. and Noda et al. with the predicted galactic background.

Many galaxies are dusty and these will produce a *far*-IR background. There are three types of IR emission: (i) the radiation from dusty disks; (ii) the "starburst" radiation coming from dust clouds; and (iii) the radiation from quasars and active galactic nuclei. In each case the associated density is very uncertain because of unknown evolutionary factors. If these are as large as inferred by Saunders *et al.* [26] from the IRAS redshift surveys, and if this is due to the source density evolving as $(1+z)^P$ with P=6.7, then the background from (i) and (ii) would be as indicated by the upper curves in Figure (3), taken from Oliver et al. [27]. In this case, the background depends on the redshift z_{max} up to which the evolution applies and the DIRBE results already constrain this to be less than 2. On the other hand, with more modest evolution the background could be several orders of magnitude smaller, as shown by the lower curves, taken from Franceschini et al. [22]: P1 and P2 give the contribution from primeval galaxies, on the assumption that these generate solar metallicity at a redshift of 2 and 4.3, respectively, with all their light being reprocessed by dust; G2 gives the contribution from (i), (ii) and (iii) with no evolution; G1 allows for evolution in (ii) and (iii).

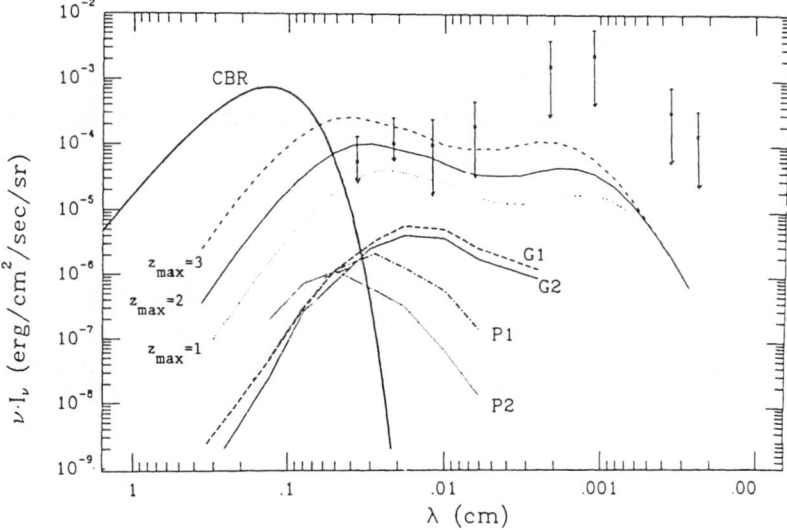

Figure (3). A comparison of the far-IR background from dusty galaxies predicted by the models of Oliver et al. [27] and Franceschini et al. [22] with the DIRBE upper limits.

3. PREGALACTIC SOURCES OF BACKGROUND RADIATION

Henceforth I will focus on pregalactic or protogalactic sources since these have been somewhat neglected in the literature. The point is that there could be several kinds of astrophysical generators of radiation in the period between z=10 and z=10^3. In what follows, I indicate the total energy density Ω_R and peak wavelength λ_{peak} associated with each background, on the assumption that the radiation propagates to us freely apart from redshift effects. The associated spectra in representative cases are summarized in Figure (1). Most of them have a dilute black-body form, on the assumption that there is not enough neutral hydrogen in the background Universe to absorb photons shortward of the Lyman cut-off. (Were this not the case, most of the radiation would come out as recombination lines.) If the backgrounds are reprocessed by dust, λ_{peak} will change but Ω_R will be roughly the same (as discussed in Section 4). More detailed calculations can be found in Bond *et al.* [28].

Many of the scenarios discussed below depend on the controversial assumption that there was a phase of pregalactic star formation. One reason for this proposal is that cosmological nucleosynthesis arguments [29] indicate that the baryonic density parameter must lie in the range 0.04 to 0.06 for a Hubble parameter of 50, whereas the density associated with visible galaxies is only 0.01. This suggests that a large fraction of the baryons must be dark. It is possible that the dark baryons are in a hot intergalactic medium [30] but, more likely, they have been processed into dark remnants through a first generation of "Population III" stars. In this case, their density could at least suffice to explain the dark matter in galactic halos, although one would still need non-baryonic dark matter to explain the critical density required by the inflationary picture, and the stars would inevitably generate background light [31]. A variety of arguments [32] show that the Population III objects must be either the black hole remnants of "Very Massive Objects" (VMOs) larger than 100 $M_\odot$ which collapse during their oxygen-burning phase [33] or "brown dwarfs" which are too small to burn hydrogen at all [34]. Even if galactic halos are non-baryonic, one could still invoke Population III stars as the precursors of the black holes in galactic nuclei or as the seeds for explosive galaxy formation.

Halo VMOs. The most striking signature of the VMO scenario would be the background radiation generated during the hydrogen burning phase. This can be predicted rather precisely since all VMOs have a surface temperature T_s of about 10^5K and generate radiation with efficiency $\epsilon \approx 0.004$. (Both quantities are almost independent of the VMO mass [35].) We assume that the VMOs have a density parameter $\Omega_* = 0.02h^{-2} = 0.1h_{50}^{-2}$, corresponding to the sort of density required to explain the dark mass in galactic halos, and that they produce black-body radiation with temperature T_s. If the radiation is affected only by cosmic redshift, it will have a peak wavelength and density

$$\lambda_{peak} = 4\left[\frac{1+z_*}{100}\right]\mu, \qquad \Omega_R = 10^{-6}h^{-2}\left[\frac{1+z_*}{100}\right]^{-1}\left[\frac{\Omega_* h_{50}^2}{0.1}\right] \qquad (3)$$

at the present epoch, where z_* is the redshift at which the VMOs burn their nuclear fuel. We can place an upper limit on z_* by noting that the main-sequence time of a VMO is $t_{MS} = 2\times10^6$y (independent of mass), so that z_* cannot exceed the redshift when the age of the Universe is t_{MS}. This implies $z_* < 240h^{-2/3}$ and hence $\lambda_{peak} < 15\mu$ and $\Omega_R > 5\times10^{-7}h^{-2}$. We can place a lower limit on z_* from UV-IR background light limits. As discussed by McDowell [36], these imply a constraint on the density of VMOs burning at any redshift z_*. If one requires $\Omega_* = 0.1h_{50}^{-2}$, this places a lower limit on z_*, mainly because one needs the radiation density to be redshifted into the near-IR band, where the background light limits are weaker. In the absence of neutral hydrogen absorption, one requires $z_* > 30$ and this implies $\lambda_{peak} > 1\mu$ and $\Omega_R < 3\times10^{-6}h^{-2}$. The peak of the VMO background must then lie somewhere on the heavy line in Figure (1) and the spectrum must lie within the broken lines. It would almost certainly be the dominant background in the near-IR and it could soon be detectable unless z_* is so large ($\gtrsim 100$) that the VMO light is pushed to 10μ, where it would be hidden by interstellar dust.

Halo Brown Dwarfs. If galactic halos comprised brown dwarfs, there would still be an IR background but it would have a much smaller density ($\Omega_R \approx 10^{-9}$-10^{-8}) and it would peak in the range 10-100μ and extend over a wider waveband. The curve in Figure (1) assumes a brown dwarf mass of 0.085 $M_{\odot}$ and is based on the calculation of Karimabadi & Blitz [37]. Note that this background is non-thermal because it is associated with the *formation* of the brown dwarfs (viz. a Hayashi phase, followed by a degenerate cooling phase) rather than their emission at the present epoch. It would almost certainly be hidden by the background from IR galaxies. The prospects of detecting individual brown dwarfs in our own halo or the collective emission from the brown dwarfs in other halos are also poor [38], although they will improve with the ISO satellite.

Black Holes in Galactic Nuclei. In order to explain quasars and active galactic nuclei, it is commonly supposed that some galaxies have black holes in their nuclei with a mass M of order $10^8 M_{\odot}$ [39]. One could expect such holes to have radiated at the Eddington limit for at least a "mass-doubling" time t_E, in which case they would have generated a background with

$$\Omega_R \simeq 10^{-7} \left[\frac{\epsilon}{0.1}\right] \left[\frac{1+z_E}{10}\right]^{-1} , \qquad \lambda_{peak} \simeq 2 \left[\frac{1+z_E}{10}\right] \mu \qquad (4)$$

as illustrated in Figure (1). Here ϵ is the efficiency with which the accreted material generates radiation and $z_E \approx 6h^{-2/3}\epsilon^{-2/3}$ is the (M-independent) redshift corresponding to the epoch when the Hubble time is t_E. The wavelength estimate assumes that one has an optically thick accretion torus at a temperature of 2×10^4K, as suggested by the models of Begelman [40].

Halo Black Holes. If galactic halos consist of VMO remnants, then they would also accrete. Although the accretion rate would be very low at the present epoch, it could have been much larger in the pregalactic era (assuming they existed then). In this period, the holes would usually accrete at the (sub-Eddington) Bondi rate from a medium with the mean cosmological gas density Ω_g and a temperature $T \approx 10^4$K. In this case, they should produce a background with density [41]

$$\Omega_R \simeq 7\times10^{-7} \left[\frac{\Omega_g}{0.1}\right] \left[\frac{\epsilon}{0.1}\right] \left[\frac{M}{10^6 M_{\odot}}\right] \left[\frac{T}{10^4 K}\right]^{-3/2} \left[\frac{1+z_*}{10}\right]^{1/2} h\Omega^{-1/2} \qquad (5)$$

(where Ω is the total density parameter) and peak wavelength

$$\lambda_{peak} \simeq \left[\frac{\epsilon}{0.1}\right]^{-1/4} \left[\frac{\Omega_g h^2}{0.1}\right]^{-1/4} \left[\frac{T}{10^4 K}\right]^{3/8} \left[\frac{1+z_*}{10}\right]^{1/4} \mu \qquad (6)$$

We have normalized M to $10^6 M_{\odot}$ since there may be dynamical evidence for halo black holes of this mass [42]. The curve in Figure (1) assumes M=$10^6 M_{\odot}$, ϵ=0.1 and z_*=40, the choice of z_* corresponding to the redshift at which most of the radiation is likely to be generated [41].

Gravitational Cooling. There will inevitably be some background generated by the cooling of gas clouds which have bound gravitationally and virialized. The characteristics of this background depend on the density of the cooling gas Ω_c, the redshift of cooling z_c and the velocity dispersion V in the final bound structures. If we normalize to the values appropriate for galaxies, we get

$$\Omega_R = 2 \times 10^{-7} \left[\frac{\Omega_c}{0.1}\right] \left[\frac{1+z_c}{5}\right]^{-1} \left[\frac{V}{300}\right]^2 , \quad \lambda_{peak} = 0.1 \left[\frac{1+z_c}{5}\right]^{-1} \left[\frac{V}{300}\right]^{-2} \mu \quad (7)$$

where V is in km/s. How Ω_c, z_c and V scale with mass depends on the particular cosmological scenario. In the standard hierarchical clustering scenario, pregalactic clouds would produce a background with a smaller density but a longer wavelength.

Pregalactic Explosions. It has been proposed that some features of large-scale cosmic structure can be explained by pregalactic explosions [43,44]. One envisages each explosive seed (a star or a cluster of stars) generating a shock which sweeps up a shell of gas. In order to explain the existence of giant voids and the form of the galaxy correlation function, the shells must eventually overlap with a characteristic radius of order 10 Mpc [45]. However, the stars which generate the explosive energy will also generate light, so one can predict a minimum background radiation density [46]: to generate density fluctuations of order 1 on a comoving scale d corresponds to a kinetic energy density

$$\rho_{exp}c^2 \simeq m_b \, n_b \left[\frac{H(z)d}{1+z}\right]^2 \quad (8)$$

where m_b and n_b are the baryon mass and number density, z is the redshift at which the shells overlap, and H(z) is the Hubble rate at that redshift. Since stars generate $\eta \simeq 100$ times as much radiation energy as explosive energy during the preceding main-sequence phase, the background radiation density must be

$$\Omega_R \simeq 10^{-4} \left[\frac{\eta}{100}\right] \left[\frac{\Omega_g h^2}{0.1}\right] \left[\frac{d}{10Mpc}\right]^2 \quad (9)$$

The spectrum should peak at the wavelength given by eqn (4). The curve in Figure (1) assumes $\eta = 0.3$, as applies for exploding stars of mass $10 M_\odot$. This density is already in conflict with the optical to IR limits unless the radiation is reprocessed by dust.

Decaying Particles. Elementary particle relics of the Big Bang would be expected to pervade the Universe and, if their mass is sufficiently large, they could have an appreciable cosmological density. In certain models, these particles would be expected to decay radiatively on some timescale τ_d. For $\tau_d < 50y$, they would contribute to the CBR, while for $50y < \tau_d < 3 \times 10^4 y$ they would distort its spectrum [47]). However, for $\tau_d > 3 \times 10^4 y$, they would just generate a background with

$$\Omega_R \simeq 5 \times 10^{-6} \, \Omega_X \left[\frac{1+z_d}{10^5}\right]^{-1} , \quad \lambda_{peak} \simeq 120 \left[\frac{1+z_d}{10^5}\right] \left[\frac{m_X}{keV}\right]^{-1} \mu \quad (10)$$

Here z_d is the decay redshift, Ω_X is the density parameter which would be associated with the particles had they not decayed, and m_X is the particle mass. Many models relate Ω_X, m_X and τ_d, so they are not necessarily independent. The curve in Figure (1) corresponds to $\Omega_X=0.01$, $m_X=1$keV and $z_d=10^3$. Note that the spectrum deviates somewhat from the black-body form. A particular variant of this scenario is that due to Sciama [48] in which $\tau_d=10^{23}$s and $m_X=26$eV, both fine-tuned in order to explain the ionization of the interstellar and intergalactic medium. In this case, one expects a background peaking at

$$\Omega_R = \Omega_X H_o \tau_d \simeq 10^{-6}, \qquad \lambda_{peak} = 0.09\mu \qquad (11)$$

although some of this will be reprocessed into recombination lines.

4. REPROCESSING OF PREGALACTIC RADIATION BY COSMIC DUST

The spectra predicted above apply only if the radiation propagates freely between us and the source. However, most of the backgrounds discussed in Section (3) are initially in the optical or UV and might therefore be absorbed by dust. In this case, they would be re-emitted at a longer wavelength, as discussed by many authors [49-65]. The dust could either be confined to galaxies (if galaxies cover the sky) or it could be uniformly spread throughout the Universe. The last situation is only likely to apply if there was a generation of pregalactic stars.

In order to determine the condition for dust absorption and the characteristics of the re-emitted radiation, I will adopt the simplistic analysis of Bond et al. [28] (BCH1) throughout this section. A more precise treatment is presented in Section (5). In BCH1 the dust cross-section for photons of wavelength λ is assumed to be

$$\sigma_d = \frac{\pi r_d^2}{1+(\lambda/r_d)^\alpha} \qquad (12)$$

where r_d is the grain radius. Thus one has a geometric cross-section for $\lambda \ll r_d$ but σ_d falls as $\lambda^{-\alpha}$ for $\lambda \gg r_d$. This implies that the optical depth for photons back to a redshift z is

$$\tau(\lambda,z) = \left[\frac{c\ \Omega_d\ \rho_{crit}}{2\ \rho_{id}\ r_d\ H_o\ \Omega^{1/2}}\right](1+z)^{3/2}\left[1 + \frac{5\ \lambda}{3r_d(1+z)}\right]^{-1} \qquad (13)$$

where ρ_{id} is the internal density of the grains and Ω_d is their mean cosmological density. Pregalactic grains would then absorb photons with $\lambda < r_d$ for

$$1+z_d > 10\left[\frac{\Omega_d}{10^{-5}}\right]^{-2/3}\left[\frac{r_d}{0.1\mu}\right]^{2/3}\left[\frac{\rho_{id}}{3g/cm^3}\right]^{2/3} h^{-2/3}\Omega^{1/3} \qquad (14)$$

where we have normalized Ω_d, r_d and ρ_{id} to the sort of values appropriate for galaxies. Since Ω_d is itself a function of z, it is useful to express the $\tau>1$ condition in terms of the (Ω_d,z) space of Figure (4). If one thinks of the mean cosmological dust density as following a trajectory $\Omega_d(z)$ in Figure (4), then photons from pregalactic sources will be absorbed by intervening dust providing there is some redshift between their emission and now at which the trajectory penetrates the shaded region. It is not clear whether this is the case. Ω_d clearly starts off above the shaded region since $\Omega_d=0$ initially; observations of distant quasars also imply that a uniform dust distribution must have $\Omega_d<6\times10^{-5}h^{-1}$ back to z=2 [66], so one certainly has $\tau<1$ at the present epoch. However, one could still have $\tau>1$ in some intermediate redshift range. For example, one would only need $\Omega_d>10^{-7}$ at z=300, the earliest epoch at which VMOs could complete their nuclear burning, and such a small abundance could certainly be generated by pregalactic stars in the hierarchical clustering scenario.

Even if there is no pregalactic grain abundance, the dust *within* galaxies could still absorb any pregalactic radiation providing two conditions are satisfied. Firstly, the *mean* dust density (i.e. the density which would apply if the galactic dust were spread uniformly throughout the Universe) must be large enough for τ to exceed 1 at the redshift of galaxy formation (z_G). The contribution of galactic dust to Ω_d can be written as

$$\Omega_d \simeq 10^{-5}\left[\frac{f_d}{0.01}\right]\left[\frac{f_g}{0.1}\right]\left[\frac{\Omega_{GB}}{0.01}\right] \qquad (15)$$

where f_g is the fraction of the galaxy's mass in gas, f_d is the fraction of the gas in dust, and Ω_{GB} is the density parameter associated with the baryons in galaxies. This just corresponds to the normalization in eqn (14). Secondly, we need the galaxies to cover the sky at z_G. Otherwise, *most* of the background photons would be unaffected. This requires

$$1 + z_G > 11 \left[\frac{R_G}{10kpc}\right]^{2/3}\left[\frac{\rho_{iG}}{10^{-24}}\right]^{2/3}\left[\frac{\Omega_{GB}}{0.01}\right]^{-2/3}\Omega^{1/3}h^{-1/3} \qquad (16)$$

where R_G is the radius of the dust-containing part of a galaxy and ρ_{iG} is the density within a galaxy in g/cm^3. Since the redshift of galaxy formation is in the range 3 to 10, it is not clear whether these two conditions are satisfied. It is certainly possible, and Ostriker and Heisler [67] have even proposed this as the explanation for why quasars cut off at a redshift of 4, but it is not necessarily the case. Note that we have normalized Ω_G, R_G and ρ_{iG} to values appropriate for galaxies like our own. In practice, these parameters will span a range of values and in some models one would expect the smallest galaxies to contribute most to the covering factor. Although galaxies might be expected to cover the sky at redshifts exceeding 4 in the Cold Dark Matter (CDM) picture [68], the analysis of Fall et al. [69] indicates that the dust-to-gas ratio in primordial galaxies may only be in the range 1/20 to 1/4 that of the Milky Way for 2<z<3, which makes condition (16) difficult to satisfy.

If the radiation is absorbed by galactic or pregalactic dust, then the thermal balance implies that the dust temperature T_d evolves as

$$T_d(z) = T_c(z) \left[1 + \left[\frac{\Omega_R}{\Omega_c}\right]\left[\frac{r_d}{0.1\mu}\right]^{-1}\left[\frac{1+z}{10^*}\right]^{-1}\right]^{1/5} \qquad (\alpha=1) \qquad (17)$$

Here $T_c(z)$ is the temperature of the CBR photons, Ω_c is the CBR density, and we have assumed $T_d \ll r_d^{-1}$, $T_c \ll r_d^{-1}$ and $T_s \gg r_d^{-1}$ in appropriate units. Thus if the radiation density is less than

$$\Omega_{crit} \simeq 2 \times 10^{-7} h^{-2} \left[\frac{r_d}{0.1\mu}\right] \left[\frac{1+z}{100}\right] \tag{18}$$

the dust temperature will just be the CBR temperature (the CBR heating alone ensuring that it never drops below this). However, if Ω_R exceeds the value given by eqn (18), the dust temperature will be somewhat larger than T_c. In this case, one expects a far-IR or submillimetre background with a spectrum peaking at a present wavelength

$$\lambda_{peak} \simeq 700 \left[\frac{\Omega_R h^2}{10^{-6}}\right]^{-1/5} \left[\frac{r_d}{0.1\mu}\right]^{1/5} \left[\frac{1+z}{10}\right]^{1/5} \mu \qquad (\alpha=1) \tag{19}$$

where we have normalized the radiation density Ω_R to the sort of value anticipated in Section (3). The appropriate value for z in eqn (19) is the redshift at which most of the dust emission occurs (i.e. the redshift of dust or radiation production, whichever is smaller). The crucial point is that λ_{peak} is very insensitive to the various parameters appearing in eqn (19) because the exponents are so small. (The exponents would be even smaller for $\alpha > 1$.) Thus one expects all the reprocessed radiation from pregalactic sources to pile up at roughly the same wavelength. In a sense, this is unfortunate since it means that the spectrum itself contains little information about the origin of the radiation. On the other hand, it is interesting because it means that one can *predict* that a background with these characteristics ought to exist.

These considerations show that, if $\Omega_R > \Omega_{crit}$, one expects the total background spectrum to have three parts: the CBR component (peaking at 1400μ), the far-IR/submillimetre dust component (peaking at λ_{peak}), and the residual source component (peaking in the optical or near-IR). If $\Omega_R < \Omega_{crit}$, the dust radiation will be superposed on the CBR, so the overall spectrum will have only two peaks. However, since the dust radiation does not have a black-body spectrum, it will still distort the CBR spectrum [50,51,54]. These considerations must be modified if the dust emission is itself absorbed. In this case, the distortion will be narrower, with the radiation being partially thermalized, so that the dust is actually generating part of the CBR. The condition for this is that the value of τ given by eqn (13) exceed 1 at the wavelength given by eqn (19). This requires

$$1 + z > 65 \left[\frac{\Omega_d}{10^{-5}}\right]^{-0.4} \left[\frac{\rho_{id}}{2}\right]^{-0.4} \left[\frac{r_d}{0.1\mu}\right]^{0.1} \left[\frac{\Omega_R}{\Omega_c}\right]^{-0.1} h^{-0.4} \tag{20}$$

which corresponds to the heavily shaded part of Figure (4). In principle, provided the $\Omega_d(z)$ trajectory passes through this double-shaded region, one could hypothesize that the *entire* microwave background derives from grains [49,53,62-64]. However, for a reasonable grain density, this requires that either the radiation be produced at a very high redshift or the grains be rather exotic.

224

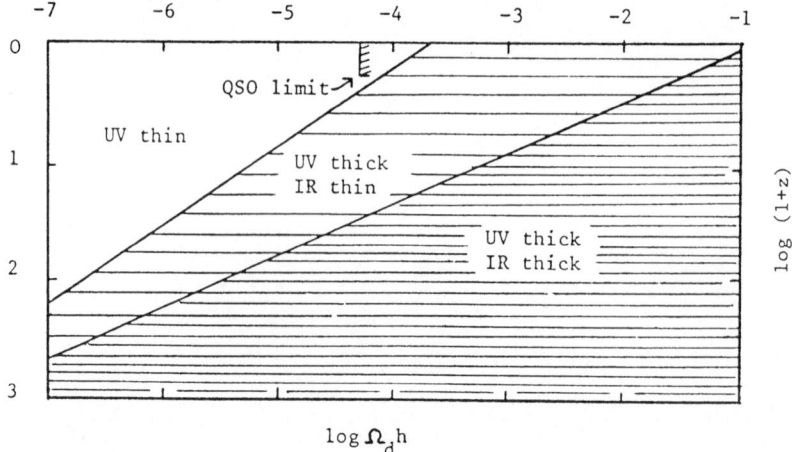

Figure (4). This shows the dust density required to absorb UV radiation at a redsdift z, thereby reprocessing it into the far-IR. In the heavily shaded region the far-IR radiation is itself absorbed, leading to partial thermalization. We assume $r_d = 0.1\mu$.

5. NUMERICAL MODELS FOR DUST ABSORPTION

The fact the Nagoya-Berkeley experiment [2] appeared to find a submillimetre excess peaking at almost exactly the wavelength predicted by eqn (19) was very exciting and prompted a much more detailed attempt to fit the data by Bond et al. [65] (BCH2). The Nagoya-Berkeley excess has now been disproved but the need for a more careful analysis remains since there must be reprocessed pregalactic light at some level. In the more sophisticated analysis of BCH2, the dust is characterized by a function $A_d(\lambda)$, related to the grain radius r_d and grain absorption cross-section σ_d by

$$A_d(\lambda) = \left[\frac{\sigma_d}{\pi r_d^2}\right]\left[\frac{3\lambda}{8\pi r_d}\right] \quad , \tag{21}$$

and we allow for the fact that the exponent α in eqn (12) could itself be wavelength-dependent. A_d is a useful parameter because it is constant over the wavelength domain in which $\alpha=1$. The grain model is then specified by three parameters rather than two: an "effective" grain radius r_s for the source radiation ($r_s=r_d$ for a geometrical cross-section), the value of A_d at 100μ (denoted as A_{d100}), and the value of α in the far-IR. We also introduce a more realistic model for the source luminosity history $L_s(z)$, allowing for both "burst" and "continuous" models. In the latter, $L_s(z)$ is assumed to evolve as $(1+z)^p$ between a turn-on redshift z_1 and a turn-off redshift z_0. We assume that the dust density $\Omega_d(z)$ is created at some redshift z_d and remains constant thereafter.

With these assumptions, the dust temperature evolves according to

$$\frac{T_d(z)}{T_c(z)} = \left[1 + \left[\frac{10}{1+z}\right]^\alpha \left[\frac{r_s A_{d100}}{100\mu}\right]^{-1} \left[\frac{\Omega_{RS}(z)}{\Omega_c(z)}\right] \right]^{1/(4+\alpha)} \quad (22)$$

where Ω_{RS} is the source radiation density at redshift z. For $r_s = r_d$ and $\alpha = 1$, this would correspond to eqn (17) but the parameter Ω_{RS} must be interpreted carefully. If the UV opacity up to the epoch corresponding to redshift z is small, $\tau_{uv}(z) < 1$, then Ω_{RS} is just associated with the integral of L_s over time until then. (This was essentially the case considered in BCH1.) However, if $\tau_{uv}(z) \gg 1$ and P<4, then the energy density at any point comes only from the region within one UV optical depth of that point. In this case, Ω_{RS} is roughly the radiation density generated over a cosmological time at redshift z divided by $\tau_{uv}(z)$. In particular, this includes the case in which the sources are completely shrouded in local dust clouds (cf. IRAS galaxies). If $\tau_{uv} \gg 1$ and P>4, then the emission from the turn-on redshift z_1 dominates and one effectively has a "burst" with $\Omega_{RS}(z)$ being the radiation density at z_1 divided by $\tau_{uv}(z_1)$ and multiplied by $\exp[\tau_{uv}(z) - \tau_{uv}(z_1)]$.

In BCH2 we consider 14 different models (corresponding to different values of α, Ω_d, Ω_R, z_1, z_0 and P) and the resulting spectra are shown in Figure (5). Also shown are the current and ultimate FIRAS limits, the 275μ Nagoya-Berkeley limit, and the limit of Gush et al. [70]. We can apply the BCH2 analysis to the VMO scenario in particular by noting that eqn (3) implies that each value of Ω_R corresponds to a specific value of z_*. Figure (5) shows that scenarios with $\Omega_R = 10^{-6}$ (eg. the VMO scenario) are viable in all models except 1, 5 and 6. For example, if $r_d = 0.01\mu$, $z_* = 100$ and $z_1 = 9$, then one is below the constraints for $\Omega_d = 10^{-6}$ but above them for $\Omega_d = 10^{-5}$. If COBE does eventually eliminate a particular pregalactic radiation-plus-dust scenario, it should be stressed that one would not necessarily expect dust reprocessing anyway.

6. ANISOTROPIES IN THE BACKGROUND FROM COSMIC DUST

One of the important implications of the above discussion is that the mere detection of a far-IR or submillimetre background will not identify its astrophysical source, since the expected spectrum depends only weakly on the epoch and mode of radiation production. However, crucial cosmological information could come from the detection of any anisotropies in the background. Indeed, since the peak of the dust background is so close to that of the CBR, interesting constraints on dust models can already be inferred from CBR anisotropy limits.

In BCH1 we argued that the dust emission can be regarded as coming from a "shell" of thickness $\sim 1h^{-1}$Gpc at a distance $6h^{-1}$Gpc. We then calculated the anisotropies on the assumption that the dust clumps in the same way as galaxies, with a statistical fluctuation in the number of galaxies within each beam. If the galaxy correlation function [71] has the form $\xi_g(x) = (x/x_0)^{-\gamma}$ with $\gamma \approx 1.8$ for $x < x_0$, then one just has Poisson fluctuations for beam sizes σ exceeding the angular scale θ_0 associated with the correlation scale x_0 but the fluctuations are enhanced for $\sigma < \theta_0$. One can infer that the rms intensity fluctuations have the form

FIGURE (5a). All these models have the same energy release $(\Omega_R h^2 = 10^{-6})$ but the quantity and character of dust varies. $\alpha = 1$ in 1 and 2; $\alpha = 2$ in 3 and 4; $\Omega_d = 10^{-5}$ in 1 and 3; $\Omega_d = 10^{-6}$ in 2 and 4. All the models have $z_1 = 9$ and $z_0 = 0$ and the energy is released in a burst.

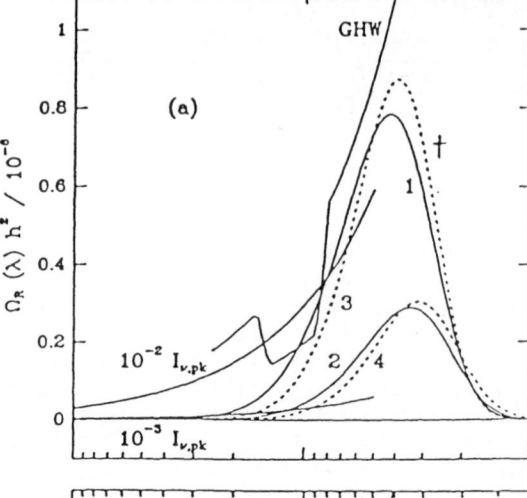

FIGURE (5b). In these models the dust properties are fixed ($\alpha = 1.5$) but the amount and characteristic redshifts are varied. $\Omega_d = 10^{-5}$ in 5 and 7; $\Omega_d = 10^{-6}$ in 6 and 8; $z_1 = 50$ and $z_0 = 25$ in 5 and 6; $z_1 = 5$ and $z_0 = 0$ in 7 and 8. All the models have $\Omega_R h^2 = 10^{-6}$. The energy is released in a burst in 7 and 8.

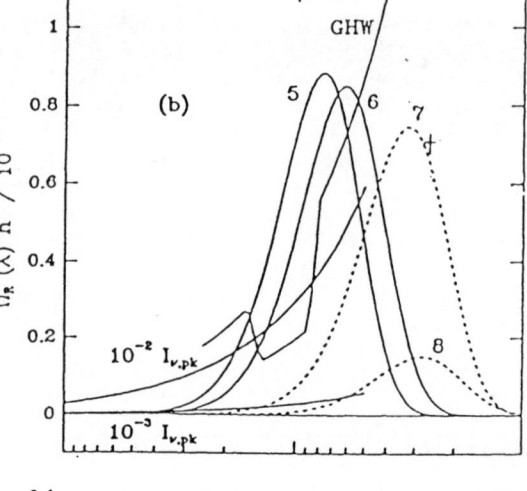

FIGURE (5c). All these models have $\alpha = 1.5$ and $\Omega_R h^2 = 10^{-7}$ but the other dust characteristics vary. $\Omega_d = 10^{-5}$ in 9, 11 and 14; $\Omega_d = 10^{-6}$ in 10 and 13; $z_1 = 50$ and $z_0 = 25$ in 9 and 10; $z_1 = 9$ and $z_0 = 0$ in 11 and 13; $z_1 = 5$ and $z_0 = 0$ in 14. The energy is released in a burst in 11 and 14.

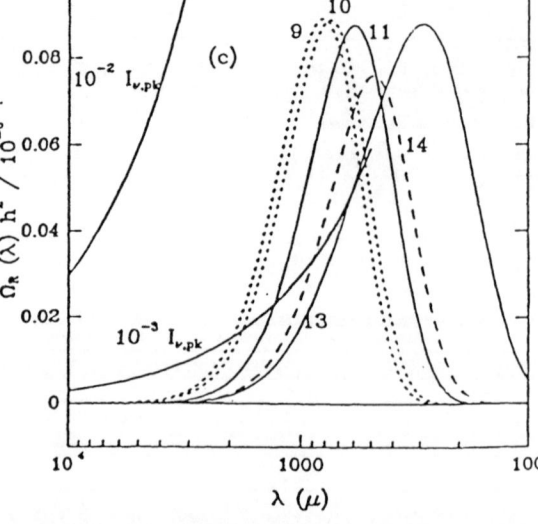

$$\left\langle \left[\frac{\Delta I}{I}\right]^2\right\rangle^{1/2} = \left\{ \begin{array}{ll} \left[\frac{x_0}{ct}\right]^{1/2}\left[\frac{\sigma}{\theta_0}\right]^{-1} & (\sigma > \theta_0) \\[2ex] \left[\frac{x_0}{ct}\right]^{1/2}\left[\frac{\sigma}{\theta_0}\right]^{(1-\gamma)/2} & (\sigma < \theta_0) \end{array} \right. \tag{23}$$

Here ct is the horizon size and x_0 is the correlation length at the epoch of dominant emission. This would be about $5h^{-1}(1+z)^{-1}$Mpc if there were no evolution of the clustering pattern, corresponding to $\theta_0 \approx 1'$, so eqn (23) gives fluctuations of a few percent on arcmin scales. The distinctive feature of the dust anisotropies is that they are wavelength-dependent and much larger than the primary anisotropies at small angular scales (where the latter are expected to be erased).

In BCH2, we confirm that the concept of an emission shell is a good one but we have refined the anisotropy estimates by allowing for fluctuations in the temperature of the emission shell as well its optical depth. (The temperature fluctuations are just associated with spatial fluctuations in the luminosity of the sources.) We have also extended the analysis to the high optical depth case, so as to include, for example, the situation with IRAS galaxies where the dust and sources are highly clumped. Our analysis includes both the linear regime, appropriate when the density fluctuations are small, and the "shot noise" regime, appropriate after galaxies form.

Some of the results of our analysis are summarized in Figure (6), where we compare the predicted anisotropies at 1300μ with the current limits from the IRAM experiment of Kreysa & Chini [72], the OVRO experiment of Readhead et al. [73] and the VLA experiment of Hogan & Partridge [74]. Figure (6) represents the ability of each experiment to probe the submillimetre anisotropy as a filter in multipole space. (Note that the ℓ-pole probes an angular scale of $3438/\ell$ arcmin.) When one multiplies this by the angular power spectrum $C_\ell(\lambda)$ of the background and integrates over $d\ln\ell$, one obtains the rms anisotropy for the experiment. The overall normalization is the rms energy density fluctuation $\sigma_I = \langle(\Delta I/I)^2\rangle^{1/2}$, which is straightforwardly related to the rms temperature fluctuation $(\Delta T/T)_{rms}$.

The figure shows the fluctuations associated with model 7 and also the primary fluctuations in the CDM scenario (cdm-sr indicates the "standard recombination" case). "PO2" is the Poisson fluctuation model in which the galaxies are placed randomly; "C1" is the $\xi \sim r^{-1.8}$ continuous clustering model, with the amplitude appropriate for observed galaxies; "cdm-d" assumes the usual CDM linear fluctuation spectrum with biasing and linear growth. One can also consider a hybrid of the last two cases in which one has both Poisson and continuous clustering contributions. The figure confirms that the dust fluctuations peak on a smaller scale ($\ell=10^3$ to 10^6) than the primordial density fluctuations. The optimal filter scale would be around 1" but the resolution achievable with submillimetre telescopes is generally above 10". One can also infer from the figure which experiments place the most interesting constraints on each dust scenario.

As a specific example, consider a CDM-like scenario with galaxies and $\alpha=1.5$ dust forming at $z_G=5$ and the burst of radiation having energy $\Omega_R h^2=10^{-6}$. The Kreysa & Chini experiment [72] at 1300μ uses a beam with $\theta_{fwhm}=11"$ and $\theta_{throw}=30"$ and gives a constraint

228

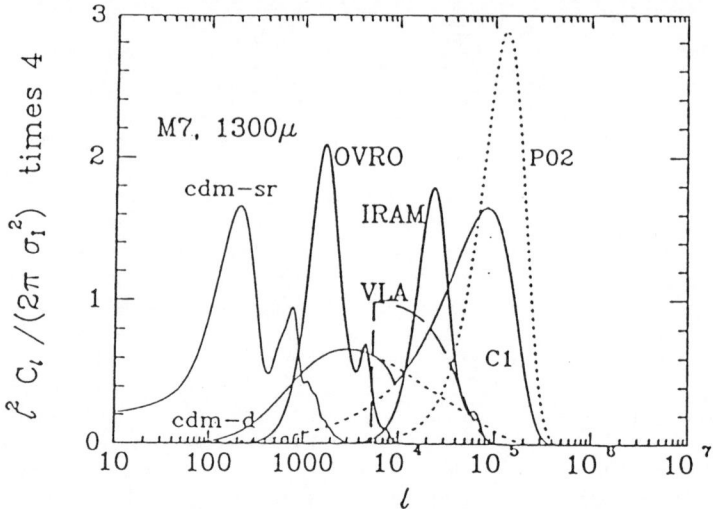

Figure (6). This shows the ability of various experiments to probe the anisotropies predicted in some of the dust scenarios discussed in the text.

$(\Delta T/T)_{rms} < 2.6 \times 10^{-4}$. This implies that the number density of the sources must satisfy $n_G{}^* > 0.06(h^{-1}Mpc)^{-3}$ and their correlation length must be $x_0 < 4h^{-1}Mpc$. If one took the background at 1300μ to have the maximum density compatible with the COBE constraints, then one would need $n_G{}^* > 1(h^{-1}Mpc)^{-3}$ and $r_0 < 0.7h^{-1}Mpc$. Future anisotropy experiments will be able to strengthen these limit on $n_G{}^*$ and r_0, thus providing a very promising probe of the formation of protogalaxies at high redshift.

ACKNOWLEDGMENT

Much of this lecture is based on work done in conjunction with Dick Bond and Craig Hogan. I thank them for an enjoyable collaboration.

REFERENCES

1. Boulanger, F., 1991, this volume.
2. Matsumoto, T. et al., 1988, Ap.J., 329, 567.
3. Rowan-Robinson, M., 1986, Mon.Not.R.Astr.Soc., 219, 737.
4. Matsumoto, T, Akiba, M. & Murakami, H., 1988, Ap.J., 332, 575.
5. Woody, D.P. & Richards, P.L., 1981, Ap.J., 248, 18.
6. Zeldovich, Ya.B. & Sunyaev, R.A., 1969, Ap.Space Sci. 4, 301.
7. Fabbri, R. & Melchiorri, F., 1979, Astr.Ap., 78, 376.
8. Birkinshaw, M. & Gull, S.M., 1984., Mon.Not.R.Astr.Soc., 206, 359.
9. Ceccarelli, C. et al., 1983, Ap.J.Lett., 275, L39.
10. Smoot, G.F. et al., 1991, Ap.J., in press; also this volume.
11. Fabbri, R., Guidi, I., Melchiorri, F. & Natale, V., 1980, Phys.Rev.Lett., 44, 1563.
12. Mather, J.C. et al., 1990, Ap.J.Lett., 354, L37.

13. Mather, J.C. et al., 1991, Ap.J., in press.
14. Lange, A.E. et al., 1991, Ap.J., in press.
15. Jones, M.H. & Rowan-Robinson, M., 1991, in *The Early Universe from Diffuse Backgrounds*, ed. B. Rocca-Volmerange (Editions Frontieres).
16. Noda, M. et al., 1991, Nagoya University Preprint.
17. Peebles, P.J.E. & Partridge, R.B., 1967, Ap.J., 148, 377.
18. Leinert, C., Mattila, K., Lemke, D. & Joseph, R.D., 1991, in *The Early Universe from Diffuse Backgrounds*, ed. B. Rocca-Volmerange (Editions Frontieres).
19. Tinsley, B., 1973, Astr.Ap., 24, 89.
20. Yoshii, Y. & Takahara, F., 1988, Ap.J., 326, 1.
21. Djorgovski, S. & Weir, N., 1990, Ap.J., 351, 343.
22. Franceschini, A. et al., 1991, Astr.Ap.Supp., in press.
23. Rocca-Volmerange, B. & Guiderdoni, B., 1991, in *The Early Universe from Diffuse Backgrounds*, ed. B. Rocca-Volmerange (Editions Frontieres).
24. Cowie, L.L., Gardner, J.P., Lilly, S.J. & McLean, I., 1990, Ap.J.Lett., 360, L1.
25. Tyson, J.A., 1990, in *Galactic and Extragalactic Background Radiation* ed. S.Bowyer & C.Leinert, p 245 (Dordrecht: Kluwer).
26. Saunders, W. et al., 1991, Mon.Not.Roy.Astr.Soc., 242, 318.
27. Oliver, S., Rowan-Robinson, M. & Saunders, W., 1991, preprint.
28. Bond, J.R., Carr, B.J. & Hogan, C.J., 1986, Ap.J., 306, 428 [BCH1].
29. Olive, K.A., Schramm, D.N., Steigman, G. & Walker, T., 1990, Phys.Lett.B., 236, 454.
30. Bartlett, J.G. & Stebbins, A., 1991, Ap.J.Lett., in press.
31. Thorstensen, J. & Partridge, R.B., 1975, Ap.J., 200, 527.
32. Carr, B.J., 1990, Comm.Ap., 14, 257.
33. Carr, B.J., Bond, J.R. & Arnett, W.D., 1984, Ap.J. 277, 445.
34. Ashman,K.M. & Carr,B.J., 1988, Mon.Not.R.Astr.Soc., 234, 219.
35. Bond, J.R., Arnett, W.D. & Carr, B.J., 1984, Ap. J., 280, 825.
36. McDowell, J.C., 1986, Mon.Not.R.Astr.Soc., 223, 763.
37. Karimabadi, H. & Blitz, L., 1984, Ap.J., 283, 169.
38. Adams, F.C. & Walker, T.P., 1991, Ap.J., 359, 57.
39. Rees,M.J., 1984, Ann.Rev.Astr.Ap., 22, 471.
40. Begelman,M.C., 1984, in *Astrophysics of Active Galaxies and Quasi-Stellar Objects*, ed. J.S.Miller (Mill Valley: University Science Books).
41. Carr,B.J., McDowell,J.C. & Sato,H., 1983, Nature, 306, 666.
42. Lacey,C.G. & Ostriker,J.P., 1985, Ap.J., 299, 633.
43. Ostriker,J.P. & Cowie,L.L., 1981, Ap.J.Lett., 273, L127.
44. Ikeuchi,S., 1981, Pub.Astr.Soc.Japan, 33, 211.
45. Saarinen,S., Dekel,A. & Carr,B.J., 1987, Nature, 325, 598.
46. Hogan,C.J., 1984, Ap.J.Lett., 284, L1.
47. Silk,J. & Stebbins, A., 1983, Ap.J., 269, 1.
48. Sciama, D.W., 1990, Comm. Ap., 15, 71.
49. Layzer, D. & Hively, R., 1973, Ap.J., 179, 361.
50. Rowan-Robinson, M., Negroponte, J. & Silk, J., 1979, Nature, 281, 635.
51. Puget, J.L. & Heyvaerts, J., 1980, Astr.Ap.Lett., 83, L10.
52. Rana, N.C., 1981, Mon.Not.R.Astr.Soc., 197, 1125.
53. Wright, E.L., 1982, Ap.J., 255, 401.
54. Negroponte, J., Rowan-Robinson, M. & Silk, J., 1981, Ap.J., 248, 38.

55. Negroponte, J., 1986, Mon.Not.R.Astr.Soc., 222, 19.
56. Hayakawa, S. et al., 1987, Pub.Astr.Soc.Japan, 39, 941.
57. Lacey, C. & Field, G., 1988, Ap.J.Lett., 330, L1.
58. Naselsky, P.D. & Novikov, I.D., 1989, Space Res.Inst. Preprint 1424.
59. Adams, F.C., Freese, K., Lenon, J. & McDowell, J.C.,1989,Ap.J.,344, 24.
60. Draine, B.T. & Shapiro, P.R., 1989, Ap.J.Lett., 344, L45.
61. de Bernardis, P., Masi, S., Melchiorri, B., Melchiorri, F. & Moreno, G.,
    1984, Ap.J., 278, 150.
62. Hawkins, I. & Wright, E.L., 1988, Ap.J., 324, 46.
63. Rees, M.J., 1978, Nature, 275, 35.
64. Hoyle, F. & Wickramasinghe, N.C.,1988, Ap.Space Sci.,18, 489.
65. Bond, J.R., Carr, B.J. & Hogan, C.J., 1991, Ap.J., 367, 420 [BCH2].
66. Wright, E.L., 1981, Ap.J., 250, 1.
67. Ostriker, J.P. & Heisler, J., 1984, Ap.J., 278, 1.
67. Heisler, J. & Ostriker, J.P., 1988, Ap.J., 332, 543.
68. Najita, J., Silk, J. & Wachter, K.W., 1990, Ap.J., 348, 383.
69. Fall, S.M., Pei, Y.C. & McMahon, R.G., 1989, Ap.J.Lett., 341, L5.
70. Gush, H.P., Halpern, M. & Wishnow, E., 1990, Phys.Rev.Lett., 65, 537.
71. Peebles, P.J.E., 1980, *The Large-Scale Structure of the Universe*,
    Princeton University Press.
72. Kreysa, E. & Chini, A., 1989, in *Proceedings of Berkeley Particle
    Astrophysics Workshop.*
73. Readhead, A.C.S. et al., 1989, Ap.J., 346, 566.
74. Hogan, C.J. & Partridge, R.B., 1989, Ap.J.Lett., 341, L29.

PRELIMINARY RESULTS FROM THE FIRAS AND DIRBE EXPERIMENTS ON COBE

E. L. Wright
UCLA Dept. of Astronomy
Los Angeles, CA 90024-1562

ABSTRACT. The FIRAS and DIRBE experiments on COBE will provide data that strongly constrain models of Population III stars and dust. This paper describes the analysis procedures necessary to process COBE data, and compares simple Population III models to the preliminary COBE data.

1. COBE Overview

The COBE satellite was launched from Vandenberg AFB in California on the morning of November 18, 1989. Its orbit is a 900 km altitude, 99° inclination circular orbit that precesses one cycle per year to maintain the orbit plane approximately normal to the Earth-Sun line. The satellite spins around an axis that always points 94° degrees from the Sun and away from the Earth. The spin rate is about 0.8 RPM. The liquid helium cryogen was exhausted on September 21, 1990. At this time (March 1991) the DMR experiment is 100% operational, and the short wavelength (1.25-4.9 μm) channels of the DIRBE continue to function. The long wavelength channels of the DIRBE required liquid helium to operate, and have thus ceased operation after covering 100% of the sky and completing >60% of a second complete sky coverage. The FIRAS instrument ceased operations when the cryogen ran out, after covering >98% of the sky.

Preliminary descriptions of COBE results can be found in Mather *et al.* (1990a, 1990b), Hauser *et al.* (1990), and Smoot *et al.* (1990, 1991).

2. COBE Data Products

The COBE instruments are all designed to produce all-sky maps. The FIRAS and DMR instruments have 7° circular beams, while the DIRBE instrument has an

M. Signore and C. Dupraz (et al.), The Infrared and Submillimetre Sky after COBE, 231–248.
© 1992 *Kluwer Academic Publishers*.

0.7° square beam. The data products exist in two forms: the time-ordered data received from the spacecraft, and the pixel-ordered data in skymaps. In order to have a consistent method for handling skymaps, a uniform pixelization scheme has been adopted. (See Chan and O'Neill, 1975). This scheme is based on projecting the sphere onto the faces of a circumscribed cube. This divides the sky into 6 square faces. The tangent plane coordinates on each face are then distorted to produce an equal-area map of the sphere. The distorted coordinates are converted into integers IX and IY with a resolution appropriate to the individual instrument. For FIRAS and DMR a resolution of 32 steps is used, giving a total of $6 \times 32 \times 32 = 6144$ pixels. For DIRBE 256 steps are used, giving $6 \times 256 \times 256 = 393,216$ pixels. This scheme was adopted to allow 64 DIRBE pixels to exactly cover a FIRAS or DMR pixel. The centers of the cube faces are the axes of the J2000 ecliptic coordinate system. The pixel numbers are constructed by interleaving the bits of the binary representations of IX and IY. The reason for doing this is to minimize the number of page faults necessary to access a compact region on the sky. For example, the 64 DIRBE pixels corresponding to a given DMR pixel are a sequence of 64 consecutive pixels.

DMR and FIRAS Pixelization Scheme

$$NPIX = 1024*NFACE + NXY$$
$$NXY_2 = msb(IY)msb(IX)\ldots lsb(IY)lsb(IX)$$

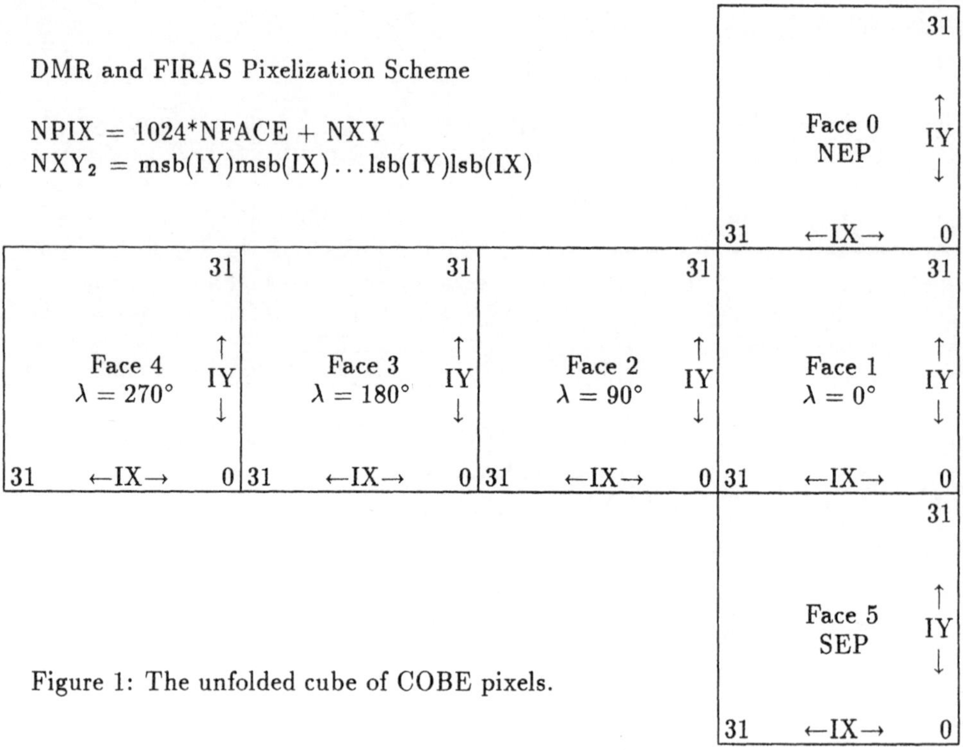

Figure 1: The unfolded cube of COBE pixels.

3. The Diffuse InfraRed Background Experiment

The DIRBE experiment on COBE has 16 channels covering ten infrared bands from 1.25 to 240 μm. The field of view is a square with sides of 0.7°. This beam is displaced by 30° from the spin axis of the satellite. The spin and orbit of the spacecraft cause the DIRBE beam to trace a cycloidal pattern on the sky.

The DIRBE instrument has an internal reference source (IRS) that is a variable temperature body viewed by reflection from the shutter when it is closed. This calibrator is designed to have a low effective emissivity to match the low optical depth of the interstellar dust and zodiacal dust. An IRS calibration run requires that the shutter be closed, then the temperature of the IRS is gradually increased. While the long wavelength detectors are calibrated, the short wavelength detectors see almost no signal. While the short wavelength detectors are calibrated, the long wavelength detectors are saturated. The IRS is viewed through the chopper, so the AC gain at 32 Hz is measured. IRS runs are typically taken 5 times per orbit. The stability of the IRS is monitored by comparing its signal levels to celestial standards. These celestial standards will also serve to define the absolute calibration of the DIRBE instrument.

3.1 DIRBE DATA ANALYSIS

The DIRBE data set is highly redundant. There are 393,216 DIRBE pixels on the sky and about 240 million DIRBE observations per year. Allowing for data interruptions due to calibration and the South Atlantic Anomaly, there are still several hundred observations per pixel in each band. The scan rate of DIRBE is about 2.6° per second, and the pixel size is about 0.32° , so a given pixel is usually only hit once in a given scan. Thus the DIRBE data set contains several hundred scans through an average piece of sky. Pixels that are at ecliptic latitudes close to ±60° are hit much more often, while pixels on the ecliptic are hit less often.

The observations of a given position on the sky vary during the year as the line-of-sight through the zodiacal dust cloud changes. This modulates the zodiacal flux by about a factor of two as the solar elongation ranges from 64° to 124° . Near the ecliptic poles the range of solar elongation is less, but the inclination of the zodiacal dust cloud and the eccentricity of the Earth's orbit combine to produce an annual variation of the polar flux. The DIRBE data analysis pipeline will create a large file called the "DIRBE Annual File", which will contain weekly averages of the brightness of each pixel. An average pixel will be observed for 26 weeks during the year, leading to a total of (393216 pixels)*(26 weeks)*(16 channels) = 160 million values. This is essentially a 4 dimensional data hypercube, with two dimensions for position on the sky, a third dimension for time of observation, and a fourth dimension for wavelength.

The annual variations in the zodiacal light will be used to adjust the parameters in a physical model of the interplanetary dust cloud. Once a model that correctly predicts the variation of the flux is found, then it will be subtracted from the data, leaving a map that contains the flux from galactic stars, interstellar dust, and a possible isotropic cosmic infrared background.

At short wavelengths, the contribution from galactic stars will be significant. I have used the star count model of Elias (1978) to predict the star counts expected at high galactic latitude at 3.5 μm wavelength, where there is a local minimum in the total flux seen by DIRBE. The star count model predicts dN/dS in stars per steradian. After all stars brighter than S_{lim} have been subtracted from the observed total flux, the correction for unresolved stars is

$$\Delta S = \int_0^{S_{lim}} \Omega \ S dN/dS dS = \Omega \ \Delta I$$

and its variance is

$$\sigma(\Delta S)^2 = \int_0^{S_{lim}} \Omega \ S^2 dN/dS ds = \Omega^2 \sigma(\Delta I)^2$$

If we consider data taken only in regions with no 12 μm IRAS sources then there shouldn't be any stars brighter than 6 Jy at 3.5 μm. The expected flux from fainter stars is 4 Jy in the 0.7° square DIRBE beam. The standard deviation of the flux in the beam is 2 Jy. This flux corresponds to 27 ± 14 kJy/sr, which is 15% of the measured flux at the South Ecliptic pole. In order to reduce the uncertainty in this correction, one needs to survey many square degrees of high latitude sky at 3.5 μm. The DIRBE filter was chosen to be a ground-based window to allow this. One should easily be able to find areas with no stars brighter than 0.66 Jy at 3.5 μm, since the expected number of such stars is one per beam, in which case the correction for unresolved stars is 17 ± 3.4 kJy/sr. Finally, for a survey limit of 0.06 Jy at 3.5 μm, six stars per DIRBE beam are expected, and after they are subtracted the correction for unsurveyed stars is 11 ± 0.9 kJy/sr. Assuming that one can cross-calibrate the ground-based and DIRBE flux scales to 10% accuracy, the limiting error will be about 1-2 kJy/sr and will require surveying enough sky to see hundreds of stars at 0.1 Jy at 3.5 μm ($L \approx 9$) in order to verify the star count models.

The analysis of the DIRBE data has not reached the point where the foregrounds can be subtracted, however. The following table shows only the total flux at the South Ecliptic Pole. The long wavelength channels have uncertain calibrations because the planets used as calibrators have temperatures much higher than the color temperature of the sky, so filter bandpass corrections are significant.

The 100 μm flux measured by DIRBE is much smaller than the flux measured by IRAS. This discrepancy has two sources: one is a constant offset in the IRAS data. Since IRAS was DC coupled and did not chop against a cold reference, such an offset is not unexpected. The second effect is a gain difference in the IRAS experiment between point sources and large scale structure. This is caused by the existence of two very different time constants in infrared photoconductors. Thus point sources, which IRAS saw at a frequency band near 1 Hz, will have a different calibration than the large scale structure seen by IRAS at 0.0003 Hz. The DIRBE experiment has a chopper which gives a known zero reference, but also converts any signal from the sky into a signal near 32 Hz. Thus the multiple time constants do not cause a major problem for DIRBE.

DIRBE MEASUREMENTS OF SOUTH ECLIPTIC POLE BRIGHTNESS

λ	νI_ν
μm	10^{-11}W cm^{-2}sr^{-1}
1.3	8.3 ± 3.3
2.2	3.5 ± 1.4
3.5	1.5 ± 0.6
4.9	3.7 ± 1.5
12	29 ± 12
22	21 ± 8
55	2.3 ± 1.0
96	1.2 ± 0.5
151	1.3 ± 0.7
241	0.66 ± 0.4

4. The Far InfraRed Absolute Spectrophotometer

The FIRAS instrument on COBE is a polarizing Michelson interferometer. It has two inputs, one of which views the sky through a sky horn that defines a 7° beam, and the other views an internal reference source through a reference horn. In addition, and external blackbody calibrator can be commanded into the sky horn. The internal reference source is close to a blackbody, with an average emissivity over the 1-20 cm^{-1} range of about 0.96. The external calibrator is designed to be a very good blackbody, with a design emissivity greater than 0.9999 and a measured reflectance less than 0.001. The temperatures of both calibrators and both horns can be independently varied during calibration observations.

There are also two outputs from the FIRAS interferometer, which have opposite phases of modulation. Each output is further split into a low frequency channel (1-20 cm^{-1}) and a high frequency channel by a dichroic filter. The detectors used in FIRAS are bolometers with a diameter of 8 mm. Non-imaging concentrators are used to give π sr of effective illumination on the detectors, leading to a throughput of 1.5 cm^{2}sr.

The path length difference range for FIRAS is commandable to be either -1.2 cm to 0.6 cm for short scans or -1.2 to 5.9 cm for long scans. The scan speed of FIRAS is either 0.8 cm/sec for slow scans or 1.2 cm/sec for fast scans.

Due to telemetry limitations, several interferogram scans are co-added before the sum is transmitted to the ground. One set of co-added interferograms is transmitted about every half minute.

4.1 FIRAS DATA ANALYSIS

The fundamental model of the FIRAS instrument is that the output is an interferogram that is the Fourier transform of a superposition of the spectra from sky plus other components of the instrument.

$$I(x) = \int \cos(2\pi\nu x + \phi_\nu)\big(C(\nu) + G(\nu)(S_\nu + \Sigma\ \epsilon_i(\nu)B_\nu(T_i))\big)\ d\nu$$

The phase correction ϕ_ν and gain $G(\nu)$ are found by inserting the external calibrator into the sky horn and adjusting its temperature over a wide range. The emissivities ϵ_i of the components are found by varying the temperatures of horns and internal reference source over a wide range. The offset term $C(\nu)$ is included for completeness, but should vanish. The calibration model has been extended to include the dihedral mirrors and the overall structure of the instrument. Neither of these is controlled, but their temperatures do vary.

The FIRAS instrument is affected by charged particles which cause large spikes or "glitches" in the detector output. There is a program in the onboard microprocessor that tries to identify glitches. It sets a glitch flag when it finds a glitch and will remove the glitch from the data if commanded to do so. The signal available for de-glitching is much larger before co-addition, but the resources available in the flight microprocessor are very limited. Since the onboard deglitcher is not very smart, a ground-based deglitching is also done. Interferograms taken under identical conditions are run through a consistency checker that compares the interferograms point by point. Ideally this consistency checking should be done before the onboard coadding, but it was not possible to include enough RAM in the flight microprocessors to allow this. The rapid changes in operating conditions caused by charged particles confusing the Mirror Transport Mechanism have made the consistency checking much more difficult.

Since the internal calibrator has a high emissivity (> 0.9) and appears with negative sign in the output, the ϵ_i for the internal calibrator is close to -1. By setting its temperature close to the temperature of the sky, the interferogram can be nulled out. Thus the FIRAS experiment makes a fairly direct measurement of the normalized deviation from a blackbody

$$\delta(\nu) = (I_\nu - B_\nu(T_c))/B_{max}$$

where T_c is the temperature of the reference blackbody, and the scaling factor that I use for B_{max} is the Planck function for 2.8 K evaluated at 6 cm^{-1}. The preliminary FIRAS results show that $\delta(\nu)$ is less that 0.01 from 1 to 20 cm^{-1}, and the ultimate sensitivity will be $|\delta| < 0.001$ or better.

4.2 PRELIMINARY FIRAS RESULTS

The first dramatic new data from the COBE satellite was published in Mather et al. (1990). The observations comparing the North Galactic Pole region to a nulled internal calibrator, and the observations comparing the external calibrator to the internal calibrator at the same temperature both showed null interferograms. This showed that any deviation of the cosmic background spectrum from a blackbody spectrum was less than 1% of the peak of the spectrum, over the 1-20 cm^{-1} region. Fitting Compton models to the spectrum showed that the Kompaneets parameter y must be less than 0.001, while fitting Bose-Einstein models to the data showed that the dimensionless chemical potential μ must be less than 0.01. Uncertainties in the thermometry of the calibrators limited the precision of the actual temperature to $T_o = 2.735 \pm 0.060$ K. Less than a month after this FIRAS result was announced, the Gush et al. (1990) data from a rocket flight confirmed the lack of distortion and gave $T_o = 2.736 \pm 0.017$ K.

The second preliminary FIRAS result was announced by Cheng et al. (1990): the spectrum of the dipole anisotropy follows $\Delta T dB_\nu/dT$, with $T_o = 2.735$ from the spectrum, and $\Delta T = 3.3$ mK, in agreement with the DMR anisotropy data at lower frequencies. This differential spectrum can be determined quite well as long as the instrument is stable for half an orbit.

The third preliminary result from the FIRAS experiment is the spectrum of the galaxy. Wright et al. (1991) have published an average spectrum of the galaxy, that clearly shows both the interstellar dust and lines from interstellar ions, atoms and molecules. The average continuum spectrum of the galaxy given by Wright et al. from 1-90 cm^{-1} is well represented by

$$g(\nu) = 0.00022(\nu/30)^2 \left(B_\nu(20.4\text{K}) + 6.7B_\nu(4.77\text{K})\right) + 5 \times 10^{-9}\nu^{-0.75}$$

where I_ν is in erg/cm^2/sec/sr/cm^{-1}, and ν is in cm^{-1}. The use of two Planck functions modified by a ν^2 emissivity law gives a convenient and versatile fitting function, but does not require the existence of two types of dust or very cold dust.

5. COBE and the Steady State

The discovery of the cosmic microwave background was a beautiful confirmation of the hot Big Bang model, and served to falsify the Steady State model. The Universe now is not producing a blackbody spectrum, so if the Steady State model held, and the Universe always looked like it does today, then there should not be a 2.735 K blackbody background. However, some scientists (Arp *et al.* 1990) have claimed that needle-shaped iron dust grains could save the Steady State. These iron needles would have frequency independent opacities of 2×10^7 cm^2/gm or more in the millimeter region, but much smaller opacities in the optical (Hoyle and Wickramsinghe, 1988). Thus their temperature now would be very close to T_o, even in the optical radiation field of the galaxy.

In the Steady State, T_{dust} is constant, so the apparent spectrum from the dust at redshift z is a blackbody with temperature $T_{dust}/(1+z)$. The range of redshifts that contribute strongly to the observed spectrum is $0 < z < 2/\tau_o$, where τ_o is the optical depth per Hubble radius. As a result, the observed spectrum is a mixture of blackbodies with a variance $\mathrm{var}(T)/T^2 \approx 1/\tau_o^2$. One way to make a Compton distorted spectrum is to mix blackbodies with a variance $\mathrm{var}(T)/T^2 = 2y$. Thus using the preliminary FIRAS limit of $y < 0.001$, we have $\tau_o > 20$. This approximate derivation of τ_o can be confirmed using the exact formula for I_ν given below. For $H_o = 50$, $\tau_o = 20$ requires an density in iron needles of 5×10^{-35} gm/cm^3, or an $\Omega_d = 10^{-5}$.

For dust grains with frequency dependent opacities, one calculates the predicted Steady State model intensity using

$$I_\nu = \int B_\nu(T_{dust}/(1+z)) d\exp(-\tau(z))$$

with

$$\tau(z) = \int \tau_o(\nu/\nu_o)^n (1+z)^n d[\ln(1+z)]$$

where a power law with index n for the frequency dependence of the dust opacity has been assumed. For n = 1 dust, this Steady State model is similar to a Bose-Einstein spectrum, and the preliminary FIRAS limit of $\mu < 0.01$ implies that $\tau_o > 20$ for $\nu_o = 1$ cm^{-1}. For n = 2 dust, the distortions are concentrated at low frequencies, and to be less than the preliminary FIRAS limit of 1% one needs $\tau_o > 13$ with $\nu_o = 1$ cm^{-1}. Satisfying the low frequency balloon and ground-based data requires higher values of τ_o for n = 2. In fact one needs $\tau_o > 160$ at $\nu_o = 1$ cm^{-1} to be within 1σ of the 4 cm datum of Kogut *et al.* (1989).

These large optical depths in the millimeter region are unlikely, because some distant sources like 3C273 are seen at 3 mm. But the Steady State proponents will say quasars are local. Clearly for the Steady State to succeed all radio sources observable at $\lambda < 2$ cm must be local. However, Stockton (1978, 1980) has seen faint galaxies near to 3C273 and other QSOs that have the same redshift as the nearby QSO. These galaxies fit nicely on the Hubble diagram for galaxies, and would seem to indicate that quasars really are very luminous distant objects.

Iron needles have higher absorptions in the millimeter than the visible, so a large millimeter optical depth need not imply a large visible optical depth.

However, the expansion of the Universe naturally cools the cosmic microwave background to which the iron needles are strongly coupled. In order to maintain T_{dust} constant as the Universe expands one needs a source of heat that supplies $4HU_{CMB}$ ergs/cm^3/sec, where U_{CMB} is the energy density of the cosmic background. The heat absorbed from the extragalactic background light is $\kappa_d\rho_d cU_{EBL}$ ergs/cm^3/sec, where $\kappa_d\rho_d$ is the absorption per cm in the visible due to the iron needles. So if these dust grains are heated by the diffuse light from galaxies, then the visible optical depth per Hubble radius must be about

$$\tau_V = 4U_{CMB}/U_{EBL} \approx 10^2$$

to provide sufficient heat to keep the dust temperature constant as the Steady State model requires. The EBL at 0.55 μm is less than about 1 kJy/sr (1 S_{10}) which gives an energy density more than 100 times less than that of the CMB (Dube, Wicks and Wilkinson 1977; Toller 1983). Thus if the dust were heated by the V band EBL, then $\tau_V > 400$ per Hubble radius. The EBL at 3.5 μm is certainly less than 0.1 MJy/sr which is about half of the flux observed by DIRBE at the South Ecliptic Pole. This upper limit on the near IR EBL is quite weak, and will improve as the zodiacal light is modeled and removed from the DIRBE flux. Even so, if the 3.5 μm EBL heats the dust we can show that $\tau > 30$ per Hubble radius at 3.5 μm, which implies an even higher optical depth in the visible. Thus if the visible or near infrared extragalactic background light were to heat the iron needles, then the optical depth in the visible must be so great that no distant galaxies could be seen, the expansion of the Universe would remain to be discovered, and neither the Big Bang nor the Steady State cosmology would exist as a topic. Thus the iron needles must be heated by a mechanism other than the absorption of starlight. No known electromagnetic background has sufficient energy density to provide adequate heating. The only bands where a large enough extragalactic electromagnetic background could be hidden are the 6-60 μm band obscured by the thermal emission from the zodical dust cloud, and the EUV and soft X-ray bands obscured by absorption in the ISM.

5.1 A MODEST PROPOSAL

The continuous creation of energy in the Steady State comes in an unspecified form. For the iron needles to produce the blackbody background, this energy must be coupled directly into the needles without going through radiation. However, why invent an *ad hoc* mechanism to heat the needles? Why not eliminate the middleman and create photons at the following rate:

$$(4\pi H_\circ/h\nu c)dB_\nu(T)/d[\ln(T)] \; \gamma/cm^3/sec/Hz$$

with T = 2.735 K? Then one would not even need the iron needles!

5.2 A CAUTIONARY NOTE:

Even though iron needles do not save the Steady State model without further *ad hoc* mechanisms, they may still have a large effect on the microwave background. An Ω_d of 10^{-7} in iron needles is perfectly compatible with standard Big Bang nucleosynthesis and will give an optical depth of 0.2 per Hubble radius. In the Big Bang model one assumes that the needles are cooling off in the expanding Universe, so no exotic heating mechanism is required. Also, the density increases with redshift so an optical depth of unity is reached at $z = 3$. These iron needles would not only erase the small-scale anisotropy produced during galaxy formation, like electron scattering in a re-ionized Universe, but would also eliminate any primordial spectral distortions.

5.3 FURTHER THOUGHTS ON IRON WHISKERS

Arp *et al.* (1990) suggest that radiation pressure will drive iron whiskers away from their natal galaxies. However, since they have much larger millimeter cross-sections they will be trapped in the CMB radiation field. The radiation drag force is

$$F = (4/3)v\sigma U/c,$$

so the time scale for slowing down a whisker is

$$\tau = v/a = mv/F = (3/4)c/(U\sigma/m) = 6 \times 10^{15}\text{sec} = 0.01/H_\circ.$$

This suggests that the whiskers will always be following the "cosmic standard of rest", and will not have peculiar velocities that will lead to small-scale anisotropy. One should note, however, that the needles will almost certainly be electrically charged and will thus be coupled directly to the magnetic field, and through the magnetic field to the ionized component of the gas, and through ion-atom collisions to the neutral component of the gas. This coupling could create small-scale anisotropy. The real test of the "cosmic fog" hypothesis lies with millimeter and centimeter observations of distant radio sources. If the Hubble law gives correct distances for QSOs, then their observed evolution falsifies the Steady State. Other radio sources such as the Sunyaev-Zeldovich effect from distant clusters of galaxies can be observed in the millimeter region.

6. LIMITS on Population III Stars and Dust

Wright (1981) presented simple models for distorted spectra produced by dust heated by an early generation of stars. (Population III stars). Reasonable fits could be made to the Woody and Richards (1981) spectrum by heating the dust at a redshift of 100 to 150 so the silicate feature was redshifted onto the peak of the cosmic microwave background spectrum. However, producing a large distortion such as the W-R distortion at high redshifts requires a large amount of energy release, since once matter is converted into radiant energy, the expansion of the Universe reduces its energy density by a factor of $1/(1 + z)$.

Now that the FIRAS experiment on COBE has shown that any distortions are small, I can turn the process around and try to place limits on any early energy release by stars. These limits are interesting once they get tight enough to rule out a release as large as the energy released by converting about 2 percent of all baryons from hydrogen to helium. For $H_o = 50$, this corresponds to the conversion of 10^{-5} of the mass of the Universe into energy. (All the models discussed here will have $\Omega = 1$).

The radiative transfer problem that needs to be solved to determine the distortion of the spectrum of the cosmic background by Population III stars and dust is quite simple, because of the isotropy of the background. Thus the usual angle dependence of the intensity drops out. Let us define an intensity variable $N(\nu, z)$ that is the "comoving mean intensity":

$$N(\nu, z) = J_{\nu(1+z)}(z)/(1 + z)^3$$

The function N is not changed by the expansion of the Universe. It differs from the distribution function by a factor proportional to ν^3, and it differs from the variable used by Wright (1981) by a constant factor of $hc/4\pi$. Note that the variable ν is always the observed frequency in these equations. The optical depth $\tau_\nu(z)$ is defined by

$$d\tau_\nu(z) = \kappa_{\nu(1+z)}\rho_d(z)cdt$$

where κ is the absorptive part of the dust opacity in cm^2/gm: scattering does not affect the spectrum of an isotropic background. The radiative transfer equation is

$$dN(\nu, z)/d\tau_\nu = N(\nu, z) - S(\nu, z)$$

where the source function $S(\nu, z)$ is given by

$$S(\nu, z) = B_\nu(T_d) + S_*$$

$$S_* = j^*_{\nu(1+z)}/\left(\kappa_{\nu(1+z)}\rho_d(z)\right)/(1 + z)^3$$

where $T_d = T_{dust}/(1 + z)$ is the "comoving dust temperature", and j^*_ν is the emissivity of the stars that are heating the dust. The equilibrium dust temperature is found using

$$\int B_\nu(T_d)\kappa_{\nu(1+z)}d\nu = \int N(\nu, z)\kappa_{\nu(1+z)}d\nu$$

Wright (1981) described the time history of the Population III stars using 2 parameters: a starting redshift z_s and a burning time t_s expressed as a fraction of the Salpeter time, t_E, which is the time constant associated with the Eddington limit: 4×10^8 yr. Massive stars radiate approximately at the Eddington limit, but convert only a fraction of a percent of their mass into radiation. Hence their lifetime is close to one percent of t_E. Wright (1981) was trying to model a large distortion and needed to radiate large amounts of energy, so the Eddington limit was a constraint. Given the closeness of the COBE data to a blackbody, only small amounts of energy can be radiated by Population III stars. I have computed some new models using the code from Wright (1981). All of these models convert f=0.0001 of the critical density for $H_o = 50$ into radiation. This is 10 times higher than the interesting limit given above, but the preliminary COBE data on spectral distortions is 10 times higher than the expected ultimate limits. The starting redshift is a parameter to be varied, with the burning time being chosen so that the Population III stars burn out at $z_f \approx 0.75 z_s$. The current dust opacity is not well known. I have used the result from Wright and Malkan (1987) that the visual extinction to $z = 1$ is 0.02 ± 0.03. Note that for a normal gas-to-dust ratio this requires $\Omega_B = 0.001$ for $H_o = 50$, which is about 100 times less than Ω_B from nucleosynthesis. To span this range in dust content I have computed models with current dust densities scaled to give 0.008, 0.02, and 0.05 for the visual extinction per Hubble radius.

The DIRBE experiment measures the sky brightness in 10 bands from 1.25-240 μm. The total sky brightness at the South Ecliptic Pole has been published by Hauser et al. (1990). Any model that exceeds the observed total brightness at any wavelength can clearly be ruled out. Modeling of the zodiacal light, galactic stars, and interstellar dust should allow extragalactic backgrounds as small as 0.1 - 0.01 of the total sky brightness to be detected. Thus comparing models to the DIRBE total sky brightness is approximately as conservative as using the preliminary FIRAS spectral distortion data.

Some models with low dust densities and low starting redshifts produce a lot of visible light. The visible sky brightness of the models has been expressed in S_{10} units: approximately 1 kJy/sr. Models producing more than 1 S_{10} can probably be ruled out, but since the current DIRBE and FIRAS limits are a factor of 10 or more higher than the hoped for ultimate sensitivity, a more consistent approach is to rule out models that are brighter than 10 S_{10}.

In the following Table, the column labeled (I_C/I_D) is the maximum ratio of model flux to observed DIRBE flux at the South Ecliptic Pole, and the next column is the wavelength at which the maximum occurs.

Population III Stars and Dust Models

$A_V(z=1)$	z_s	S_{10}	$100\delta(max)$	$(I_C/I_D)_{max}$	λ
0.008	10	7.79	0.33	0.074	1.25
0.008	20	0.152	0.32	0.115	3.5
0.008	40	0.000	0.63	0.066	3.5
0.008	80	0.000	0.45	0.000	—
0.02	10	6.3	0.71	0.069	1.25
0.02	20	0.09	0.60	0.099	3.5
0.02	40	0.000	0.87	0.007	3.5
0.02	80	0.000	0.47	0.000	—
0.05	10	4.2	1.36	0.060	2.2
0.05	20	0.000	1.37	0.036	3.5
0.05	40	0.000	0.88	0.000	—
0.05	80	0.000	0.40	0.000	—

The conclusion is that if an early generation of massive stars converts 2 percent of the baryonic density allowed by standard Big Bang nucleosynthesis from hydrogen to helium, the energy released will create a detectable EBL in the visible if the stars are bright at $z_s < 10$. For $z_s \approx 20$, the EBL may be detectable at 3.5 μm, the short wavelength hole in the zodiacal light. But this will require the identification and removal of at least 99% of the zodiacal and galactic foregrounds from the DIRBE data. The FIRAS instrument should see deviations from a blackbody that are close to its quoted sensitivity of $\delta = 0.001$, and these deviations vary by only $\pm$ a factor of 2 as the dust density and redshift are varied.

Wright (1982) considered whether one could make all of the cosmic background well after the Big Bang, using needle-shaped grains to thermalize the spectrum. This requires a much greater release of energy than the expected nuclear burning considered earlier. To minimize the required energy one must reduce z_s, which requires a large dust opacity in the millimeter region. Iron whiskers can provide this opacity. The main limitation on such models is provided by the short wavelength DIRBE and visible light limits on the EBL. Since the observed EBL has an energy density > 10-100 times smaller than the CMB, a large fraction of the starlight must be absorbed by dust. In order to convert 99% of the light into the far IR the optical depth in the near IR and visible must be > 4.6. Since iron whiskers have larger opacities in the millimeter than they do in the visible, this amount of dust is adequate to thermalize the background. For example, a model with $z_s = 40$ and $A_V = 0.02$ per Hubble radius that converts 0.004 of the critical density for $H_o = 50$ into radiation has $|\delta(max)| < 0.0001$ but produces 37% of the 3.5 μm flux observed by DIRBE at the SEP. For this example I assumed that the millimeter opacity of the iron needles was 10 times higher than the visible opacity. Arp et al. (1990) suggest the millimeter opacity could be 10^3-10^4 times higher than the visible opacity, but while that makes $\delta(max)$ even smaller it does not reduce the short wavelength flux that actually limits the model.

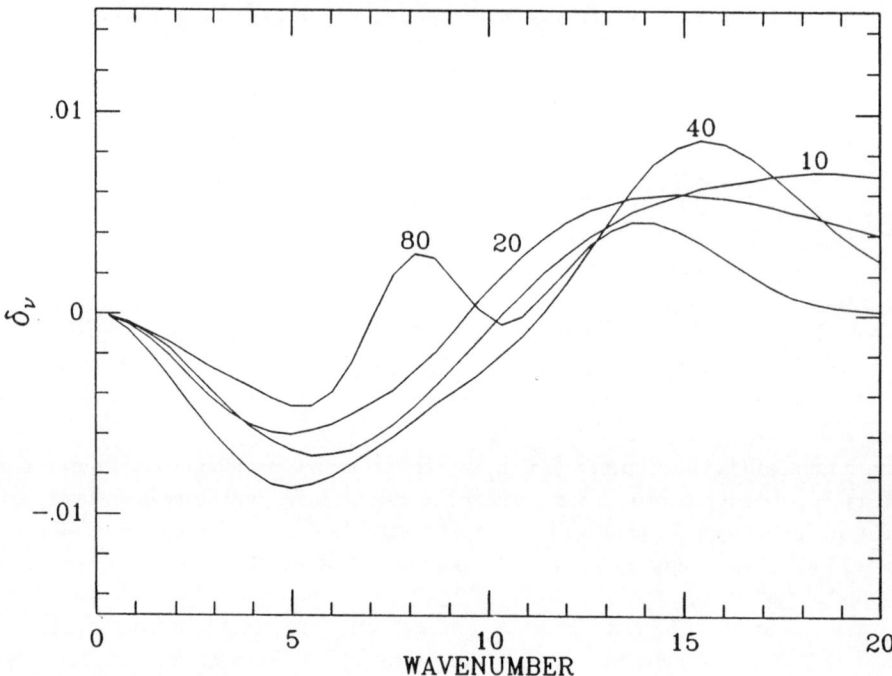

Figure 2: The spectral distortions produced for different starting redshifts, all with an extinction per Hubble radius of 0.02 and all converting 20% of the baryon mass from hydrogen to helium.

7. Local Dust

The FIRAS spectrum of the galaxy can used to estimate the order of magnitude expected distortion in high latitude spectra. The IRAS 100 μm all-sky data, when analyzed with the same spatial averaging used on the FIRAS data, gives 70 ± 14 MJy/sr. Since the gradient of a csc $|b|$ fit to the IRAS data is about 3.4 MJy/sr (de Bernardis et al. 1988) one might expect about 5% of the FIRAS spectrum to be present at the galactic poles. If I take a spectrum given by

$$I_\nu = B_\nu(2.735K) + 0.05g(\nu)$$

and adjust T_c to minimize the maximum value of $\delta(\nu)$, the minimax δ is 0.0012. Thus achieving the FIRAS sensitivity goal will require an understanding of the high latitude galactic dust. A comparison of this local dust distorted spectrum with a Compton distorted spectrum is shown in Figure 3. This Compton model has $y = 0.0001$ and can fit most of the variation in $\delta(\nu)$, so without an understanding of galactic dust a determination of $y < 0.0001$ will be impossible.

8. Evolving IR Galaxies

The IRAS 60 μm galaxy counts show a deviation from a non-evolving model. Hence the luminosity or the density of infrared bright galaxies must have been higher in the past. The redshift will move the peak flux of these distant IR galaxies into the band observed by FIRAS. The IRAS data only sample low redshift objects ($z \approx 0.1$ or less), but have been modeled by Beichman and Helou (1991) using an evolution in the comoving density of IR galaxies that varies like $(1+z)^n$ for $z < z(max)$. Since the total power radiated by a comoving volume diverges as $z(max) \rightarrow \infty$ for $n > 1.5$, this formulation requires a cutoff redshift. I have modified the Beichman and Helou evolution model to one following an exponential in cosmic time, $\exp(\gamma(1 - t/t_o))$. This model is usually applied to luminosity evolution instead of density evolution, but since only the luminosity density matters in the calculation of the spectrum seen by FIRAS and DIRBE, I will ignore this concern. For low redshifts the exponential and power law models are similar if $\gamma = n/1.5$. The exponential evolution model has the advantage that no cutoff redshift is required, so a single parameter γ can describe the strength of the evolution. Since there is no evidence yet for a distortion in the background spectrum, I feel that one adjustable parameter is the most one should consider. Figure 3 shows the result of adding the contribution of evolving IR galaxies with $\gamma = 3$ to the blackbody spectrum. In this case the minimax δ is 0.0011. With this luminosity density history, the energy released in the IR alone corresponds to turning 0.9% of the baryons allowed by Big Bang nucleosynthesis from hydrogen to helium. Unless these IR galaxies are all quite dusty with $L_{IR}/L_B >> 1$ the near infrared and visible light from stars will violate the DIRBE and visible EBL limits.

Finally, galactic dust and distant IR galaxies give spectral distortions that are quite indistinguishable. Figure 3 shows the $\gamma = 3$ evolving IR galaxy model fit to an adjustable amount of galactic dust. The maximum deviation between the two models is only $\delta = 0.00026$. The best way to distinguish between dust in our galaxy and dust in an evolving population of IR galaxies is to extend the IRAS source counts to the flux level where the flux integral has converged, which is about 1 mJy at 60 μm. This will be possible using the SIRTF telescope.

ACKNOWLEDGEMENT

The National Aeronautics and Space Administration/Goddard Space Flight Center (NASA/GSFC) is responsible for the design, development, and operation of the Cosmic Background Explorer (COBE), under the guidance of the COBE Science Working Group. GSFC is also responsible for the software development through to the final processing of the space data.

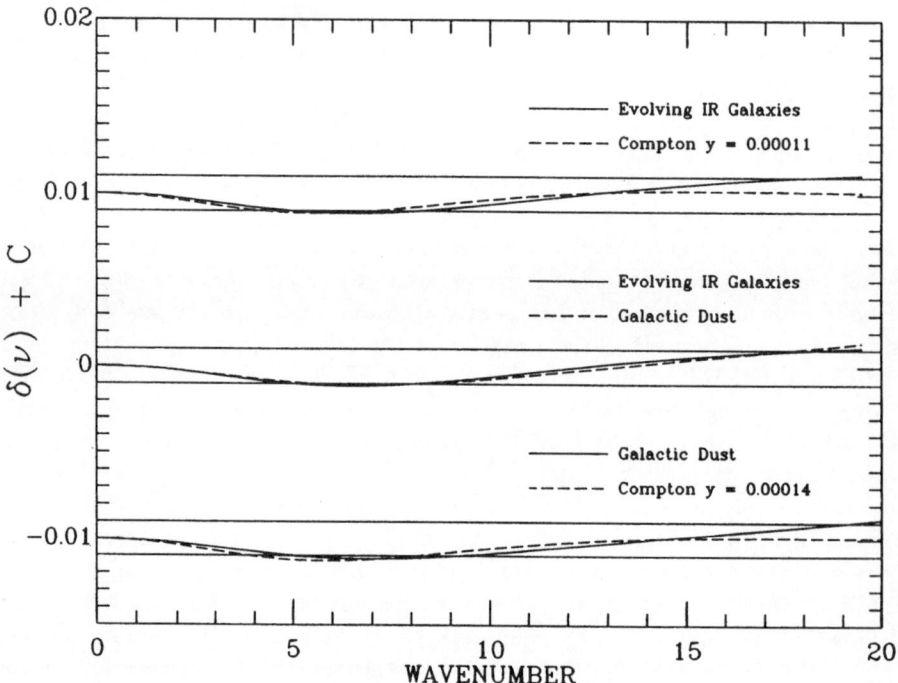

Figure 3: Distortion spectra produced by local galactic dust and/or evolving IR galaxies compared to each other and to Compton models

REFERENCES

Arp, H. C., Burbidge, G., Hoyle, F., Narlikar, J. V., and Wickramasinghe, N. C. 1990, Nature, 346, 807-812.

Beichman, C. A., and Helou, G. 1991. "What COBE Might See: The Far-Infrared Cosmological Background", ApJL, 370, L1.

Chan, F. K., and O'Neill, E. M. 1975, "Feasibility Study of a Quadrilateralized Spherical Cube Earth Data Base", Computer Sciences Corp, March 1975, EPRF Technical Report 2-75 (CSC).

Cheng, E., Mather, J., Shafer, R., Meyer, S., Weiss, R., Wright, E., Eplee, R., Isaacman, R., and Smoot, G., 1990, Bulletin of the APS, 35, 971.

de Bernardis, P., Masi, S., Melchiorri, F., Moreno, G., Vannoni, R., and Aiello, S. 1988, ApJ, 326, 941-946.

Dube, R. R., Wicks, W. C., and Wilkinson, D. T. 1977, ApJL, 215, L51.

Elias, J. 1978, ApJ, 223, 859.

Hauser, M. G., Kelsall, T., Moseley, S. H. Jr., Silverberg, R. F., Murdock, T., Toller, G., Spiesman, W., and Weiland, J., 1991, in "After the First Three Minutes" Conference, Univ. of Md., eds. S. S. Holt, C. L. Bennett, and V. Trimble, AIP Conference Proceeding TBD.

Hoyle, F. and Wickramsinghe, N. C. 1988, Ap. & Sp. Sci., 147, 245-256.

Kogut, A., Bensadoun, M., de Amici, G., Levin, S., Smoot, G. F., and Witebsky, C. 1989, ApJ, 355, 102-113.

Mather, J. C., Cheng, E. S., Eplee, R. E., Jr., Isaacman, R. B., Meyer, S. S., Shafer, R. A., Weiss, R., Wright, E. L., Bennett, C. L, Boggess, N. W., Dwek, E., Gulkis, S., Hauser, M. G., Janssen, M., Kelsall, T., Lubin, P. M., Moseley, S. H. Jr., Murdock, T. L., Silverberg, R. F., Smoot, G. F., and Wilkinson, D. T. 1990, ApJL, 354, L37-L41.

Smoot, G. F., Bennett, C. L., Kogut, A., Aymon, J., Backus, C., De Amici, G., Galuk, K., Jackson, P. D., Keegstra, P., Rokke, L., Tenorio, L., Torres, S., Gulkis, S., Hauser, M. G., Janssen, M., Mather, J. C., Weiss, R., Wilkinson, D. T., Wright, E. L., Boggess, N. W., Cheng, E. S., Kelsall, T., Lubin, P., Meyer, S., Moseley, S. H., Murdock, T. L., Shafer, R. A., and

Silverberg, R. F., 1991, ApJL, 371, L1.

Stockton, A. 1978, "The Nature of QSO Redshifts", ApJ, 223, 747-757.

Stockton, A. 1980, "Association between QSOs and Groups of Galaxies", in "Objects of High Redshift", Proceedings of IAU Symp. 92, eds. Abell, G. O., and Peebles, P. J. E., Dordrecht : Reidel, pp 89-97.

Toller, G. N., 1983, ApJL, 266, L79.

Woody, D. P., and Richards, P. L. 1981, ApJ, 248, 18.

Wright, E. L. 1981, ApJ, 250, 1.

Wright, E. L. 1982, ApJ, 255, 401.

Wright, E. L., Cheng, E. S., Dwek, E., Bennett, C. L., Boggess, N. W., Mather, J. C., Shafer, R. A., Hauser, M. G., Kelsall, T., Moseley, S. H. Jr., Silverberg, R. F., Smoot, G. F., Eplee, R. E., Isaacman, R. B., Meyer, S. S., Weiss, R., Gulkis, S. G., Janssen, M., Lubin, P. M., Murdock, T. L., and Wilkinson, D. T., 1990, "Preliminary Millimeter and Sub-Millimeter COBE Observations of the Milky Way Galaxy with a 7 Degree Beam" BAAS, 22, 874.

Wright, E. L., Mather, J. C., Bennett, C. L., Cheng, E. S., Shafer, R. A., Fixsen, D. J., Eplee, R. E. Jr., Isaacman, R. B., Read, S. M.,Boggess, N. W., Gulkis, S., Huaser, M. G., Janssen, M., Kelsall, T., Lubin, P. M., Meyer, S. S., Moseley, S. H. Jr., Murdock, T. L., Silverberg, R. F., Smoot, G. S., Weiss, R., and Wilkinson, D. T., "Preliminary Spectral Observations of the Galaxy with a 7° Beam by the Cosmic Background Explorer (COBE)", 1991, to be published in ApJ.

Wright, E. L., and Malkan, M. A., 1987, "An Upper Limit on Quasar Reddening", Bull.AAS, 19, 699.

SUBMM EMISSION FROM GALAXIES AND QUASARS

Rolf Chini
Max-Planck-Institu̧ c für Radioastronomie
Auf dem Hügel 69
D-5300 Bonn 1
Federal Republic of Germany

ABSTRACT. This lecture discusses the new field of extragalactic submm astronomy that has been developed during the last five years since large submm telescopes started their operation. Starting from some physical and technical aspects of submm observations in general, a variety of extragalactic sources like normal spirals, active galaxies and radio-quiet and radio-loud quasars are discussed with respect to their submm properties. It turns out that the FIR/submm regime of the spectra is dominated by thermal emission from dust. This even holds for the bulk of quasars. Being a good tracer for the total amount of gas the optically thin emission from dust at 1300µm is used to derive gas masses and, in a second step, luminosity-to-gas ratios for the various classes. The ratio L_{IR}/M_{gas} clearly defines the stage of activity in extragalactic objects.

1. Introduction

The submm continuum emission from galaxies and quasars is a rather new field in astronomy. Due to their intrinsic energy distributions, which attain a minimum around the wavelength of 1mm, most objects are extremely faint and require large telescopes and sensitive detectors in order to be measured. On the other hand, the few existing data are promising and corroborate the suggestion that this wavelengths range will play a key role in determining important physical properties. Interestingly, one must notice that it is just the submm range, where the two dominant radiation mechanisms, namely non-thermal Synchrotron radiation and thermal emission from heated dust grains, overlap and thus can be distinguished from each other by corresponding measurements. Having separated these components one may derive global quantities like source sizes, electron energies and magnetic field strengths for the non-thermal case and dust temperatures, total gas masses and star forma-tion efficiencies for the thermal case.

The following lecture summarizes the present stage of extragalactic submm research and starts with the physical mechanisms of submm contin-uum radiation (Section 2). After that, some technical details concerning submm telescopes and detectors as well as some observing modes are given

M. Signore and C. Dupraz (et al.), The Infrared and Submillimetre Sky after COBE, 249–262.
© 1992 *Kluwer Academic Publishers.*

(Section 3) in order to provide a better understanding of the submm data under discussion and to evaluate their significance and reliability. Then, different classes of extragalactic objects are presented with an emphasis on their submm properties. As it will turn out, thermal emission from heated dust grains dominates the spectra of normal spirals (Section 4) and active galaxies (Section 5) and thus allows the determination of total gas masses. In the case of quasars (Section 6), generally to be thought of as template sources of pure Synchrotron emission, it also turns out that most objects show evidence for a thermal component of varying strength relative to the non-thermal component. Some final remarks (Section 7) summarize the derived results and compare them for the different classes of objects. The luminosity-to-mass ratio turns out to be a major quantity, suited to serve as a new definition for the stage of activity in extragalactic objects.

2. Radiation Mechanisms at Submm Wavelengths

At submm wavelengths basically three components may contribute to the observed flux density. Thus, for an interpretation of the measurements a careful separation of them is necessary.

2.1 Free-Free Emission

One important thermal process is the emission from ionized gas. While it dominates the millimeter spectrum of star-forming regions its contribution at submm wavelengths is generally masked by emission from dust. Likewise, it is not clear, what fraction of free-free emission is present in the spectra of active galactic nuclei as there, it is difficult to separate from possible non-thermal radiation. In any case, it is important to take into consideration a possible contamination before interpreting submm data in terms of dust or synchrotron radiation.

2.2 Non-Thermal (Synchrotron) Emission

The radio spectrum of active galactic nuclei is dominated by Synchrotron emission which is produced by relativistic electrons spiralling in a magnetic field. It is likely that this radiation mechanism also contributes to the submm and FIR regime. For details concerning the Synchrotron emission from compact radio sources I refer to a review by Marscher (1987). For the present purpose it is sufficient to remember that the energy losses of electrons in a magnetic field due to Synchrotron radiation produce a characteristic spectrum where the optically thin part is described by

$$S_\nu \propto B^{(s+1)/2} \nu^{-(s-1)/2} \tag{1}$$

Here B is the magnetic field and s the index of the power law distribution of relativistic electron energies. At some frequency the source will become optically thick due to self-absorption and will radiate like a black body at the appropriate temperature. The resulting spectrum of the emitted energy is then of the form

$$S_\nu \; \alpha \; \nu^{5/2} \tag{2}$$

regardless of the electron energy distribution s. Thus $\alpha = 2.5$ is the steepest theoretical spectral index while the observational record is of the order of 1.7 for Synchrotron sources. Recent observations of steeper than canonical FIR turnovers in the spectra of several radio-quiet quasars (see below), however, have induced new calculations (DeKool and Begelman, 1989) which use Synchrotron vacuum theory and a concave double electron power law distribution in order to rescue the non-thermal model and to produce steeper than 2.5 spectra.

2.3 Thermal Emission From Dust

The most important thermal emission process at submm wavelengths is the re-radiation of high energy photons absorbed by dust grains. Many sources like star-forming regions or circumstellar disks remain optically thick even out to 100μm so that one sees only the surface of the emitting volume. Submm radiation, however, is optically thin for most objects and thus one samples the volume, i.e. the total mass of the emitting grains. The observed flux density S_ν correlates with the mass of dust M_d and its temperature T_d as

$$S_\nu = M_d \; k_\nu \; B_\nu(T_d) \; D^{-2} \tag{3}$$

where k_ν is the dust opacity and D the distance to the source. The spectral slope in the Rayleigh-Jeans region is ν^{2+m}, where m is the wavelength dependence of the dust opacity $k_\nu \; \alpha \; \nu^m$. In principle equation (3) then allows the determination of the total gas mass M_{gas} by adopting a certain gas-to-dust ratio R. In practice, however, there remain many uncertainties, particularly with the determination of k_ν and R (see e.g. Krügel et al. 1989).

3. Continuum Observations With Bolometers

Although ground-based submm observations started already in the early 80's, the astronomical results have been rather sparse at that time. This was mainly due to two disadvantages: i) Astronomers who wanted to observe in this wavelengths range had to use optical telescopes whose collecting area and spatial resolution was comparatively poor. ii) A second limitation was given by the detectors which lacked sensitivity because of their inadequate cooling systems. Fortunately, the interest in this last unexplored gap of the electromagnetic spectrum available from the ground increased rapidly and meanwhile a new generation of telescopes and detector systems has started its operation and further developments are on their way. The following Section is supposed to introduce some technical concepts of submm continuum observations but restricts to the framework that is needed in the present course.

3.1 Telescopes

The largest telescope operating at wavelengths around 1300μm and shorter is the IRAM 30m dish (MRT) on Pico Veleta, Spain, which has proven its capabilities at 870 and 1300μm. Most of the extragalactic results obtained so far and presented in the following come from this telescope. On Mauna Kea, Hawaii, two other submm dishes are operating, the 15m James Clark Maxwell Telescope (JCMT) and the 10m CalTec Submm Observatory (CSO), both covering all atmospheric windows from 350 to 1300μm. Finally, there is the Swedish ESO Submm Telescope (SEST) on La Silla, Chile, which has also been used so far at 870 and 1300μm. Three identical 15m telescopes are forming an interferometer on Plateau de Bure but there are currently no plans to use them below a wavelength of 1300μm. Finally, a 10m Submm Telescope (SMT) is under construction on Mt. Graham, Arizona, which will have the most accurate surface of all submm dishes, making it ideally suited for the shortest wavelengths of 350μm.

3.2 Detectors

At submm wavelengths, direct detection of continuum radiation is more sensitive than heterodyne detection. Due to their intrinsically unlimited bandwidth bolometers are the best choice at these wavelengths, whereas for observations shortward of 200μm, at least for low backgrounds, photoconductors are more sensitive. The passband of a bolometer system is defined by a filter either matched to atmospheric windows for ground based operation or chosen as a compromise between sensitivity and good definition of the effective wavelength for space instruments. A fundamental sensitivity limit is given by the photon noise of the thermal background radiation.

The MPIfR bolometers (Kreysa, 1990) are ^{3}He-cooled Ge-bolometers operating at 0.27K. There are several single channel systems in separate cryostats, each of them adapted to one of the atmospheric windows at 350, 870 and 1300μm. Generally, the systems work in the diffraction limit. This design has the advantage that all components, including the bolometers can be optimized for one wavelength. Under these conditions the background on the bolometer is minimal and highest sensitivity for point sources can be achieved. Likewise, this configuration yields the highest spatial resolution obtainable at a given telescope. Another advantage of the diffraction limited system is that a short piece of cylindrical waveguide can be used as a very efficient high pass filter. The low pass filters are produced at MPIfR for wavelengths as short as 30μm and will be implemented in the ISOPHOT project. They consist basically of capacitive meshes embedded in polyethylen or free-standing resonant meshes, i.e. metal foils with numerous cross-shaped apertures.

3.3 Observing Modes

Ground-based observations require signal modulation in order to cancel atmospheric emission fluctuations. Undoubtedly, chopping with the subreflector is the best way of doing this. The necessity of performing differential measurement becomes very convincing if one imagines that during average observing conditions, e.g. with the MRT on Pico Veleta, a variable flux density of about 1E3Jy is contained in the 11" beam at 1300µm, whereas the faint objects one wishes to observe emit only 1E-3Jy. The practical observation of a point source is identical to classical IR techniques, i.e. performing ON-ON measurements at the position of the source. After a certain amount of integration time, typically 10sec, the telescope is moved by the amount of the beam separation to center the object into the second beam. In this way the object is observed for equal intervals of time in the two beams, hopefully cancelling out fluctuations and drifts and a possible imbalance of the system. Likewise, the background of the source is measured on opposite locations at the sky. The procedure of changing the object from one beam into the other and integrating for a few seconds is repeated until one has achieved the desired signal to noise ratio.

This technique is strictly applicable to point sources only, i.e. to sources whose angular diameter is small compared to the HPBW. As soon as the object is larger than the HPBW but still smaller than the chopper throw one looses flux for obvious reasons. If, finally, the object is even larger than the beam separation an additional loss comes from the fact that the OFF-beam doesn't see a zero level with respect to the source but measures a gradient on the source. The only way to determine total flux densities and, what is of equal importance, the structure of extended objects is then to map the source, i.e. covering (with one or both beams) a field, large enough to contain the entire object and to assure that the edges of the map are on a constant, relative zero level. The mapping technique is very time consuming and insensitive compared to ON-ON measurement because one spends a considerable fraction of the time outside the source. The technical description by Kreysa (1990) contains a lot more detailed information on this subject.

Finally one has to calibrate the incoming flux to an absolute scale, generally provided by observations of the planets. The atmospheric transmission is determined by so-called "skydips" which is basically a procedure that measures the thermal emission of the atmosphere at different elevations.

4. Dust Emission From Normal Spiral Galaxies

The gas content of a galaxy is one of its basic parameters as it determines the star formation rate and the ratio of infrared to optical luminosity. The gas mass can basically be derived from CO lines (molecular hydrogen) and from the 21cm line of HI (atomic hydrogen). Both phases can be traced by means of optically thin dust emission at mm/submm wavelengths.

4.1 Nearby Spirals

The class of bright well-known optical spirals like M51 has turned out to be extremely difficult to be observed in the submm range. Although their total 100µm flux densities amount to several hundred Jy which implies 1300µm fluxes of the order of several 100mJy for dust temperatures between 20 and 30K, it is impossible to detect them e.g. at the MRT by means of conventional techniques in an 11" beam. This means that i) The submm emission must be extended. ii) The gradients must be smaller than the typical sensitivity which allows to detect sources of several mJy in rather short integration times.

Chini et al. (1986) have tried to overcome these technical disad-vantages by observing a sample of extended Sb and Sc spirals with a 90" beam and a chopper throw of 300" at 1300μm at the 3m IRTF. Combining these results with IRAS observations, they derived a spectral fit to the FIR/submm data by two dust components, a warm one of 48K and a rather cold one of about 17K. Eales et al. (1989), using a 136" chopper throw and slightly smaller beams, found for a similar sample of extended spirals 350 and 1100μm fluxes that, when extrapolated to 1300μm, are an order of magnitude smaller than our observed values. As a consequence, Eales et al. did not find any evidence for a cold dust component but claim that the FIR/submm emission in these galaxies is dominated by warm dust of 30 to 50K. Such high temperatures decrease the involved dust and gas masses drastically. Finally Devereux and Young (1990) suggested that in spiral galaxies 90% of the dust mass belongs to a component which is so cold that it is not detected by IRAS. Similar positions are taken by Stark et al. (1989) and Eales et al. (1989) who suggest a cold component of about 10K. These remarks show how uncertain our knowledge about the dust and gas content in bright nearby spirals presently is.

4.2 Small Distant Spirals

In order to extend our investigation about the presence of cold dust in normal spirals and to overcome the observational difficulties concerned with the limited beam throw in largely extended galaxies we selected a complete flux limited sample of small spirals with optical diameters smaller than 200" and 100μm flux densities larger than 10Jy. The north-ern hemisphere part of this sample comprises 80 galaxies mainly classi-fied as Sb and Sc, whose energy distributions rise from 12 to 100μm. Their distances range between 10 and 90Mpc. The beam size for the measurements was 11"; the beam separation, provided by the wobbling secondary, was 60". In this way we approached a situation where the OFF-beam was located outside or at least in the outer parts of most galaxies, hopefully providing the maximum possible signal. So far, 33 objects could be observed at the MRT at a wavelengths of 1300μm with an encouraging detection rate of 76%. The obtained flux densities at 1300μm range between 10 and 35mJy. Due to the fact that there exist not yet any measurements concerning the spatial extent of the objects at submm wavelengths one must regard these values as lower limits for the total 1300μm emission.

Combining the new 1300μm fluxes with the IRAS data one obtains energy distributions which attain their maximum at wavelengths longer than 100μm. There is no doubt that the origin of the FIR and submm emission is due to dust grains heated by stellar UV photons and re-radiating their energy at longer wavelengths. Therefore, one can inter-pret the observed spectra as thermal emission from dust and may derive formal dust temperatures from flux density ratios at different wavelengths. The coldest dust component clearly comes from the observations at 100 and 1300µm. Due to the absence of any other submm data we adopt a wavelength dependence of dust opacity m=2 and calculate a formal dust temperature according to Equ. (3). This yields an average value of T_d = 27.4 ± 2.5K. According to the remarks above this temperature would decrease in case the 1300µm emission is extended beyond the observational beam of 11". A 1300µm flux density twice as high as the observed value would e.g. result in a temperature of 23.6K and increase the M_{gas} by a factor of 2.2. For the individual galaxies the 60µm data are consistent with the derived dust component, very often they are even underestimated.

4.3 The L/M_{gas} Ratio of Non-Active Galaxies

The optically thin emission at 1300µm is now used to estimate the mass associated with the coldest observed dust component. From the data of Chini et al. (1986), concerning the bright nearby spirals, one may derive a relation between the luminosity and the gas mass of the form

$$\log L = (0.85 \pm 0.18) \log M_{gas} + (1.91 \pm 0.82) \tag{4}$$

which was interpreted as to origin from a proportionality between the star formation and the available gas content of a galaxy. The average luminosity-to-gas ratio is of the order of 4.

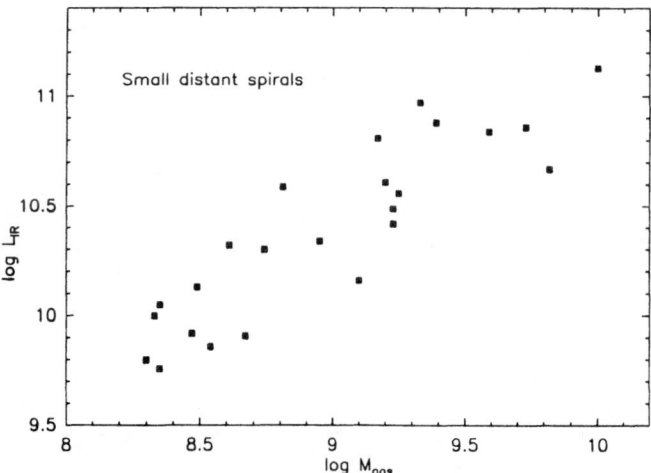

Fig.1: Luminosity vs. gas mass for a sample of normal spirals

In analogy we interpret the results for the sample of small distant spirals. According to Equ. (3) we convert the observed flux densities into a gas mass and derive values that range from 2E8 to 1E10M⊙, typical for spiral galaxies. Integrating the energy distributions at the IRAS bands yields luminosities between 6E9 and 2E11L⊙. Fig. 1 compares the obtained values of L and M_{gas} and shows a good correlation of both quantities. A least square fit yields

$$\log L = (0.71 \pm 0.08) \log M_{gas} + (4.01 \pm 0.71) \tag{5}$$

Like in the case of nearby spirals we interpret this proportionality of luminosity and gas mass as a result of the star formation. The more gas available the more stars are to be formed. The average luminosity-to-gas ratio of these galaxies is 29.1 ± 14.7, a factor of six higher than in the nearby spirals. As mentioned above, all numerical results must be treated as preliminary values due to the unknown submm extent of these distant spirals. Undoubtedly, they are first limits which in case of extended dust emission rapidly evolve towards the properties derived for nearby spirals, i.e. the temperature will decrease, the gas mass will increase and thus the luminosity-to-mass ratio also decreases.

5. Active Galaxies

In active galaxies the interstellar material is closely related to the physical processes of activity itself, either via star formation or via the inflow of gas onto a massive relativistic object. The study of the interstellar matter in these sources is therefore an interesting field. Today one distinguishes between a variety of activities like Seyferts, liners, star bursts, HII region galaxies and quasars. It is a very attractive idea that all types of activity have a common origin and the varying appearance of the individual objects merely reflects evolution- ary stages or special conditions of the line-of-sight orientation.

5.1 Markarian Galaxies

The First Byrakan Spectral Sky Survey is a catalog of 1500 active galax- ies (Markarian et al. 1981) exhibiting a strong UV continuum and an emission line spectrum. The Second Byrakan Spectral Sky Survey (Markar- ian et al. 1985a,b) shows that 10% of these galaxies are Seyferts and most of the remaining objects are so-called star burst and HII region galaxies. Krügel et al. (1988a,b) and Chini et al. (1989a) have selected an FIR flux density limited sample of about 50 Markarian galaxies in order to study their submm continuum. The broad band energy distribu- tions of these galaxies are strikingly similar in the sense that they steadily increase from optical wavelengths to about 60µm and turn over towards 100µm. Spectra, including the submm data at 870 and 1300µm, obtained at the IRAM 30m MRT, and IRAS data from 12 to 100µm are shown by Chini et al. (1989a).

From the steep average spectral index of about 2.8 and from the common turnover of the spectra around 100µm one must conclude that the radiation at FIR and submm wavelengths is most likely due to thermal emission from dust. Having established the thermal nature of the emis- sion, it is possible to fit the data from 100 to 1300µm by a single dust component of 30 ± 5K. In many cases this dust component even fits the data point at 60µm. It is evident, however, that additional thermal components of higher temperatures are required in order to explain the energy distribution at shorter wavelengths. In these simple fits of the form $\nu^m B_\nu(T_d)$ the wavelengths dependence of dust opacity m and the dust temperature T_d were left as free parameters. For m we find a value of 2.0 ± 0.2 in accordance with the value observed in the diffuse inter- stellar medium of our Galaxy. This finally means, that although we know very little about galactic dust, the dust in extragalactic systems seems not to be considerably different.

The comparatively high temperature of 30K for the coldest dust component corroborates our assumption that also the IRAS fluxes are contained within the inner 11" of the galaxies: In case of thermal emission, the longest wavelengths trace the lowest temperatures and thus the most outer parts of the emitting area. If the emission at 1300μm is confined to a region of 11", it is evident that the radiation at IRAS wavelengths cannot be more extended than this.

5.2 The L/M_{gas} Ratio of Active Galaxies

Although the spectra are rather simple, there is even more information contained in these few data points. It is a general behavior that the coldest dust component carries the bulk of mass (> 90%) whereas the warmer dust components emit the bulk of luminosity (>70%). In that sense, one may use the FIR/submm data to determine both quantities fairly accurate.

Because the emission at 1300μm from these objects is optically thin one may use Equ. (3) to determine the mass of dust and hence, by adopt- ing the galactic gas-to-dust ratio of 100, the total gas mass of the galaxies; the corresponding values reach from 1E8 to 5E10$M_\odot$ as to be shown later. At this point one may ask: How reliable are gas masses derived from the 1300μm emission? In order to check the validity of the method applied, we went the "classical" way of determining gas masses and observed our sample of active galaxies at the CO(1-0) and (2-1) transitions (Krügel et al. 1990, 1991), the latter one with the identi- cal beam size as used for the dust measurements. Comparing the results of both methods, i.e. dust masses obtained from optically thin emission at 1300μm and from the CO(2-1) transition, we find reasonable agreement. Being aware of all uncertainties associated with both methods it seems that the global properties of the diffuse interstellar medium in extrag- alactic systems cannot be dramatically different than those observed in our Galaxy.

As explained above, the observations between 60 and 1300μm of active galaxies can roughly be interpreted in terms of one single dust component. With this result at hand, it is now possible to determine dust and gas masses of active galaxies also in the absence of corre- sponding submm data: One simply determines the dust temperature from the IRAS data at 60 and 100μm and extrapolates to the flux density at 1300μm. The expected 1300μm flux may then be converted into a dust and gas mass. In order to enlarge our data base on active galaxies we derived gas masses for another 140 Markarians by using the IRAS fluxes to determine the dust temperature T_d, extrapolating to the expected flux at 1300μm and to convert this value into a dust and gas mass according to Equ. (3).

The bulk of luminosity in these galaxies can be derived by integrating the energy distribution between 12 and 100µm; this yields values
between 2E10 and 3E12L₀. Again we find a proportionality between
luminosity and gas mass (Fig. 2) of the form

$$\log L = (0.74 \pm 0.07) \log M_{gas} + (4.53 \pm 0.66) \qquad (6)$$

very similar to that derived for distant spirals. In analogy we interpret the relation as due to star formation in the sense that the larger
the reservoir of gas the higher the formation of new stars. We will come
back to this point in Section 7.

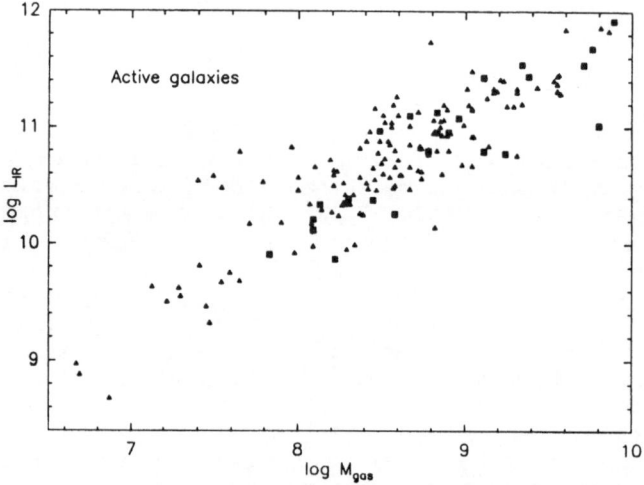

Fig.2: Luminosity vs. gas mass for Markarian galaxies; squares denoted
galaxies with submm data, triangles denote IRAS data only.

6. Quasars

There is no doubt that the submm regime is an important spectral region
for the interpretation of quasar spectra. It is the last unexplored gap
between IR and radio wavelengths and seems to play a key role in discriminating between thermal and non-thermal components. In particular since
IRAS data have become available for a number of quasars (Neugebauer et
al. 1986, Sanders et al. 1989) a controversial discussion about the
origin of FIR emission has started.

6.1 Radio-Quiet Quasars

Although quasars were originally discovered as bright radio objects,
meanwhile about 90% of all known quasars don't show any or only
extremely week radio emission. The spectra of most radio-quiet quasars

are identical to those of normal spirals or active galaxies, i.e. they rise at least from NIR wavelengths to about 100µm. From the absence of any significant radio emission in the cm regime it follows that the spectral turnover must occur between 100µm and 1mm like also known from galaxies.

In order to determine the spectral shape at submm wavelengths and to elucidate the origin of the FIR emission Chini et al. (1989b) observed all northern radio-quiet quasars which were detected at least in 3 IRAS bands. The observations were performed at 1300µm at the MRT and are by far the most sensitive measurements ever done in this wavelengths range. The 25 objects were integrated down to a 3σ limit of approximately 3mJy - a value being a compromise between telescope time and the implications following from an upper limit of that order. Five of the objects could be positively detected above the 3σ level. All spectra are given by Chini et al. (1989b) from X-ray to 1300µm.

The most important quantity that can be derived from those data is the spectral index α between 100 and 1300µm, defined as $S_\nu \propto \nu^\alpha$. The average value of α for this sample of radio-quiet quasars corresponding to a straight line that connects the 100µm points and the 1300µm upper limits is 2.18 ± 0.32. In reality this spectral index must be even larger because i) Most values at 1300µm are upper limits and ii) Most spectra rise from 60 to 100µm indicating that the turnover point lies beyond 100µm at a still higher flux density. Both effects will increase α considerably. Taking into account that four of the quasars already exceed the limit of 2.5 and another 8 come rather close to it one is tempted to rule out Synchrotron radiation as the origin of the FIR emission. For a more detailed discussion of the data see Chini et al. (1989b).

As already addressed above, the similarity of the energy distributions of radio-quiet quasars with those of normal and active galaxies indicates that thermal emission might be responsible for the spectral turnover beyond 100µm. In order to fit the spectrum over the entire wavelengths range, one faces the well known fact that several components are required to do that job. In a thermal model, where an energy source in the center of the host galaxy heats the interstellar or circumnuclear dust, this is easy to understand because one expects a temperature gradient.

Following the interpretation outlined during Sections 4 and 5 we start to calculate the bulk of the material by determining the coldest dust component. Due to the upper limits at 1300µm one may obtain only lower limits for the dust temperature when combining 100 and 1300µm data. The 60 and 100µm data, on the other hand, give upper limits of T_d, implying that the entire range from 60 to 1300µm is dominated by one single dust component. In this way one derives mean limits for the coldest dust ranging between 24 and 36K. The first limit of 24K seems to be rather low taking into account that we are dealing with the most luminous objects in the universe. The upper limit of 36K is slightly higher than the coldest dust found in active galaxies and seems to be a better approach. We therefore adopt in the following that a single dust component is a good approximation to explain the spectra from 60 to 1300µm and use the temperatures derived from the 60 and 100µm data to describe the cold dust component in radio-quiet quasars.

From the temperatures $T_{60/100}$ we can extrapolate again to the 1300µm flux density which should be observed if all energy at that wavelength is emitted from heated dust. As it turns out the corresponding values are generally below 1mJy and explain our failure to detect most of the sources. Following Equ. (3) one may derive the gas masses associated with the cold component. The results show that the involved gas masses range from 7E7 to 2E10M$_\odot$, comparable to the masses of normal spiral galaxies (Chini et al., 1986) and active galaxies (Krügel et al. 1988a,b). This agreement of deriving reasonable gas masses and dust temperatures from FIR flux densities can be taken as further evidence for the thermal origin of the FIR radiation. An independent corroboration for the thermal model comes from 2µm polarization measurements by Sitko and Zhu (1991). They find that the extremely low degree of polarization in their sample of quasars can be at most due to scattering from dust particles and not origin from Synchrotron emission. All observations favour a picture where the central engine heats a dusty disk in its environment. The dimensions of such a disk may be of the order of 0.1pc to several kpc and its emission dominates the spectrum from NIR wavelengths to about 1300µm (e.g. Sanders et al. 1989).

6.2 Radio-Loud Quasars

Finally we want to investigate the submm properties of radio-loud quasars. Their energy distributions from X-rays to 100µm are roughly identical to those of the objects discussed so far, i.e. they are steadily increasing towards longer wavelengths. The radio spectrum, on the other hand, divides these quasars into two groups: i) The very compact sources whose core dominates the total emission and produces a flat Synchrotron radio spectrum. ii) The steep spectrum sources whose emission originates from extended radio lobes.

Chini et al. (1989c) have performed 870 and 1300µm observations for all radio-loud quasars from the list of Neugebauer et al. (1986) with two or more detections in the IRAS bands. The spectra, displayed from the X-ray to the radio range (Chini et al. 1989c) indicate that the observed submm emission is an extension of the steep spectrum radio component and thus non-thermal in origin. Simultaneously, the submm data clearly demonstrate the existence of a second component shortward of 100µm. Even a large number of classical flat spectrum sources show a steep decrease towards submm wavelengths dividing the energy distribution into two components. For both groups of radio-loud quasars the spectral index between the FIR and the X-ray range is similar to that seen in radio-quiet quasars. Likewise, the spectral index between 100µm and 870 or 1300µm is rather steep for some objects so that a thermal interpretation for the FIR emission is again rather tempting. As discussed by Chini et al. (1989c) a thermal model yields reasonable result concerning dust temperatures and gas masses. Therefore, we favour an interpretation where the basic concept for quasars involves accretion onto a massive compact object. The orientation of the active nucleus (jet, disk, radio lobes) with respect to the observer is vital for its appearance as a radio-loud or radio-weak object. For small angles

between the jet axis and the line of sight the relativistic jet dominates and can swamp the entire isotropic emission. For larger angles,
the Synchrotron radiation becomes weaker and, if not dominated by steep
spectrum radio lobes, the thermal emission from the gaseous disk
increases relative to the radio component and dominates the FIR regime.

6.3 The L/M_{gas} Ratio of Quasars

In order to accomplish the investigation of the luminosity-to-mass ratio
in extragalactic objects, Chini et al. (1989b) have calculated the
dependence of L_{IR} and M_{gas} in quasars under the very plausible
assumption that the FIR range is dominated by thermal emission from
dust. The temperature of the dust is determined from the IRAS data as
described above, the mass was calculated according to the procedure
explained above. Most objects follow a relation of the form

$$\log L = (0.66 \pm 0.23) \log M_{gas} + (5.99 \pm 2.10) \qquad (7)$$

where the slope is similar to those in the L/M_{gas} relations for galaxies. This time, however, individual L/M_{gas} ratios range between 5E2 and
4E3, i.e. a factor of 5 to 40 higher than in active galaxies.

7. Conclusions

This lecture has demonstrated that the FIR range of most extragalactic
sources is dominated by thermal emission from dust. Even most quasars
show a thermal bump around $100\mu m$ originating from a circumnuclear disk
heated by the active nucleus. The optically thin emission from dust
allows fairly accurately the determination of the bulk of gas contained
in the individual objects. It could be shown that there exists a tight
proportionality between M_{gas} and the luminosity. Surprisingly this
proportionality is similar for all classes of objects in the sense that
an increase of the mass by an order of magnitude implies a luminosity
that is roughly five times higher. There is a difference in the temperature for the bulk of dust increasing from about 20K in normal spirals to
30K in active galaxies and to 35K in quasars. This difference in temperature is closely related to the luminosity-to-mass ratio; objects of a
given L/M_{gas} ratio heat the dust to a fairly well determined minimal
temperature. Chini et al. (1989a) derived an empirical relation

$$\log L/M_{gas} = -1.44 + 0.12 \, T_d \qquad (8)$$

which in principle reduces the difficult determination of L/M_{gas} to a
measurement of the coldest dust temperature. This may be of practical
use for classifying unknown sources to be discovered during future FIR
and submm surveys.

Likewise, the ratio L/M_{gas} describes how efficiently gas is
converted into luminosity: i) Normal star formation, like known from our
own galaxy, result in an L/M_{gas} ratio of 5 to 30. ii) Violent star
formation as it occurs in star burst galaxies produce an L/M_{gas} ratio of

262

about 100, reflecting that the same amount of gas is more efficiently used to form stars. This can either be achieved by increasing the star formation rate or by shifting the IMF towards more massive stars. iii) The high L/M_{gas} ratio of about 1000 as observed in quasars requires an even more efficient process of turning gas into luminosity. This is probably achieved by the accretion of mass onto the central compact object. In this sense, the L/M_{gas} ratio is a direct measure for the activity of the objects discussed in this lecture.

References

Chini, R., Kreysa, E., Krügel, E., Mezger, P.G.: 1986, Astron. Astrophys. **166**, L10

Chini, R., Krügel, E., Kreysa, E., Gemünd, H.-P.: 1989a, Astron. Astrophys. **216**, L7

Chini, R., Kreysa, E., Biermann, P.L.: 1989b, Astron. Astrophys. **219**, 87

Chini, R., Biermann, P.L., Kreysa, E., Gemünd, H.-P.: 1989c, Astron. Astrophys. **221**, L3

DeKool, M., Begelman, M.C.: 1989, Nature **338**, 484

Devereux, N., Young, J.: 1990, Ap.J. **350**, L25

Eales , S., Wynn-Williams, G., Duncan, D.: 1989, Ap.J. **339**, 859

Kreysa, E.: 1990, in "From Ground-Based to Space-Borne Submm Astronomy", Proc. 29th Liège International Astrophysical Colloqium July 1990, Eds. N. Longdon, B. Kaldeich, ESA SP-314, p. 265

Krügel, E., Chini, R., Kreysa, E., Sherwood, W.A.: 1988a, Astron. Astrophys. **190**, 47

Krügel, E., Chini, R., Kreysa, E., Sherwood, W.A.: 1988b, Astron. Astrophys. **193**, L16

Krügel, E., Steppe, H., Chini, R.: 1990, Astron. Astrophys. **229**, 17

Krügel, E., Steppe, H., Chini, R.: 1991, Astron. Astrophys. (submitted)

Markarian, B., Lipovetskij, V., Stepanian, D.: 1981, Astrofiz. **17**, 619

Markarian, B., Erastova, L., Lipovetskij, V., Stepanian, D., Shapovalova,A.: 1985a, Astrofiz. **22**, 215

Markarian, B., Stepanian, D., Erastova, L.: 1985b, Astrofiz. **23**, 43 9

Marscher, A.P.: 1987, in "Proceedings of Workshop on Superluminal Radio Sources", eds. A. Zensus and T.J. Pearson, Cambridge University Press

Neugebauer, G., Miley, G.K., Soifer, B.T., Clegg, P.E.: 1986, Ap.J. **308**, 815

Sanders, D.B., Phinney, E.S., Neugebauer, G., Soifer, B.T., Matthews, K.: 1989, Ap.J. **347**, 29

Sitko, M.L., Zhu, Y.: 1991, Ap.J. **369**, 106

Stark, A.A., Davidson, J.A., Harper, D.A., Pernic, R., Loewenstein, R., Platt, S., Engagiola, G. and Casey, S., 1989, Ap.J. **337**, 650

INFRARED EMISSION FROM THE GALAXY

F. BOULANGER

Radioastronomie, Ecole Normale Supérieure,

24 rue Lhomond, F-75005 Paris, France

F.X. DESERT

DEMIRM, Observatoire de Meudon,

F-92195 Meudon Principal Cedex, France

ABSTRACT: This lecture summarizes empirical knowledge on the infrared emission from the Galaxy in the plane and at high Galactic latitude. In discussing observations we convey to the reader a basic understanding of the origin of the infrared emission of a normal galaxy such as the Milky Way and of the composition of interstellar dust. We analyze the infrared emission from the different components of the interstellar medium: neutral atomic, ionized and molecular gas, and the contribution of dust grains, small particles and large molecules to the emission from the near-infrared to 1mm. Most observations analyzed here were obtained with the Infrared Astronomy Satellite (IRAS) which surveyed the sky at 4' resolution at 12, 25, 60 and 100μm. The Cosmic Background Explorer (COBE) has recently extended these observations to both shorter and longer wavelengths. We briefly discuss how these observations, when analyzed, will complement and test current knowledge on interstellar matter. A motivation for studying the Galactic emission is the necessity to subtract it before one may discuss the existence of an infrared extragalactic background. Since the school was mostly focused on cosmology special attention is given to the infrared emission seen in the region of minimum emission including its structure on small scale which complicates the observation of faint extragalactic objects.

1. Introduction

Small pieces of solid particles and large molecules present in the interstellar medium

M. Signore and C. Dupraz (et al.), The Infrared and Submillimetre Sky after COBE, 263–289.

absorb light from stars and radiate the corresponding energy at infrared and sub-mm wavelengths. This emission was surveyed over the whole sky by the Infrared Astronomy Satellite (IRAS) at 12, 25, 60, and $100\mu m$. Emission within a few degree of the Galactic Plane and from some bright nearby molecular complexes had been previously measured from balloons (see lecture by Puget 1985 in a former Les Houches school). At 100 μm, for the Solar Neighborhood radiation field, the sensitivity of the IRAS survey corresponds to the emission from a column density of matter of a few $10^{19} H cm^{-2}$; that of DIRBE (Diffuse Infrared Background Experiment) is one order of magnitude better. The IRAS sensitivity was sufficient to detect emission from the Galaxy over most of the sky. These observations have been used to probe and compare the spatial distribution of dust and gas outside the Galactic plane, to study the size distribution of interstellar dust grains, to measure the radiation field within cloud and to discuss the origin of the infrared emission from the Galaxy. The study of the IR emission from the Galaxy is the basis of our understanding of the IR emission from galaxies which is used to discuss large scale properties of star formation within and among galaxies. The most recent reviews on the Galactic infrared emission are Cox and Mezger (1989) for the emission in the plane and Boulanger (1989) for the emission at high Galactic latitude.

Since the main focus of the school was cosmology this lecture concentrates mainly on presenting knowledge which is useful for subtracting the Galactic contribution to the sky emission. With this perspective in mind we only briefly describe the infrared emission from the Galactic Plane. Readers interested in understanding the detailed derivation of the conclusions presented here are referred to Cox and Mezger and references therein.

At high latitude the line of sight goes through the Galaxy along a short path within which, in most cases, the radiation field is approximately constant. Wherever this is the case the infrared brightness scales as the column density of dust and, for constant dust abundance and optical properties, as the gas column density. Direct correlation of the distribution of interstellar gas allows to verify this expectation and to measure the infrared emissivity (brightness per unit column density of matter) for each of the components of the interstellar medium: atomic, molecular and ionized. These emissivities may be used to estimate the absolute brightness of the Galactic emission in the regions of minimum Galactic emission and discuss the existence of an extragalactic background. The fact that interstellar dust is not made of one single type of particle complicates this simple picture. In a given radiation field the emission spectrum depends on the size, the chemical composition and the structure (their surface to volume ratio, i.e. compact or fluffy particles) of the particles. The smallest particles contributing to the emission are made of only a few tens of atoms, the largest, billions. The smallest particles are so tiny that their heat capacity is small compared to the energy of a stellar photon. Such particles have not a fixed temperature, they reach high temperatures after a photon absorption and thereafter cool on a time scale shorter than the mean frequency of heating absorptions. For example a UV photon heats a molecule of 50 carbon atoms to a peak temperature of ~ 1000 K; such absorptions occur once a year while the cooling time-scale at 1000 K is one second. Temperature fluctuations are significant up to sizes of 10 nm or a few 10^5 atoms. For larger sizes there is at any time a statistical balance between absorbed and emitted energy which sets the particle at a fixed equilibrium temperature. This temperature depends on the ratio between the absorption cross section in the UV and visible and that in the far-infrared. This ratio depends on the size and the chemical nature of the particles but also much on their structure. Fluffy particles have much larger emissivities in the far-infrared than compact particles of the same mass and composition, and consequently are at lower temperatures. This short discussion shows the importance of understanding the dust composition in order to model the infrared emission from the Galaxy. This is why we devote a significant fraction of this lecture to a description of what we know about interstellar dust from extinction, scattering, and infrared emission.

The lecture has two main sections one discusses the infrared emission at high Galactic latitude from an empirical approach based on the comparison of the spatial distribution of the infrared emission with that of interstellar matter (section 2). The other relates the spectral distribution of the emission to the dust composition (section 3). The following sections bear on more specific points. Section 4 gives a brief summary of studies of the emission in the Galactic plane. Section 5 describes statistical studies of the spatial structure of the high latitude emission. Section 6 concludes the lecture by presenting some current questions to which COBE observations are expected to contribute.

2. Galactic Emission outside the Galactic Plane

The galactic emission has a complex morphology with filamentary structures named IR cirrus after their resemblance to clouds in the terrestrial atmosphere. In a first sub-section we discuss global properties of the emission from latitude profiles. In a second one we study the correlation between infrared cirrus and interstellar clouds, atomic and molecular, seen in maps of H I, CO emission, of optical nebulosity and visual extinction. A special sub-section is devoted to the contribution of diffuse ionized gas. We end this section by estimating the emission in the regions of the sky which are the most devoid of interstellar matter and are therefore the most suitable to look for an extragalactic background. A budget of the infrared emission from the Solar Neighborhood is given in Table 2.

2.1 Latitude Profiles

Figure 1 from Boulanger and Pérault (1988, hereafter BP) presents Galactic latitude profiles of the infrared emission obtained after subtraction of zodiacal light. Data in the direction of the Magellanic Clouds, the main nearby molecular clouds and OB associations, and a few bright point sources were discarded when computing these profiles. An H I profile built over the same parts of the sky as the infrared profiles is also presented in figure 1. The fact that the latitude profiles follow a cosecant law proves the existence of Galactic emission for all b at all IRAS wavelengths. The subtraction of the zodiacal light is particularly critical for the detection of the the Galactic emission at 12 and 25 μm which at $|b| > 10°$ represents less than 1% of the zodiacal light in the ecliptic plane. The absence of emission above 30° in the 12 and 25 μm profiles results directly from the procedure used by BP to model the zodiacal light and does not mean that there is no Galactic emission at 12 and $25\mu m$ all the way to the poles. From the slopes of the cosecant laws BP derived the average spectrum of the IR emission at high Galactic latitude listed in Table 1. This spectrum is extended to the near-IR by the $3.3\mu m$ balloon measurement of Giard et al. (1989) at a few degrees above the molecular ring and to the submillimeter waveband by the DIRBE measurements at the southern ecliptic pole (Hauser et al. 1990).

The question of the origin of the infrared emission measured by the cosecant

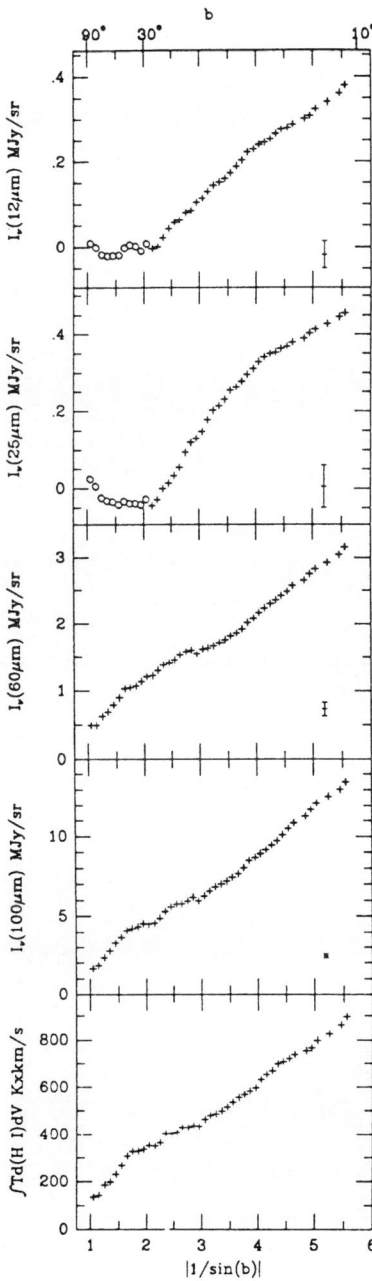

Fig. 1: Profiles of infrared and H I emission for absolute latitudes larger than 10°. A representative error bar is plotted on the bottom-right corner of the infrared profiles.

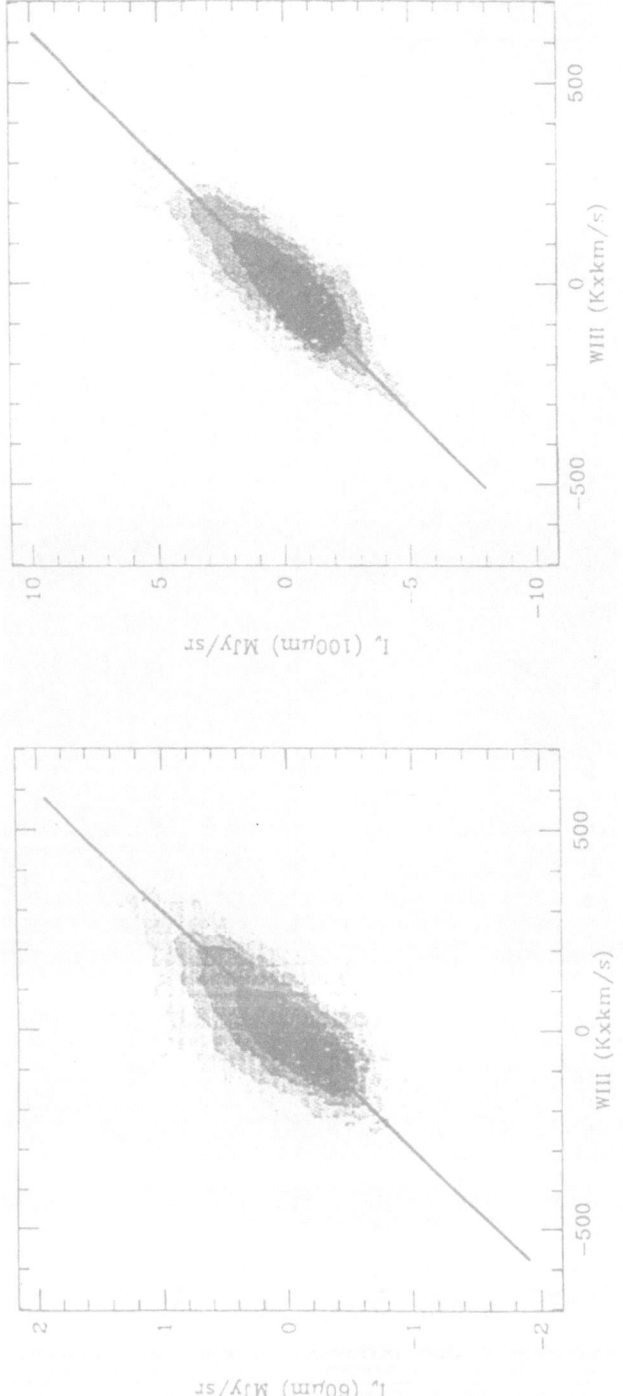

Fig. 2: Correlation between infrared and H I emission at $|b| > 10°$. For each pixel in the 60, 100μm, and H I all-sky maps, we computed the difference between the map value and the average emission at the Galactic latitude of the pixel. The two gray-scale plots show the density distribution of these differences in a 60μm − H I and a 100μm − H I diagram. The straight lines superposed on the gray-scale plots go through the origin and have slopes equal to the ratios of 60 and 100μm to H I emission derived from the cosecant laws. Pixels in the Magellanic Clouds and in the vicinity of the main nearby molecular clouds and OB associations were not included in these plots.

laws can be addressed by comparing its spatial structure about the average latitude profile with the distribution of interstellar matter. BP computed for each $1/2° \times 1/2°$ pixel all-sky maps of 60, 100 μm and H I emission the difference between the observed emission and the average brightness at the Galactic latitude of the pixel. A pixel-by-pixel comparison of these differences is presented in Figure 2. This figure shows a clear correlation between the structure in the infrared and H I maps. The straight lines on the $60\mu m - H \ I$ and $100\mu m - H \ I$ diagrams represent the infrared emissivity per H atom derived from the cosecant laws (see Table 1). The fact that these lines approximately follow the major axis of the distribution of points in the IR-H I diagrams implies that the far-infrared emission at high galactic latitude arises mainly from dust associated with neutral atomic gas or from some other gas component which spatial distribution is correlated with that of H I. Uncertainties in the zodiacal light subtraction is too large to allow a pixel-by-pixel correlation between the 12 and 25 μm emission and the distribution of interstellar dust over the whole sky. BP proved that the 12 and $25\mu m$ emission comes from the interstellar medium by showing that stars and circumstellar dust-shells account for only a small fraction of the observed emission. Photospheres and circumstellar dust- shells account for about 10% of the $12\mu m$ and 2.5% of the $25\mu m$ emission seen at $|b| > 10°$.

2.2 Infrared Cirrus

In the first series of papers published on IRAS observations Low et al. (1984) reported the discovery of filamentary structures which they describe as infrared cirrus. In this early report Low et al. gave several examples of infrared cirrus which coincided spatially with H I clouds but also pointed out that many extended structures did not have any clear counterpart in H I maps. Since this early report much work has been done in associating infrared cirrus with interstellar clouds seen in H I, CO, or as nebulosity or extinction patches in optical plates. We show in the following that most structures in the $100\mu m$ emission on scales larger than a degree are correlated with the distribution of H I gas. A large number of infrared cirrus with sizes up to several degrees and smaller structures within cirrus which are predominantly atomic do not have an $H \ I$ emission counterpart but these regions account for a small fraction of the $100\mu m$ emission and cover a small part of the sky. In order to study the extragalactic sky they may be discarded and the $100\mu m$ emission modeled from the H I data. This conclusion is not true for shorter wavelengths and ignores a possible contribution from the diffuse ionized gas which is investigated in the next section.

2.2.1 The $100\mu m$-H I correlation

Boulanger et al. (1985), Terebey and Fich (1986), de Vries et al. (1987) and BP have investigated the correlation between far-IR and H I data over several regions of the sky typically a few tens of a degree in size. Within each of the fields analyzed by these authors, there is a tight correlation between the far-IR and H I emission wherever there are no molecular clouds. As an example we present in Figure 3 a pixel-by-pixel comparison between the $100\mu m$ and H I emission in the northern and southern polar cap. The good correlation observed within each of the fields proves that, on scales of the order of 100 pc, the interstellar radiation field and the dust abundance are reasonably uniform. On larger scales significant variations in the infrared emission per H atom are observed. These variations are the main source of scatter in Figure 2. A systematic analysis of the $100\mu m$ - H I correlation was carried out by Désert, Bazell, and Boulanger (1988, hereafter DBB) who built a map of the $100\mu m$ emissivity per H atom from the results of linear correlations within cells of $20°$ in longitude and $8°$ in latitude. At $|b| > 20°$ away from the Sco- Cen and Orion OB associations the $I_\nu(100\mu m)/N_H$ ratio is rather uniform and remains within $\sim 50\%$ of the cosecant law value, 0.86 MJy/sr for $10^{20} H cm^{-2}$. More important variations are observed at lower latitudes and close to OB associations. At low latitude in the direction of the outer parts of the Galaxy the $I_\nu(100\mu m)/N_H$ ratio is about a factor 2 lower than the cosecant value (Terebey and Fich 1986). In the Lupus region, close to the Sco-Cen OB association, BP measured a ratio 3 times higher than the cosecant value. The fact that high values of the far-IR emissivity are observed in the vicinity of OB associations, while low values are observed towards the external parts of the Galaxy suggests that changes in the infrared emission per H atom result mainly from variations in the intensity of the interstellar radiation field rather than from changes in the dust-to-gas ratio.

2.2.2 Molecular Gas at High Galactic Latitude

The launch of the IRAS satellite and the observation of infrared cirrus coincided

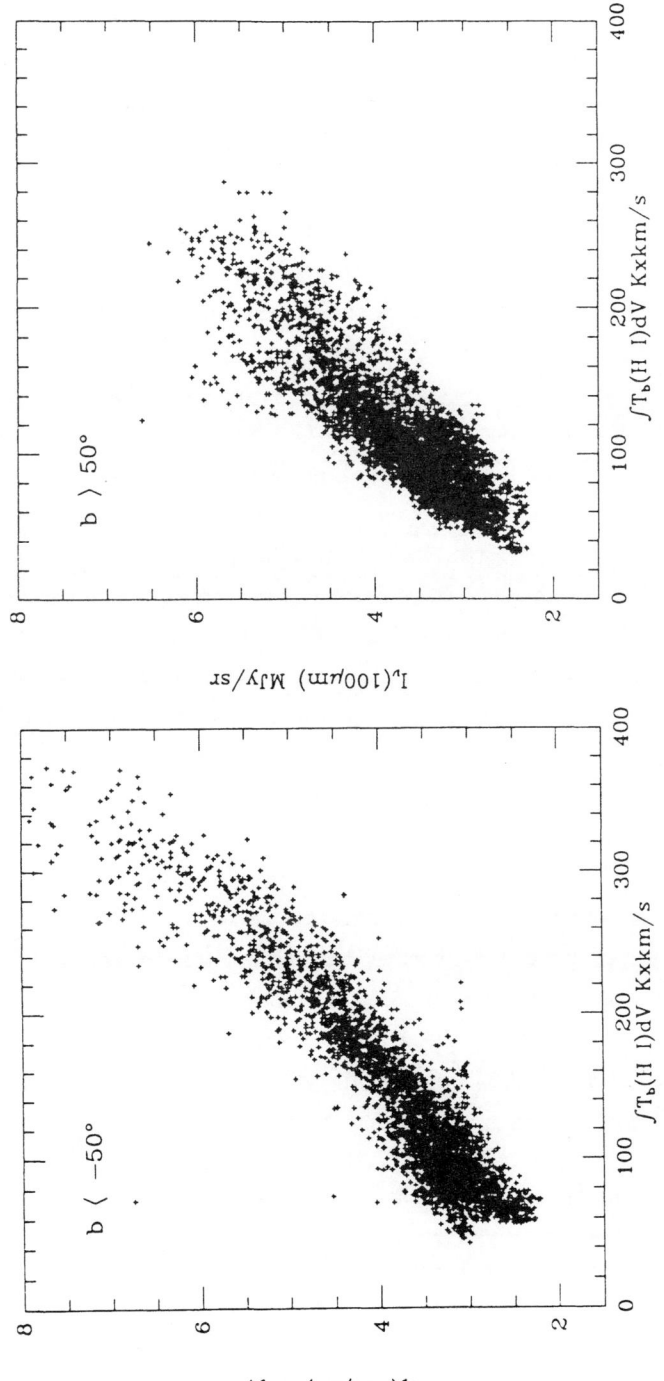

Fig. 3: Pixel-by-pixel comparison between 100μm and H I emission in the northern and southern polar caps. To avoid contamination by stray radiation, we used for this comparison the H I survey made with the horn reflector of the Bell Laboratories. The 100μm data were smoothed to the lower resolution of this survey; each pixel represents a 1.5° x 1.5° pixel.

with the discovery of small molecular clouds at high Galactic latitude by Blitz et al. (1984) and Magnani et al. (1985). Early after several high latitude molecular clouds known from CO observations were identified with infrared cirrus (Weiland et al. 1986; de Vries et al. 1987). The IRAS data have since been used to guide CO observations looking for more clouds. The systematic comparison of $100\mu m$ and H I data of DBB was carried out with the purpose of establishing a catalog of regions with an excess IR emission from what is expected from gas seen in emission in $H\ I$. This analysis led to a list of 578 clouds with excess infrared emission distributed over the whole sky. The low resolution of the analysis, which is limited by that of the H I surveys (30'), precludes the detection of excess regions with extents smaller than about a degree among which some clouds detected in CO by Magnani et al. DBB suggested that most of these clouds correspond to high latitude molecular clouds. However, Blitz et al. (1990) surveyed these clouds for CO emission and detected only 13% of the clouds. Clouds not detected in CO could be molecular but without CO (see Lada and Blitz 1988) or made of cold neutral atomic gas. One surprising result of the analysis of DBB is the existence of a number of regions with a deficit of infrared emission compared to the emission expected from H I gas as large as the number of excess clouds. There is no obvious correlation between these deficit regions and the presence of H I gas at intermediate velocities. Globally, regions with excess or deficit IR emission with respect to the IR-H I correlation represent only a small fraction, $\sim 2\%$, of the total IR emission. Using the high latitude CO survey of Magnani et al. (1985) and the brightness ratio, $I_\nu(100\mu m)/I(CO)$, of 1.0 ± 0.5 (MJy/sr)/(Kkm/s) measured by Weiland et al. and de Vries et al. BP estimated the contribution of small molecular clouds observed at high latitude to less than 1% of the $100\mu m$ emission integrated over the same latitude range than that of the CO survey. This result is not surprising since these clouds account for a small fraction of the gas surface density: $\sigma(H_2) \sim 0.2 M_\odot/pc^2$ while $\sigma(H\ I + H\ II) \sim 6 M_\odot/pc^2$ (Magnani et al. 1986, Reynolds 1989).

2.2.3 Emission at Wavelengths Shorter than $100\mu m$

On scales larger than a degree over most of the high latitude sky, the $100\mu m$ brightness may be accurately estimated from the observed H I emission. This conclusion does not apply at shorter wavelengths. Numerous studies of the IR emission of interstellar clouds heated by the average interstellar radiation field of the Galaxy

(Leene 1986; Weiland et al. 1986; Heiles, Reach and Koo 1988; Laureijs, Mattila, and Schnur 1987; Laureijs, Chlewicki, and Clark 1988 and Boulanger et al. 1990) have shown than the ratio between the mid-IR (12 and $25\mu m$) and the $100\mu m$ emission varies by more than one order of magnitude from cloud to cloud, and within clouds on scales as small as the IRAS resolution. These color variations tend to average out on scales larger than 1 degree but are visible up to regions 10 degrees in size in the IRAS low- resolution maps. It is thought that they trace variations in the composition of dust and not excitation effects related to changes in the radiation field. Unlike the large particles emitting at $100\mu m$, the small particles which make the emission at shorter wavelength are non-uniformly distributed in the interstellar medium.

2.2.4 Optical Images of Infrared Cirrus

The presence of interstellar dust at high Galactic latitude was inferred from nebulosity in Schmidt plates many years before the IRAS launch (see e.g. de Vaucouleurs and Freeman 1972, Sandage 1976). Shortly after the report of the discovery of infrared cirrus de Vries and Le Poole (1985) looked for the optical counterpart of two cirrus clouds and showed that they could be seen as regions of faint diffuse emission in the blue plates of the ESO/SERC sky survey. For a peak $100\mu m$ brightness between 5 and 15 MJy/sr these clouds have a maximum optical brightness between 24 and 25 mag/arcsec^2. From this analysis de Vries and Le Poole concluded that, with a detection limit of 27 mag/arcsec^2, the southern optical survey has a sensitivity to high latitude cirrus similar to that of IRAS. Optical properties of cirrus have since been analyzed in several papers (Chlewicki and Laureijs 1987; Laureijs, Mattila, and Schnur 1987; Guhathakurta and Tyson 1989). Observed brightnesses from the U to the V band generally agree with what is expected from reflection of Galactic light by dust in the clouds. In the R and I band the emission is generally larger than what is expected from dust scattering. The excess emission is presently interpreted as luminescence from the small particles which produce the near-IR emission from the interstellar medium (Section 3.2).

2.3 Contribution from Diffuse Ionized Gas

The interstellar matter seen at $|b| > 20°$ is mostly atomic gas, about 3/4 of it is neutral and 1/4 ionized (see Reynolds 1989). Unless there is a difference in dust abundance between neutral and ionized gas, the ionized component of the nearby interstellar medium must account for $\sim 25\%$ of the infrared emission measured by the cosecant laws. Unless the distribution of neutral and atomic gas are spatially correlated the $I_\nu(100\mu m)/N_H$ ratio derived from the scattered diagrams in Figure 3 measures the $100\mu m$ emission from the neutral gas alone. Note that this $I_\nu(100\mu m)/N_H$ ratio measures the IR emissivity of individual H I clouds because there is hardly any variation of the background emission with Galactic latitude at $|b| > 50°$. If we scale the latitude distribution of H I emission by the $I_\nu(100\mu m)/N_H$ ratio derived from Figure 3 we get a latitude distribution which differs from the profile of infrared emission by much less than the 25% contribution expected from diffuse ionized gas (see Figure 4). The questions which arises then is: Did IRAS measure any $100\mu m$ emission from the diffuse ionized gas?

As the H I and $100\mu m$ profiles in Figure 4 agree closely at all latitudes this problem cannot be solved by simply changing the zero level of the Galactic emission because the column density of ionized gas increases with the cosecant of the latitude (Reynolds 1984). The close agreement seen between the two profiles in Figure 4 could be coincidental if the IR emissivity of the H I gas was decreasing with latitude just so to compensate for the increase of the emission of the ionized gas. But analyses of the $100\mu m$ - H I correlation (BP, DBB) show on the contrary that $I_\nu(100\mu m)/N_H$ increases slightly from the polar caps to $|b| = 20°$. Therefore, we see only two possible explanations to the coincidence of the two profiles in Figure 4: (1) the distribution of ionized gas is correlated with the distribution of the H I gas, (2) the diffuse ionized gas is deficient in dust (at least in the large grains emitting at $100\mu m$). In the first case the $100\mu m$ emissivities derived from the slope of the $100\mu m$- H I correlation will include the contribution of ionized gas. Some degree of correlation between the spatial distribution of neutral and ionized gas is expected in the three phases picture of the interstellar medium where ionized gas is found in transition regions between H I clouds and hot coronal gas (McKee and Ostriker 1977). Reynolds (1980) has pointed out that some of the structures seen in $H\alpha$ maps appear spatially and kinematically correlated with H I filaments and clouds but not all of the diffuse ionized gas can be at the surface of H I clouds since the scale height of the ionized gas is much greater than that of cold neutral clouds

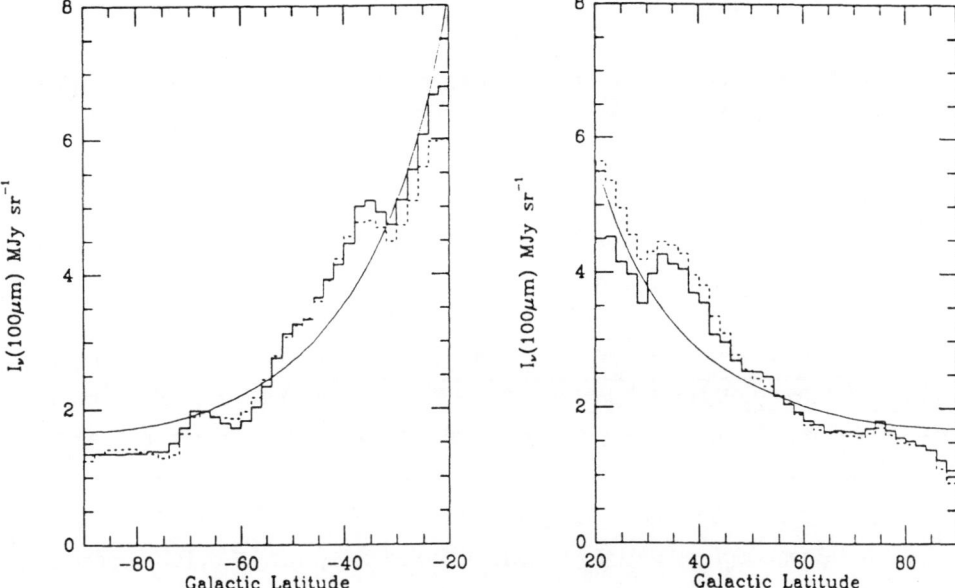

Fig. 4: Galactic latitude profiles of the $100\mu m$ emission at $b < -20°$ and $b > 20°$. In each plot the histogram drawn with a solid line represents the $100\mu m$ profile. The dotted histogram represents the profile of the integrated H I emission derived from the survey of the Bell Laboratories and scaled by the $I_\nu(100\mu m)/WH\ I$ ratios derived from the $100\mu m - H\ I$ correlation at $b > 50°$ and $b < -50°$ (Figure 3). For the southern profiles only the part of the sky covered by the Bell Laboratories survey was used. In each plot the continuous line represents a least-squares fit of the $100\mu m$ profile by a cosecant law plus a constant. These cosecant laws are explicitly formulated in BP.

(Lockman 1984, Reynolds 1989). Larger maps are necessary in order to clarify the degree of correlation between $H\alpha$ and H I emission. Due to a lower density the rate of destruction of grains by interstellar shocks is much higher in the warm phases (ionized and neutral) than in the cold phase of the interstellar medium (McKee et al. 1987). If the cycling time of the gas between these two phases is larger than the time scale for dust destruction in the warm phase, $\sim 3 \times 10^7$yrs, this phase could be significantly depleted in dust compared to the cold phase which could be an explanation for proposition (2) above.

The $10° \times 12°$ $H\alpha$ map centered at $l = 144°$, $b = -21°$ presented by Reynolds (1980) indicates that the diffuse ionized gas has a complex morphology made of sharp filaments and some more diffuse structures. Reynolds estimates that features in these maps correspond to column densities of $\sim 5 \times 10^{19}$cm^{-2} which should be detectable in the IRAS data if they have an IR emissivity similar to that of H I clouds. It is thus conceivable to look for a contribution from diffuse ionized gas to the infrared emission by correlating structures in $H\alpha$ emission with structure in infrared maps. However this is not easy to do because one needs, first, to remove from the infrared maps the dominant contribution from the neutral gas.

2.4 Looking at the Extragalactic Sky

Gray-scale plots of the distribution of the $100\mu m$ emission around the northern and southern Galactic poles down to $|b| = 30°$ have been presented by BP. Regions of low IR emission can be identified on these plots. If we extrapolate the $100\mu m$-H I diagram of figure 3 to zero H I emission one finds a residual emission of $1.8MJy/sr$. This emission is not meaningful because it is within the known uncertainties on the zero level calibration of the IRAS data. Ignoring this offset and assuming that the Galactic emission is zero for zero H I emission, one finds from the IRAS data that 20% of the sky at $|b| > 50°$ has a $100\mu m$ brightness smaller than 1.3 MJy/sr which corresponds to a visual extinction A_v of 0.08 mag (scaling relation 15.9 MJy/sr per magnitude for low A_v). The region of lowest H I emission in the sky covers about 10 deg^2 in the Ursa Major constellation, around $\alpha = 10h40m$, $\delta = 58°$. It has been investigated in details by Lockman, Jahoda, and McCammon (1986) and Jahoda et al. (1990) and coincides with a region of minimum infrared emission in the $100\mu m$

TABLE 1

Energy Distribution of IR Emission at High Galactic Latitude

λ	$4\pi\nu I_\nu$ $(10^{-31}W/H$
3.3 μm^1	1.9 ± 0.4
12 μm	1.1 ± 0.3
25 μm	0.7 ± 0.2
60 μm	1.1 ± 0.06
100 μm	3.2 ± 0.12
100 μm	2.0 ± 1.9^2
151 μm	3.9 ± 2.5^2
241 μm	2.3 ± 1.4^2

(1): Intensity at the peak of the 3.3μm feature from Giard et al. (1989). The full width at half maximum of the feature is 0.05μm. The total intensity in the feature is 2.9×10^{-33}W/H. The level of the continuum emission around the feature is unknown.

(2):COBE Measurements at the South Ecliptic Pole after subtraction of the zodiacal light and scaling by N(H I)= 3.510^{20} cm^{-2} (see Table 3).

TABLE 2

Budget of the Infrared Emission

ISM Component	Solar Neighborhood		Inner Galaxy (1)	
	$\sigma_g(M_\odot/pc^2)$	$L_{IR}(L_\odot/pc^2)$	$M_g(10^8 M_\odot)$	$L_{IR}(10^9 L_\odot)$
H_2	1.3	1.5	10.	2.5
H I	5.	10.	8.3	5.6
H II Regions	-	3.	-	2.8
Diffuse H II	1.2	< 3(2)	-	(2)

(1): 2kpc $< R < R_\odot = $ 10kpc. For other $R_\odot$ luminosities and masses should be scaled by $(R_\odot/10kpc)^2$.

(2): The contribution from diffuse ionized gas may be included in that of the H I gas.

IRAS data. The H I emission in this region corresponds to a column density of $5 \times 10^{19} cm^{-2}$. For the high latitude $100 \mu m$ to H I emission ratio, the corresponding $I_\nu(100\mu m)$ is 0.4MJy/sr. If there were as much ionized gas in the holes as on average at high latitude, assuming an emissivity similar to that for the H I gas, the $100\mu m$ emission from ionized gas would be $\sim 0.8 MJy/sr$. However, one Hα measurement in this region led to an upper limit one order of magnitude lower than the average emission at high latitude (Jahoda et al.). This suggests but does not prove, due to the dependence of the Hα emission on clumping, that the H I hole in Ursa is also a hole of ionized gas. Highly ionized gas has been detected in the Ursa hole in CIV and OIII lines; the corresponding column density is $\sim 10^{18} Hcm^{-2}$ (Martin and Bowyer 1990). This is one order of magnitude smaller than the H I column density but still within the detection limit of DIRBE.

3. Composition of Interstellar Dust

3.1 Extinction Curve

The extinction of stellar light by interstellar dust gives the dust optical depth τ_λ as a function of wavelength – it can be separated into the scattering and the absorption parts, the albedo (between 0 and 1) being the scattering fraction of the extinction. Initially, the composition of interstellar dust was initially inferred from extinction measurements.

The contribution of dust grains to the extinction curve depends on their size unless the particles are very small. Particles with sizes much smaller (typically by a factor 10) than the wavelength have negligible albedo (little scattering compared to absorption) and a small optical depth to the radiation (see textbook by van de Hulst 1981). Their extinction cross section is proportional to their mass. In practice, even in the UV, the contribution of particles smaller than $\sim 10nm$ to the extinction and scattering is independent of their size and shape. Dust models constructed on the basis of extinction and scattering data (Mathis, Rumpl, and Nordsieck 1977, Draine and Lee 1984) had therefore an arbitrary lower limit to the diameter of particles of $10nm$. Similarly in the UV and the visible, the contribution to the extinction of a set of particles does not differ much if they are either independent or connected in a fractal structure with little mutual shadowing. In emission in the infrared, small particles may be distinguished because the smaller they are the hotter they can get. Fractal grains may be distinguished from compact grains because they have much larger emissivities in the far-infrared and are therefore colder (Wright, this book) These uncertainties set aside, the slope of the extinction curve from the UV to the near-IR may be fitted with a size distribution (number of particles per size interval) following a power law of index -3.5 with an uncertain lower cut-off and an upper cut-off $\sim 0.2 \mu m$ constrained by observations of reflected light in the near-infrared.

Compared to solar abundances, almost all of the elements are observed to be depleted in the interstellar medium. From abundances and depletions the basic constituents of interstellar dust are inferred to be C, Si, O and some metals. There are only three spectroscopic features in the extinction curve which allow to be more specific about dust composition. Two of them at 9 and $18 \mu m$ are characteristic of silicates. The third one, the $220 nm$ bump, is assumed to come from graphite or some amorphous carbonaceous compound.

3.2 Emission of Interstellar Dust

Since the first models of interstellar dust were proposed on the basis of extinction curves, data on the infrared emission have widened and complicated our understanding. In the spectrum listed in Table 1 one sees that the power radiated by interstellar dust is distributed over a wide range of wavelengths from the near-IR to the submillimeter. Since dust at high latitude is illuminated by a rather homogeneous radiation field the emission must originate from different types of particles emitting at different temperatures. The IRAS data, therefore, confirmed the discovery, first made in reflection nebulae (Sellgren, Werner, and Dinerstein 1983), of the existence in the interstellar medium of small particles transiently heated to high temperatures by the absorption of single photons.

In an attempt to reproduce the spectrum in Table 1 (Fig. 5) at the same time as the extinction curve Désert, Boulanger, and Puget (1990) concluded that there must be at least three different dust components two of which being small particles emitting through temperature fluctuations: (1) Polycyclic Aromatic Hydrocarbons

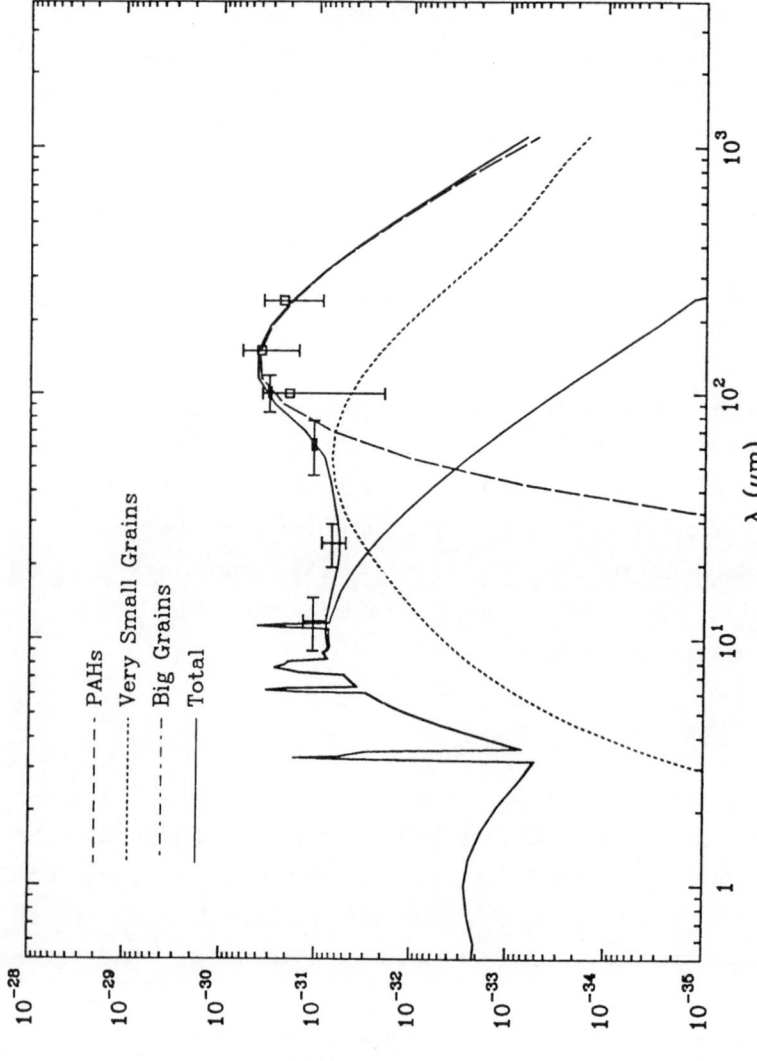

Fig. 5: Spectrum of Galactic emission outside the Galactic plane. The continuous spectrum is the calculated emission for the dust model of Désert et al. (1990). The contribution of each of the components of the model are shown. The four IRAS fluxes and three COBE (open squares) fluxes from Table 1 are plotted.

molecules (PAHs) made of about 50 to 200 carbon atoms which account for the near-IR emission and part of the mid- IR, (2) Very Small Grains (VSGs), 3d particles with sizes between 1.5 and 5 nm contributing to the 25 and $60\mu m$ emission measured by IRAS, and (3) large dust grains in thermal equilibrium with the radiation field emitting in the far-IR and the submillimeter. For the Solar Neighborhood radiation field, PAHs and VSGs account each for $\sim 20\%$ of the total emission, large grains for the remaining 60%. Désert et al. have investigated the variation of the emission spectrum with the intensity and the spectral shape of the radiation field.

Obviously, this model does not fully describe interstellar dust. It was conceived to analyze properties of the infrared emission and, in particular, to look for connections between emission and extinction properties. The model is based on many arbitrary assumptions but I would rather say that it is opened to test many hypothesis on the absorbing and emitting properties of particles. An important uncertainty is that no observations yet permit to ascribe a specific contribution to the extinction curve to any of the dust components. The PAHs are the only particles which are specifically described. Their identification is based on a characteristic set of emission features at 3.3, 6.2, 7.7, 8.6 and $11.3\mu m$ which have been seen in Galaxies including ours and in reflection nebulae in the solar neighborhood (Puget and Léger 1989). The natures of the very small and big grains are unknown. Since grains are destroyed by shock waves in the diffuse interstellar medium (McKee et al. 1987) on time-scale of a few 10^8years, smaller than the replenishment time-scale by stars and supernovae by at least one order of magnitude, grains must grow in the interstellar medium by accreting atoms and molecules, and through coagulation. This random growth is expected to lead to composite grains, possibly very fluffy and with fractal shapes. Observational evidence for evolution of interstellar dust in the interstellar medium (Draine 1985) comes from infrared spectroscopy which shows the formation of ices on grains within molecular clouds (Roche 1989), and the spectacular variations of the extinction curve and the spectral distribution of the infrared emission from place to place in the solar neighborhood (Fitzpatrick and Massa 1988, Boulanger et al. 1990). In section 6, we will see how the COBE data may complement the present knowledge of interstellar dust.

4. Emission in the Galactic Plane

In the Galactic Plane the infrared emission is made of a diffuse component and localized sources accounting for 3/4 and 1/4 of the infrared luminosity of the Galaxy. The sources are all associated with known star forming regions. The diffuse emission in the plane is maximum for longitudes within 35° of the Galactic centre. This range of longitudes coincides with that of the molecular ring which corresponds to a concentration of molecular clouds and young stars. The origin of the diffuse emission has been for a long time a debated issue. Is it truly diffuse in nature or does it come from a large number of sources non-separated by the observations? Does it come from the molecular, the neutral atomic or ionized components of the interstellar medium? Is the source of energy the old stellar population or recently formed stars? These questions bear on the relation between star formation and the infrared emission of galaxies.

Historically, the debate over these questions was confused by the ignorance of the existence of small particles emitting at high temperatures even in regions of low radiation field (see section 3). The spectrum suggested that a large fraction of the diffuse emission comes from warm dust ($T > 40\,K$). Assuming that all emitting particles are in thermal equilibrium one then reaches the conclusion that a large fraction of the diffuse emission comes from dust close to hot stars, i.e. near OB associations. On the other hand the overall luminosity of ionized stars as derived from radio studies of H II regions did not seem sufficient to account for the observed infrared luminosity even in the extreme case where all of their luminosity is absorbed by dust. The IRAS data clarified this debate in two ways. First, it established the widespread existence of small particles experiencing temperature fluctuations. The large amount of emission observed shortward of $100\mu m$ did not need to come from grains in the vicinity of hot stars. Second, data over a wider range in latitudes allowed to separate the contribution from atomic and molecular gas on the basis of their different scale-heights. Longitude profiles a few degrees from the plane allowed to measure the radial variation of the infrared emission of dust associated with the atomic gas independently of the contribution of molecular clouds. Once this emission is known, the contribution from molecular clouds may be derived from longitude profiles in the plane, after subtraction of the emission associated with the atomic gas. In the plane, within 35° in longitude from the Galactic Center (i.e. in the direction of the molecular ring), dust in atomic gas contributes for about 1/3 of the observed emission.

The various groups which have analyzed the IRAS data (Pérault 1987, Cox and Mezger 1988, Sodroski et al. 1989, Bloemen et al. 1990) now agree on the following conclusions. The budget of interstellar matter and infrared emission from the Galaxy within the solar circle is summarized in Table 2.

(1) Most of the diffuse emission comes from interstellar matter, molecular and atomic, not directly associated with the formation of high mass stars, i.e. not in the vicinity of OB associations. Although there is slightly more molecular gas than atomic gas within the solar circle this last component contribute to 70% of the diffuse emission. There is a difference of about a factor 3 to 4 in the ratio of infrared luminosity to mass of gas between atomic and molecular clouds. Much dust within molecular clouds is shielded from stellar radiation and contributes little to the observed luminosity.

(2) The heating of dust is dominated by O, B and A stars. Older stars contribute only 30% of the overall luminosity of the Galaxy (sources included). Ionizing stars, O and early B, contribute 40%. A large fraction of the luminosity of dust is provided by stars of intermediate mass, late B and A stars with ages in the range 10^7 to a few 10^8 yrs. This is an important conclusion which implies that the infrared emission provides an information on star formation, distinct from that supplied by radio observations which measure the thermal emission from H II regions. From the infrared data, one may derive the radial distribution cumulative luminosity of O, B and A from the Solar Neighborhood to the molecular ring. This radial distribution follows that of ionizing stars derived from radio observations. Thus, the distribution of late B and A stars, a few 10^8 years old, follows that of O stars. The enhanced rate of star formation currently observed in the molecular ring has thus been lasting for at least a few 10^8 years: if it is a burst it is not a short one.

5. Statistical Analysis of the Angular Distribution of the Emission

IRAS images of high latitude cirrus have been used to study the power spectrum of high latitude cirrus (Gautier et al. 1991). The shapes of the power spectra were found to be well represented by a power law. Two dimensional analyses yield indices near -3 from 0.125 degree^{-1} to the Nyquist limit of the data 0.5 arcmin^{-1}. The spectral index was generally found to be roughly independent of azimuth in frequency space. The power spectrum analysis made on images was extended to lower frequencies by one dimensional analyses which yielded spectral indices near -2 down to a frequency of 0.07 degree^{-1}. Below this frequency the statistics is too poor to say anything. The difference of about unity between the one and two dimensional analyses results from the dimension change in the Fourier transform.

The power spectra were used to measure the second order structure function: the rms of brightness differences between two apertures versus the angular separation of the two apertures. Using sky maps, Gautier et al. found that the brightness fluctuations on a scale of 5 arcminutes are non-gaussian. For example, the 99.7% quantile of the distribution (i.e. level below which are 99.7% of the points in the distribution) is 5 times the 67% quantile while, for a gaussian distribution, the 67% quantile is equal to the rms of the distribution and the 99.7% quantile is only 2.75 times this value. Brightness fluctuations in the Galactic emission may eventually limit the ability of future observations in detecting faint extragalactic sources. Gautier et al. estimate that on average at $|b| > 50°$ the confusion limit will be 1 and 0.5 mJy at $100\mu m$ for the two next IR space missions: ISO and SIRTF equipped with cooled telescope of diameter 60 and 80 cm. The expected sensitivity of ISO is just above this confusion limit.

In the region of lowest $100\mu m$ emission IRAS maps do not show any measurable emission in the power spectra. One of these region is the region of lowest H I emission in the sky at 10h45m, 57° where N(H I) is estimated to $5.510^9 cm^{-2}$ (section 2.4). Lockman, Jahoda and McCammon (1986) have measured the structure function of the H I emission in this area. The possibility that the auto-correlation function of galaxies could be biased by dust extinction was raised during the school. In particular people wondered if dust extinction could explain the break in the auto-correlation function measured by Maddox et al. (1990) at 3°. This question has been since then investigated by Harmon (private communication). The median extinction at $|b| > 50°$ derived from the $100\mu m$ IRAS data is $A_v = 0.10$ mag with a dispersion around this value of the same order (BP). Using the IRAS data scaled to τ_v, Harmon measured the autocorrelation function of extinction over the exact area used by Maddox et al. The amplitude of the extinction over this region, and consequently of the autocorrelation function, is much too small to have any effect on the measured property of the distribution of galaxies. The slope of the autocorrelation function of τ_v is -0.6 from 1 to 10° with no break point.

6. Prospects with the COBE Data

6.1 Dust Composition

One can easily list a few questions that will certainly be addressed through COBE observations. They derive from our very coarse knowledge of the sub–millimeter spectrum of the dust emission.

1) Are there one or two components of IS large grains? Either silicates and graphite grains separately or carbon-coated silicate grains?

2) What is the large grain emissivity law (ϵ_ν vs. frequency ν) in the submillimeter domain? Observations have so far reached a limit where $\epsilon_\nu \propto \nu^\alpha$ with $1 \leq \alpha \leq 2$. The corollary of this question is the temperature distribution. It is not certain that the emissivities and the temperatures can be completely disentangled. A first attempt has already been done by Pajot et al. (1986) (see also Taolong Xie et al. 1991).

3) In the spectrum of Figure 5, big grains which are assumed to be compact have a temperature of 18 K. Is there an additional dust component of very cold fractal grains contributing to the submillimeter spectrum (Wright, these proceedings)? Wright et al. (1991) say that the FIRAS (Far Infrared Absolute Spectrophotometer) spectrum of dust emission is well fitted by the addition of two components with a ν^2 emissivity law: one with temperature 20.4 K and a colder one with temperature 4.8 K. But the spectrum is equally well fitted with a single component of emissivity $\alpha\nu^{1.65}$ and $T = 23K$. Additional uncertainties on the physical meaning of these fits comes from the fact that in the Galactic plane, regions with a wide range of radiation field intensities contribute to the observed emission. With that respect, the spectrum of the diffuse emission at high latitude will be easier to interpret. Further, the knowledge and homogeneity of the heating radiation allows to tie together emissivity laws and temperatures, which should help select plausible fits among all possible ones. Very small grains spend most of the time at low temperatures close to but without reaching the cosmic bacground temperature because their cooling time at these temperatures ($\sim$ 6K) is longer than the mean time between photon

absorptions. If they have large emissivities, they could contribute to the sub-mm emission (Désert et al. 1990). In that case, a correlation between mid-IR and submillimeter emission would be expected.

One can understand how these questions are relevant to the subtraction technique that will have to be used to fully exploit the COBE data in order to get a reliable value of any extragalactic background.

6.2 Measuring an Infrared Extragalactic Background

Table 3 lists the measured brightness of COBE at the South Ecliptic Pole together with estimates of the zodiacal and Galactic contribution to the measured fluxes. These estimates account for the observed flux within the quoted error bars but one should bear in mind all of the uncertainties attached to them. The numbers in Table 3 are instructive in giving the order of magnitude of the contributions and their wavelength dependence. The zodiacal light was estimated from the $100\mu m$ IRAS data assuming that the pole brightness is equal to the slope of the cosecant law followed by the zodiacal brightness for $|\beta| > 30°$. Depending on the geometry of the zodiacal cloud this assumption could either underestimate or overestimate the zodiacal contribution. At longer wavelength the zodiacal light is assumed to follow a Rayleigh-Jeans spectrum. The Galactic emission was derived from the $100\mu m$ to H I emission ratio at high Galactic latitude and an H I column density of $3.5 \times 10^{20} cm^{-2}$. The $100\mu m$ flux is extrapolated at longer wavelength using the dust model of Désert, Boulanger and Puget (1990). This estimate ignores emission from the diffuse ionized gas and the possibility that some of the gas contributing to the H I emission has an infrared emissivity different from that measured for clouds by correlating structure in the $H\ I$ and infrared maps. In particular the warm and cold neutral medium may not have the same dust abundance since dust is destroyed by shock waves in the diffuse component only.

Will COBE be able to improve on the present uncertainties on the separation of the zodiacal, Galactic and Extragalactic contribution to the infrared brightness of the sky? Probably yes because the three contributions have different wavelength dependence and COBE has a more extensive wavelength coverage than IRAS. A specific spectroscopic signature of the Galactic emission is the C II line at $158\mu m$. Unfortunately, the FIRAS sensitivity is insufficient to detect the C II line in the regions of minimum Galactic infrared emission. For a line to continuum ratio of 8×10^{-3}, the 1 σ detection limit for one year, one pixel corresponds to a $100\mu m$ brightness of 0.6 MJy/sr.

6.3 Interstellar Matter in the Galaxy

Measurements by DIRBE of the Galactic Plane emission at longer wavelengths than IRAS will facilitate the separation of the atomic and molecular contributions to the diffuse infrared emission of the Galaxy and, hopefully, will confirm the conclusions reached with the IRAS data. The DIRBE sensitivity is sufficient also to detect the emission of PAHS in the $3.3\mu m$ at high Galactic latitude. Such a detection would demonstrate the presence of PAHs in nearby cold interstellar matter. The only comparable observations are that of Giard et al. (1989) in the plane. Observations by FIRAS of the main cooling lines of gas in the Galaxy (Wright et al. 1991) are a significant step in the investigation of interstellar matter. They complement dust observations in establishing the thermal budget of the Galaxy and constrain physical conditions in molecular clouds and H II regions.

TABLE 3

Emission at the Southern Ecliptic Pole

λ	I_ν MJy/sr (1)	Zodiacal MJy/sr (2)	Galactic MJy/sr (3)
100 μm	4.0 ± 1.7	2.2	3.
151 μm	6.5 ± 3.5	1.0	4.5
241 μm	5.6 ± 3.2	0.4	2.

(1): Observed Brightness from Hauser et al. (1990).

(2): Zodiacal Light estimated at $100\mu m$ from the cosecant law in the IRAS data at $|\beta| > 30°$ and a solar elongation of 90°. The extrapolation at longer wavelengths follows a Rayleigh-Jeans spectrum.

(3): Galactic Contribution estimated from the $100\mu m$ to H I emission ratio measured by IRAS at high Galactic Latitude. The scaling to longer wavelengths is based on the dust model of Désert, Boulanger and Puget (1990). Note that comparison with COBE indicates that IRAS brightness for diffuse emission are somewhat overstimated (Hauser et al. 1990).

References

Blitz, L., Magnani, L., and Mundy, L. 1984, Ap. J. (Letters) 282, L9.

Blitz, L., Bazell, D., and Désert, F.X. 1990, Ap. J. 352, L13.

Bloemen, J.B.G.M., Deul, E.R., and Thaddeus, P. 1990, Astr. Ap. 233, 437.

Boulanger, F., Baud, B., and van Albada, G.D. 1985, Astr. Ap. 144, L9.

Boulanger, F., and Pérault, M., 1988, Ap. J. 330, 964 (BP).

Boulanger, F. 1989, in Galactic and Extragalactic Background Radiation, eds. S. Bowyer and C. Leinert (Dordrecht: Reidel).

Boulanger, F., Falgarone, E., Puget, J.L.,and Helou, G. 1990, Ap. J. 364, 136.

Chlewicki, G., and Laureijs, R.J. 1987, Polycyclic Aromatic Hydrocarbons and Astrophysics, eds. A. Léger, L. d'Hendecourt, and N. Bocarra (Reidel), p. 335.

Cox, P. and Mezger, P.G. 1989, Astr. Ap. Rev. 1, 49.

Désert, F.X., Bazell, D., and Boulanger, F. 1988, Ap. J. 334, 815 (DBB).

Désert, F.-X., Boulanger, F., and Puget, J.-L. 1990, Astr. Ap, 237, 215.

Draine, B.T. 1985, Protostars and Planets II, eds. D.C. Black and M.S. Matthews, The University of Arizona Press, p. 621.

Draine, B.T., and Lee, H.M., 1984, Ap. J., 285, 89.

Draine, B.T., and Anderson, N. 1985, Ap. J., 292, 494.

de Vaucouleurs, G., and Freeman, K. C. 1972, Vistas Astron., 14, 163.

de Vries, C.P., and Le Poole, R.S. 1985, Astr. Ap., 145, L7.

de Vries, H.W., Heithausen, A., and Thaddeus, P. 1987, Ap. J., 319, 723.

Fitzpatrick, E.L. and Massa, D. M. 1988, Ap. J 328, 734.

Gautier, T.N.G., Boulanger F., Pérault, M., and Puget J.L. 1991, Astr. J. in press.

Giard, M., Pajot, F., Lamarre, J.M., Serra, G., and Caux, E. 1989, Astr. Ap. 215, 92.

Guhathakurta, P., and Tyson, J.A., 1989, Ap. J., 346, 774.

Hacking, P., and Houck, J.R. 1987, Ap. J. Suppl., 63, 311.

Hauser, M.G., et al. 1991, in After the first Three Minutes, University of Maryland, eds. S.S. Holt, C.L. Bennett, and V. Trimble, AIP Conference Proceeding

Heiles, C., Reach, W.T., and Koo, B.C. 1988, Ap. J., 332, 313.

Jahoda, K., Lockman, F.J., and McCammon, D. 1990, Ap. J. 354, 184.

Lada, E., and Blitz, L. 1988, Ap. J. 326, L69.

Laureijs, R.J., Mattila, K., and Schnur, G. 1987, Astr. Ap. 184, 269.

Laureijs, R.J., Chlewicki, G., Clark, F.O. Astr. Ap. 192, L13.

Leene, A. 1986, Astr. Ap., 154, 295.

Lockman, F.J. 1984, Ap. J., 283, 90.

Lockman, F.J., Jahoda, K., and McCammon, D. 1986, Ap. J. 302, 432.

Low, F.J., et al. 1984, Ap. J. (Letters), 278, L19.

Maddox, S.J., Efstathiou, G., Sutherland, W.J. and Loveday J. MNRAS 242, 43P.

Magnani, L., Blitz, L., and Mundy, L. 1985, Ap. J. 295, 402.

Magnani, L., Lada, E.A., and Blitz, L. 1986, Ap. J. 301, 395.

Martin, C., and Bowyer, S. 1990, Ap. J. 350, 242.

Mathis, J.S., Rumpl, W., and Nordsieck, K.H. 1977, Ap. J., 217, 425.

McKee, C., and Ostriker, J.P. 1977, Ap. J., 218, 148.

McKee, C.F., Hollenbach, D.J., Seab, C.G., Tielens, A.G.G.M. 1987, Ap.J. 318, 674.

Pajot, F., Boissé, P., Gispert, R., Lamarre, J.M., Puget, J.L. 1986, Astr. Ap. 157, 393.

Pérault, M. 1987, Thèse d'Etat (PHD), Université Paris VII.

Puget, J.L. 1985, Birth and Infancy of Stars, Les Houches Summer School 1983, A. Omont and R. Lucas eds., North Holland Publisher.

Puget, J.L., and Léger, A., 1989, Ann. Rev. Astr. Ap. 27, 161.

Reynolds, R.J. 1980, Ap. J., Ap. J. 236, 153.

Reynolds, R.J. 1984, Ap. J. 282, 191.

Reynolds, R.J. 1989, Ap. J. 339, L29.

Roche, P.F. 1989, Eslab Symposium on Infrared Spectroscopy in Astronomy, B.H. Kaldeich ed., ESA Publications Division, p. 79.

Sandage, A. 1976, A. J., 81, 954.

Sellgren, K., Werner, M.W., and Dinerstein, H.L. 1983, Ap. J. (Letters), 271, L13.

Sodroski, T.J., Dwek, E., Hauser, M.G., and Kerr F.J. 1989, Ap. J. 336, 762.

Taoling Xie, Goldsmith, P.F., and Weimin Zhou 1991, Ap. J. 371, L81.

Terebey, S., and Fich, M. 1986, Ap. J. (Letters), 309, L73.

van de Hulst, H.C. 1981, Light Scattering by Small Particles, Dover Publications Inc.

Weiland, J.L., Blitz, L., Dwek, E., Hauser, M.G., Magnani, L., and Rickard, L.J. 1986, Ap. J., 306, L101.

Wright, E.L. et al. 1991, Ap. J. in press.

LINEAR AND NON-LINEAR GRAVITATIONAL EFFECTS ON THE CMB ANISOTROPY

E. MARTINEZ-GONZALEZ
Instituto de Estudios Avanzados en Física Moderna y
Biología Molecular,
CSIC-Universidad de Cantabria,
Facultad de Ciencias, Dept. Física Moderna
Avda. Los Castros s.n.,
39005 Santander,
Spain.

ABSTRACT. A review is given on the gravitational effects produced by linear and non-linear structures on the anisotropies of the microwave background. We study the effect of single prominent structures with different physical characteristics that can be present in our surrounding universe. Also it is considered the non-linear effect of a statistical distribution of matter and is compared with the linear Sachs-Wolfe and Doppler effects.

1. Introduction

In the standard scenario of the evolution of the universe the cosmic microwave background (CMB) radiation, first discovered in 1965 by A.A. Penzias and R.W. Wilson, is interpreted as a relic of a very hot and dense phase of our early universe. The special properties of this radiation, respect to its spectrum and isotropy, have become very relevant to understand the past and present states of the universe (for a recent review on the implications of these properties see Sanz and Martínez-González 1990 and also Sanz 1991). In the standard picture the radiation decouples from the matter at a redshift of $z_r \approx 1000$ and the only interaction between both components afterwards is through the gravitational field. Then, the primary anisotropies appearing at z_r consist of a combination of the following effects: photon number fluctuations at recombination, the Doppler shift due to the motion of the last scatters (Doppler effect) and the gravitational fluctuations at recombination (Sachs-Wolfe effect, Sachs and Wolfe 1967).

The recombination process is not instantaneous implying a thickness for the cosmic photosphere of $\Delta z \approx 10^2$ (Wyse and Jones 1985). This introduces a scale $\Theta_r \approx 4' \Omega^{1/2}$ below which fluctuations are smoothed out. Another interesting scale

M. Signore and C. Dupraz (et al.), The Infrared and Submillimetre Sky after COBE, 291–301.

is the angle subtended by the horizon at recombination $\Theta_H \approx 2°\Omega^{1/2}$ above which the fluctuations are preserved from the physical processes operating at that time. In the case that the universe is reionized at later times this last scale can be as large as $12°\Omega^{1/2}$ for $z_r = 30$ (for $z < 30$ photon-electron scatterings become very rare due to the low mean density of the barionic component in the universe). Consecuently, the primary temperature anisotropies on scales above Θ_H, where the Sachs-Wolfe effect dominates, can be observed undisturbed at present (we will see in section 4 that this is not strictly true and that the gravitational effect of non-linear structures can compete with the Sachs-Wolfe effect on those large angular scales at least for some cosmological scenarios). On scales below Θ_H the Doppler effect usually dominates down to Θ_r; around this last scale the fluctuations in the photon number can also be relevant.

Secondary anisotropies can arise in the later history of the universe due to a number of phenomena: i) Reionization of the matter in the universe erases primary anisotropies and produces new ones through the coupling of velocity and density fluctuations of the electrons (Vishniac effect, Vishniac 1987). ii) Heated dust present at high redshift, related to the process of structure formation, generates spectral distortions and anisotropies (Bond et al. 1991). iii) Contribution from discrete sources can also be relevant for a wide range of angular scales and wavelengths (Franceschini et al. 1989). iv) The gravitational effect of the growing non-linear structures (Martínez-González, Sanz and Silk 1990).

In this paper we extensively study the generation of anisotropies due to linear and non-linear structures. In section 2 we describe the main characteristics of the gravitational effects. Section 3 summarizes studies on the gravitational redshift produced by single structures and section 4 contains very recent results on the different gravitational redshifts produced by a statistical distribution of matter.

2. Gravitational Effects

At the time of decoupling of matter and radiation the density fluctuations producing the gravitational field were very small, and linear theory can account for the gravitational redshift felt by the photons. Later on, the fluctuations grow via gravitational pulling and non-linear effects dominate the final steps of the collapse.

The total redshift suffered by the photons in their way from recombination till us is summarized in a recent result given by Martínez-González, Sanz and Silk (1990) for a flat Friedmann universe

$$\frac{\Delta T}{T} = \frac{1}{3}\phi_r - \vec{n} \cdot \vec{v} + 2 \int_r^o dt \frac{\partial \phi}{\partial t}(t, \vec{x}) \quad . \tag{1}$$

(the units are such that $d_{H_0} = 3t_0 = 1$, $c = 8\pi G = 1$, d_{H_0} being the horizon distance at the present time, c the speed of light and G the gravitational constant). In this formula $\vec{n}$ is the direction of observation, $\vec{v}$ the velocity of the emitters and ϕ the gravitational potential at a given point, the integral is calculated from recombination till the observer and $\vec{x}$ is the comoving distance from the observer to the photon at time t. The first two terms is what comes out of linear theory and were first found by Sachs and Wolfe (1967). The first term represents the gravitational contribution of all the structures at the recombination time and is therefore dominated by the linear structures near the decoupling surface (Sachs-Wolfe effect, SWE); and the second one is a Doppler effect (DE) due to the fall-in of the last scatters into the gravitational potential. The third term appearing in the equation is an integrated effect that accounts for the work performed by the photon in its way towards us against the non-static gravitational potential created by all the non-linear structures placed in its path. This integrated gravitational effect (IGE), except for the factor 2 that comes from general relativity in the same manner than in the Einstein deflection of a photon by a spherical mass, can be understood in the framework of Newtonnian mechanics (Peebles 1980).

We will discuss in the following sections the gravitational redshift produced by different isolated structures that can be present either in our surroundings or near the horizon of the universe, and also the redshift of the photons propagating in several cosmological scenarios where the density fluctuations are distributed in different statistical ways.

3. Gravitational Redshift Produced by Single Structures

Isolated structures present in our universe may leave their imprint on the maps of the CMB temperature. Analysis of these maps can shed light on their existence and properties. Many works have appeared in the literature to study the CMB temperature anisotropy produced by either linear or non-linear structures via their gravitational interaction with the microwave photons, being this interaction, more precisely, the Sachs-Wolfe effect and the integrated gravitational effect, respectively. Usually the spherical symmetry has been assumed for the objects studied, although other more realistic models have also been considered in some cases and, in particular, N-body simulations have recently been performed by van Kampen and Martínez-González (1991) to calculate the integrated effect produced by cosmic voids and great attractors. We summarize those works in the following subsections.

3.1. LINEAR STRUCTURES

Linear structures, i.e. fluctuations whose density is $\delta \ll 1$, affect the temperature of the microwave photons by means of the Sachs-Wolfe effect. The dominant effect comes from those structures which are near the decoupling surface (Argüeso et al. 1989, Argüeso and Martínez-González 1989). Additionally, linearity is the common feature of the matter density at the epoch of recombination of the universe, being the r.m.s. density fluctuation at that epoch and at the relevant scales for this effect $\delta_r \leq 10^{-3}$.

The generic signatures of a spherical dominant structure on the microwave maps, once the monopole and dipole have been subtracted, are hot and cold rings and spots (spots can also appear in homogeneous and isotropic Bianchi models due to geodesic focusing, Barrow et al. 1983). These features depend on the following parameters: radius of the structure, its density contrast, distance from the structure to the observer and the possible intersection with the cosmic photosphere (Argüeso et al. 1989). In the non-intersecting case, the temperature profile is obviously independent of the density profile and, as an example, 3σ peaks in an unbiased CDM scenario produce temperature anisotropies of the order of $\simeq 10^{-5}$. For a structure intersecting the cosmic photosphere the result will depend on its density profile, and the maximum temperature fluctuation produced by a 3σ homogeneous peak is also of the order of $\simeq 10^{-5}$.

The Sachs-Wolfe effect due to homogeneous spheroidal structures has been studied by Argüeso and Martínez-González (1990). They calculate the quadrupole moment of the temperature fluctuations, which is greater for spheroids than for the equivalent spheres (spheres with the same volume and density) when the structures are near us. However beyond the distance $c/\sqrt{5}$ (c being the major semiaxis) spheroids and spheres give similar effects, producing the maximum quadrupole contribution when they are placed on the photosphere. In general, the Relict upper limit on the quadrupole constrains the density contrast of the spheroids being the constraint more stringent for disk-like than for the equivalent filament-like objects (same volume and eccentricity) and for objects that are close to the last scattering surface.

Very recently, Bertschinger (1990) has calculated the Sachs-Wolfe effect for the local gravitational potential field within $\sim 6000 kms^{-1}$ from the Local Group (reconstructed from the peculiar motions of several hundred of elliptical and spiral galaxies and groups within that distance, after smoothing the velocity data and applying to it the Zeldovich approximation; see Bertschinger et al. 1990), as seen from an observer placed at our horizon. Although the calculation considers a very realistic situation, however linear theory does not apply rigurously to this case. The maximum anisotropy produced by the Great Attractor as seen by such observer, or equivalently considering that the Great Attractor is placed on our last scattering surface, is found to be $\simeq 10^{-5}$ on angular scales of $1°$.

3.2. NON-LINEAR STRUCTURES

Non-linear structures affect the temperature of the microwave photons via the "integrated gravitational effect" (IGE) discussed above, through their gravitational potential changes acting on the path of the photons. We study below the IGE produced by cosmic voids and great attractors whose existence has recently been claimed in the literature.

3.2.1. *Cosmic Voids.* A void of galaxies in the thin-shell approximation is represented by an empty spherical region surrounded by a thin-shell where the matter that would have been inside the void is distributed, and the external universe is Einstein-de Sitter (see Berstchinger 1985). The radius of the shell expands according to the law $R \propto t^\gamma$ with $\gamma \simeq 0.13$ either for collisionless gas with adiabatic compression or collisional gas with rapid cooling. This compensated model has been recently considered in the literature to represent voids like Bootes (Thompson and Vishniac 1987; Scaramella 1989; Martínez-González, Sanz and Silk 1990; Martínez-González and Sanz 1990). The integrated gravitational effect for such a void gives a maximum fluctuation at the center with respect to the edge of $\Delta T/T(0) \simeq -3 \times 10^{-7}$, being the angular scale of the coldspot of $\simeq 15°$. On the other hand, if we assume the existence of: i) voids like Bootes at higher redshifts ($z > 1$), the $\Delta T/T(0)$ value slightly increases but the angular scale is drastically reduced to $\leq 1°$, ii) a supervoid (i.e. an empty region with radius $\simeq 100h^{-1}Mpc$ and placed at a distance of $\simeq 150h^{-1}Mpc$) will give a maximum fluctuation $\Delta T/T(0) \simeq -10^{-5}$ with an angular scale for the coldspot of $\simeq 50°$.

More recently, van Kampen and Martínez-González (1991a) have simulated isolated voids (as -2σ density fluctuations) in an unbiased CDM $\Omega = 1$ scenario and calculated the integrated effect for 64^2 photons propagating through the evolving N-body distribution; the result for $\Delta T/T$ is $\simeq 10^{-7}$ for a void of radius $5h^{-1}Mpc$ placed at $z \simeq 0.6$. Finally, the statistical effect of many voids randomly distributed has been calculated by Thompson and Vishniac (1987), they use voids with the same size in the thin-shell approximation starting at some redshift and the $\Delta T/T$ value is $\leq 10^{-6}$ for the most favorable case of a radius of $\simeq 30h^{-1}Mpc$ and a filling factor of 1.

Summing up, considering all of these partial results one can conjecture that for a $\Omega = 1$ universe the real statistical distribution of voids in the universe (unless it is quite different from the one really observed) will produce a level of fluctuation $\Delta T/T$ that is $\leq 10^{-6}$, i.e. below the expectations of present observations.

3.2.2. Great Attractors.

A strong concentration of galaxies like the Great Attractor GA (Lynden-Bell et al. 1988), with a mass of $\simeq 5 \times 10^{16} M_\odot$ at a distance of $\simeq 40h^{-1}Mpc$, or the Shapley concentration SC (Scaramella et al. 1989, Raychaudhury 1989), with a mass of $\simeq 3 \times 10^{17} M_\odot$ at a distance of $\simeq 140h^{-1}Mpc$, can be represented by a Swiss-cheese model (i.e. a homogeneous sphere of matter is replaced by a concentric homogeneous internal sphere of matter leaving a shell without any matter). An estimation of the $\Delta T/T$ profile based on time delay was done by Rees and Sciama (1968), further studies using the framework of general relativity were done by Dyer (1976) and Nottale (1984). We have recently studied the integrated effect in a clear way through equation (2) (Martínez-González, Sanz and Silk 1990) and applied it to both concentrations of matter mentioned above (Martínez-González and Sanz 1990). The result is a maximum fluctuation at the center $\Delta T/T \simeq -3 \times 10^{-6}$ for the GA with physical characteristics taken from Dressler (1988), and $\Delta T/T \simeq -3 \times 10^{-5}$ for the SC, being the angular scale of the coldspot of $\simeq 30°$ and $20°$ respectively. Moreover, if we take the GA and place it at a high redshift the effect decreases and the angular scale is drastically reduced. Finally, the statistical effect of many lumps, represented by Swiss-cheese models randomly distributed, has been considered by Dyer and Ip (1988) finding that the level of fluctuation is $(\Delta T/T)_{rms} \leq 10^{-6}$ (except for a couple of models where the mass of the lumps is $3 \times 10^{19} M_\odot$ and the present number density $3.2 \times 10^{-3} Mpc^{-3}$, respectively, and $H_0 = 70 km s^{-1} Mpc^{-1}$).

More recently, van Kampen and Martínez-González (1991b) have simulated isolated lumps (from 2σ to 5σ peaks) in an unbiased CDM $\Omega = 1$ universe to represent clusters and GA-like structures. The code used was an adapted version of the Barnes and Hut (1986) treecode, with initial conditions produced by a code deviced by Bertschinger (1987) based on a path integral method which allows to constrain the initial density field in order to assure the presence of a peak of a given highness inside the simulated volume. Following Bertschinger et al. (1990), such GA was now defined as an object having a peak amplitud of $\delta = 1.2$ when averaged over a radius of $14h^{-1}Mpc$ (this would corresponde to a 5.5σ peak in a CDM scenario). They have estimated the integrated effect for 64^2 photons propagating through the evolving N-body distribution representing such a big lump and the result is a fluctuation $\Delta T/T$ of a few times 10^{-6}, almost independent of redshift of the structure for $0 \leq z \leq 2$.

The Great Wall (GW) found in the CfA redshift survey (Geller and Huchra 1989) is a large concentration of galaxies distributed in a sheet-like form containing a mass of $\approx 8 \times 10^{16} M_{\odot}$. This concentration has a mean density contrast of $\simeq 5$ and therefore its non-static gravitational field will produce a gravitational redshift to the microwave photons crossing the structure. Atrio-Barandela and Kashlinsky (1991) have calculated the IGE due to the Great Wall by modelizing it with an oblate spheroid. The result for photons passing through the center is $10^{-6} \leq \Delta T/T \leq (a few) \times 10^{-5}$ depending on the physical characteristics of the spheroid an on Ω.

Summarizing, considering all of these partial results one can conjecture that, for a $\Omega = 1$ universe, the real statistical distribution of GA-like structures in the universe will produce a level of fluctuation that is below the level $\simeq 10^{-5}$, whereas for SC-like and GW-like structures the level could be above this value, i.e. maybe large-scale observations with the appropiate angular resolution could reveal in the near future the presence of such very large-scale features.

4. Gravitational Redshift Produced by a Statistical Distribution of Matter

We now consider that the matter is statistically distributed according to different distributions. We assume the standard picture where the density fluctuations generated after inflation are Gaussian distributed and consider two cases for the density correlation function: In the first case the density fluctuations are correlated with a constant value up to a certain scale with a cut-off beyond that scale; in the second case we consider the same correlation which is observed for the actual distribution of galaxies. An interesting question we want to answer here is which of the three effects appearing in equation (1), i.e. IGE, SWE and DE, is the dominant one. The expresions for these effects may be written separately as:

$$\left(\frac{\Delta T}{T}\right)_{IGE} = 2 \int_{r}^{o} dt \frac{\partial \phi}{\partial t}(\vec{t}, \vec{x}) \quad , \tag{2}$$

$$\left(\frac{\Delta T}{T}\right)_{SWE} = \frac{1}{3}\phi_r \quad , \tag{3}$$

$$\left(\frac{\Delta T}{T}\right)_{DE} = -\vec{n} \cdot \vec{v} \quad . \tag{4}$$

For the case of the IGE we calculate this effect in the second order theory for the density fluctuation.

Perturbation theory up to the 2nd order in an Einstein-de Sitter universe gives the following value for the density fluctuation (Peebles 1980)

$$\frac{\delta_2}{a} = \delta + a(\frac{5}{7}\delta^2 + \frac{1}{6}\vec{\nabla}\delta \cdot \vec{\nabla}\phi + \frac{1}{126}\phi_{,ij}\phi^{,ij}) \quad , \quad \Delta\phi = 6\delta \quad , \tag{5}$$

where all the quantities in the r.h.s. of both equations are taken at the present time (i.e. $\delta = (1 + z_r)\delta_r$, δ_r is the density fluctuation at recombination) and the scale factor $a(t)$ is normalized to the present time $a_0 = 1$. Therefore, the integrated effect -as given by equation (2)- yields to (Martínez-González, Sanz and Silk 1991)

$$\left(\frac{\Delta T}{T}\right)_{IGE} = 2 \int_r^0 da \Psi\big(\vec{x}(a)\big) \quad , \quad \vec{x}(a) = \vec{n}\big(1 - a^{1/2}\big) \quad , \tag{6}$$

$$\Delta\Psi = 6\frac{\partial}{\partial a}\left(\frac{\delta_2}{a}\right) = \frac{30}{7}\delta^2 + \vec{\nabla}\delta \cdot \vec{\nabla}\phi + \frac{1}{21}\phi_{,ij}\phi^{,ij} \quad . \tag{7}$$

In order to get a first estimation of the relevance of the IGE as compared with the other two effects, we have made the calculation only for the δ^2-term. A justification of the use of only that term come from the Zeldovich aproximation since this is the only term which appears in that approximation at the 2nd order perturbation, although with a multiplicative coefficient which is 1 instead of 5/7. In fact all the results will be given for the coefficient 1.

Details of the calculation will be given elsewhere (Martínez-González, Sanz and Silk 1991). Here we simply summarize the main results for the two cases considered and compare them with the Relict, Tenerife and COBE (Klypin et al. 1987, Watson et al. 1989 and Smoot et al. 1991, respectively) experiments. The characteristics of these experiments which are of relevance for this calculation are related in the contribution by Martínez-González and Cayón 1991 of these proccedings. For the first case considered of a constant correlation with a Gaussian cut-off we may precisely write $\xi_l(r)$ as

$$\xi_l(r) = B \exp\left(-\frac{r^2}{2R^2}\right) \quad , \tag{8}$$

where the subindex l refers to linear theory. The relation between the linear correlation ξ_l and the 2nd order theory one ξ_{nl} considering only the δ^2-term is $\xi_{nl}(r) = 2\xi_l^2(r)$. The value of the constant B is obtained by imposing on $\xi_l(r)$ the standard requirement that the typical mass fluctuation be 1 on a scale of $8h^{-1}Mpc$. The results for the temperature autocorrelation function for the IGE, SWE and DE effects as measured by the COBE experiment (the beam-width of the antenna for this experiment is 7° FWHM) are shown in figure 1 for a cut-off $R = 10h^{-1}Mpc$. We have previously elliminated the monopole and quadrupole contributions from the correlation. We see that in this case the IGE clearly dominates over the SWE and the DE ones, the difference being more than an order of magnitude with respect to the SW and even more respect to the DE. When one consideres the double differences of the Tenerife experiment (beamwidth of 5° FWHM and beamthrow 8°.2) , see figure 2, the IGE is again greater than the SWE and DE effects although the differencies are smaller for this type of measurements than for the COBE ones. The quadrupole moment is for the IGE $(\Delta T/T)_2 = 3.3 \times 10^{-4}$, SWE $(\Delta T/T)_2 = 8.7 \times 10^{-5}$ and DE $(\Delta T/T)_2 = 6.4 \times 10^{-6}$, which clearly violate the upper limit given by the Relict, $(\Delta T/T)_2 \leq 3 \times 10^{-5}$. Although these theoretical results also violate the limits for the anisotropy given by Tenerife and COBE, we want to remark the interesting point that the IGE is greater than the SWE, the effect which is normally considered dominant on these angular scales.

To test whether the results already obtained for the simple case of a constant correlation are generic or not, we consider a second case of a more realistic correlation of the form

$$\xi_{nl}(r) = \left(\frac{r_0}{r}\right)^2 \exp\left(-\frac{r}{R}\right) \quad , \qquad (9)$$

where in order to fit that observed for galaxies we take $r_0 = 5h^{-1}Mpc$ and the exponential cut-off $R = 10h^{-1}Mpc$. The linear correlation is in this case

$$\xi_l(r) = \frac{1}{\sqrt{2}}\left(\frac{r_0}{r}\right) \exp\left(-\frac{r}{2R}\right) \quad . \qquad (10)$$

The results for the COBE measurements are now shown in figure 3. The correlations for the IGE and the SWE are now very similar. The major discrepancy appears at $\alpha = 0°$ where the SWE dominates, but the difference is not very significant $< 15\%$. The correlation for the DE is much smaller on these large angular scales and it is very close to 0 for all the α range. The $\Delta T/T$ values for the Tenerife experiment can be seen in figure 4, where the SWE effect dominates over the IGE and DE ones and the differencies at $\alpha = 8°.2$ are $\approx 50\%$. Notice that at approximately this angular scale the IGE curve exceeds the DE one. In this case the quadrupole moment is for the IGE $(\Delta T/T)_2 = 5.1 \times 10^{-5}$, SWE $(\Delta T/T)_2 = 5 \times 10^{-5}$ and DE $(\Delta T/T)_2 = 3.7 \times 10^{-6}$. We would like to comment that, although the mean temperature anisotropy is above the observed upper limits, these results do not contradict the observations if we take in account the possible bias of the galaxies respect to the matter distribution and also the uncertainties given by the probability distribution function of the temperature anisotropy (see Martínez-González and Cayón 1991 in these proccedings).

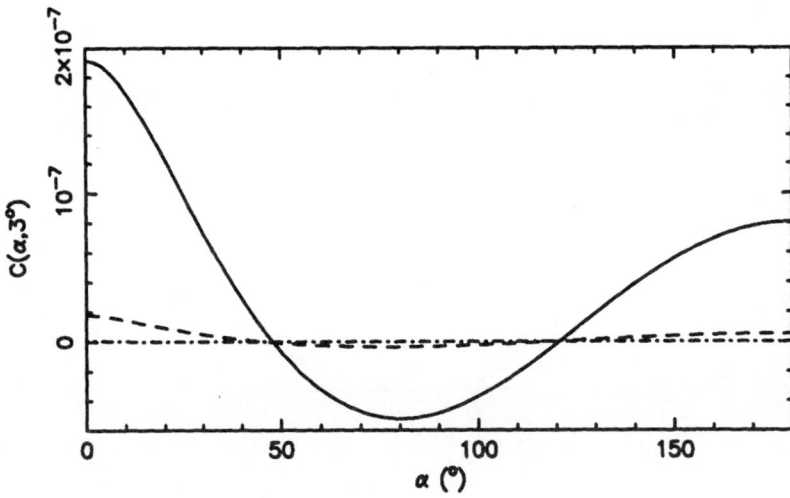

Fig.1. *Temperature autocorrelation function with* $\sigma = 3°$ *(COBE) for different effects, considering a constant correlation with a Gaussian cut-off at* $R = 10h^{-1}Mpc$. *Solid line corresponds to IGE, dashed line corresponds to SWE and the dashed-dotted line is for DE.*

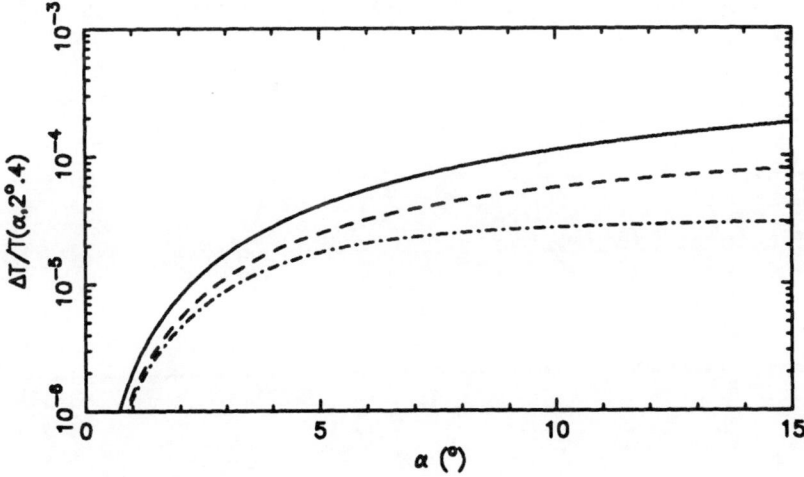

Fig.2. *Temperature fluctuations with* $\sigma = 2°.4$ *(Tenerife) for different effects, considering a constant correlation with Gaussian cut-off at* $R = 10h^{-1}Mpc$. *Different lines correspond to the effects indicated in fig.1.*

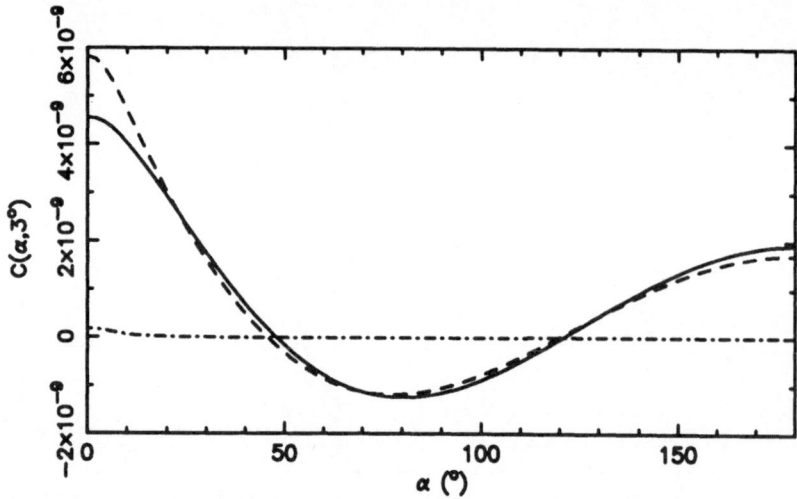

Fig.3. *Temperature autocorrelation function with* $\sigma = 3°$ *(COBE) for different effects, considering a power law correlation with an exponential cut-off at* $R = 10h^{-1}Mpc$. *Different lines correspond to the effects indicated in fig.1.*

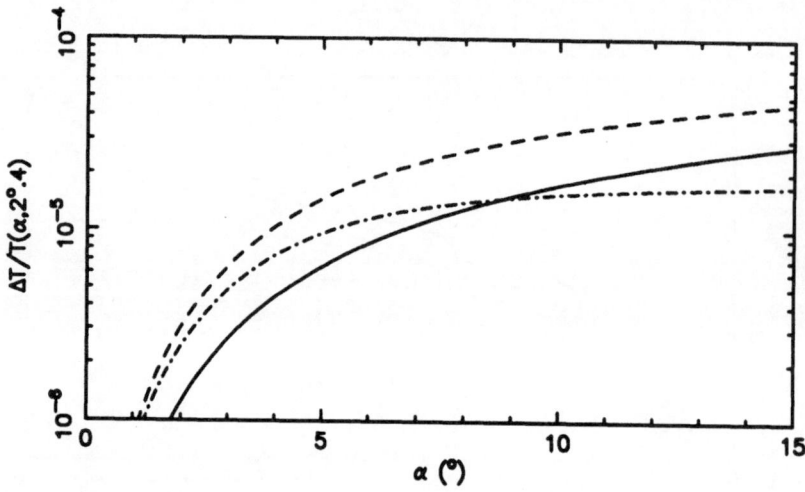

Fig.4. *Temperature fluctuations with* $\sigma = 2°.4$ *(Tenerife) for different effects, considering a power law correlation with an exponential cut-off at* $R = 10h^{-1}Mpc$. *Different lines correspond to the effects indicated in fig.1.*

References

Argüeso, F. & Martínez-González, E. 1989, *Mon. Not. R. ast. Soc.*, **238**, *1431*.
Argüeso, F., Martínez-González, E. & Sanz, J.L. 1989, *Ap. J.*, **336**, *69*.
Atrio-Barandela, F. & Kashlinsky, A. 1991, preprint.
Barnes, J. & Hut, P. 1986, *Nature*, **324**, *446*
Barrow, J.D., Juszkiewicz, R. & Sonoda, D.M. 1983, *Nature*, **305**, *397*.
Berstchinger, E. 1985, *Ap. J. Supplement Series*, **58**, 1.
Bertschinger, E. 1987, *Ap. J. Lett.*, **323**, *L103*.
Bertschinger, E. 1990, *Nature*, **345**, *507*.
Bertschinger et al. 1990, *Ap. J.*, **364**, *370*.
Bond, J.R., Carr, B.J. & Hogan, C.J. 1991, *Ap. J.*, **367**, *420*.
Dressler, A. 1988, *Ap. J.*, **329**, *519*.
Dyer, C.C. 1976, *Mon. Not. R. ast. Soc.*, **175**, *429*.
Dyer, C.C. & Ip, P.S.S. 1988, *Mon. Not. R. ast. Soc.*, **235**, *895*.
Franceschini, A., Toffolati, L., Danese, L. & De Zotti, G. 1989, *Ap. J.*, **344**, *35*.
Geller, M.J. & Huchra, J.P. 1989, *Science*, **246**, 897.
van Kampen, E. & Martínez-González, E. 1991a, *IInd Rencontres de Blois: Physical Cosmology*, Pergamon Press, in press.
van Kampen, E. & Martínez-González, E. 1991b, *submitted to Mon. Not. R. ast. Soc.*
Klypin, A.A., Shazin, M.V., Strukov, I.A. & Skulachev, D.P. 1987, *Sov. Astron. Lett.*, **13**, 104.
Lynden-Bell, D., Faber, S.M., Burstein, D., Davies, R.L., Dressler, A., Terlevich, R. & Wegner, G. 1988, *Ap. J.*, **326**, *19*.
Martínez-González, E. & Cayón, L. 1991, in these proceedings.
Martínez-González, E. & Sanz, J.L. 1990, *Mon. Not. R. ast. Soc.*, **247**, *473*.
Martínez-González, E., Sanz, J.L. & Silk, J. 1990, *Ap. J. Lett.*, **355**, *L5*.
Martínez-González, E., Sanz, J.L. & Silk, J. 1991, in preparation.
Nottale, L. 1984, *Mon. Not. R. ast. Soc.*, **206**, *713*.
Peebles, P.J.E. 1980, The Large Scale Structure of the Universe, Princeton: Princeton Univ. Press.
Raychaudhury, S. 1989, *Nature*, **342**, *251*.
Rees, M.J. & Sciama, D.W. 1968, *Nature*, **517**, *611*.
Sachs, R.K. & Wolfe, A.M. 1967, *Ap. J.*, **147**, *73*.
Scaramella, R., Baiesi-Pillastrini, G., Chincarini, G., Vettolani, G. & Zamorani, G. 1989, *Nature*, **338**, *562*.
Smoot et al., 1991, *Ap. J.*, *submitted*.
Thompson, K.L. & Vishniac, E.T. 1987, *Ap. J.*, **313**, *517*.
Vishniac, E.T. 1987, *Ap. J.*, **322**, *597*.
Watson et al. 1989, *Large Scale Structure and Motions in the Universe*, p.133, eds.
    Mezzetti, P. et al., Kluwer, Dordrecht.
Wyse, R.G. & Jones, B.J. 1985,*Astr. Astrophys.*, **149**, 144.

STATISTICS OF THE CMB MAPS

E. MARTINEZ-GONZALEZ* & L. CAYON**
*Instituto de Estudios Avanzados en Física Moderna y
Biología Molecular,
CSIC-Univ. Cantabria,
Facultad de Ciencias, Dept. Física Moderna
Avda. Los Castros s.n.,
39005 Santander,
Spain.
**Dpto. Física Moderna,
Universidad de Cantabria,
Facultad de Ciencias,
Avda. Los Castros s.n.,
39005 Santander,
Spain.

ABSTRACT. The maps of the cosmic microwave background CMB temperature contain information on the anisotropy and on the topology of the microwave background. We first study the distribution of the temperature autocorrelation function and its implications on the cosmological models of the universe. Secondly, we analyse the topological descriptors that have been proposed to characterize the microwave temperature fluctuations.

1. Introduction

In this review we focus on the statistics of the CMB temperature fluctuations. These temperature fluctuations are considered to be gaussian distributed, based on the inflationary paradigm that provides the initial gaussian matter density perturbations present at recombination. These perturbations are of the adiabatic type and have a nearly flat variance when they cross the horizon (the Harrison-Zeldovich spectrum, $P(k) \propto k$). However it is usual to consider a more general power law for the spectrum $P(k) \propto k^n$. We also consider the standard scenario for the evolution of the microwave photons, where it is assumed an instantaneous decoupling of matter and radiation and the only effects acting on the photons are the linear ones (Sachs and Wolfe 1967) which preserve the gaussian character of the fluctuations (see Martínez-González 1991 for a review on the non-linear gravitational effects).

303

M. Signore and C. Dupraz (et al.), The Infrared and Submillimetre Sky after COBE, 303–314.

A large number of studies have been dedicated to calculate the anisotropies expected in the CMB temperature for cosmological scenarios dominated by baryons (B) either with adiabatic or isocurvature perturbations (Gouda et al. 1987, 1989; Efstathiou and Bond 1987), hot dark matter (HDM) (Bond 1988, Sugiyama et al. 1989) and cold dark matter (CDM) (Fukugita et al. 1990, Efstathiou and Bond 1986, Bond 1987). An exhaustive study on the constraints of these scenarios is given in Holtzman (1989) and Suto et al. (1990). Summarizing all these results for the case of standard recombination and normalizing the spectrum by imposing that the present typical mass fluctuation is 1 at $8h^{-1}Mpc$, we can say that barionic models with $\Omega \leq 1$, $-1 \leq h \leq 1$ and either adiabatic or isocurvature fluctuations produce anisotropies which are above the upper limits given on large or intermediate angular scales. HDM and CDM models with isocurvature fluctuations are ruled out by the large angular scale limits whereas for adiabatic fluctuations they are aceptable for values of $\Omega \sim 1$ and $n \sim 1$. However HDM models can have other problems related to the epoch of structure formation and its distribution in the universe. In the case of the pure barionic models, one way of lessening those constraints is by assuming that a significant fraction of the matter is reionized after recombination (Efstathiou 1989, Gorski and Silk 1989), erasing the primary anisotropies and generating new ones due to the coupling of the bulk flow of the electrons with their density fluctuations (Ostriker and Vishniac 1986, Vishniac 1987). Also the introduction of a cosmological constant help to loosen the constraints on low density universe models (Sugiyama et al. 1990).

However in order to properly reject a model we have previously to consider confidence intervals for the temperature anisotropies. In section 3 we show how to calculate the probability distribution for the temperature autocorrelation function.

The maps of the temperature of the microwave sky contain more information than the temperature autocorrelation function and its multipoles. There exit topological descriptors that describe the texture of the CMB maps and that have already been studied by several authors (see for instance Vittorio and Juszkiewicz 1987 and Gott et al. 1990). In section 4 we review the present state of those studies, emphasizing their relevance in the analysis of the observational data and in discriminating between the different cosmological scenarios. In section 5 we give some topological results for several experiments.

2. Recent experiments on large and intermediate angular scales

There are different kinds of experiments to measure the CMB temperature on intermediate and large angular scales. One of the main problems is the contribution of the earth and atmosphere radiation at these frecuencies. To solve these problems some of the experiments work aboard satellites (COBE, Relict) or balloons (Melchiorri et al., 1981). The groundbased Tenerife and OVRO experiments use an special technique, based in a double subtraction to suppress the atmospheric contribution.

-COBE was launched on November 18th 1989 and is working at present, it has three differential microwave radiometers with two receivers each of them having two antennas of 7° of diameter (FWHM) separated 60°. It has already measured the spectrum of the CMB radiation finding a blackbody spectrum with deviation $< 1\%$ of the peak intensity (Mather et al., 1990). Upper limits on the correlation function (between 15° and 165°), $|C(\alpha, \sigma)| \leq 10^{-9}$, and multipoles $(\Delta T/T)_2 \leq 3 \times 10^{-5}$, $(\Delta T/T)_l \leq 4 \times 10^{-5}$ $(l \geq 2)$, at 95% confidence level, have recently been given (Smoot et al. 1991). After 2 years of observations it will obtain a map of the sky, at the level $\leq 10^{-5}$, in three wavelengths 3.3, 5.7 and 9.6 mm.

-The soviet experiment, Relict-1, aboard the satellite Prognoz 9 (July 1983-February 1984) measured the temperature differencies between two beams separated 90° with a FWHM of 5.5° and operating at $\lambda = 8$mm. The result is a map covering 70% of the sky with the data previously smoothed with a gaussian window of dispersion $\sigma = 5°$. In this way they have restricted the quadrupole $a_2 \leq 4.8 \times 10^{-5}$ (at 95%) and the correlation function between 20° - 150° at the 99% confidence level (Klypin et al., 1987 and Strukov and Skulachev, 1988).

-The Tenerife experiment (Jodrell-IAC collaboration), used two antennas of resolution 8°.5, separated 8°.2 and operating at $\lambda = 3$ cm (at present another similar experiment is working at $\lambda = 2$ cm). The signal is received through a conmutant reflector. This strategy gives measurements by the difference between the signal received in a central beam and the mean of the signals observed in two lateral beams separated 8°.2 from the central one. With these measurements an upper limit of $\Delta T/T \leq 3.7 \times 10^{-5}$ is given at declination 40° (Davies et al., 1987).

-The 40 m radiotelescope at OVRO has provided measurements on intermediate scales (between 4' and 2°) operating at $\lambda = 1.5$ cm. The data have been obtained with a technique of double conmutation (equivalent to the Tenerife experiment one), with resolution 1'.8 (FWHM) and separation 7'.15. The result is an upper limit of the temperature fluctuations of the CMB of $\Delta T/T \leq 2.1 \times 10^{-5}$ at 95% confidence level (Readhead et al., 1989).

-The balloon-borne experiment performed by Melchiorri et al. (1981) worked with an antenna of beamwidth $\sigma = 2°.2$ and obtained data at an angular scale of 6° using a single subtraction technique (beam-switching experiment). This experiment gave an upper limit on the CMB fluctuations of $\Delta T/T \leq 5.4 \times 10^{-5}$.

3. Distribution of the temperature anisotropy

3.1. THEORETICAL RESULTS

We expand the temperature fluctuations of the CMB in a sum of spherical harmonics:

$$\frac{\Delta T}{T}(\vec{x}, \vec{n}) = \sum_{l=1}^{\infty} \left(\sum_{m=-l}^{l} a_l^m(\vec{x}) Y_{lm}(\vec{n}) \right) \tag{1}$$

where, $\vec{x}$ is the observer position, $\vec{n}$ is a unit vector which indicates the observation direction around the observer. The coefficients a_l^m of this expantion are gaussian random variables, independent among them, with mean value zero and variance $\langle |a_l^m(\vec{x})|^2 \rangle = \sigma_l^2$. For a power law spectrum of the primordial density perturbations, $P(k) = Ak^n$, we have (Fabbri, Lucchin and Matarrese 1987):

$$\sigma_l^2 = \frac{2^{n-1}A}{r_0^{(n+3)}} \frac{\Gamma(3-n)}{\Gamma[(4-n)/2]^2} \frac{\Gamma[(2l+n-1)/2]}{\Gamma[(2l+5-n)/2]} \tag{2}$$

(we consider units such that $d_h = 8\pi G = c = 1$, where d_h is the horizon distance and c the speed of ligth).

In order to observe the temperature fluctuations we use an antenna whose response is normally represented by a Gaussian function, giving rise to the following correlation function after averaging over the whole sky:

$$C(\alpha,\sigma) = \langle \frac{\Delta T}{T}(\vec{n}_1,\sigma)\frac{\Delta T}{T}(\vec{n}_2,\sigma) \rangle_{sky}$$

$$= \frac{1}{4\pi}\sum_{l=2}^{\infty}(2l+1)a_l^2 P_l(\cos\alpha)\exp(-(l+\frac{1}{2})^2\sigma^2) \tag{3}$$

$\vec{n}_1$ and $\vec{n}_2$ are the two observation directions with $\vec{n}_1\vec{n}_2 = \cos\alpha$, and $\sigma << 1$ is the dispersion of the antenna. The antenna introduces a exponential cut-off in the sum over all multipoles and then only a finit number of them are necesary to be considered. If we want to calculate the mean value of the correlation function we have to do an average over all the possible realizations of the CMB:

$$\langle C(\alpha,\sigma) \rangle = \frac{1}{4\pi}\sum_{l=2}^{\infty}(2l+1) < a_l^2 > P_l(\cos\alpha)\exp(-(l+\frac{1}{2})^2\sigma^2) \tag{4}$$

and $< a_l^2 >=< |a_l^m|^2 >= \sigma_l^2$.

We now calculate the probability density function of the random variable $Z \equiv C(\alpha,\sigma)$. The sum of the squares of $(2l+1)$ independent gaussian random variables $(|a_l^m|)$ is distributed with a χ^2 density function with $(2l+1)$ degrees of freedom. Thus, the probability density function of the correlation is given as a convolution of the density functions of each term of the sum in eq.(3). Moreover, the characteristic function of Z is given as the product of the characteristic functions of random variables which are of the form of a constant times a χ^2 distributed variable. Finally, the probability density function of the temperature correlation can be obtained by the invers Fourier transform of the characteristic function (this result has very recently been obtained by Cayón, Martínez-González and Sanz):

$$f_Z(t) = \frac{1}{2\pi}\int_{-\infty}^{+\infty} dx \left(\prod_{l=2}^{m}[1-2ix\bar{\sigma}_l]^{\frac{-(2l+1)}{2}}\right)\exp(-itx) \tag{5}$$

where $\bar{\sigma}_l = \frac{\sigma_l^2}{4\pi}P_l(\cos\alpha)exp[-(l+1/2)^2\sigma^2]$.

All of the formalism used above for a single beam experiment can be applied to a beam-switching experiment and a double beam-switching one, defining the new variables Z_2 and Z_3 being $< Z_2 >^{\frac{1}{2}} = \left(\frac{\Delta T}{T}\right)_2(\alpha, \sigma)$, $< Z_3 >^{\frac{1}{2}} = \left(\frac{\Delta T}{T}\right)_3(\alpha, \sigma)$. The probability density function for Z_2 and Z_3 is again given by (4), where now $\bar{\sigma}_l$ is for the beam-switching experiment:

$$\bar{\sigma}_l = \frac{1}{2\pi}\sigma_l^2 \left[1 - P_l(\cos \alpha)\right] \exp(-(l + \frac{1}{2})^2\sigma^2) \qquad (6)$$

and, for the double beam-switching experiment:

$$\bar{\sigma}_l = \frac{1}{4\pi}\sigma_l^2 \left[\frac{3}{2} - 2P_l(\cos \alpha) + \frac{1}{2}P_l(\cos 2\alpha)\right] \exp(-(l + \frac{1}{2})^2\sigma^2) \qquad (7)$$

3.2. IMPLICATIONS ON COSMOLOGICAL MODELS

In this section we compare our results with three types of experiments. These experiments differ in the strategy used to measure the temperature anisotropies.

In figure 1 we have plotted the probability density function of the normalized temperature correlation for the COBE experiment at zero lag, $C(0, 3°)/ < C(0, 3°) >$, in the case of a Harrison-Zeldovich spectrum for the density fluctuations. This probability is asymmetric with a long tail towards large correlations. This asymmetry is also present on other angular scales different from 0. The 95% confidence interval of $C(\alpha, 3°)/ < C(0, 3°) >$ as a function of the angular scale α is shown in figure 2. At zero lag the 95% confidence interval goes from $\sim 50\%$ above to $\sim 30\%$ below the mean correlation and widens for angular scales near 180°.

The implications of considering confidence intervals for the temperature autocorrelation obtained in several cosmological models are related in Cayón, Martínez-González and Sanz (1991). Considering the COBE upper limit for the correlation between angular scales 15° and 165°, $|C(\alpha, \sigma)| \leq 10^{-9}$, we show in table 1 the predictions of several cosmological models on an angular $\alpha = 20°$ (the models are normalized in the standard way, i.e. the r.m.s. density fluctuation is 1 at $8h^{-1}Mpc$; in the calculations $h = 0.5$ is considered). The HDM model with $\Omega = 0.5$ predicts a mean correlation above that limit but it can not be rejected at the 95% confidence level. Therefore the only model that can be rejected at that confidence level is the pure barionic one with $\Omega = 0.1$. On the other hand, for experiments involving 1-beam (Melchorri et al. 1981) or 2-beams (Davies et al. 1987) subtractions the 95% interval becomes very narrow and therefore the uncertainties with respect to the mean correlation becomes very small (see tables 3 and 4 of Cayón, Martínez-González and Sanz 1991). As a result, the two models considered with $\Omega < 1$, HDM and B, with standard normalization predict 95% confidence intervals above the observed upper limits.

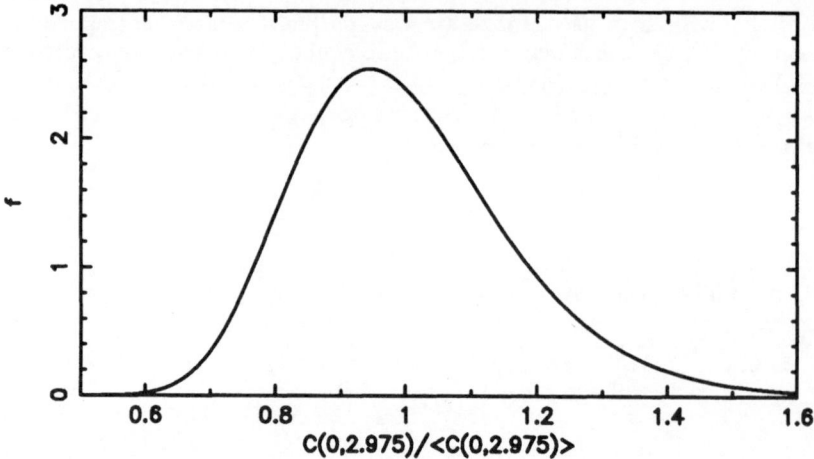

Fig.1. *Probability density function of the normalized correlation at* $\alpha = 0°$ *and* $\sigma = 2°.975$ *(COBE)*.

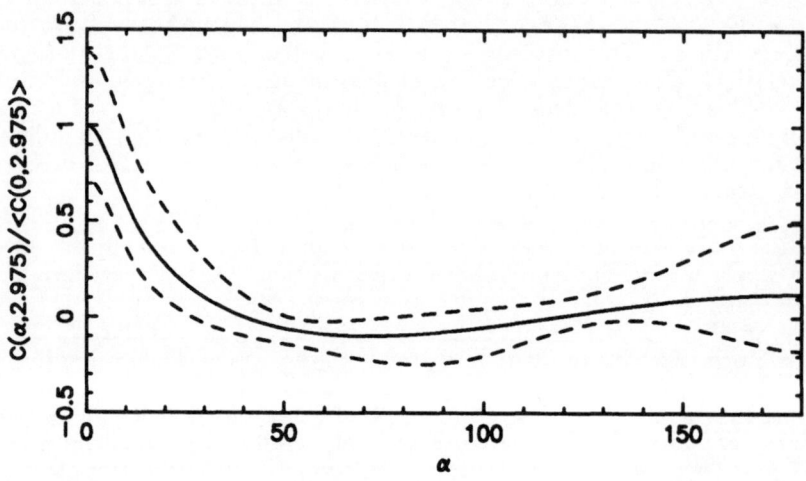

Fig.2. *Confidence intervals (at 95%) for the normalized correlation function with* $\sigma = 2°.975$ *(COBE)*.

TABLE 1. 95% confidence intervals of $C(20°, 2°.975)$ (COBE) for different scenarios.

Scenarios	$C(20°, 2°.975) \times 10^{10}$	95% conf. interval
CDM	0.22	(0.066,0.46)
HDM	0.31	(0.093,0.65)
HDM ($\Omega = 0.5$)	16.85	(5.1,35.4)
B	3.8	(1.15,7.97)
B($\Omega = 0.1$)	379.5	(114.7,797.0)

Finally, there are other uncertainties related to the normalization of the power spectrum that should be considered. One is the bias parameter $b \approx 1 - 2.5$ that tells us about the discrepancy between the distribution of matter and galaxies. Another posible source of uncertainty comes from using linear theory to normalize the fluctuations at scales of $10h^{-1}Mpc$. Martínez-González and Sanz (1991) have shown that there is a factor of ~ 2 difference between linear theory and the Zeldovich approximation. Both uncertainties result in a decrease of a factor ~ 10 in the expected temperature correlation as well as in the confidence interval, concluding that the only model that can be rejected , in a conservative way, by the large angular scale experiments is the pure barionic B with $\Omega = 0.1$.

4. Topology of the CMB fluctuations

Additionally to the statistical information contained in the temperature autocorrelation function and its distribution discussed before, there are other statistical descriptors which can be used to characterize the topology of the CMB maps. Some of these descriptors like the expected number and size of regions above a certain level have already been studied by a number of authors for gaussian (Sazhin 1985, Zabotin and Nasel'skii 1985, Bond and Efstathiou 1987, Vittorio and Juszkiewicz 1987, Martínez-González and Sanz 1989) and several non-gaussian temperature fluctuations (Coles and Barrow 1987). In the case that the temperature fluctuations correspond to a homogeneous and isotropic gaussian random field on the sphere with correlation function $C(\alpha)$, analytical expressions can be obtained for the expected number of spots $< N_s >$ over the whole sphere and the expected area of the spots $< A >$ in the 2D maps, and for the expected number $< N_p >$ and expected size $< \Theta >$ of the pips at fixed declination in the 1D scans, above the level $\nu\, C(0)^{\frac{1}{2}}$:

$$< N_s >= \frac{2}{\pi\lambda^2} \frac{e^{-\nu^2}}{erfc(\nu/\sqrt{2})} \quad , \quad < A >=< \Theta >^2 \tag{8}$$

$$< N_p >= \frac{1}{\lambda} e^{\frac{-\nu^2}{2}} \cos\delta \quad , \quad < \Theta >= \pi\lambda e^{\frac{\nu^2}{2}} erfc(\nu/\sqrt{2}) \tag{9}$$

where $\lambda \equiv [-\frac{C(0)}{C''(0)}]^{\frac{1}{2}}$ is the coherence angle of the field.

Although these statistical quantities give a useful insight into the behaviour of the temperature field, they are not very sensitive to discriminate between gaussian and non-gaussian fields (Coles and Barrow 1987). More recently other topological quantities have been used to characterize the texture of the CMB (Coles 1988, Gott et al. 1990): fractional area of the excursion regions a, contour length s and total curvature of the contours or "genus" G. Of special interest in the genus G which is rotational invariant, unlike the integral geometric characteristic Γ (Coles 1988, Adler 1981), and it does not require the compatness of the Euler-Poincare characteristic χ (Adler 1981). These three statistics agree if the excursion domains do not cross the boundaries or if the boundaries are periodic. In this case the genus G is the total number of the excursion regions minus the total number of holes in them. G has been proved to be a sensitive statistic to test the gaussian nature of the CMB temperature field (Gott et al. 1990).

For a homogeneous and isotropic gaussian random field the mean values of a, s and G can be given as

$$< a >= \frac{1}{2} erfc(\nu/\sqrt{2}) \quad , \quad < s >= \frac{1}{2\lambda} e^{\frac{-\nu^2}{2}} \tag{10}$$

$$< G >= erfc(\nu/\sqrt{2}) + (\frac{2}{\pi})^{\frac{1}{2}} \frac{1}{\lambda^2} \nu e^{\frac{-\nu^2}{2}} \tag{11}$$

5. Topological results for several experiments

We now calculate the topological quantities described above for several experiments, assuming a gaussian random field for the temperature fluctuations and a scale-invariant power spectrum. The experiments considered differ in the geometry used to measure the CMB temperature: Relict and COBE obtain maps with direct measures of the temperature, Melchiorri et al. (1981) measure single subtraction whereas the Tenerife experiment measures double subtraction of the temperature. The characteristics of the experiments are explained in section 2.

Figure 3 shows the mean number of spots $< N_s >$ for the four experiments as a function of the level ν. The number of spots over the whole celestial sphere is higher for the single and double subtraction experiments (Melchiorri et al. (1981) and Tenerife) than for the Relict and COBE ones. For the mean size of the spots $< \Theta >$ the result is just the opposite, the spots found in the temperature maps are bigger in size than the ones found in the maps of the temperature gradients (see figure 4). N_s as a function of the beamthrow α is shown in figure 5 for the gradient maps. The second difference map contains more spots than the first difference one for the same α.

The expected total curvature of the contours for the whole celestial sphere $< G >$ is given in figure 6 as a function of the level ν. It is clear from this figure that this topological statistic is quite sensitive to the antenna beamwidth (compare the bottom curves) and also the geometry of the experiment. Moreover, it is also known that the genus is a good discriminator between gaussian and non-gaussian random fields (Coles 1988). In particular , it has been successfully applied to discriminate the non-gaussian nature of the temperature field produced in the string scenario (Gott et al. 1990) from the gaussian one.

In a recent work, Gutiérrez et al. 1991 have used some of the topological statistics to analyse the data of the Tenerife experiment (Davies et al. 1987). This experiment covers 70° in rigth ascension at fixed declination $\delta = 40°$ and the measures are the result of second differences with antenna beamwidth $\sigma = 3°.6$ and beamthrow $\alpha = 8°.2$. The data are binned into 1° intervals of rigth ascension. By simply joining the mean values of the data with straight lines we are able to calculate the number of pips N_p from the 1D scan which are above different levels and also their mean size $< \Theta >$ (it is important to remark that these statistical quantities N_p and $< \Theta >$ obtained from the data are almost independent of the way we interpolate the data. The N_p and $< \Theta >$ calculated for a cubic spline interpolation give very similar results). N_p and $< \Theta >$ so obtained are compared with the values expected in gaussian temperature distributions. The first conclusion is that it is not possible to reproduce the observed N_p and $< \Theta >$ with the only presence of the cosmic signal. It is therefore required the existence of some noise, the instrumental noise, which is assumed to be white. Another conclusion is that with a r.m.s. amplitud for the noise of 0.22 mk and $< C(0,\sigma) >^{1/2} = 0.16$ mk for a cosmic signal with scale invariant power spectrum it is possible to reproduce the observed N_p and $< \Theta >$ at the 95% confidence level (the value of $< C(0,\sigma) >^{1/2} = 0.16$ mk is the one obtained by applying the maximum likelihood method, Vittorio et al. 1989).

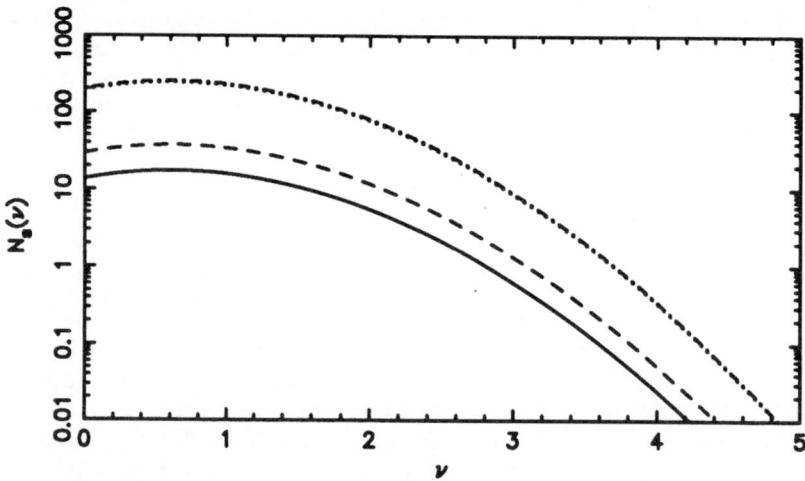

Fig.3. *Mean number of spots* $< N_s >$ *as a function of the level* ν. *The solid line is for Relict; the dashed line is for COBE; the dashed-dotted line corresponds to the Melchiorri et al. (1981) experiment and the dotted one is for the Tenerife experiment.*

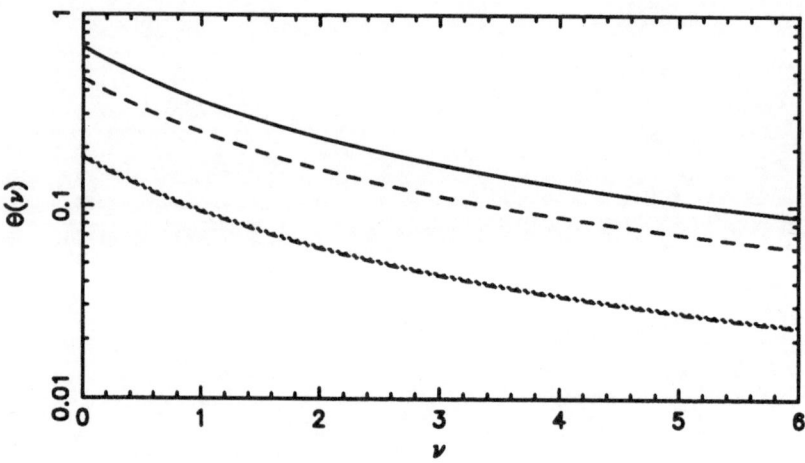

Fig.4. *Mean size of the spots* $< \Theta >$ *as a function of the level* ν. *Different lines correspond to the same experiments indicated in fig.3.*

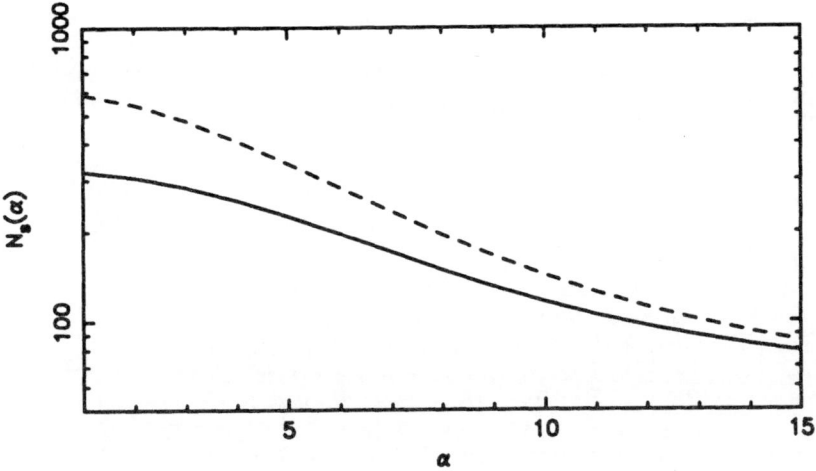

Fig.5. Mean number of spots N_s as a function of the beamthrow α (in deg). The solid line corresponds to the Melchiorri et al. (1981) experiment and the dashed one is for Tenerife.

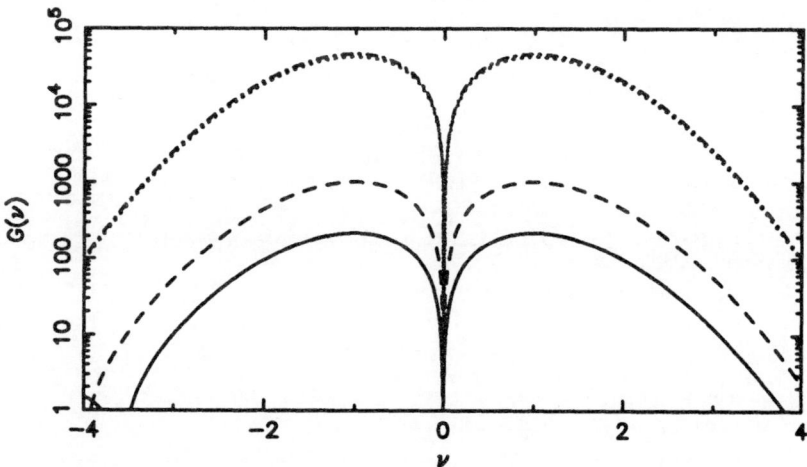

Fig.6. The mean value of the genus $< G >$ as a function of the level ν. Different lines correspond to the same experiments indicated in fig.3.

314

References

Adler, R.J. 1981, *The Geometry of the Random Fields*, (New York: John Wiley).
Bond, J.R. & Efstathiou, G. 1987, *Mon. Not. R. ast. Soc.*, **226**, *655*.
Bond, J.R. 1987, in NATO ASI "The Early Universe", ed. W.G. Unruh and
    G.W. Semenoff. (Dordrecht:Reidel).
Bond, J.R. 1988, in "Large Scale Structure of the Universe", Audouze et al.
    (eds.), Reidel, Dordrecht.
Cayón, L., Martínez-González, E. & Sanz, J.L. 1991, *Mon. Not. R. ast. Soc.*,
    submitted.
Coles, P. & Barrow, J.D. 1987, *Mon. Not. R. ast. Soc.*, **228**, *407*.
Coles, P. 1988, *Mon. Not. R. ast. Soc.*, **234**, *509*
Davies, R.D., Lasenby, A.N., Watson, R.A., Daintree, E.J., Hopkins, J., Beckman,
    J., Sánchez-Almeida, J & Rebolo, R. 1987, *Nature*, **326**, *462*.
Efstathiou, G. & Bond, J.R. 1986, *Mon. Not. R. ast. Soc.*, **218**, *103*.
Efstathiou, G. & Bond, J.R. 1987, *Mon. Not. R. ast. Soc.*, **227**, *33p*.
Efstathiou, G. 1989, Proceedings of the Vatican Conference "Large Scale Motions
    in the Universe".
Fabbri, R., Lucchin, F. & Matarrese, S. 1987, *Ap. J.*, **315**, *1*.
Fukugita, M., Sugiyama, N. & Umemura, M. 1990, *Ap. J.*, **358**, *28*.
Gorski, K.M. & Silk, J. 1989, *Ap. J. Lett.*, **346**, *L1*.N
Gott et al. 1990, *Ap. J.*, **352**, *1*.
Gouda, N., Sasaki, M. & Suto, Y. 1987, *Ap. J. Lett.*, **321**, *L1*.
Gouda, N., Sasaki, M. & Suto, Y. 1989, *Ap. J.*, **341**, *557*.
Gutiérrez, C., Martínez-González, E., Rebolo, R. & Sanz, J.L.
    1991, *in preparation.*
Holtzman, J.A. 1989, *Ap. J. Supplement Series*, **71**, 1.
Klypin, A.A., Shazin, M.V., Strukov, I.A. & Skulachev, D.P. 1987,
    Sov. Astron. Lett., **13**, 104.
Martínez-González, E. 1991, *in this volume.*
Martínez-González, E. & Sanz, J.L. 1989, *Mon. Not. R. ast. Soc.*, **237**, *939*.
Martínez-González, E. & Sanz, J.L. 1991, *Ap. J.*, **366**, *1*.
Mather et al. 1990, *Ap. J.*, **354**, *137*.
Melchiorri, F., Melchiorri, B.O., Ceccarelli, C. & Pietranera, L. 1981,
    Ap. J. Lett., **250**, *L1*.
Ostriker, J.P. & Vishniac, E.T. 1986, *Ap. J. Lett.*, **306**, *L51*.
Readhead, A.C.S., Lawrence, C.R., Myers, S.T., Sargent, W.L.W., Hardebeck,
    H.E. & Moffet, A.T. 1989, *Ap. J.*, **346**, *566*.
Sachs, R.K. & Wolfe, A.N. 1967, *Ap. J.*, **147**, *73*.
Sazhin, M.V. 1985, *Mon. Not. R. ast. Soc.*, **216**, *short communication, 25p.*
Smoot et al. 1991, *Ap. J.*, *submitted.*
Strukov, I.A. & Skulachev, D.P. 1988, *Sov. Sci. Rev. E. Astrophys Space Phys.*,
    6, 145.
Sugiyama, N., Sasaki, M. & Tomita, K. 1989, *Ap. J. Lett.*, **338**, *384*.
Sugiyama, N., Gouda, N. & Sasaki, M. 1990, *Ap. J.*, **365**, *432*.
Suto, Y., Gouda, N. & Sugiyama, N. 1990, *Ap. J.Supplement Series*, **74**, *665*.
Vishniac, E.T. 1987, *Ap. J.*, **322**,*597*.
Vittorio, N. & Juskiewicz, R. 1987, *Ap. J. Lett.*, **314**, *L29*.
Vittorio, N., de Bernardis, P., Masi, S. & Scaramella, R. 1989, *Ap. J.*, **341**, *163*.
Zabotin, N.A. & Nasel'skii, P.D. 1985, *Sov. Astron.*, **29(6)**, 614.

CMB AND GALACTIC MAPS

IN THE MILLIMETRIC REGION

Paolo de Bernardis
Dipartimento di Fisica
Università La Sapienza, Roma, Italy

Silvia Masi
Dipartimento di Fisica
Università La Sapienza, Roma, Italy

Nicola Vittorio
Dipartimento di Fisica
Università de L'Aquila, L'Aquila, Italy

ABSTRACT: We present low angular resolution, full sky maps of the expected millimetric diffuse emission, in order to illustrate the level of Galactic contamination in CMB (Cosmic Microwave Background) large scale anisotropy experiments. The maps are obtained by simple modelling the diffuse emission of our Galaxy at mm wavelengths, including dust and radio continuum emission. CMB maps expected in cold dark matter (CDM) and baryonic isocurvature scenarios are compared to the Galactic anisotropy pattern. We present the results of a complete analysis of the Quadrupole anisotropy, which, in the case of COBE-DMR, seems to be the most useful tool for cosmological tests.

1. INTRODUCTION

The sensitivity of Cosmic Microwave Background (CMB) anisotropy experiments is continuously increasing, and the detection of CMB anisotropies is of crucial importance in testing galaxy formation scenarios. A serious contamination of these measurements on large angular scales is due to our own Galaxy. Local diffuse emission is present at high galactic latitudes, due to either dust particles (heated by the interstellar radiation field and emitting thermal radiation), and to electrons (moving both in Galactic magnetic fields, thus emitting synchrotron radiation, and in a high temperature plasma, thus emitting free-free radiation). The peak wavelength of dust thermal emission lies between 100 and 200 μm, while radio continuum emission has a power law spectrum steeply decreasing with frequency: so, in the mm region, there is a window with minimal Galactic emission. However, both dust and synchrotron emission show significant angular structures also at high

315

M. Signore and C. Dupraz (et al.), The Infrared and Submillimetre Sky after COBE, 315–330.

galactic latitudes: a cosec(b) law is a quite good first order description of their large scale distribution; moreover 'cirrus' patchy structures were ubiquitously detected by IRAS (Hauser et al. 1984), and high latitude 'radio loops and spurs' are present in the 408 MHz map (Haslam et al. 1982). When compared to the faint intrinsic fluctuations expected in the CMB, these structures become quite important also in the mm region, and can strongly contaminate the cosmological signal. This problem has been already addressed in the past (see e.g. Weiss 1980, Fixsen et al. 1983, Lubin et al. 1985, Banday and Wolfendale, 1990), but a comprehensive analysis is still missing and is particularly important in view of the forthcoming results of the COBE-DMR experiment and, later, for the RELIKT-II mission.

We present here full sky maps of the expected Galactic emission in the mm region; we will compare qualitatively these maps to CMB anisotropies generated in flat, biased, cold dark matter (CDM) models and in open, reionised, pure baryonic universes. We also carry out a quantitative analysis of the Quadrupole anisotropy, by taking into account simultaneously the Galactic contamination, its uncertainty and the detectors noise.

2. MAPS OF THE GALAXY

We are far from having a complete and comprehensive mapping of our Galaxy in the frequency range of interest here, although the COBE DIRBE and FIRAS experiments will greatly improve the situation in the near future. So, we base our analysis on the extrapolation of the available data for dust and radio continuum emission. For each frequency of interest, we have obtained a diffuse Galactic emission map by adding the brightnesses extrapolated from low and high frequency maps (due to synchrotron emission, I_s, free-free emission, I_f, and dust emission, I_d):

$$T_G(\nu) = \frac{I_s(\nu) + I_f(\nu) + I_d(\nu)}{BB(T_{CMB}, \nu)} \cdot T_{CMB} \cdot \frac{e^x - 1}{x e^x} \quad ; \quad x = \frac{h\nu}{k T_{CMB}}. \quad (1)$$

Here T_G is the thermodynamic temperature of a Black Body producing the same emission of our Galaxy, BB is the Planck function, T_{CMB} is the thermodynamic temperature of the CMB. We refer to de Bernardis, Masi and Vittorio (1991; hereafter dBMV) for details. Here we want to give the basic assumptions.

2.1 Dust emission

We are interested in dust emission in the Rayleigh-Jeans tail of the thermal spectra, at wavelengths larger than typical dust grains dimensions ($\lesssim 0.1\mu m$). We use a simple single - temperature thermal spectrum with a power law spectral brightness:

$$I_d(\nu) = A \cdot I_{IRAS} \cdot \left(\frac{\nu}{\nu_{IRAS}}\right)^\beta \cdot \frac{e^{x_{IRAS}} - 1}{e^x - 1} \quad ; \quad x = \frac{h\nu}{k T_d} \quad (2)$$

We use this spectrum for extrapolating to longer wavelengths the dust emission observed in the $100\mu m$ IRAS map (I_{IRAS}). We will assume the following parameters for the dust : $T_d = 22K$; $\beta = 1.5$; $A = 0.6$, (see Masi et al., 1991), in agreement with the recent results of COBE-DIRBE, recalibrating IRAS diffuse measurements (Wright, 1991, this volume). Obviously, we must take into account the errors on the best fit parameters and propagate them in the extrapolation. This will be determined quite poorly at long wavelengths: we have a 1 sigma relative error in the extrapolation of about 35 % at $\lambda = 1mm$ and more than 50 % at $\lambda = 1cm$.

2.2 Synchrotron emission

We used the 408 MHz map of Haslam et al. (1982) to monitor the synchrotron emission of our Galaxy. This approach has been used quite successfully in the past to model Galactic emission at higher frequencies. We assumed a simple power law

$$I_e(\nu) = I_e(408MHz) \cdot \left(\frac{\nu}{408MHz}\right)^{\gamma} \tag{3}$$

with $\gamma = -0.8$, which marginally fits the NGP antenna temperature $T_a = 200\mu K$ at 25 GHz obtained by Fixsen et al. (1983).
The error propagation, including both spectral index variations and the 5% relative errors in the 408 MHz map values, produces a 1 sigma relative uncertainty in the extrapolated emission of about 15% at $\lambda = 1cm$ and 25% at $\lambda = 1mm$.

2.3 Thermal Bremsstrahlung

Free-free emission is generated by hot, ionized gas $(T \gtrsim 10^4 K)$, present both in HII regions in the Galactic plane and in diffuse form above the disk of our Galaxy (Reynolds 1989, Bowyer 1990). A detailed map of this contribution is missing, and we have modeled it at high Galactic latitudes $(|b| > 5°)$ by using a simple cosec(b) law, as in Weiss (1980):

$$I_f(\nu) = I_f(\nu_o)(\nu/\nu_o)^{-0.1} \frac{1}{sin(|b|)} \tag{4}$$

We normalized this spectrum in such a way to obtain the sum of synchrotron plus free-free emission consistent with the NGP emission measured by Fixsen et al. (1983). So we get, for free-free only, 32 mK (at NGP) at $\nu = 1GHz$. This component is the most uncertain; a high galactic latitude survey of H_α emission should be very useful in this respect. Also, the 19 GHz map of Boughn et al. (1990), should be sensitive to both the free-free and synchrotron components, and should improve significantly the precision of the extrapolation of the radio continuum at mm wavelengths.

In conclusion, the extrapolated maps presented here can have a pixel to pixel uncertainty of about 50% at 1 mm. On the other hand, if one assumes that the fluctuations of the spectral parameters in Eq. (2), (3), (4) are uncorrelated on large scales, this uncertainty is considerably reduced, as we will show in the case of the Quadrupole anisotropy.

3. MAPS OF THE CMB

The temperature fluctuation field as observed from position $\mathbf{x}$ in the direction $\hat{\gamma}$ with an antenna of beam σ_B (the dispersion of a Gaussian approximating the angular response of the antenna) can be conveniently expanded in spherical harmonics (see, e.g., Scaramella and Vittorio 1990):

$$\frac{\Delta T}{T}(\mathbf{x}, \hat{\gamma}, \sigma_B) = \sum_{\ell=2}^{\infty} a_\ell^m(\mathbf{x}) \exp\left[-\frac{1}{2}(\ell + \frac{1}{2})^2 \sigma_B^2\right] Y_\ell^m(\hat{\gamma})$$

Here we omit the dipole term, which is dominated by the observer's present peculiar velocity. The large scale pattern of the CMB temperature distribution in our own sky (e.g. $\mathbf{x} = 0$) is then uniquely determined by the set of coefficients $\{a_\ell^m(0)\}$. In the framework of the gravitational instability scenarios, each coefficient $a_\ell^m(\mathbf{x})$ is assumed to be a stochastic variable of the position $\mathbf{x}$, to obey to a Gaussian distribution function with zero mean, $\langle a_\ell^m(\mathbf{x}) \rangle = 0$, and with a rotationally invariant variance $\langle |a_\ell^m(\mathbf{x})|^2 \rangle \equiv a_\ell^2$. Here the symbol $\langle \rangle$ indicates an ensemble average, or, equivalently, an average over all the possible observer positions.

In a CDM scenario, CMB anisotropies on large angular scales are generated by gravitational potential fluctuations on the last scattering surface (Sachs and Wolfe, 1967). In the CDM scenario, the amplitude of the $\{a_\ell^2\}$ has a simple scaling (Peebles, 1982):

$$a_\ell^2 = \frac{A}{2\pi^2}\pi\frac{H_0^4}{c^4}\frac{1}{2\ell(\ell+1)}\frac{1}{b^2}$$

Here A is the normalization constant of the density fluctuations power spectrum and b is the biasing factor. The spectrum is usually normalized by requiring that the rms mass fluctuation in a sphere of $8h^{-1}Mpc$ is b^{-1}

In BDM scenarios there are not large scale potential fluctuations on the last scattering surface: matter density inhomogeneities are strongly damped on scales larger than the matter radiation Jeans length at decoupling. The CMB anisotropy is due to primordial entropy inhomogeneities, gradually transferred in radiation energy density fluctuations because of the constant curvature constraint. For open BDM models, the curvature of the universe should be in principle taken into account when one propagates the CMB radiation brightness from last scattering up to the present (Górski and Silk 1990). However, for the isocurvature models considered here (Peebles, 1987, see Table 1), with a substantial reheating of the intergalactic medium up to the present epoch, the last scattering surface is at a distance much smaller than the curvature radius. Then it is possible to write (Scaramella and Vittorio, 1991):

$$a_\ell^2 = \frac{\pi}{4}\frac{1}{2\pi^2}\int k^2 dk P_{rad}(k) j_\ell^2(kr_{LS})$$

where r_{LS} is the distance of the last scattering surface, and

$$P_{rad} \sim (8/5)^2 A k^n exp(-1.5k^2/k_J^2)$$

is the radiation power spectrum (cf. Gorski and Silk, 1990). In the previous expression k_J is the comoving wavenumber associated with the matter–radiation Jeans length at decoupling. The normalization of the density fluctuation power spectrum is done as in the CDM model, but imposing $b = 1$.

Having fixed the second moment of the a_ℓ^m distribution, we can build CMB maps, by randomly extract the coefficients of the harmonic expansion, $\{a_\ell^m(\mathbf{x})\}$. It is possible to show that the behaviour of the a_l^2 vs. ℓ in the BDM models is much flatter than in the CDM scenario. In the latter case the CMB pattern is then more correlated than in the former one (Scaramella and Vittorio 1991; see the maps in the panels 07-12). We can also evaluate the magnitude of the expected CMB quadrupole anisotropy, i.e. $\langle Q_2^2 \rangle = 5a_2^2$ (see Table 2).

4. SIMULATED MM SKY MAPS

We have averaged the maps in 4584 pixels, each $3° \times 3°$ wide. This resolution is higher than, *e.g.* the COBE-DMR resolution by a factor $\lesssim 3$. The results are displayed in panels 7-12, and must be compared to the COBE-DMR maps published by Smoot et al. (1991), where the sky is dominated by the dipole anisotropy, and the Galaxy is barely visible outside the Galactic plane. There the color scale corresponds to a thermodynamic temperature ranging linearly from around -4 mK (full red) to +4 mK (full blue). Such maps visualize the optimal choice of the observation wavelengths of COBE-DMR for the study of CMB anisotropies: the Galactic contribution is really minimal, being relevant only in the Galactic Plane. In our simulated maps the dipole is removed, the Galactic plane is masked (at $|b| < 5°$), and the scale is expanded to the range -127 μK (full red) +127 μK (full blue). In this way we amplify the small, intrinsic CMB fluctuations at scales smaller than the dipole, and the Galactic fluctuations at high Galactic latitudes. In the panels 07 and 08 the frequency is 90 GHz, in panels 09 and 10 is 53 GHz, in panels 11 and 12 is 31 GHz. The six maps are organized as follows: the galactic dust contribution is plotted at top left, the synchrotron contribution at center left, the free-free contribution at bottom left; the CMB anisotropy map at top right, the sum of Galactic and CMB anisotropies at center right and the sum of Galactic and CMB emission, and detector noise at bottom right.

Due to the large pixel size, only a few sources are visible in the maps: Orion and the Large Magellanic Cloud in the dust map; LMC and the North Galactic Spur in the radio map. The CMB maps have been simulated as outlined in section 3, using a CDM model with 15 μK total rms temperature fluctuations (panels 7,9,11) and a Baryonic dark matter model with $\Delta T_{rms} = 50\mu K$ (panels 8,10,12).

In order to take into account the receiver noise, we consider as an example the 6 radiometers of COBE-DMR, using the following parameters: observation frequencies $\nu_1 = 90 GHz$, $\nu_2 = 53 GHz$, $\nu_3 = 31 GHz$; bands $\Delta\nu_1/\nu_1 = 0.94\%$, $\Delta\nu_2/\nu_2 = 1.57\%$, $\Delta\nu_3/\nu_3 = 1.67\%$; noise equivalent temperature $NET_1 = 23mK/\sqrt{Hz}$, $NET_2 = 16mK/\sqrt{Hz}$, $NET_3 = 42mK/\sqrt{Hz}$ (Smoot et al. 1990). The noise per pixel in the final map was estimated assuming a uniform sky coverage and two years of continuous operation; the results are $\mathcal{R}_1 = 140\mu K$, $\mathcal{R}_2 = 100\mu K$, $\mathcal{R}_3 = 250\mu K$. Note that the full dynamical range of the map is comparable to the rms detector noise per pixel: so the high Galactic latitude features visible in the maps are quite faint with respect to the level of detector noise.

5. QUADRUPOLE ANISOTROPY

The less uncertain component of the Galactic anisotropy is the Quadrupole (which is obtained from a large number of independent pixels): for this reason we concentrate our quantitative analysis on the Quadrupole in the following. As a first step, we searched for the quadrupole component of the Galactic map, obtained at different frequencies using Eq.(1) and the mean values of the parameters in

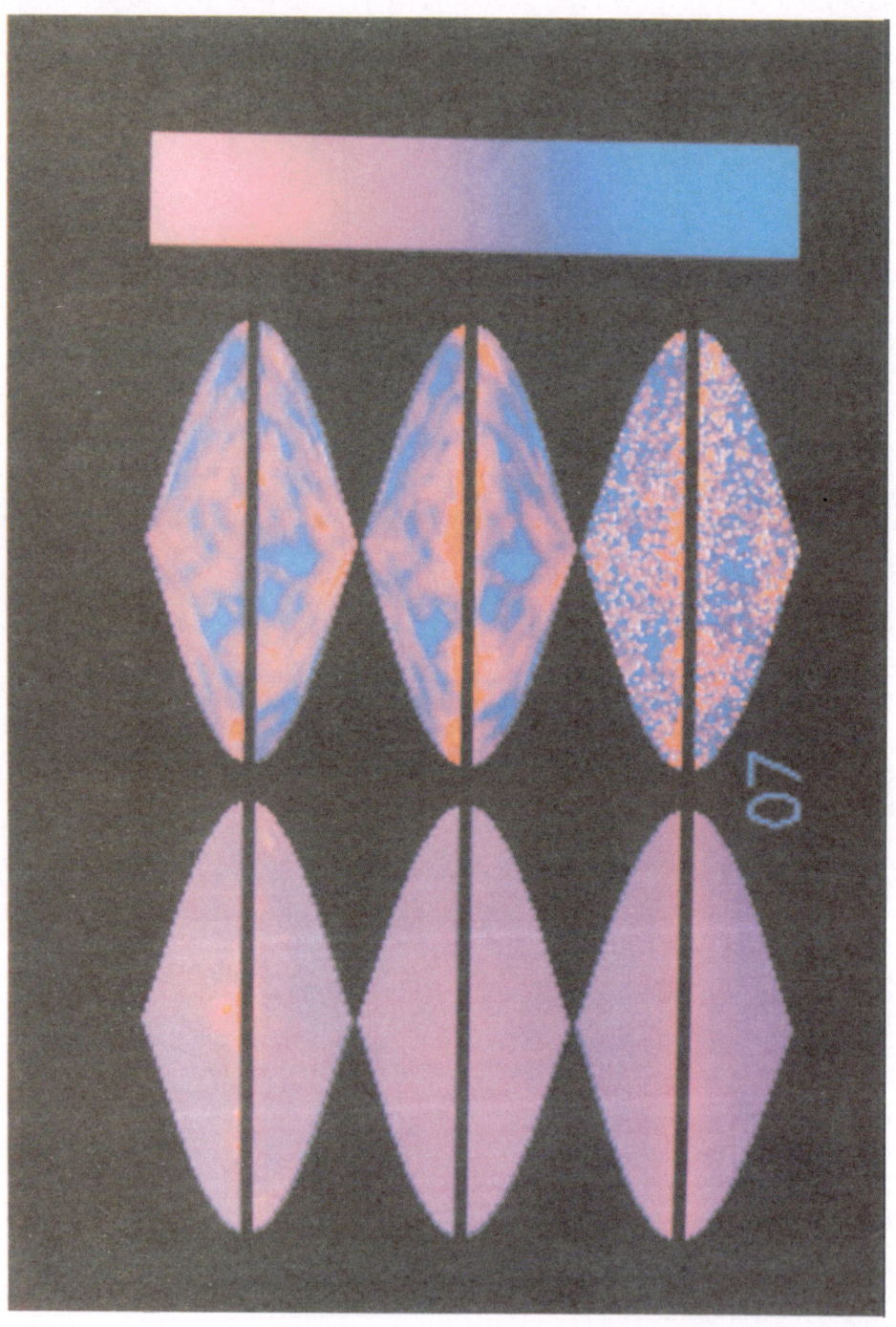

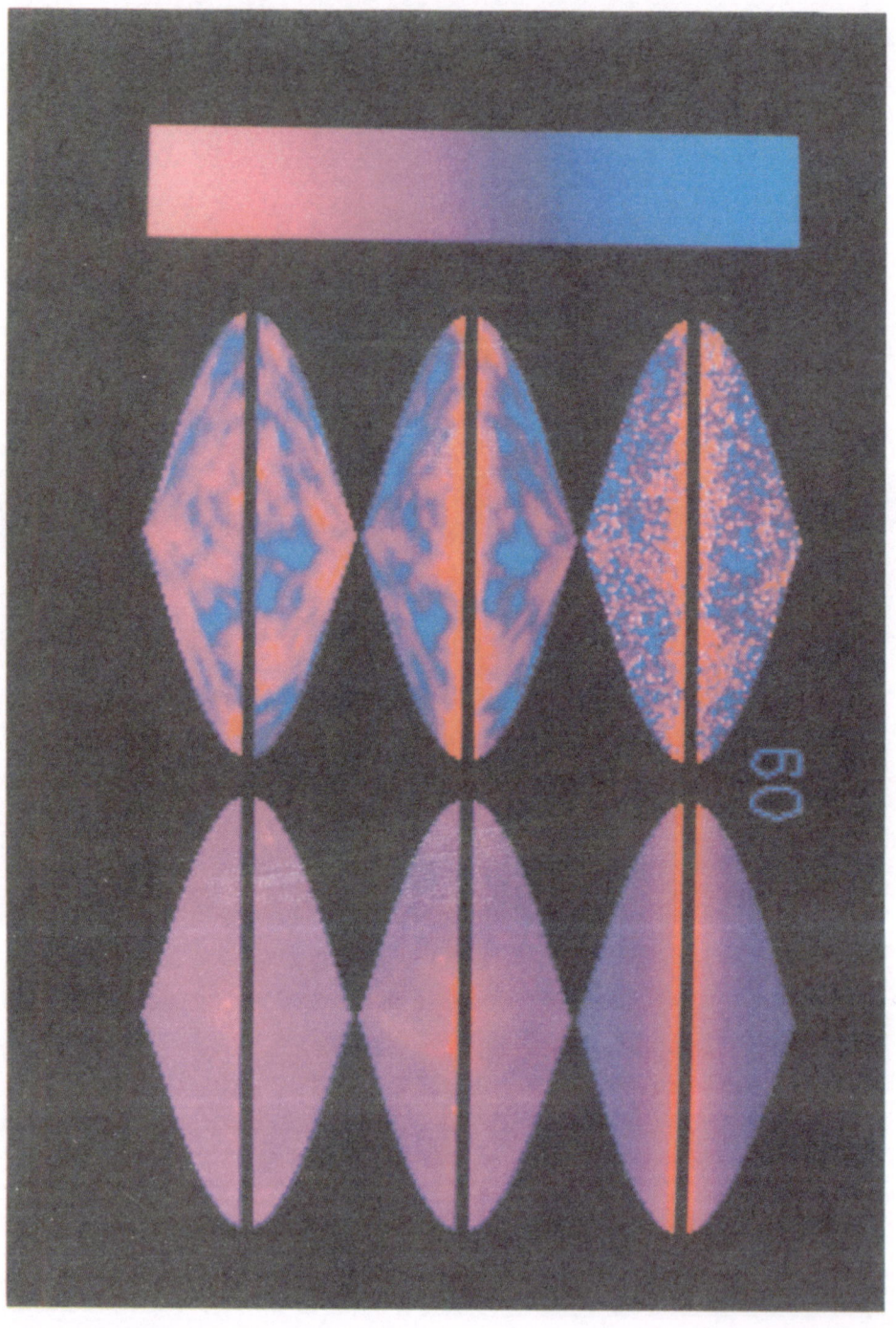

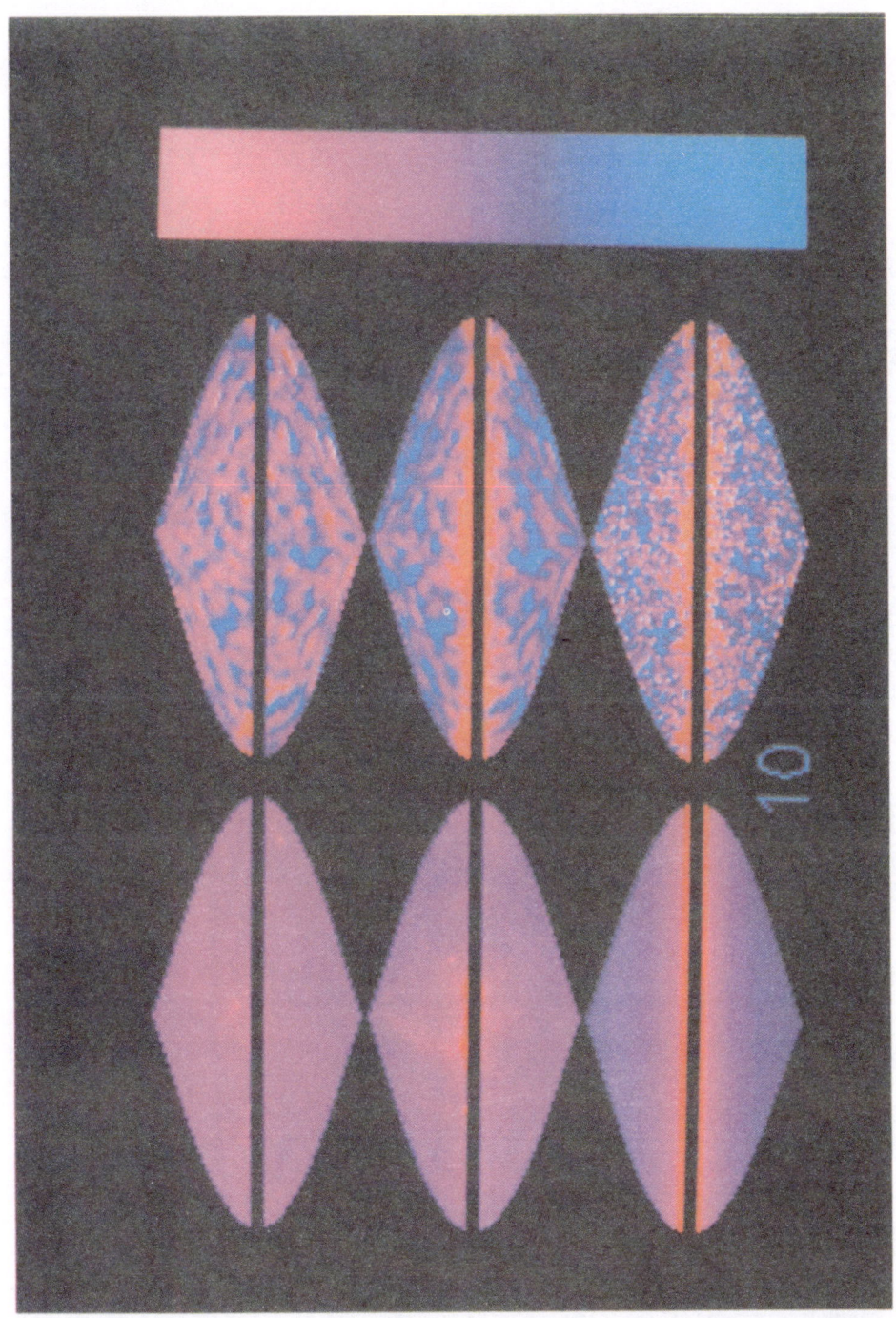

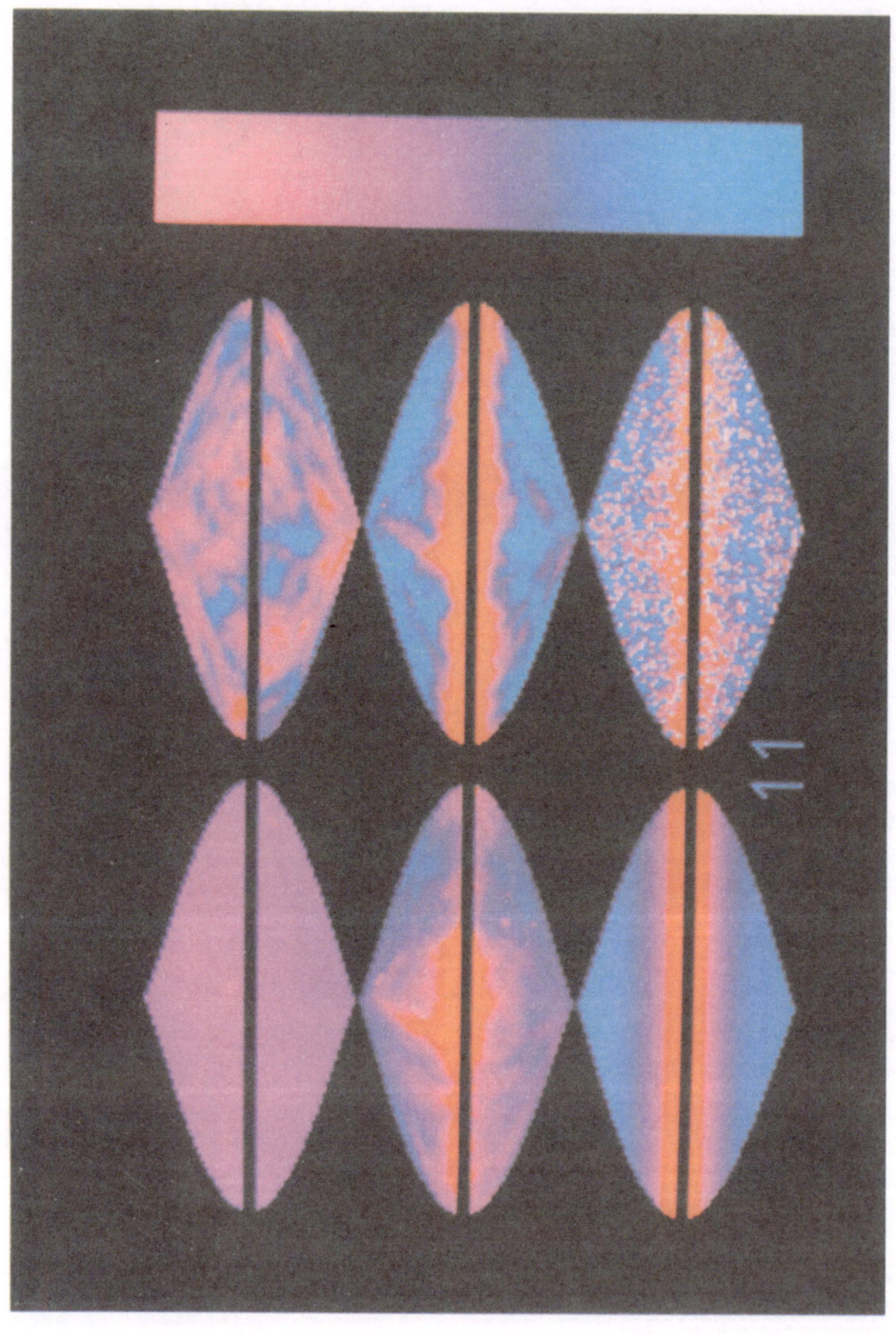

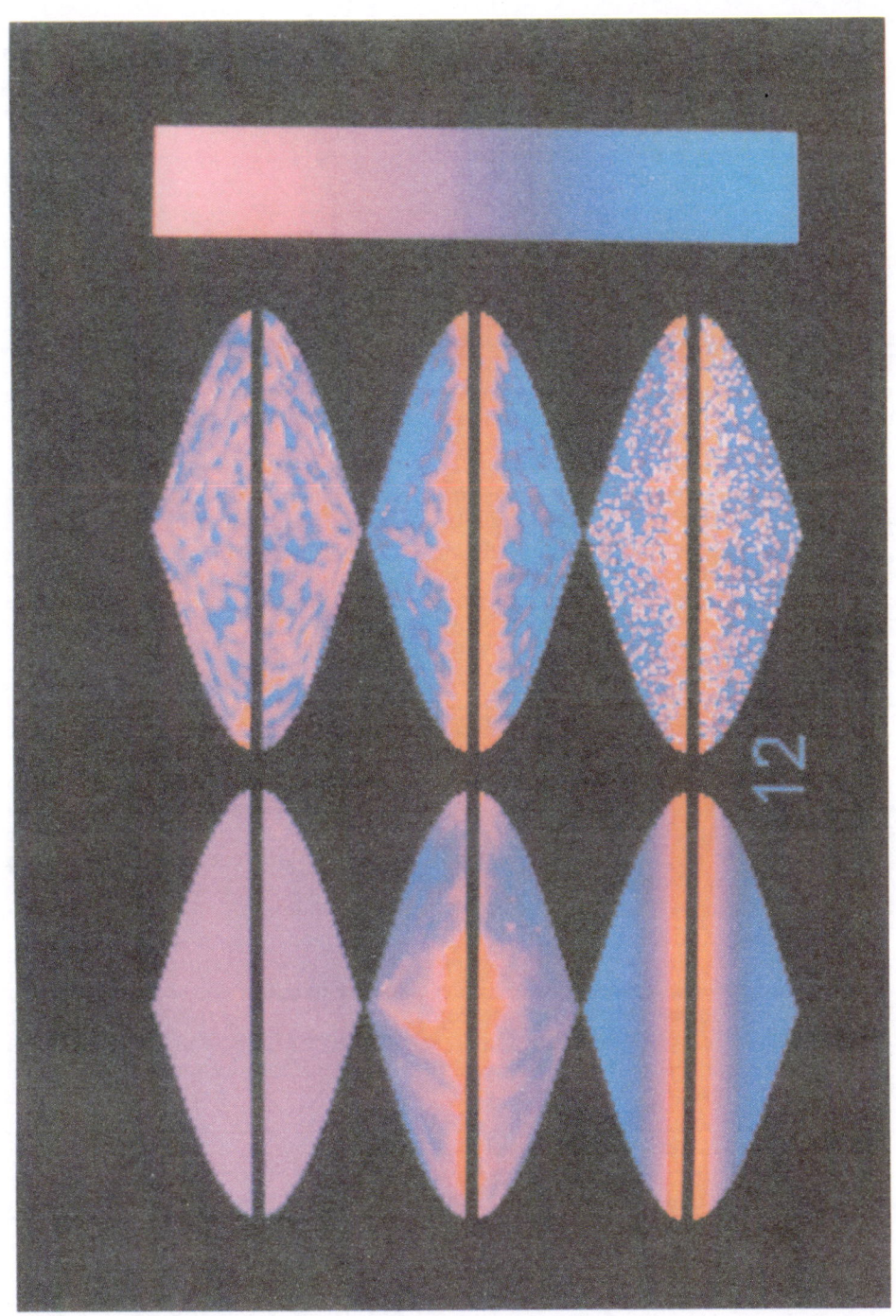

Eq.(2) and Eq.(3). We use the quadrupole components Q_i, of the sky temperature fluctuations, expressed in the so-called Berkeley notation (Lubin et al. 1983). The quadrupole of the sky temperature distribution can then be written as $Q = \sum_{i=1}^{5} Q_i X_i$, where $X_1 = (3\sin^2 \delta - 1)/2$, $X_2 = \sin 2\delta \cos \alpha$, $X_3 = \sin 2\delta \sin \alpha$, $X_4 = \cos^2 \delta \cos 2\alpha$, and $X_5 = \cos^2 \delta \sin 2\alpha$. Here α and δ are Right Ascension and Declination, respectively.

As expected, the best fit quadrupole has two hot spots coincident with the Galactic center and anti-center respectively. Its frequency dependent amplitude resembles the dust and radio continuum spectra (see Fig.1), with a minimum at $\lambda \simeq 3\div 4mm$. We use only the pixels with $|b| > b_t$, with $b_t = 5°$ or $15°$. These two values define the range where a good compromise among sky coverage and galactic contamination is achieved: the pixels used in the two cases are 4104 and 3396 (out of 4584), and the values of the normalized correlation matrix between the quadrupole and dipole components for the cut map are all much less than 1%.

In order to take into account pixel to pixel fluctuations of the parameters in Eq. (1), (2), (3), we simply added to the extrapolated Galactic map a white gaussian noise component, ΔT_G, with dispersion given by the total extrapolation uncertainty. We built 10,000 independent realizations of these maps. For each realization we performed a best fit using a quadrupole pattern (5 coefficients, Q_i^G), and two separate offset terms for the two Galactic hemispheres (de Bernardis et al. 1988). Averaging over all the simulations provides an estimate of the Galactic quadrupole components, $\langle Q_i^G \rangle$, and of the corresponding errors, $\sigma(Q_i^G)$, due only to the uncertainties in the knowledge of the extrapolation parameters. We list the results of this analysis in Table 1: it is evident that Galactic emission contaminates significantly the measurements at $\lambda < 1mm$ and at $\lambda > 10mm$. A preliminary analysis of the 19 GHz full sky map with a Galactic latitude cut, $|b_c| > 5°$, provide $Q_1 = 50\mu K$, $Q_2 = 360\mu K$, $Q_3 = 210\mu K$, $Q_4 = -400\mu K$, and $Q_5 = 60\mu K$, with standard deviations $\approx 20\mu K$ (Boughn et al., 1990). This is in quite good agreement with the predictions of our crude modelling of the Galaxy (see row 4 in Table 1).

The expected measurement error in the quadrupole components Q_i due to the detector noise only can be found by means of the standard error propagation formulae: $\sigma(Q_1^N)_j = \mathcal{R}_j\sqrt{5/\mathcal{N}}$; $\sigma(Q_i^N)_j = \mathcal{R}_j\sqrt{15/4/\mathcal{N}}$ (i=2,5), where $\mathcal{N}$ is the total number of independent sky pixels ($\mathcal{N} = 4584$ in our case). We obtain $\sigma(Q_1^N)_j = 5,3,9\mu K$ and $\sigma(Q_i^N)_j = 4,3,8\mu K$ (i=2,5), at $\lambda_j = 3.3, 5.7, 9.7mm$ respectively; these errors will be reduced by a factor 0.82 in the case of 3 years of continuum operation of COBE-DMR. Here we assume a uniform sky coverage, while, due to a satellite typical scan strategy, the weight to be assigned to the pixels varies from the poles to the equator. This may, for example, improve the COBE DMR sensitivity to medium scale anisotropies (a polar cap region is observed more accurately), but is not important in the case of the quadrupole analysis, where a large sky coverage is essential. We verified that all our results change by less than 10 % when the integration time per pixel changes a factor 3 from the equator to the poles, maintaining the same total integration time.

The level of the Galactic quadrupole components has a minimum value $Q_i^G \approx 5\mu K$ at around $\lambda \sim 4mm$. This result is well above the uncertainty due to our scanty knowledge of the Galactic emission $[\sigma(Q_i^G) \lesssim 0.2\mu K]$. The total uncertainty in the measured Q_i^G is obviously given, at the 1 sigma level, by $\sqrt{\sigma^2(Q_i^G) + \sigma^2(Q_i^N)}$.

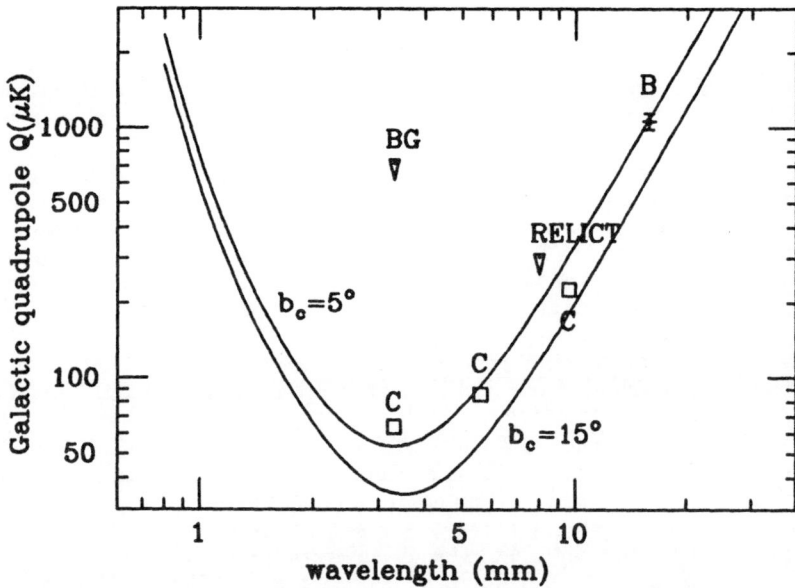

Figure 1: The Galactic quadrupole amplitude Q^G as a function of the wavelength; the Strukov et al. (1990) at 8 mm (RELIKT) , the Berkeley Group (BG) at 3 mm (Lubin et al. (1985a) upper limits, and the detections (B) of Boughn et al. (1990) at 15.6 mm and of COBE-DMR (C, data given by G. Smoot at this school) are also shown. The two curves are labeled with the value of the Galactic latitude cut.

λ (mm)	$\langle Q_1^G \rangle \pm \sigma(Q_1^G)$ (μK)	$\langle Q_2^G \rangle \pm \sigma(Q_2^G)$ (μK)	$\langle Q_3^G \rangle \pm \sigma(Q_3^G)$ (μK)	$\langle Q_4^G \rangle \pm \sigma(Q_4^G)$ (μK)	$\langle Q_5^G \rangle \pm \sigma(Q_5^G)$ (μK)	$\langle Q^G \rangle \pm \sigma(Q^G)$ (μK)
1.2	-44 ± 3	119 ± 2	104 ± 4	-122 ± 4	13 ± 3	370 ± 10
3.3	-2 ± 1	18 ± 1	13 ± 1	-19 ± 1	0 ± 1	53 ± 1
5.7	3 ± 1	32 ± 1	19 ± 1	-36 ± 1	-1 ± 1	95 ± 1
9.7	13 ± 2	101 ± 2	66 ± 2	-121 ± 2	0 ± 1	315 ± 3
15.8	40 ± 6	337 ± 5	239 ± 6	-419 ± 6	19 ± 4	1080 ± 11

Table 1: Galactic Quadrupole anisotropy at millimetric wavelengths. The five components of the Quadrupoles are listed together with the estimeted error due to our poor knowledge of Galactic emission at high Galactic latitudes. Five observation wavelengths are considered : 1.2 mm; the three wavelengths of COBE-DMR, and the wavelength of the 19 GHz sky survey of Boughn et al. (1990). All the sky pixels at Galactic latitudes higher than 5° have been included in the Quadrupole best fit of the extrapolated sky maps. The units are thermodynamic temperature fluctuations of the CMB.

6. CONSTRAINTS ON GALAXY FORMATION MODELS

Since the Galactic quadrupole seems to be well determined from our simple Galactic model, we can investigate what happens when a cosmological quadrupole component, Q_i^{CMB} is present. Each of the measurable coefficients Q_i is given by the sum of four terms: $Q_i = Q_i^{CMB} + Q_i^S + Q_i^{FF} + Q_i^D$, describing the contribution to the quadrupole anisotropy due to intrinsic CMB anisotropies, and Galactic synchrotron, free-free and dust emission. The obvious goal is to disentangle the different effects and a good analysis can be carried out by taking advantage of multi-frequency mapping. In fact, an accurate choice of observation bands allows for measuring the local disturbances, removing them, and isolating the cosmological information. This procedure should also self consistently take into account any systematic deviations from the simple Galactic model we have outlined. Several different approaches are possible for evaluating each of the four contributions to the quadrupole anisotropy of the mm sky. dBMV have shown that, assuming the simple Galactic model outlined here, the best method consists in considering only two of the COBE-DMR channels (the more channels one uses, the larger is the noise in the resulting cleaned data), and removing from the measured data the estimates for synchrotron and dust emission found by means of Eq.(2) and (3). A simple system of two equations in two unknowns has to be solved in order to find the 'cleaned' data for the CMB and the free-free components. The uncertainty in the CMB quadrupole obtained in this way is dominated by the detector noise, and is obviously higher than in the case of a single frequency measurement. The use of the 100 μm data is more effective than the 3 mm channel in cleaning the data from dust emission, and the free-free emission is obviously recovered much better using the 9.7 mm channel: for this reason the COBE-DMR 5.6 mm and 9.7 mm channels should be used. In this way, the amplitude of the true CMB quadrupole can be extracted from the data with a rms accuracy of roughly 20 μK. This conclusion rests of course on the assumed Galactic model. However, the final uncertainty on the CMB quadrupole is mainly due to receiver noise. A better estimate will be possibly reached using the COBE-FIRAS spectra of dust emission.

Let us first consider the possibility that an experiment as COBE-DMR is able to set only an upper limit on the quadrupole anisotropy. The expected upper limit that COBE-DMR should be able to set on the rms value of the CMB quadrupole anisotropy is found in the framework of a classical hypothesis test. We require that the distributions of quadrupoles under hypothesis $\mathcal{H}_0$ (Detector and Galactic noises only) and $\mathcal{H}_1$ (CMB Quadrupole plus noises) are clearly distinct (only the two 5% tails overlap). Quantitatively we must define a critical value, Q_*^2, such that

$$P[Q^2 < Q_*^2 | \mathcal{H}_0] = 95\%; \qquad P[Q^2 < Q_*^2 | \mathcal{H}_1] = 5\% \qquad (5)$$

Only in this way we can clearly accept or reject theoretical models. Since the distributions of Q^2 can be easily expressed in terms of χ_5^2 distributions, it easy to find out the solution of the system (5), which is providing an upper limit with a confidence level larger than the usual 95% if the power of the test is less than 95%. dBMV estimate $\langle |Q|^2 \rangle^{1/2} < \sigma_Q \simeq 80 \mu K$, for a Galactic latitude cut at $|b_c| < 5°$. Increasing the integration time to 3 years will reduce the upper limit to $\sim 65 \mu K$. These upper limits must be compared to the level of the rms Quadrupole anisotropy expected in several theoretical models. The comparison is summarized in Table 2. It is evident that these upper limits are not enough for constraining most of the popular models for galaxy formation. Only open, completely re-

model	Ω_0	Ω_b	x_e	n	$\langle[Q^{CMB}]^2\rangle^{1/2}$ (μK)	f(2 y) (%) $b_c = 5°$	f(2 y) (%) $b_c = 15°$	f(3 y) (%) $b_c = 5°$	f(3 y) (%) $b_c = 15°$
CDM	1	0.03	standard	1	27	40	32	53	44
CDM	1	0.2	standard	1	34	52	44	66	58
CDM-Λ	0.2	0.03	standard	1	83	97	95	100	100
CDM-Λ	0.4	0.03	standard	1	60	89	84	98	98
CDM-Λ	0.6	0.03	standard	1	51	82	75	94	91
BDM	0.1	0.1	1	-1	167	100	100	100	100
BDM	0.1	0.1	1	0	12	11	10	14	12
BDM	0.1	0.1	0.1	-1	41	68	60	80	73
BDM	0.1	0.1	0.1	0	4	6	5	6	6
BDM	0.4	0.4	1	-1	112	99	98	100	99
BDM	0.4	0.4	1	0	15	13	11	18	15
BDM	0.4	0.4	0.1	-1	35	52	44	66	58
BDM	0.4	0.4	0.1	0	5	6	5	6	6
minimal a	1	-	-	1	32	49	41	63	55
minimal b	1	-	-	1	39	66	57	78	71

Table 2: Quadrupole anisotropy expected in different models for Galaxy formation (*rms* values, column 6). Ω_0 and Ω_b are the total and baryonic density parameters, while n is the density fluctuation primordial spectral index. The ionization fraction x_e follows its standard evolution in the CDM models, while is fixed to a constant value in the baryonic (BDM) models. The minimal models discussed by Gorski (1990) are normalized to: a) $390km/s$ on a scale of $R = 4000km/s$ and b) to $327km/s$ on a scale of $R = 6000km/s$ (see Bertschinger et al., 1990). The fraction of Cosmic Observers detecting a quadrupole anisotropy using a COBE-DMR like experiment is given by f. Columns 7 and 8, and columns 9 and 10 refer to 2 and 3 years of integration, respectively.

ionised, pure baryonic universes with $n = -1$ and $\Omega = 0.1 \div 0.4$, or a low density ($\Omega_0 \lesssim 0.2$) flat (because of a suitable cosmological constant) CDM model would be ruled out.

On the other hand, one can evaluate the fraction of cosmic observers that would *detect* a quadrupole anisotropy in different theoretical scenarios, with a COBE-DMR experiment. This fraction is also listed in table 2. It is not negligible for most of the models and is unconfortably low only for BDM models with white noise primordial power spectrum. At the expected level of sensitivity, COBE-DMR could have a $\lesssim 50\%$ chances of detecting a quadrupole anisotropy, even in the flat, biased CDM scenario.

The analysis presented here neglects systematic errors in the forthcoming data. Also, the quadrupole analysis, on which we concentrated here, is only one among the possible on the COBE-DMR data: better information can probably be extracted from correlation analysis at smaller angular scales (see *e.g.* Scaramella and Vittorio 1990), but the Galactic contamination (and its uncertainty) is expected to be much more important than in the case of the Quadrupole anisotropy.

REFERENCES

Banday A.J. and Wolfendale A.W., 1990, *M.N.R.A.S.*, **254**, 182.

Boughn, S., Cheng, E.W., Cottingham D., 1990, *in preparation*

Bowyer S., 1990, in *The Galactic and Extragalactic Background radiation*, Bowyer S. and Leinert C. eds., IAU Symp. 139, pg. 171-183.

de Bernardis P., Masi S., Melchiorri F., Moreno G., Vannoni R., 1988, *Ap.J.*, **326**, 941

de Bernardis P., Masi S., Vittorio N., *CMB and Galactic Quadrupoles of the mm sky*, 1991, *Ap.J.*, in press

Fixsen D.J., Cheng E.S., Wilkinson D.T., 1983, *Phys. Rev. Lett.*, **50**, 620

Gorski K., and Silk J. 1989, *Ap.J.Lett.*, **346**, L1

Haslam C.G.T., Salter C.J., Stoffel H., Wilson W.E., 1982, *Astron. Ap. Suppl. Ser.*, **47**, 1

Hauser M.G., and IRAS team, 1984, *Ap.J.Letters*, **278**, L15.

Lubin P., Epstein G., Smoot G, 1983, *Phys. Rev. Letters*, **50**, 616.

Lubin P., Villela T., Epstein G. and Smoot G., 1985, *Ap.J.Letters*, **298**, L1.

Masi S., de Bernardis P., De Petris M., Epifani M., Gervasi M., Guarini G., 1990, *Ap.J.Letters*, **366**, L51.

Peebles P.J.E., 1982, *Ap.J.Letters*, **263**, L1.

Peebles P.J.E., 1987a, *Nature*, **327**, 210.

Peebles P.J.E., 1987b, *Ap.J.Letters*, **315**, L73.

Reynolds R.J., 1989, *Ap.J.Letters*, **339**, L29-L32.

Sachs, R.K., and Wolfe, M.A., 1967, *Ap.J.*, **147**, 73.

Scaramella R. and Vittorio N., 1990, *Ap.J.*, **353**, 372.

Scaramella R. and Vittorio N., 1991, in preparation

Smoot G., et al., 1990, *Ap.J.*, **360**, 685.

Smoot G., et al., 1991, *Ap.J. Lett.*, **371**, L1.

Weiss R., 1980., *Ann. Rev. Astron. Astroph.*, **18**, 489.

COBE DMR RESULTS AND IMPLICATIONS

GEORGE F. SMOOT
Lawrence Berkeley Laboratory and Space Sciences Laboratory,
Bldg 50-351 - 1 Cyclotron Road
University of California, Berkeley, 94720

ABSTRACT. This lecture presents early results obtained from the first six months of measurements of the Cosmic Microwave Background (CMB) by Differential Microwave Radiometers (DMR) aboard *COBE*[*] and discusses significant cosmological implications. The DMR maps show the dipole anisotropy and some galactic emission but otherwise a spatially smooth early universe. The measurements are sufficiently precise that we must pay careful attention to potential systematic errors. Maps of galactic and local emission such as those produced by the FIRAS & DIRBE instruments will be needed to identify foregrounds from extragalactic emission and thus to interpret the results in terms of events in the early universe. The current DMR results are significant for Cosmology.

1. INTRODUCTION/MOTIVATION

The DMR major scientific objectives are to investigate the large scale structure and evolution of the universe. This means probing the geometry and dynamics of the universe and making a map of the early matter and energy distribution. Goals are to determine the isotropy and symmetry of the universal expansion, to determine the rotation of the universe, and to check causal connectedness of the early universe. Another important goal is to test the standard model of cosmology - in particular the inflationary paradigm.

The method is to make full-sky microwave surveys at frequencies of 31.5, 53 and 90 GHz (wavelengths of 9.5, 5.66, and 3.3 mm respectively.) These frequencies were chosen to be in a region where the CMB intensity is 1000 times the galactic emission. A fourth frequency was originally in the DMR complement but it was moved from the DMR to a 19 GHz balloon-borne system (Cottingham *et al.* 1991).

1.1. PRODUCTION OF CMB ANISOTROPY

Anisotropy in the CBR is erased nearly completely at the surface of last scattering. Thus we can treat the surface of last scattering as a clean slate. We then can use the Liouville (or

---

[*] The National Aeronautics and Space Administration/Goddard Space Flight Center is responsible for the design, development, and operation of the Cosmic Background Explorer. GSFC is also responsible for the software development through to the final processing of the space data. The *COBE* program is supported by the Astrophysics division of NASA's Office of Space Science and Applications.

M. Signore and C. Dupraz (et al.), The Infrared and Submillimetre Sky after COBE, 331–344.
© 1992 *Kluwer Academic Publishers.*

Einstein-Louiville) theorem that for a radiation field, such as the cosmic background radiation, the ratio I/v^3 is a constant. Thus if we know the Intensity I at the surface of last scattering then we only need to calculate the frequency shift from the surface of last scattering to the observer to determine the present intensity.

1.1.1. *Primordial Intensity Variation.* Any original intensity variation present at the surface of last scattering will manifest itself to the DMR unless there is a cosmic conspiracy that produces just the right frequency shift to compensate for it. For example, we can consider the possibility that we are displaced from the rough center of the inflation bubble we might inhabit. Then it is possible that an entropy gradient could exist and produce (1) a dipole, (2) generally a comparable quadrupole, and (3) probably higher order terms. We can expect in the inflationary scenario that the present size of the inflation bubble is likely to be extremely much larger ($>10^{15}$) than our present horizon size and the entropy gradient is likely to be very small. Generally, it is assumed that we do not occupy a special position in the universe and that the intensity of the CBR is the same across the surface of last scattering except for the primordial perturbations. There are two orthogonal types of primordial fluctuations: curvature (adiabatic) and isocurvature (isothermal). Adiabatic fluctuations are variations in the number (energy) density and thus the spatial curvature. All species participate equally, $\delta n_x/n_x = \delta n_\gamma/n_\gamma = \delta I/I = 3\delta T/T$. The temperature fluctuation is 1/3 the number fluctuation. Isocurvature perturbations do not represent a variation in energy density and thus no variation in spatial curvature. They are instead a variation in the relative number densities of species that conspire to keep the energy density constant. For a number density variation of species x, there is a compensating temperature fluctuation on the order of $\delta T/T \sim -1/3\ \delta n_x/n_x$ at matter-radiation equality.

Map of Present Universe with Entropy Gradient

Present Inflation
Bubble Size

Our Location

and current horizon size

Terra Incognita
Here Be Dragons

1.1.2. *Mechanisms Causing Frequency Shift* and thus a change in the observed intensity or CMB temperature

Mechanism

$$\frac{\Delta T}{T} = \begin{array}{l} R_e / R_0 = 1/(1+ z) \quad\quad \text{Cosmological Redshift} \\[2mm] \dfrac{(\phi_0 - \phi_e)}{3c^2} \quad\quad\quad \text{Gravitational Redshift (Sachs-Wolfe Effect)} \\[2mm] + \ \mathbf{n}\cdot(v_0 - v_e)/c \quad\quad \text{Doppler Shift} \\[2mm] + \ 2\int_e^0 dt\ d\phi(t,x)/dt/c^2 \quad\quad \text{Induction (Rees-Sciama Effect)} \end{array} \quad (1)$$

1.1.3. *Later Time Material and Energy*.

Since we are observing the cosmic background radiation it must pass through all the evolving matter and energy in the universe on its way to be observed by us. Thus any interactions and added energy can perturb the intensity of incoming radiation. Examples of these effects are:

A. Sunyaev-Zeldovich Effect: The scattering of the CMB photons by hot electrons, e.g. by hot gas in clusters:

$$(\Delta T/T)_{RJ} = -2y \quad\quad y = \int n_e k T_e / m_e c^2 \ \sigma_T \ dl$$

B. Cold Dust: Cold dust in the early universe will absorb the CMB and any early starlight and will reemit the energy adding back to the CMB.

C. Radio sources: There are many extragalactic radio sources and radio-active galaxies in the universe and they add their emission to the incoming CMB. These sources are discussed by Franceschini *et al.* (1989). For the DMR their contributions are generally below the 10^{-6} level of the CMB.

D. Other effects: include the galaxy and other local sources covered in section 2.3.

1.2. THE DOPPLER EFFECT

The Doppler effect deserves special mention. Can we distinguish between local motion of the observer and the motion of the distant surface of last scattering? Or the equivalent intrinsic dipole variation of the temperature at the surface of last scattering? The answer is yes, in principle. The Doppler Shift that produces the kinematic dipole anisotropy also produces a well-defined kinematic quadrupole.

$$T(\theta) = T_0 \frac{(1 - \beta^2)^{1/2}}{1 - \beta\cos(\theta)} = T_0 [\ 1 + \beta\cos(\theta) + \frac{1}{2}\beta^2\cos(2\theta) + O(\beta^3)\ ] \quad (2)$$

Melchiorri's lecture covers the quadrupole in more detail. An intrinsic dipole must be a larger than current horizon-size isocurvature fluctuation (also equivalent to an entropy gradient mentioned above) and will in general produce a quadrupole which is less than the dipole by the ratio of the horizon size to the fluctuation size modulated by the phase of the curvature at our location (Turner 1991). This can be seen more clearly as the dipole is the gradient of the fluctuation and the quadrupole is the second derivative (see section 3.2.3).

1.3 POLARIZATION

While the last scattering erases anisotropy very effectively, it does not erase polarization nearly as effectively. In fact any intensity anisotropy with non-zero quadrupole or higher-order moment at last scattering will be converted into a polarization by Thomson scattering. If the universe were inhomogeneous, then $P \approx \Delta T/T$ on angular scales corresponding to the thickness of last scattering. For larger angular scales the polarization to anisotropy ratio decreases with the square of the angle. Thus large angular scale polarizaton is typically small - on the order of 1 to 10% of the intensity anisotropy - unless the universe was ionized in the redshift regime of $10 < z < 100$. For this medieval time ionization the polarization can be as large as the intensity anisotropy (Rees 1968, Anile 1974, Basko & Polnarev 1980, Negroponte & Silk 1980, Kaiser 1982, Tolman 1984).

The galaxy can contribute a foreground polarized signal through the mechanisms of synchrotron radiation and polarized dust emission and scattering. All of these are tied to large scale organized magnetic fields in the galaxy. Such fields might exist in the universe as a whole.Thus far only upper limits exist for potential CMB polarization. Melchiorri discusses observations in his lecture.

2. DMR STATUS AND RESULTS

The DMR instrument is working well and continues to take data. It has made full sky maps. The best maps are at 53 GHz - a previously unobserved frequency. The maps at all three frequencies are quite good and are shown in the color plates at the front of this book.

2.1. SUMMARY OF MAJOR RESULTS

The measured dipole anisotropy parameters are in agreement with previous results and will be the most precise when our full calibration is completed. The DMR and FIRAS results are consistent with a thermal dipole anisotropy and just what would be expected from the Doppler effect. The galactic plane emission is at the level expected. At this stage in the data processing there is no evidence for a quadrupole anisotropy, $\Delta T/T < 3 \times 10^{-5}$ (95% CL). There is no evidence for gaussian fluctuations, $\Delta T/T < 4 \times 10^{-5}$ (95% CL). There is no evidence for any other effects at $\Delta T/T < 10^{-4}$.

2.2. SENSITIVITY AND SYSTEMATICS

The sensitivity per pixel is about $\Delta T/T = 10^{-4}$ per pixel which matches what is expected from the instrument. This projects to roughly $\Delta T/T = 3 \times 10^{-6}$ per spherical harmonic coefficient. Why then is the quadrupole limit some ten times above this level? The answer is at this time we are limited by our knowledge of potential and real systematic errors. We have divided these potential systematic errors into three major categories: (1) foreground emission such as signal from the *COBE* shield, the earth, the moon, the sun, the planets, and the galaxy, (2) effects in the instrument, and (3) data processing effects. In his lecture Janssen covers many of these that relate to instrumental and data processing effects. Here I discuss briefly the foreground signals and especially one fundamental foreground signal - emission from the galaxy.

2.3. FOREGROUND EMISSION

Much of the foreground emission can be avoided by careful experimental design or cutting out data when the field of view is obscured. Working outward from the DMR instrument:

2.3.1. The COBE Ground Shield. To avoid much of the signal and thermal flux from the spacecraft, earth, and other objects a large ground shield/sun shade surrounds the *COBE* instruments. The ground shield is carefully designed to exclude unwanted radiation. It has a low emissivity inner surface to minimize emission to the DMR, and sufficient thermal capacity to minimize variation in emission. The shield excludes radiation from the earth while contributing less than 10^{-5}K and is stable to better than 10%.

2.3.2. Radio Frequency Interference. The DMR only utilizes data where the earth is below the ground shield and there are no spikes in the data. This procedure is to avoid any intermittent RFI. We also search the data for spikes and correlations either with ground location or geostationary satellite location. We have found no evidence for any RFI and estimate the potential effect as below 10^{-5}K.

2.3.3. Diffracted Earth Emission. The DMR only utilizes data where the earth is below the ground shield. The combination of antenna gain and the ground shield keeps the earth emission signal below a few times 10^{-5}K, when the earth limb is below the plane of the ground shield. We are studying the earth signal and testing to determine if a more restrictive cut or modeling is needed.

2.3.4. Lunar and Planetary Emission. The DMR cuts out data where the moon or planets (only Jupiter at this time) produce a signal too large to ignore or remove. A typical planet contributes only a few to tens of μK when in the beam. The interplanetary dust is negligible because its emissivity in the microwave region is much less than 10^{-7} and it temperature is less than 300K giving a total signal much less than a μK.

2.3.5. Solar Emission. The sun is always shielded by the *COBE* ground shield/sun shade. The combined effect of diffraction over the shield and good antenna off-axis rejection keeps the solar emission below 10^{-6}K.

2.3.6. Galactic Emission. One important foreground that may well ultimately limit the detection and measurement of the quadrupole and other large angular scale anisotropies is galactic emission. The rms quadrupole amplitude including galactic emission (DMR 31, 53, and 90 GHz data and the Relict 37 GHz data) decreases with increasing frequency. However, since the CMB intensity begins to decrease with frequency, the DMR 53 and 90 GHz channels bracket the apparent best observing frequency range. Including the galactic plane produces a quadrupole amplitude at a minimum of 50 μK. To obtain the ultimate DMR sensitivity we need a level 20 to 30 times lower. How much does simply cutting the galactic plane from the fitting achieve? Tests using the Haslam 408 MHz (a check of the galactic synchroton) and *IRAS* 100 micron (a check of dust emission) maps indicate that reasonable galactic latitude cuts gain roughly a factor of 5 to 10. We will have to obtain the remaining factor of 2 to 4 from careful modeling.

3. INTERPRETATION

A major DMR goal is investigating the large scale structure and evolution of the universe.

3.1. GEOMETRY OF THE UNIVERSE

A major assumption in cosmology, called the Copernican Principle, is that we are not at a special place (or time) in the universe. What we see is on large scales is what a typical observer sees. This implies that on large scales the universe is homogeneous. The assumption of homogeneity implies that the universe can be described by the 9 Bianchi types of homogeneous geometries (Bianchi 1897, Taub 1951, Estabrook, Wahlquist, & Behr 1968, Ellis & MacCallum 1969, Ryan & Shepley 1975).

3.1.1. *Metric.* The CMB photons follow geodesics as given by the action principle for wave propagation. They also obey the Liouville equation so that I/v^3 is constant. The DMR observations of full sky CMB isotropy imply that the large scale geometry of the universe is given by the Robertson-Walker metric (Ehlers, Geren, & Sachs 1968) with perturbations of order $\Delta T/T$:

$$ds^2 = c^2\, dt^2 - R(t)^2 \left[\, dr^2 + \left\{ \frac{\sinh r}{r} \atop \sin r \right\}^2 (d\theta^2 + \sin^2\theta\, d\phi^2) \right] \tag{3}$$

for $r \ll 1$ and $dR/dt \ll 1$, this expression simplifies to the Minkowski metric. Thus from elementary particle experiments and the large angular scale anisotropy measurements we know that the universe is well-described by a Riemannian manifold from 10^{-16} to 10^{28} cm. Its metric is very close to the Robertson-Walker.

3.1.2. *Gravity Waves.* One potential local metric perturbation is gravitational radiation.
3.1.2.1 *Local gravity waves*: Long-wavelength gravitational waves propagating in this region of the universe distort the metric and produce a quadrupole temperature variation in the CMB. For a single plane wave the resultant CMB anisotropy is (Burke 1975)

$$\Delta T/T = \frac{1}{2}(A_r - A_e)(1 - \cos(\theta))\cos(2\phi)) \tag{4a}$$

where A_r and A_e are the proper strains at the freely falling emitter and receiver, respectively. The gravity wave energy density for wavelength λ_{gw} is

$$\varepsilon_{gw} \sim \frac{c^4}{32\pi G} A_{gw}^2 /\lambda_{gw}^2 = \frac{1}{12}\left[\frac{c/H_0}{\lambda_{gw}}\right]^2 A_{gw}^2\, \rho_c c^2 = \frac{1}{3}\left[\frac{L_0}{\lambda_{gw}}\right]^2 \left[\frac{\Delta T}{T}\right]^2 \rho_c c^2 \tag{4b}$$

The limits $\Delta T/T < 3 \times 10^{-5}$ for quadrupole imply the energy density of single plane wave is less than

$$\Omega_{gw} < 3 \times 10^{-10}\left[\frac{c/H_0}{\lambda_{gw}}\right]^2 = 3 \times 10^{-5}\left(\frac{\lambda_{gw}}{10\ \text{Mpc}}\right)^{-2} h^{-2} \tag{4c}$$

where Ω_{gw} is the energy density of the radiation divided by the critical density, λ_{gw} is the wavelength at the current epoch, and h is the Hubble constant in units of 100 km s^{-1} Mpc^{-1}. A chaotic superposition of gravity waves has energy density 8/5 greater (Burke 1975).

3.1.2.2 *Gravity waves on the surface of last scattering* will generate chaotic fluctuations. The DMR is sensitive primarily to gravitational waves with scale sizes $\geq 7°$ at the surface of last scattering, or ~400 Mpc today. The finite thickness of the surface of last scattering is expected to wash out the effect of gravity waves with comoving wavelengths less than about 100 Mpc ($\theta \approx 1°$) (Linder 1988). The measured anisotropy, $\Delta T/T$, is just half the strain of the gravity waves at the surface of last scattering. To determine the present energy density of the gravity waves, we must propagate those gravity waves from the last scattering epoch to the present. The strain of a gravity wave remains constant until it crosses the horizon, then the energy density scales as $(1+z_{hor})^{-4}$, where z_{hor} is the redshift that the gravity wave enters the horizon. In a matter dominated universe, which we believe is the case since the surface of last scattering, the total energy density scales as $(1+z_{hor})^3$ so the gravity wave energy density scales as $(1+z_{hor})^{-1}$, times the energy density calculated for the measured strain and comoving wavelength. (Note that gravity waves entering the horizon during the radiation dominated phase have the energy density calculated for their strain and comoving wavelength but divided by $(1+z_{eq})$, where z_{eq} is the redshift of matter radiation equality.) Since the horizon crossing redshift is inversely proportional to the wavelength squared in a matter dominated universe, the gw energy density limit is independent of wavelength and depends only on the anisotropy measured on that scale:

$$\Omega_{gw\text{-}ls} < 1/3 \left[\frac{\Delta T}{T}\right]^2 \approx 3 \, x \, 10^{-10} \qquad (4d)$$

Figure 1 summarizes the DMR limits on gravitational wave energy density.

3.1.3. *Topological Defects.* While we know the metric for large scales very well, we have not determined the overall topology of the universe. In the inflationary model our observable universe and a large region around it are spherical. What lurks outside in the region marked Terra Incognita could be more bubbles or any order of complicated geometries. For example, the universe might be cylindrical with radius much less than the current horizon size so that we can see around the universe a number of times. This model easily accounts for periodic structure - one sees the same structure over and over.

For a working hypothesis a spherical symmetric universe is appropriate; however, we can then ask the question: Are there topological defects in the universe? In the inflationary and standard particle physics models the universe passes through phase transitions. The vacuum phase transitions are through vector fields so there is the possibility of stable and unstable topological defects of 0, 1, 2, and 3 dimensions. The zero-dimensional defects are called monopoles and inflation makes magnetic monopoles rare.

A. *Cosmic Strings* are stable one-dimensional defects and are actually very thin lines of the higher energy vacuum state. Cosmic strings are characterized by their mass per unit length μ. Their large energy density and relativistic velocities produce CMB anisotropies through relativistic boosts and the Sachs-Wolfe effect (gravitational lensing alone does not produce anisotropy in an otherwise isotropic background). Many authors have calculated the anisotropy produced by various configurations of cosmic strings, with typical values (Vilenkin 1985, Stebbins 1988, Stebbins, *et al.*, 1987, see lecture by Bouchet).

$$\Delta T/T \sim 8\pi\beta\gamma\frac{G\mu}{c^2} \qquad (5)$$

The DMR experiment limits the existence of large-scale cosmic strings to $G\mu/c^2 < 10^{-5}$.

B. *Domain Walls* are stable two-dimensional defects and are actually very thin sheets of the higher energy vacuum state. Domain walls cause major perturbations to the metric. Their effect is so large that if they persist from early times the domain walls dominate the universe. Thus we only consider late time phase transitions in which the energy density per unit area, σ, in the walls is relatively small. CMB photons passing through a domain wall undergo a frequency shift that depends upon the time that they go through the walls, the peculiar velocity of the walls and the impact angle. The general pattern is a rotationally symmetric cusp with maximum anisotropy amplitude about $\pi G\sigma/H_0$ (Goetz & Notzold 1990, Turner, Watkins, & Widrow 1990). The lack of apparent cusps in the DMR maps provides a limit on $\sigma < 10$ MeV^3 ($\approx 4.6 \times 10^{-5}$ gm/cm^2).

C. *Textures* are the unstable three-dimensional defects and their effects come from thin sheets where there is a high gradient caused by the change from one phase to another of the zero energy vacuum state (Turok 1989). The metric perturbations produced by collapsing knots of global texture is calculated to be a collection of various sizes of red and blue shifted disks covering about 10% of the sky with typical amplitude $\Delta T/T = 10^{-5}$ (Turok & Spergel 1990). At this stage these textures would be just at the level of detectability.

D. *Other Topological Effects of higher dimensions* are thought not to be stable and would dominate the universe excessively.

There is no evidence for topological defects such as cosmic strings, domain walls, and textures. The large scale geometry of the universe appears to be uniform and without defects.

3.1.4. *Causality.* For a Robertson-Walker space the proper distance particle horizon size is the integral over photon path line element in comoving coordinates r, θ, ϕ, scaled to the epoch of interest t.

$$d_{Hor}(t) = R(t) \int_0^t cd\tau/R(\tau) = [R(t)/R_0]^{3/2} 2c/H_0 = 2L_0/(1+z)^{3/2} \qquad (6)$$

The last equality holds for a flat, matter-dominated universe. The comoving horizon diameter is then $2L_0/(1+z)^{1/2} \approx 200\text{Mpc/h}[(1+z)/1100]^{1/2}$. Allowing for curvature and radiation the comoving size is approximately $120\text{Mpc}/[\Omega_0 h^2(1+z)/1100]^{1/2}$.

The horizon at the surface of last scattering subtends an angle of about

$$\theta_{dec} \approx 1.7° \, \Omega_0^{1/2} \, [(1+z_{dec})/1100]^{-1/2} \qquad (7)$$

If decoupling takes place at a nominal redshift of 1070, the horizon subtends an angle of about 2°. If the universe remains fully ionized, the mean depth of last scattering $z_{dec} > 10$ and the horizon subtends an angle less than $\sim 10°$. Thus any structures that the DMR discovers most surely come from regions that are not causally connected except at very early times and were imprinted on the universe during the period of quantum cosmology. The surface of last scattering, observed in the microwave today, contains over 10^3 causally disconnected regions which nevertheless have their temperature equal to a part in ten thousand.

Note that the angle subtended by a comoving size L at redshift z is given as

$$\theta_L \approx 0.956° \, \Omega_0 h \, L/100\text{Mpc} \qquad \text{for } z \gg 1. \qquad (8)$$

3.2. DYNAMICS OF THE UNIVERSE

3.2.1. *Rotation of the Universe.*The quadrupole and higher-order anisotropies limits constrain global shear and vorticity in the early universe. If the universe were rotating (in violation of Mach's Principle), the resultant metric causes null geodesics to spiral. In a flat universe, the anisotropy is dominated by a quadrupole term (Collins & Hawking 1973; Barrow, Juskiewicz, and Sonoda 1985). The DMR results limit the global rotation of the universe to $\omega/H_0 < 10^{-6}$, or less than one ten-thousandth of a turn in the last ten billion years. This is less than is needed as primordial vorticity to cause the rotation of galaxies.

Conservation of angular momentum $L = 2/5 \; M \; R^2 \; \omega_r = 8\pi/15 \; (\rho_r + \rho_m)R^5\omega_r$ implies that during the radiation dominated phase $R\omega_r$ is constant and ω scales as $1/R$; while during the matter dominated phase $R^2\omega_r$ is constant and ω scales as $1/R^2$. Energy conservation gives the dependence of H. Assuming that pressure, shear and vorticity are not important contributors to the energy density then H is proportional to R^{-2} for radiation dominated and to $R^{-3/2}$ for matter dominated epochs. If the shear or vorticity is important, then H scales as ω or σ. The ratios ω/H and σ/H scale as R for radiation dominated era and as $R^{-1/2}$ for the matter dominated era. Some modes decay more quickly than implied by conservation of angular momentum, so that depending upon the model assumed, the limits on ω/H_0 and σ/H_0 vary over several orders of magnitude and can be as low as 10^{-11}.

There are two other limits on shear and vorticity. Big Bang Nucleosynthesis sets a limit that the energy density in shear (and vorticity) must be less than about 10% of the radiation energy density or the expansion would have been too fast (this is the same argument that limits the number of neutrino species). The 10% of the CMB energy density limit implies that the ratios ω/H and σ/H were less than 0.3. Thus the current value of ω_0/H_0 and σ_0/H_0 must be less than $3 \; x \; 10^{-3}$. Shear in the universe would have been dissipated by Thomson scattering creating a distortion in the CMB spectrum. The limits on spectral distortions mean that the energy dissipated has been less than 1% of the CMB energy density which implies that the ratio σ/H was < 0.1 and the current value is $\sigma_0/H_0 < 10^{-3}$.

3.2.2. *Expansion of the Universe.* If the expansion of the universe were not uniform, the expansion anisotropy would lead to a temperature anisotropy in the CMB of similar magnitude. The large-scale isotropy of the DMR results indicate that the Hubble expansion is uniform to significantly better than one part in 10^4. This provides additional evidence for hot big bang models of cosmology, and indicates that the currently observed expansion of the universe can be traced back at least to the radiation-dominated era.

3.2.3. *Very Large Scale Structure.* Inhomogeneities on scales outside the present horizon induce gradients and shear across the region within our horizon, and these cause anisotropy in the CMB. Grischuk and Zel'dovich (1978) showed that a very long wavelength density ripple in a spatially flat universe generates a quadrupole anisotropy in the CMB.

$$(\frac{\Delta T}{T})_{\text{quadrupole}} \approx L_0^2 \; \nabla^2 \delta\phi = (c/H_0)^2 \; \nabla^2 \delta\phi \approx \frac{1}{2}[\frac{L_0}{L}]^2 \; (\frac{\delta\rho}{\rho})_L \qquad (9)$$

There is no dipole term because both the photons and the observer are in free fall. The effect that matters is the potential gradient associated with the inhomogeneities. If the universe is flat, the DMR limits on the quadrupole term constrain $(\delta\rho/\rho)_L < 10^{-4} \; (L/L_0)^2$. So we actually know that the region labeled Terra Incognita on the first figure is not too inhomogeneous; the present inflation bubble or well-behaved region must extend much more than shown in the sketch. Any dragons must be lurking at a greater distance. Inspite of what we might have thought, we can tell something about structure outside our present horizon up to of order 100 times the present horizon.

3.3. OBSERVATIONAL TESTS OF INFLATION

The observed isotropy of the universe on large angular scales presents a major problem for cosmology. At the time of primordial nucleosynthesis the presently observable universe was divided into about 10^{25} causally independent regions. The observed uniformity of light elements implies a baryon density uniformity of $\delta\rho/\rho < 3$ during synthesis (Yang *et al.* 1984). At the surface of last scattering the horizon size subtends ~2° when viewed from here. Regions separated by more than 2° were not in causal contact; consequently, DMR measures some 10^4 causally disconnected regions of the sky. Standard models of cosmology fail to explain why causally unconnected regions are the same to order unity, much less the 10^{-4} isotropy implied by the DMR observations. Inflationary scenarios provide one solution. In these models, the universe undergoes a spontaneous phase transition ~10^{-32} seconds after the Big Bang, causing a period of exponential growth in which the scale size increases by 30 to 40 orders of magnitude. The entire observed universe would then originate from a small pre-inflationary volume in causal contact with itself, eliminating the problem. In the simplest inflationary models, the pre-inflationary matter and radiation fields are diluted to zero along with any pre-existing anisotropies. The process of inflation, however, generates scale-free anisotropies with a Harrison-Zel'dovich spectrum which result in small but detectable CMB anisotropies in the present universe (Abbott & Wise 1984, Bond & Efstathiou 1984). During inflation zero point quantum fluctuations produce density fluctuations and gravitons (scalar and tensor fields) at levels comparable to the vacuum energy. Thus at the horizon wavelength scales they should produce comparable CMB anisotropies. The CMB anisotropy limits are now just pushing down to the level which is needed for observed velocities/gravitational potentials and beginning to limit the energy scale of inflation at a natural level. If we were optimistic, we could hope to see anisotropies at the 10^{-5} level, giving us the density fluctuations we need and telling us the inflation energy scale.

The inflationary paradigm is central to our modern theory of cosmology. How can it be tested? There are a number of predictions of inflation that are central to the paradigm. For example, that the universe should be flat. One original motivation is the overall isotropy of the CMB. Mike Turner covers the features of inflation in his lecture. The isotropy limits have improved about a factor of ten since inflation was first proposed as a solution to the large scale isotropy and the flatness of the universe and thus remains a valid explanation today. However, inflation does predict that there should be some perturbations that appear and some that do not. I summarize three below:

3.3.1. *Rotation - vorticity and shear.* In order for the inflationary epoch to occur the energy density in the universe in vorticity and shear must be less than the Planck scale E_p^4, which is about the kinetic energy density. As the universe expands the energy density in shear and vorticity decrease as R^2. During the non-inflationary phase the universe expands by about $E_p/kT_0 \approx 5 \times 10^{31}$. Without inflation the energy density in vorticity and shear decreases to less than $[E_pkT_0]^2$. The kinetic energy of expansion is decreased the same amount that the shear and vorticity are so that ω/H and σ/H maintain their (primordial) original ratio. The inflationary phase provides another factor of at least 10^{26} in the expansion factor. This reduces the ratio of these values to below 10^{-7}. The DMR limits in that sense support inflation and Mach's Principle.

3.3.2. *Relic Gravitons*. Inflationary models predict the zero point quantum fluctuation production of relic gravitons with a Harrison-Zeldovich spectrum of strain with amplitude $A_{graviton} = 2/\sqrt{\pi} \, H / M_{Planck}$. These gravitons will produce CMB temperature fluctuations of order $\Delta T / T \sim H / M_{Planck}$, where H is the Hubble parameter during inflation (a period characterized by a constant Hubble parameter). As a result, the possible values of the Hubble parameter at the inflationary stage are significantly restricted by the anisotropy limits. The relationship between the inflation Hubble parameter and the vacuum energy density is

$$H^2 = 8\pi/3 \; G \; \rho_{vacuum} = 8\pi/3 \; G \; M^4 = 8\pi/3 \; M^4 / M_{Planck}^2 \quad (10)$$

where M is the energy scale of inflation and the vacuum energy density is $\rho_{vacuum} = M^4$ $(=c^5 M^4/h^3)$. The energy density of gravitons (see Figure 1) is

$$\Omega_{graviton} = 10^{-4} \; \rho_{vacuum}/ \; M_{Planck}^4 = 10^{-4} \; (M/M_{Planck})^4 \quad (11)$$

for wavelengths significantly smaller than the horizon, rising to $\rho_{vacuum}/ M_{Planck}^4 = (M/ M_{Planck})^4$ at horizon size. The quadrupole CMB anisotropy and smaller angular scales limits imply $H_{inflation} < 3 \; x \; 10^{-5} \; M_{Planck}$, the inflation energy scale of interest for our observable universe is far removed from the Planck scale ($M < 10^{17}$ GeV), and the energy density of the vacuum is $\rho_{vacuum} = M^4 < 10^{-9} \; M_{Planck}^4$.

3.3.3. *Power Spectrum of Initial Perturbations*. The process of inflation generates scale-free potential fluctuations with a Harrison-Zel'dovich spectrum. If these potential fluctuations account for the structure and large scale velocity flow observed in the present universe, they must produce small but detectable CMB anisotropies in the present universe (Abbott & Wise 1984, Bond & Efstathiou 1984, Kolb & Turner 1990, Gorski 1991, Schaefer 1991, Starobinsky 1991). During inflation zero point quantum fluctuations produce the scale-free spectrum of perturbations which effectively remain frozen until they cross the horizon. Once the perturbation crosses the horizon physical processes can operate to cause evolution of the perturbation. While the perturbation is radiation dominated it continues to participate in the expansion of the universe and the perturbation remains essentially as an acoustic perturbation. Once the universe reaches the matter dominated stage (scale size $\sim 10 h^{-1}$ Mpc) then the perturbation can begin earnest evolution. The matter in the perturbation contracts just enough to maintain constant potential to conserve energy. The potential remains constant even once the perturbation is inside the horizon due to conservation of energy and can change only if there is dissipation of the energy from the region of interest. The primary means of energy dissipation is through free-streaming of photons or hot particles out of the perturbation. Thus the Silk damping of the baryon perturbations and the streaming out of photons causes the rms variation of the potential to decrease on small scales. Hot dark weakly interacting matter is named so because it can free-stream even more effectively than photons and thus structure formation is even more suppressed on small scales. However, on scales above about 10 to 30 Mpc the primordial spectrum created by the early quantum fluctuations should remain intact. The normalization at cluster and larger scales predicts minimal rms potential fluctuations on the order of 10^{-5} and thus corresponding $\Delta T/T$ fluctuations at 1/3 that level and a very flat spectrum over the whole angular range sampled by the DMR.

3.4. GROWTH OF STRUCTURE IN AN EXPANDING UNIVERSE

A second major problem in cosmology is the growth of structure in the universe. The largest structures in the current universe (walls and voids) are observed to have density fluctuations $\delta\rho/\rho$ of order unity on scale sizes ~50 Mpc. Structures of this size are at the horizon scale at the surface of last scattering; consequently, the primordial density fluctuations are small and most of the growth is in the linear regime. The assumption of linear growth requires peculiar velocities ~ 0.01c in order to move the matter the required 10^8 light years of co-moving distance in the ~10^{10} years estimated to have elapsed since the surface of last scattering, nearly an order of magnitude greater than the peculiar velocity inferred from dipole anisotropy. To explain the observed structure without violating limits on CMB anisotropy, and to generate the critical density required by inflationary models, many astrophysicists have turned to cosmological models in which most of the matter in the universe (> 90%) is composed of weakly interacting massive particles (WIMPs). The dynamical properties of this "dark matter" allow it to clump faster than the baryonic matter, which later falls into the WIMP gravitational potential wells to form the structures observed today. The gravitational potential and motion of these particles produce CMB anisotropy whose amplitude depends on the angular scale size. For scale size ~ 10°, most reasonable models predict $\Delta T/T \sim 10^{-5}$, depending on the average density of the universe. Although current observations do not provide significant limits to these models, we anticipate that the DMR will provide a stringent test of such models as it continues to accumulate data.

4. CONCLUSIONS

A year after launch, the *COBE* DMR instrument is working well and continues to collect data. The FIRAS and DMR data show the dipole anisotropy, consistent with a Doppler-shifted thermal spectrum. Galactic emission is present at levels close to those expected prior to launch, and is largely confined to the plane of the galaxy. The results are currently limited by instrument noise and upper limits to potential sources of systematic error. There is no evidence for any other large-scale feature in the DMR maps. The DMR results limit CMB anisotropies on all angular scales >7° to $\Delta T/T < 10^{-4}$. The results are consistent with a universe described by a Robertson-Walker metric and show no evidence of anisotropic expansion, rotation, or defects (strings, walls, texture). As DMR sky coverage improves and the instrument noise per field of view decreases, we anticipate improved calibration, better estimates of potential systematics, and increasingly sensitive limits to potential CMB anisotropies.

5. ACKNOWLEDGEMENTS

The *COBE* DMR instrument and satellite have been working well and continue to collect data due to the excellent work by the staff and management of the *COBE* Project. The *COBE* research is a team effort including the 19 original Science Working Group members and other scientists and professionals (see for example, Smoot *et al.*). This work is supported by NASA and in part by the Director, Office of Energy Research, Office of High Energy and Nuclear Physics, Division of High Energy Physics of the U.S. Department of Energy under Contract No. DE-AC03-76SF00098. I thank Charles Lineweaver and Al

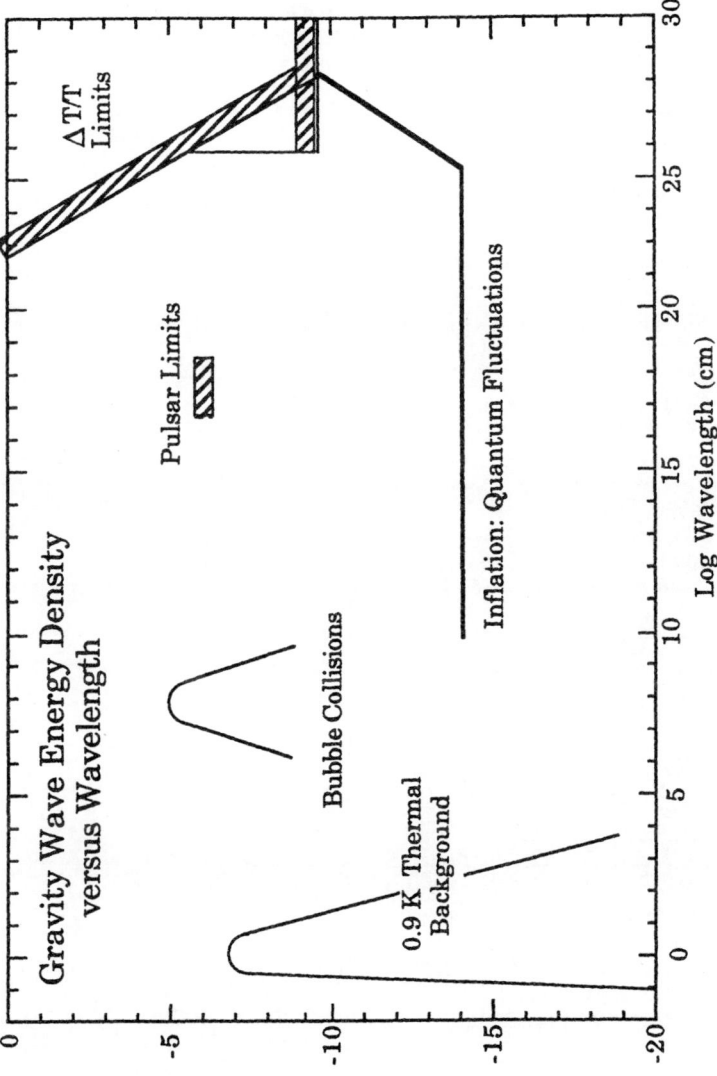

Figure 1: Limits on the energy density of long wavelength gravity waves. The DMR experiment sets two limits based upon the ΔT/T and strain to energy density relationships shown here and upon the formulas from Linder (1988): one is for the quadrupole and the other the surface of last scattering. The quantum fluctation curve is the predicted spectrum from inflation at maximum allowed normalization. This maximum curve corresponds to an energy scale of inflation slightly less than 10^{17} GeV. The potential inflation bubble collision spectrum is from Turner and Wilczek (1990). The normalization on the predictions depend upon the choice of elementary particle theory.

Kogut for their comments on this manuscript. We thank Monique Signore, Christophe Dupraz, and the Les Houches staff for the delightful and formidable school.

6. REFERENCES

Abbott, L.F., and Wise, M.B., *Ap. J. Lett.*, **282**, L47 (1984).
Anile, A.M., 1974 *Astrophysics and Space Sciences*, **29**, 415-426.
Barrow, J.D., Juskiewicz, R., and Sonoda, D.H., *MNRAS*, **213**, 917 (1985).
Basko, M.M. & Polnarev, A.G. 1980, *MNRAS*, **191**, 207.
Bianchi, L. 1897 *Mem. Soc. It. Della. Sc. (Dei.XL)(3)* **11**, 267.
Bond, J.R. and Efstathiou, G, *Ap. J. Lett.*, **285**, L45 (1984).
Burke, W.L., *Ap. J.*, **196**, 329 (1975).
Collins, C.B., and Hawking, S.W., *MNRAS*, **162**, 307 (1973).
Cottingham *et al.* 1991 submitted to *Ap. J.*.
Estabrook, F.B., Wahlquist, H.D., Behr, C.G. 1968 *J.Math.Phys.* **9**, 497.
Ehlers, J., Geren,P. & Sachs, R. 1968 *J.Math.Phys.* **9**, 1344.
Ellis, G.F.R., & MaCallum, M.A.H. 1969 *Comm.Math.Phys.* **12**,108.,
Franceschini, A., Toffolatti, L., Danese, L., & De Zotti, G. 1989 *Ap. J.*, **344**, 35-45.
Goetz, G. & Notzold, D. 1990, Fermi Lab preprint.
Gorski, K. 1991, *Ap. J. Lett.* **370**, L5.
Grischuk, L.P., and Zel'dovich, Ya. B., *Sov. Astron.*, **22**, 125 (1978).
Kaiser, N. 1982 *MNRAS*, **1xx**, 3.
Negroponte, J.& Silk, J. 1980 *Phy.Rev. Let.*, **44**, 1433.
Raychaudhuri, A.K. 1979, "Theoretical Cosmology, Oxford.
Ryan, M.P., & Shepley, L.C. 1975 "Homogeneous Relativistic Cosmologies", Princeton.
Rees, M.J. 1968 *Ap. J. Lett.*, **153**, L1.
Sachs, R.K., and Wolfe, A.M., *Ap. J.*, **147**, 73 (1967).
Schaefer 1991, *"After the First Three Minutes"*, ed. Holt, Bennett, Trinble, AIP.
Smoot, G.F., *et al.*, *Ap. J.*, **360**, 685 (1990).
Smoot, G.F., *et al.*, *Ap. J. Let.*, **371**, L1-L5 1991A
Smoot, G.F., *et al.*, *Adv. Space Res.* **11**, No2, pp.(2)193-(2)205, 1991B.
Starobinsky, A.A. 1991 *"Observational Tests of Inflation"*, ed. Shanks et al. Dordrecht.
Stebbins, A., *Ap. J.*, **327**, 584 (1988).
Stebbins, A., *et al.*, *Ap. J.*, **322**, 1 (1987).
Taub, A.H. 1951, *Ann. Math* **53**, 472. Tolman, B. W., 1984 *Ap. J.*, **290**, 1-11
Turner, M.S.,Watkins, R. & Widrow, L.M. 1990, Fermi Lab preprint.
Turner, M.S. 1991 submitted Physical Review.
Turok, N. 1989, *Phy. Rev. Let.*, **63**, 2625.
Turok, N. & Spergel, D. 1990, *Phy. Rev. Let.*, **64**, 2763.
Vilenkin, A., *Physics Reports*, **121**, 263 (1985).
Yang, J., *et al.*, *Ap. J.*, **281**, 493 (1984).

COSMIC BACKGROUND ANISOTROPIES IN THE MILLIMETRIC REGION

S. BOTTANI, P. DE BERNARDIS, M. DE PETRIS, S. MASI,
B. MELCHIORRI*, F. MELCHIORRI & P. TANZILLI
*Dept.of Physics, University of Rome   * IFA-CNR, Rome*
Piazz.le Aldo Moro 2, 00185 Rome, Italy

ABSTRACT. We discuss the observational situation of CBR anisotropies in the millimetric and sub-millimetric region

talks given by F. Melchiorri and P. de Bernardis

1. Introduction

The main goal of theoretical cosmology is that of writing and solving the Boltzmann Transport Equations (BTE) for the cosmological fluids (photons, barions, Cold Dark Matter, (CDM), neutrinos etc.) and their perturbations, starting from the Planck Era ($t \simeq 10^{-43}$ s) to now. This is a multi-step process in which one has to decide the source terms for BTE (for instance, Bremsstrahlung, Compton scattering etc. for photons), select an appropriate gauge, write down the unperturbed and perturbed Einstein Equations, identify a set of hypersurfaces where the relevant quantities have to be evaluated, expand in spherical harmonics the perturbation, solve the equations and list the solutions in terms of wavenumbers and, finally, compare them to the observations. This procedure is valid when the perturbations are small and the relevant equations can be linearized: in terms of redshift it corresponds to $z \geq 10 - 20$. For shorter redshifts we enter into the *nearby* universe characterized by structures like galaxies, clusters of galaxies, etc. and the corresponding physics is much more complicated.

Almost all these points have been discussed by various authors during the School. Here, we are interested in the *observational approach* to Cosmology. This is again a multistep program that involves the study of the transparency of the Universe to electromagnetic radiations, the analysis of the effects of the "local" backgrounds (Earth's Atmosphere, zodiacal dust scattering and emission, Galactic dust and gas emission), the choice of the best wavelength of observation, the choice of the best angular scale (spatial resolution) and, finally the comparison between the observed quantities and the theoretical models.

It is a widely diffused opinion that the comparison between theories and observations should be simpler as long as the analysis is limited to the *linear regime*, i.e. for redshifts $z \geq 10 - 20$. Unfortunately only one radiation source, namely the Cosmic Background

M. Signore and C. Dupraz (et al.), The Infrared and Submillimetre Sky after COBE, 345–384.
© 1992 *Kluwer Academic Publishers.*

Radiation (CBR), is believed to carry out information from these large redshifts. Therefore we have decided to concentrate our attention on the properties of this radiation.

The study of the distant universe ($z \geq 10 - 20$) is characterized by some peculiarities: cosmologists are no longer interested in maps or atlases, as in the case of the nearby universe. The primordial fireball is believed to be uniform enough to be described by a few statistical properties, which are collectively called *Anisotropies*.

2. Anisotropies: what are they?

In the framework of the Robertson Walker metric the density of the universe is increasing with redshift. It follows that also the density of galaxies increases with z. If we disregard the evolution, at a redshift of the order of $z = 20 - 40$, the average distance among galaxies becomes comparable with their mean size. Tidal interactions would have a strong influence in these conditions: therefore, it is usually believed that galaxies merge together at such large redshifts. From this simple consideration it follows that the early universe cannot be explored by powerful telescopes resolving single objects: there are no single objects at $z \geq 20 - 40$! Astronomers are already accustomed with "confusion" problems, when too many objects are inside the field of view and the properties of each of them are lost. In this sense, the astronomy of the early universe shears this confusion problem, being the observer unable to disentangle the single perturbation: special instruments have to be invented and different terminologies must be introduced to describe this peculiar situation. The *typical* instrument employed in Observational Cosmology is the *isotropometer* and the objects it is looking for are called *anisotropies*.

An isotropometer consists of a differential radiometer that measures the difference of the *Brightness* I_b of two sky regions separated by an angle α (called the modulation angle). Besides the modulation angle, a radiometer is characterized by its collecting area A, its beam size $\tilde{\sigma}$ (called the angular scale), its wavelength of operation λ, the bandwidth $\Delta\lambda$, and other more technical details, like the frequency of switching between the two fields f, the operating temperature T_R etc.

The product $A \cdot \Omega$ between the radiometer collecting area A and the solid angle Ω is called *Throughput* and it converts the differential Brightness of the sky (which is measured in $W\ cm^{-2}\ sr^{-1}$) into a power (W) that is received by the detector (a radio-receiver or a bolometer) and measured as a differential signal ΔV. As we move our radiometer across the sky the output ΔV will change . Let us assume that we have selected a certain path across the sky and computed the rms value of ΔV, say $\Delta V_{r.m.s.}$, along this path. If this quantity turns out to be independent on the selected path we may conclude that it contains information on some statistical property of the sky roughness. In this case we may speak of Anisotropies: the distribution of Brightness differences in the sky has a characteristic statistical distribution; $\Delta V_{r.m.s.}$ is taken as a measurement of the Sky Anisotropy at the given wavelength λ and angular scale $\tilde{\sigma}$.

To get complete information about this situation, we have to repeat our measurements with different beam sizes $\tilde{\sigma}$ and modulation amplitudes α. After these measurements have been performed, and if we know the mean Brightness I_B, we can plot the function

$$(\frac{\Delta I_B}{I_B})_{r.m.s.} = F(\alpha, \tilde{\sigma}, \lambda) \qquad [1]$$

which completely characterizes the sky anisotropies.

To better illustrate the difference between an anisotropy measurement and a conventional sky map let us apply the above technique to the study of the infrared sky at 100 microns, as provided by IRAS data set. These data are usually plotted as sky maps, where the Galactic plane represents the strongest source. In Figure 1 we have reported the result when the paths across the sky are selected in such a way so as to explore several clean sky regions, well apart from the Galactic plane. The function [1] is then plotted versus the amplitude of modulation for three fixed beam sizes. It turns out that the result depends on the beamsize but is independent of the selected path. This figure also shows a peculiar feature of a typical anisotropy: the r.m.s. amplitude increases with the modulation amplitude and saturates at a certain level. We interpretate this effect as due to a decrease of correlation between the two beams of the radiometer: at small modulation amplitudes the two beams are observing parts of the same sky spot, while for very large modulations the two beams are observing anisotropies arising from well independent sky patches. In the last case the two beams are completely uncorrelated. The modulation amplitude at which this effect occurs depends on the statistical properties of the anisotropy pattern: the flattening can even not occur if large scale gradients are present.

In Figure 2 the paths are still more than 30 degrees apart from the Galactic plane, but they start to cross important sources. Figure 2 reveals that the result is now strongly dependent on the selected trajectory in the sky: it follows that it is possible to speak of anisotropies only in the regions explored in Figure 1, while one has to produce a conventional sky map to study the situation of Figure 2. Figures 1 and 2 are what we need to characterize the anisotropies of the IRAS sky. It is clear that this characterization becomes meaningless if we extend our analysis beyond the "clean regions" selected for the paths of Figure 1.

Even in the case of Figure 1 the usefulness of the idea of anisotropy depends on the real distribution of the irregularities in the sky. Only for a Gaussian distribution the quantity $\left(\frac{\Delta I}{I}\right)_{r.m.s.}$ is a good estimator of the sky variance. If other distributions occur, then the information given by the variance of the anisotropy must be completed by other statistical properties, like the number of peaks that lie beyond a certain level and so on (for a discussion of this point see the lecture of Gonzalez-Martinez). In short, the transition from an anisotropy to a conventional sky map is determined by the topological richness of the sky .

In the framework of the Big Bang cosmology many theories for galaxy formation anticipate very peculiar anisotropies, which are Gaussian and are represented by tiny fluctuations of the Cosmic Background Temperature ΔT, so that the quantity $\left(\frac{\Delta T}{T}\right)_{r.m.s.}$ depends only on α and $\tilde{\sigma}$, and not on the wavelength of operation. These are usually called CBR Anisotropies. Their dependence on the beam size is called CBR Anisotropy Spectrum. To give a precise meaning to this one should measure the anisotropy for modulation angles large enough to guarantee that the two beams are uncorrelated: since this angle depends on the statistical properties of the anisotropies it will be different in different theoretical scenarios: that is one of the reasons why the result of the same measurement has different impacts in the various theories, because it could correspond to completely uncorrelated or to partially correlated beams.

One of the main goals of theoretical cosmology is that of connecting the spectrum of CBR anisotropies to the theories of galaxy formation, thereby offering the possibility of an experimental test (see the lecture of Silk for this point). The main goal of observational

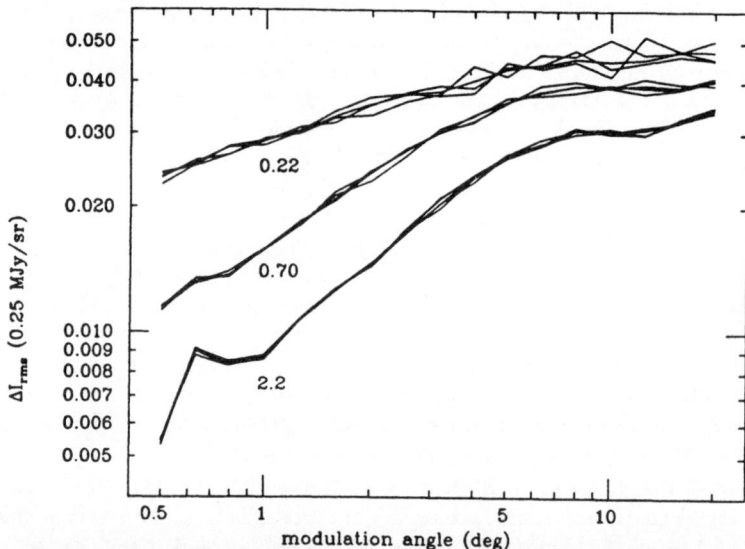

Figure 1: sky brightness anisotropy ΔI_{rms} at $\lambda = 100\mu m$, as measured by the IRAS satellite. The rms value has been computed over the 'clean sky regions' defined by the condition that the measured brightness in both the beams is within 20 % of the minimum brightness. The rms anisotropy $\Delta I_{rms}(\alpha, \sigma)$ is plotted versus the modulation amplitude α for three values of the beamsize σ (the different curves are labeled with the beamsize in degrees). Five realizations of 10000 differences heve been used for each beamsize, in order to test the reliability of these estimates.

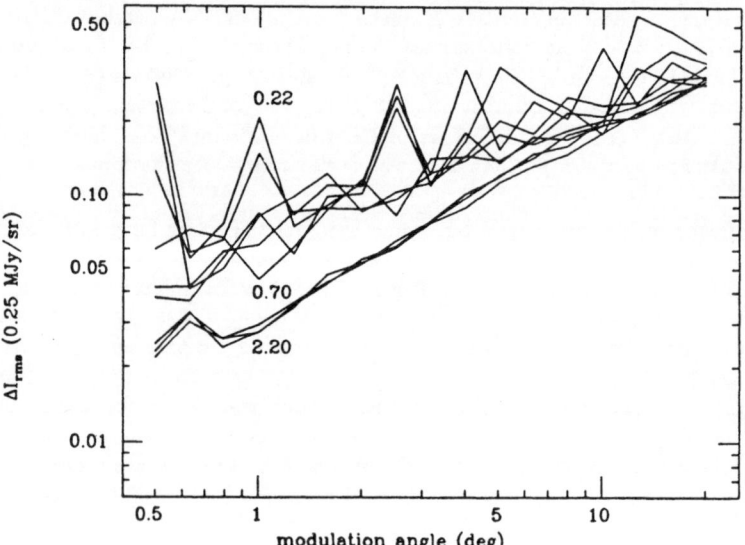

Figure 2: same as figure 1, in all the sky regions with $|b| > 30°$: the presence of sources does not allow to plot reliable general estimates of the functional dependence of anisotropy from modulation amplitude, especially in the case of small beamsizes.

cosmology is that of identifying the best wavelength and angular scale of observations to reach the greatest level of sensitivity.

The first point to underline is that, despite the relevant properties of $\left(\frac{\Delta T}{T}\right)_{r.m.s.}$ anticipated by theoreticians, it is not this quantity that a radiometer would measure: as said at the beginning, a radiometer is sensitive to the Brightness gradients ΔI_B: they are related to $\frac{\Delta T}{T}$ by the equation

$$\Delta I_B = I_B \frac{x \cdot e^x}{e^x - 1} \frac{\Delta T}{T}; \qquad x = \frac{h\nu}{kT} \qquad [2]$$

The spectrum of ΔI_B is no more Planckian, as shown in Figure 3., its maximum occurs around a wavelength of 1-2 mm. Unfortunately, this does not mean that the choice of the wavelength of operation is easy. One has to worry about the possible spurious anisotropies due to "local emissions", like the already illustrated Galactic anisotropies. The observer must carefully consider the properties of the various local backgrounds. A simple but rather approximate way to do it is just to plot the spectra of the local backgrounds in units of the CBR anisotropies (see Figure 4). However this procedure is largely misleading. Firstly the various backgrounds have different levels and spectra, depending on the sky region under observation. Secondly, one should consider the anisotropies of these backgrounds and not their absolute fluxes. The "typical" level of these anisotropies is again dependent on the direction in the sky and on the angular scale and on the modulation amplitude. We may conclude that we have, for a given wavelength, an optimum angular scale of observation and, viceversa, for a given angular scale, an optimum wavelength of operation. The idea that the best wavelength of operation is just where the local backgrounds have a minimum has to be considered as a very rough approximation.

3. The Environmental Background

In the following we discuss to some extent the properties of the local backgrounds and how they could affect and contaminate the observations of CBR anisotropies.

3.1. THE EXPERIMENTAL PROBLEM

CBR gradients in the millimetric region represent only a fraction (sometimes a tiny fraction) of the gradients in the sky: Galactic emission gradients and Atmospheric gradients (for balloon-borne experiment) are also present. Let us assume that we are working at a given α, $\tilde{\sigma}$ and wavelength λ_1: the signals we observe can be expressed as

$$\Delta S_1 = R_1 \cdot (\Delta I_{gal,1} + \Delta I_{atm,1} + \Delta I_{CBR}) \qquad [3]$$

where R_1 is the responsivity of our system and the other symbols are obvious. To extract the cosmological term ΔI_{CBR} we need to calibrate our instrument (measurement of R_1) and to measure in some other way the terms related to the Galactic and Atmospheric gradients.

To deal with this problem observations are usually performed at several different wavelengths. We can employ, for instance, a second channel operating at a much shorter wavelength $\lambda_2 \leq 500\mu m$ where CBR anisotropies are negligible. At this new wavelength we have

$$\Delta S_2 = R_2 \cdot (\Delta I_{gal,2} + \Delta I_{atm,2}) \qquad [4]$$

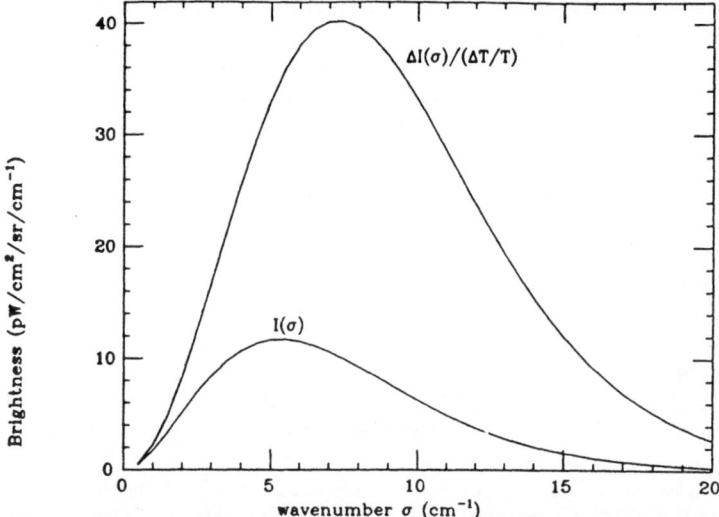

Figure 3: Brightness ($I(\sigma)$) and Brightness anisotropy ($\Delta I(\sigma)$) spectra for a 2.735 K blackbody. σ is the wavenumber in cm^{-1}.

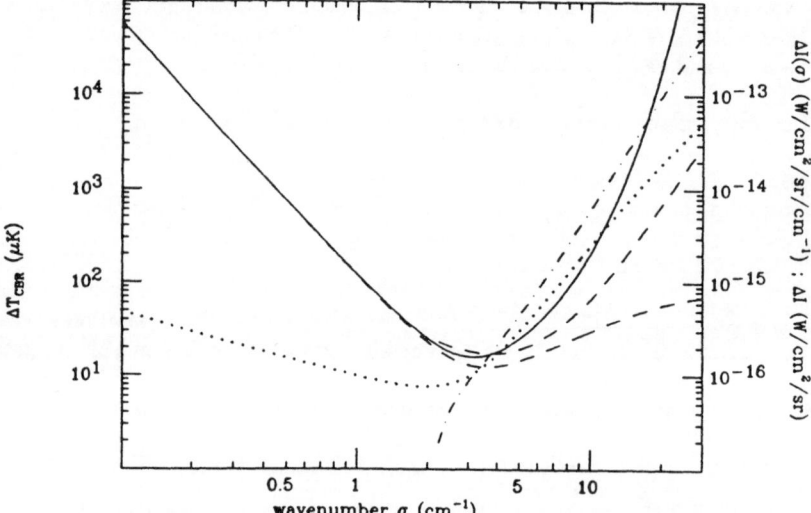

Figure 4: Diffuse Galactic emission at the North Galactic Pole. The continuous line represents the Galactic emission in terms of Thermodynamic temperature fluctuations of the CBR (left vertical axis) as measured by a monochromatic receiver; the higher dashed line refers to the same quantity expressed in antenna temperature. The lower dashed line refers to the thermodynamic temperature fluctuations measured by a wide band receiver starting at 2 cm^{-1}; the dotted-dashed line refers to the same observable in $W/cm^2/sr$. The dotted line is the Brightness measured by a monochromatic detector (in $W/cm^2/sr/cm^{-1}$).

In order to extract the cosmological anisotropies we may combine the equations [3] and [4] to cancel the contaminations; one needs

$$R_1(\Delta I_{gal,1} + \Delta I_{atm,1}) - K \cdot R_2(\Delta I_{gal,2} + \Delta I_{atm,2}) = 0$$

In theory one can hope to select two channels so that

$$\frac{\Delta I_{gal,1}}{\Delta I_{gal,2}} = \frac{\Delta I_{atm,1}}{\Delta I_{atm,2}} \equiv K \qquad [5].$$

and then cope with the precedent requirement. The real situation is much more complicated, however. What we collectively call "Galactic emission" is in fact the sum of at least three contributes, namely sinchrotron and free-free emission at long wavelengths, Galactic dust emission and zodiacal dust emission.

3.2. THE SPECTRUM OF GALACTIC EMISSION

Zodiacal dust forms a relatively uniform, low emissivity cloud embedding the inner Solar System. Its thermal far IR emission has been mapped extensively by the IRAS satellite at 12, 25, 60 and 100 μm. The temperature of the cloud is of the order of 300 K, slowly decreasing with the distance from the sun. The emission spectrum observed from a given line of sight is a weighted mean of black-bodies, with an emissivity law characteristic of relatively large ($\sim 10\mu m$) dust grains. The resulting spectrum depends on the elongation angle of the line of sight, but has in general a maximum at $\lambda \sim 20\mu m$ and a long wavelength spectral index ~ -2 (see Temi et al. 1989 for details).

Dust emission at high Galactic latitudes is due to several different components: very small grains, transiently heated at several hundreds of K, 'normal grains' with a temperature of a few tens kelvins and, perhaps, very large 'fractal' or 'needle' grains with a temperature $\lesssim 10K$. IRAS has not mapped all these components, since it is not sensitive to dust temperatures lower than 20 K. However, for the components mapped by IRAS the dust temperature is remarkably uniform at high Galactic latitudes (Boulanger and Pérault 1988, de Bernardis et al. 1988), with a typical temperature of the dust clouds around 20 K. This finding is reinforced by the analysis of the few available sub-mm measurements of dust emission: in spite of the different observed sky regions and detection geometries, all these measurements follow a simple common spectral trend, once normalized to the IRAS emission (see Masi et al. 1991, and references therein). This is consistent with a single - temperature thermal spectrum with a power law emissivity:

$$I^D(\nu) = A \cdot I_{IRAS} \cdot \left(\frac{\nu}{\nu_{IRAS}}\right)^\beta \cdot \frac{e^{x_{IRAS}} - 1}{e^x - 1} \quad ; \quad x = \frac{h\nu}{kT_d} \qquad [6]$$

The values of the best fit parameters are: $T_d = 26K$; $\beta = 1.25$; $A = 0.75$ (see Masi et al., 1991 for a complete discussion on the uncertainties). These values are also consistent with the recent redetermination of high Galactic latitude dust emission made by COBE-DIRBE (Hauser et al., 1990).

Synchrotron emission is generated by electrons moving in the Galactic magnetic field: it is well described by a simple power law spectrum

$$I^S(\nu) = I^S(\nu_o) \cdot \left(\frac{\nu}{\nu_o}\right)^\gamma \qquad [7]$$

The spectral index $\gamma = -0.8 \pm 0.1$ has been estimated by comparing the 408 MHz map with the 1420 MHz map (Reich and Reich, 1986, 1988).

Free-free emission is generated by the hot, ionized gas $(T \simeq 10^4 K)$, present both in HII regions in the Galactic plane and in diffuse form above the disk of our Galaxy (Reynolds 1989). The spectrum is well fitted by

$$I^{FF}(\nu) = I^{FF}(\nu_o) \left(\frac{\nu}{\nu_o}\right)^{-0.1} \tag{8}$$

Here the spectral index is determined by the theory of thermal bremsstrahlung: the only free parameter is the overall amplitude.

To give an idea of the real situation one has to combine the results of Figure 3 with the spectrum given by eqs.[6],[7],[8]. The results are shown in Figure 4.

3.3 ATMOSPHERIC EMISSION

The same line of reasoning has to be applied to the Atmospheric emission, which is a combination of various components: Ozone and Water vapour being the most important emitters at balloon altitudes and in the far infrared.

The angular dependence of Atmospheric fluctuations is almost unknown. There are indications that the anisotropies are time-dependent, with a time scale ranging from minutes to hours, and usually increasing with beam size and modulation amplitude.

By comparing this last point with the results for Galactic anisotropies one derives the conclusion that a minimum in spurious anisotropies should occur at angular scales ranging from 1 to 20 degrees.

3.4. THE CORRELATION METHOD

Once we have selected the "best" angular scale, some attempts can be made to search for a reasonable correlation between two or more photometric channels. However, the choice of the "near infrared" channel is quite limited: the Zodiacal dust emission becomes important at wavelengths shorter than 200 μm, thereby degrading the correlation. At wavelengths longer than 500 μm the CBR anisotropies are still present and one has to worry not to cancel them together with the spurious ones.

Let us briefly show how we have proceeded in practical cases. For the near ir channel we selected a bandwidth 48-53cm^{-1}. The Galactic and Atmospheric fluctuations are taken as $\delta I_{gal} = 10^{-2} I_{gal}$ and $\delta I_{atm} = a \cdot I_{atm}$, where the factor a is chosen so that both contaminations are similar on the spectral domain containing the CBR maximum:

$$\int_3^{15} \delta I_{gal} d\nu = \int_3^{15} \delta I_{atm} d\nu \tag{9}$$

This is quite an arbitrary choice, but it corresponds to the worse situation, where both the spurious effects (Galactic and Atmospheric) are of comparable amplitudes. For the Galactic emission I_{gal} we have employed the model fitting the COBE-FIRAS data (Wright et al. 1991) with the following parameters: $T_d = 23.3 K$, $\beta = 1.65$, $\gamma = -0.75$. The spectrum of these fluctuations is shown in Figure 5. In Figure 6 we compare the Galactic emission fluctuations to the Atmospheric emission fluctuations. In Figure 7 we plot the two ratios of eq. [5] as function of the cutoff frequency of the long wavelength channel, which is assumed to have a flat response starting from $3 cm^{-1}$ up to the cut-off wavenumber. The two curves have crosspoints between 20 and 30 cm^{-1}. Thus for two filters with bandwidth $3\text{-}20 cm^{-1}$ and $48\text{-}53 cm^{-1}$ there is the best correlation between the contaminations in the two channels and both Galactic and Atmospheric contributions can be removed simultaneously.

To evaluate the efficiency of this method we plotted in Figure 8 the residuals after the correction: $\delta I_{atm_1} + \delta I_{gal_1} - K_{1,2} \cdot (\delta I_{atm_2} + \delta I_{gal_2})$, where K_1 is the ratio between the Galactic emission in the two filter bandwidths and K_2 refers to Atmospheric emission. This suggests that a good cancellation of both Galactic and Atmospheric emissions can be obtained by a careful choice of the operating bandwidth of the far infrared filter. *Due to the dependence in both cases from the Atmospheric lines the curves show the same behaviour.* The correction with K_2 is about one order of magnitude better than the one with the Galactic coefficient. This is essentially due to the more erratic behaviour of the atmsopheric emission versus the Galactic one. The correcting term $K_2 \cdot \delta I_2$ can therefore match more easily the signal δI_2 of the first channel.

At the frequency of best correlation, as both ratios are equal, both curves of the correction residues should go to zero. In Figure 8 we see that this might occur but only at narrow intervals. Thus a complete correction is quite difficult to obtain. Nevertheless in the neighborhood of ν_0, where both ratios are close in value, there is a distinct valley where the correction is improved. The shape and width of this region depends largely on the Atmospheric model and on where the two curves cross.

The increase in the correlation is not necessarily more advantageous than the decrease of Galactic and Atmospheric contamination obtained by working at longer wavelengths, and critically depends on the choice of the filters. We should stress that our analysis is correct only if the Atmospheric and Galactic models are those selected, and if the Atmospheric fluctuations are comparable with the Galactic ones.

In conclusion, one can hope to correct simultaneously for the spurious Galactic and Atmospheric anisotropies by a careful choice of the central wavelength and bandwidth of the far infrared filter: the practical advantage of this procedure should be tested through observations, being the degree of correlation strongly dependent on poorly known parameters, like stratospheric emission and Galactic dust spectrum.

4. Large Scale Anisotropies: The Dipole and Quadrupole Kinematic Anisotropy

4.1. HISTORY AND THEORY

In 1907 Kurd von Mosengeil, in his Ph.D. Thesis entitled "Theory of stationary electromagnetic radiation in uniformly moving vacuum cavity", carried out under the tutoring of M. Planck, has studied the problem of an observer moving at a speed v inside an oven, the walls of which radiate like a blackbody at a temperature T. The purpose of this curious work was that of applying for the first time the Lorentz transformation to the case of a diffuse Doppler effect, a problem that had already been studied "classically" by F. Hasenohrl in 1905. His main result can be easily recovered by recalling that the Brightness divided

354

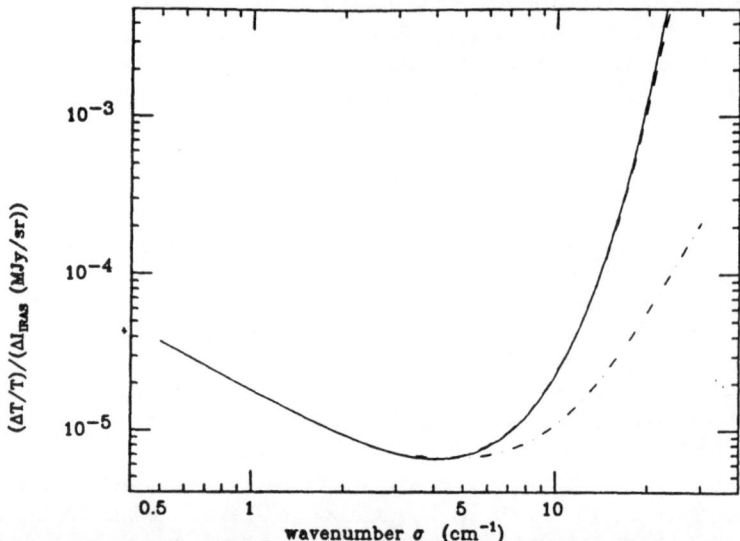

Figure 5: Galactic anisotropy expected at high latitudes in the mm region, according to the model of Masi et al. (1991). The fluctuations are normalized to IRAS data anisotropies in MJy/sr. The rms fluctuations expected from the IRAS 100 μm data are about 0.2 MJy/sr. The continuous line refer to a monochromatic receiver; the dashed line to a band with $\Delta\sigma/\sigma = 0.3$; the dashed - dotted line to a receiver with cut-on at 3 cm^{-1} and cut-off at σ.

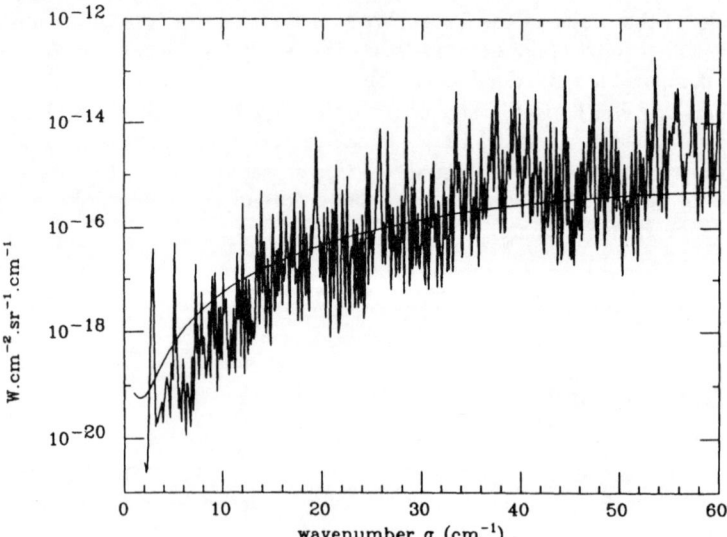

Figure 6: Spectra of Galactic and stratospheric emission fluctuations (as expected at balloon altitude) in the sub-mm region.

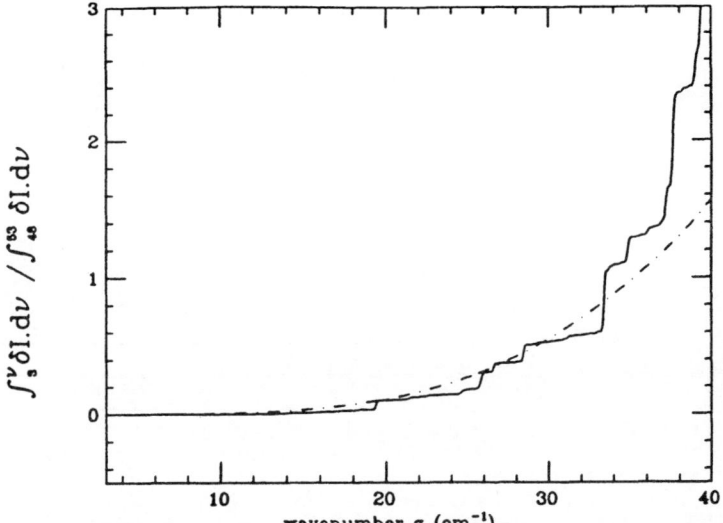

Figure 7: Ratio between the signals obtained in two photometric channels observing the same sky regions from balloon altitude. Band 1 is a broad band with cut-on at 3 cm^{-1} and cut-off at σ cm^{-1}; the band 2 is a band with cut-on at 48 cm^{-1} and cut-off at 53 cm^{-1}. The continuous line refers to atmospheric emission fluctuations; the dotted - dashed line refers to Galactic anisotropies.

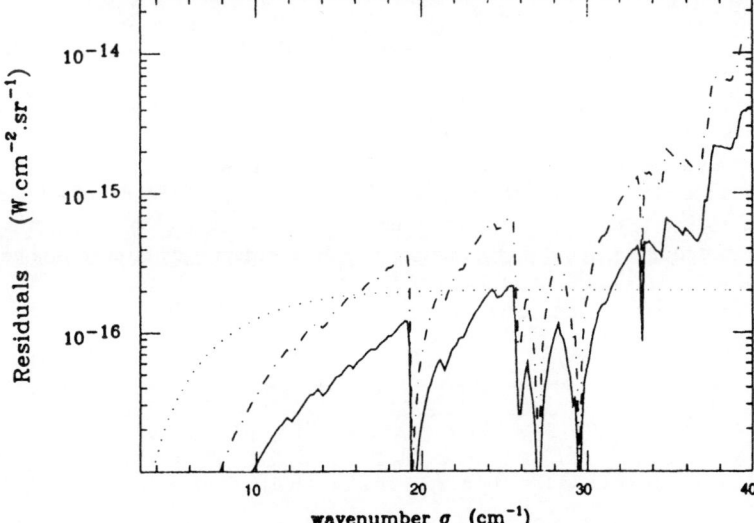

Figure 8: Residuals $\int_3^\nu \delta I.d\nu\prime - K(\nu).\int_{48}^{53} \delta I.d\nu\prime$ with $K(\nu)$ the ratio of the emission in the two channels, for the atmosphere (continuous line) and for the galaxy (dash dotted line). The dotted line indicates the contribution of CBR fluctuations at the level $\frac{\delta T}{T} = 10^{-6}$ in the FIR channel.

by the cube of the frequency is a relativistic invariant (Liouville Theorem). So we have

$$\frac{I_{obs}(\nu_{obs})}{\nu_{obs}^3} = \frac{B(\nu)}{\nu^3} \qquad [10].$$

Here $B(\nu)$ is the blackbody Brightness measured by an observer at rest inside the oven, while $I_{obs}(\nu_{obs})$ is the Brightness measured by the moving observer. But

$$\nu_{obs} = \nu \frac{1 - \beta \cos \Theta}{(1 - \beta^2)^{1/2}} \qquad , \qquad \beta = \frac{v}{c} \qquad [11],$$

where Θ is the angle between the direction of observation and that of the motion, as measured in the reference frame of the source. It follows that

$$I_{obs}(\nu_{obs}) = B(\nu, T) \frac{(1 - \beta \cos \Theta)^3}{(1 - \beta^2)^{3/2}} \qquad [12]:$$

so we get finally

$$I_{obs}(\nu_{obs}) = \frac{2h}{c^2} \frac{\nu_{obs}^3}{\exp\left[\frac{h\nu_{obs}(1-\beta^2)^{1/2}}{T(1-\beta \cos \Theta)}\right] - 1} \qquad [13].$$

The main conclusion is that a blackbody of temperature T appears to a moving observer like a blackbody of temperature

$$T_{obs} = T[1 - \beta \cos \Theta](1 - \beta^2)^{-1/2} \qquad [14]$$

The linear dependence of the observed temperature on the cosine of the angle of observation justifies the name of *"Dipole Anisotropy"* given to this effect.

Obviously the dipole character is verified as far as we measure the temperature of the radiation: this is true in the Rayleigh - Jeans region, where the Brightness of a blackbody is proportional to the temperature. If we operate in the Wien region of the spectrum, the non linear dependence between Brightness and temperature introduces higher order multipoles in the spatial distribution of the radiation.

Mathematically the problem is better solved if the spectral index $\alpha = \frac{dlnI}{dln\nu}$ is considered. In the case of a blackbody we have

$$\alpha = \frac{\nu}{I} \frac{dI}{d\nu} = 3 - \frac{x \cdot e^x}{\cdot e^x - 1} \qquad [15]$$

Up to the second order we have (de Bernardis et al. 1990a):

$$I_{obs}(\nu_{obs}) = I(\nu) \frac{(1 + \beta \cos \Theta)^3}{(1 - \beta^2)^{3/2}} \simeq$$

$$\left[I(\nu_{obs}) + \left(\frac{dI}{d\nu}\right)_{\nu=\nu_{obs}} (\nu - \nu_{obs}) + \frac{1}{2} \left(\frac{d^2I}{d\nu^2}\right)_{\nu=\nu_{obs}} (\nu - \nu_{obs})^2 \right] \frac{(1 + \beta \cos \Theta)^3}{(1 - \beta^2)^{3/2}}$$

$$[16].$$

In the case of a blackbody we get

$$I_{obs}(\nu_{obs}) \simeq I(\nu_{obs}) \left[1 + (3 - \alpha)\beta \cos\Theta + \frac{1}{2}(3 - \alpha)\beta^2 + \frac{1}{2}\beta^2 \cos^2\Theta(3 - \alpha)(2 - \alpha - x/2) \right]$$
[17]

In this equation we note at the right side the dipole term $(3 - \alpha)\beta \cos\Theta$, a term constant in Θ (which is the equivalent of the transverse Doppler effect in Special Relativity) and the quadrupole term, proportional to $\cos^2\Theta$.

These speculative questions assume relevance in the framework of Big Bang cosmology. We can imagine of substituting the walls of the Mosengeil's oven with the last scattering surface (LSS): a measurement of the Dipole Anisotropy of CBR would allow finding our peculiar velocity with respect to this distant matter. This possibility was pointed out soon after the discovery of CBR independently by D. Sciama [1967] and P.J.E. Peebles and D.T. Wilkinson [1968]. To be more precise, Sciama pointed out the interest in measuring the peculiar velocity of our galaxy in order to compare it with the infall velocity towards Virgo Cluster, as derived by optical observations: he also rose the interesting question of wheter the Virgo Cluster itself would show up some peculiar velocity. Sciama however did not compute the angular dependence of the anisotropy and from his paper one can derive the impression that he suggested measuring just the "maximum" of the anisotropy, a rather difficult task. Peebles and Wilkinson (1968) correctly addressed the question of the angular dependence of the anisotropy.

4.2. WIEN AND ABERRATION EFFECTS

In the case of Big Bang cosmology, in order to get a correct result, one has to take into account that the formulas above derived refer to the angle Θ as measured in the rest reference frame of the source (the LSS in our case). It is more realistic to plot the signal versus the angle Θ_{obs} as measured by the observer. In such a case we must take into account the effect of the light aberration. The transformation rule is

$$\cos\Theta_{obs} = \frac{\cos\Theta + \beta}{1 + \beta\cos\Theta}$$
[18]

The aberration effect is quite negligible to the first order. It transforms eq.[10] into that obtained by Peebles and Wilkinson (1968)

$$T_{obs} = \frac{T_0}{\gamma[1 - \beta\cos\Theta_{obs}]} \simeq T_0[1 + \beta\cos\Theta_{obs} + \beta^2\cos^2\Theta_{obs} - \frac{1}{2}\beta^2 + ...]$$
[19]

Therefore the effect of light aberration manifests itself with the presence of a Quadrupole Temperature pattern having the same orientation of the Dipole and an amplitude scaled by the factor β. Smoot et al. (1991) have pointed out that COBE-DMR should reach a sufficient sensitivity to detect this temperature pattern in about two years of integration. This would be an important confirmation of the kinematic nature of the observed Dipole anisotropy.

One should bear in mind that detectors are sensitive to the Brightness of radiation or to the Antenna Temperature, and not to the thermodynamic temperature of the source. The

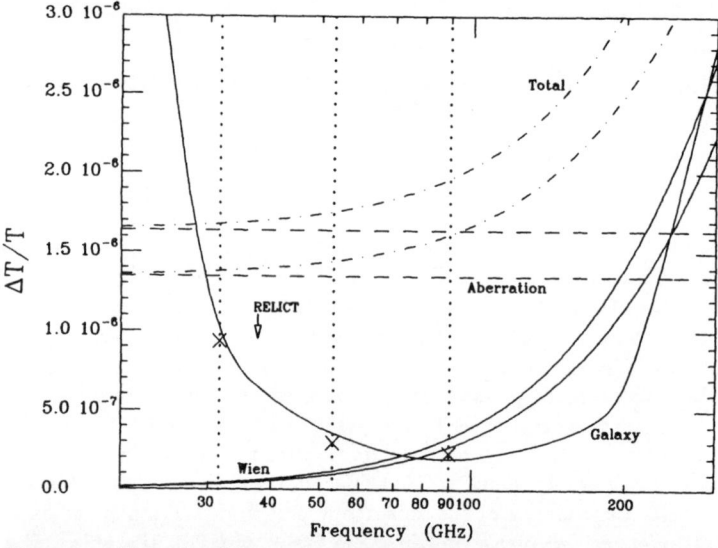

Figure 9: Aberration, Wien and total Quadrupoles (Thermodynamic Temperature fluctuations $\Delta T/T$) as a function of the frequency: each pair of curves take into account the actual uncertainty in the value of $\beta = 0.00122 \pm 0.00006$ (Smoot et al. 1991). The operating frequencies of the COBE's radiometers are indicated. Galactic Quadrupole data from COBE (x), Relikt I (upper limit), are also plotted, together with a model of Galactic contamination (see de Bernardis et al. 1991a). Both the data and the model are multiplied by 10^{-2}.

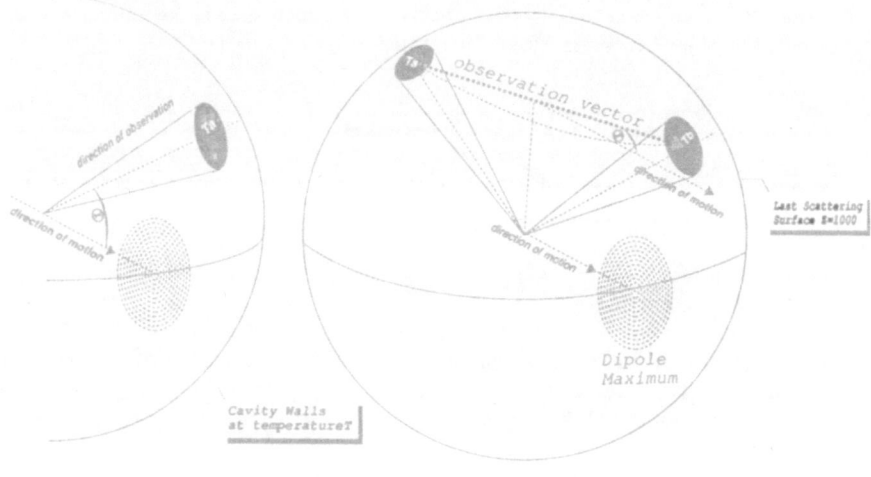

Figure 10: Measurements of the Kinematic anisotropy of the CBR, due to the Doppler effect generated by the motion of the observer with respect to the Last scattering surface restframe. Both absolute (left) and differential (right) measurements are considered.

two quantities are exactly proportional to each other only in the Rayleigh-Jeans limit. We have already seen that in the Planck formula the non linear relation between Brightness and Temperature has the consequence to produce a Quadrupole pattern even if the distribution of the temperature in the sky is that of a pure dipole: this Quadrupole, which is stronger in the Wien region, has nothing to do with the aberration effect. It follows that we have to disentangle the *Aberration Quadrupole* from the spurious term, which for the sake of simplicity we call in the following *Wien Quadrupole*.

If we develop the spectrum, seen by a moving observer, to the second order and correct for light aberration using eq. [18], we get

$$I_{obs} = I_0(\nu_{obs}) \left[1 + \frac{1}{2}(3 - \alpha)\beta^2 + (3 - \alpha)\beta\cos\Theta_{obs} + (3 - \alpha)(3 - \alpha - x/2)\beta^2\cos^2\Theta_{obs} \right]$$
[20].

Thus the Total Quadrupole is

$$I_Q = \frac{1}{2}I_0(\nu_{obs})(3 - \alpha)(3 - \alpha - x/2)\beta^2\cos^2\Theta_{obs}$$
[21];

this will result from the sum of the Wien Quadrupole and the Aberration Quadrupole. We recall from eq. [17] that the Wien Quadrupole is

$$I_W = \frac{1}{2}I_0(\nu_{obs})(3 - 2\alpha)(2 - \alpha - x/2)\beta^2\cos^2\Theta_{obs}$$
[22].

The Quadrupole connected with the aberration is obtained by subtracting the Wien Quadrupole amplitude from that of the Total Quadrupole expressed in eq. [21]:

$$I_A = \frac{1}{2}I_0(\nu_{obs})(3 - \alpha)\beta^2\cos^2\Theta_{obs}$$
[23].

We point out that I_W tends asymptotically to zero in the Rayleigh-Jeans region of the spectrum, when the Brightness tends to be proportional to the temperature, while I_A tends to an asymptotic value of $I_0(\nu_{obs})\beta^2\cos^2\Theta_{obs}$.

In Figure 9 we have plotted the amplitude of the two Quadrupoles as well as the Total Quadrupole. Unfortunately, the long-wave radiometers, where the Aberration Quadrupole dominates, are also the most contaminated by Galactic emission, which is clearly shown by Smoot et al. (1991). To give an idea of the effect of Galactic contamination we have also plotted in Figure 9 the results of a Galactic quadrupole model obtained by using the IRAS 100 micron map for the distribution of dust emission and the 408 MHz map for synchrotron emission (de Bernardis et al. 1991a). For comparision we also plot the results of COBE and of Relikt I.

We may conclude that COBE could determine the true nature of the Dipole Anisotropy : perhaps the final limit will be posed by Galactic contamination. If this could be modelled and subtracted at about 1% level, the sensitivity would be determined, among the various systematic effects, by the total integration time. An observing time longer than the already scheduled two years is strongly recommended in view of the cosmological relevance of the observation of the Aberration Quadrupole.

4.3. EXPERIMENTAL RESULTS

The search for the Dipole Anisotropy has been underway since 1968, more with the aim of proving the cosmological nature of CBR than with the goal of deriving information on the cosmic velocity fields, almost unknown at that time. From an observational point of view the search posed several new interesting problems: the first point is that one has to search for a *difference* in temperature (or in Brightness). This fact significantly reduces the problems related to the absolute measurements of the sky temperature. One does not need a precise absolute calibration, while different regions of the sky are compared with each other. In contrast with this advantage observers had to manage much smaller signals, the dipole component being about one thousand of the total CBR Brightness. This last point raises the question of possible contamination by spurious radiations at a much deeper level than in the case of CBR spectrum. Conventional radio and infrared systems have very broad angular response and the spill-over at large angles is such that ground radiation can easily contribute at a level of a few millidegrees in the beam. This serious instrumental problem was finally solved by the introduction of the *corrugated horns* in radio equipment and of the *Winston cones* in incoherent millimetric radiometers.

As shown in the second half of Figure 10, what an experiment is measuring is the difference in Brightness between two sky zones separated by an angle Θ. During the early observations few limited zones of the sky were explored so that one had to think of how to combine the data to enhance the detection of the dipole component. Smoot (1977) was the first suggesting to employ a graphic representation in which the data are averaged with respect to the angle between the "observation vector" TbTa of Figure 10 and the direction of the maximum in the Dipole Anisotropy. A typical plot in "Smoot angle" is shown in Figure 11, where we report the results of the Berkeley airborne experiment that first clearly detected the Dipole Pattern.

As more and more sky regions were covered, this representation was abandoned in favour of an analysis based on the determination of the *Spherical Harmonics coefficients*. If we indicate with α and δ the Right Ascension and the Declination of a sky map, a generic Dipole can be represented as the sum of three terms

$$\Delta T_D = T_x \cos \delta \cos \alpha + T_y \cos \delta \sin \alpha + T_z \sin \delta \qquad [24],$$

while a generic Quadrupole can be decomposed into 5 terms

$$\Delta T_Q = Q_1(3\sin^2\delta-1)/2 + Q_2 \sin 2\delta \cos \alpha + Q_3 \sin 2\delta \sin \alpha + Q_4 \cos^2 \delta \cos 2\alpha + Q_5 \cos^2 \delta \sin 2\alpha$$
$$[25].$$

A more "mathematical" definition for the Quadrupole would involve the complex spherical harmonics $Y_{l,m}$, that is:

$$Q_{tot} = \sum_{m=-2,2} a_{2,m} Y_{2,m} \qquad [26].$$

Here the $a_{2,m}$ have all the same *r.m.s.* value, $a_{r.m.s.}$; $Q_{1,r.m.s.} = \sqrt{\frac{5}{4\pi}} a_{r.m.s.}$ and for the remaining Q_i we have $Q_{i,r.m.s.} = \sqrt{\frac{3}{4}} Q_{1,r.m.s.}$.

Depending on the sky coverage, the various terms can be determined almost independently of each other. At the present time four experiments have an almost total sky

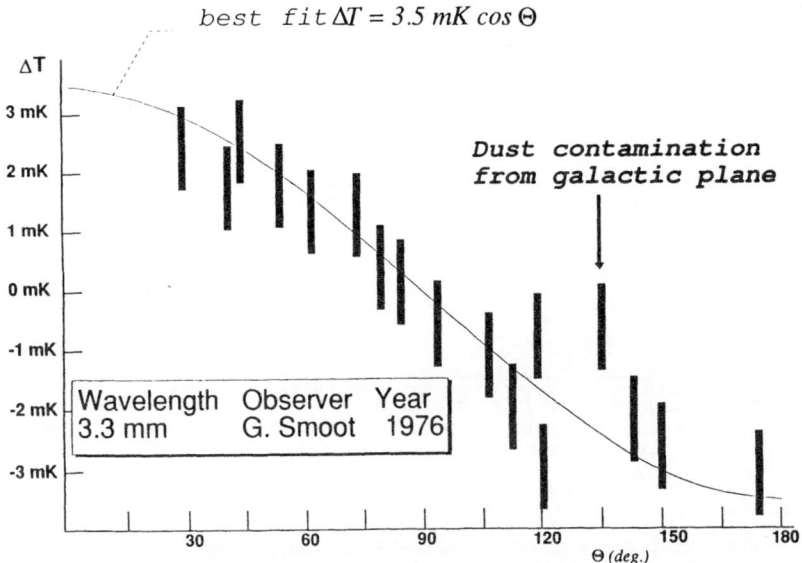

Figure 11: Dipole anisotropy detected in the Berkeley group early observations: in the abscissa is the so called 'Smoot Angle'.

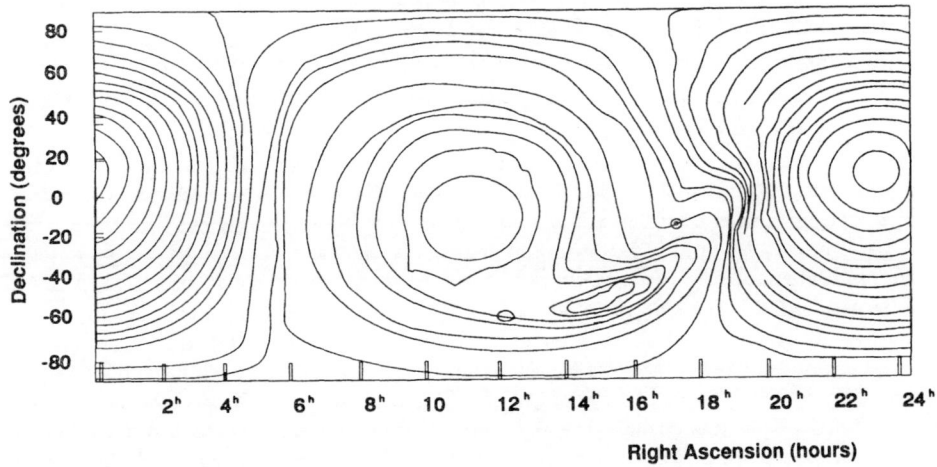

Right Ascension (hours)

Figure 12: Isophotes map of the full sky measured by the RELIKT-I satellite experiment at 37 GHz. The CBR Dipole anisotropy pattern is strongly contaminated by radio emission in the Galactic plane.

coverage, namely that of Berkeley (Lubin et al. 1985), Princeton (Fixsen et al. 1983), IKI-Moskow (Strukov et al. 1988) and COBE (Smoot et al. 1991).

From eq. [24] we get for the total dipole amplitude

$$(\Delta T_D)^2 = T_x^2 + T_Y^2 + T_z^2 \tag{27}$$

From Eqs.[25] and [26] different definitions for the total quadrupole amplitude have been used in literature. Theoreticians adopt the quantity

$$(\Delta T_Q)^2 = \frac{16\pi}{15} \left(\frac{3}{4}Q_1^2 + Q_2^2 + Q_3^2 + Q_4^2 + Q_5^2 \right) \tag{28}.$$

while Strukov et al. (1988) and also Smoot et al. (1991) have adopted the quantity

$$(\Delta T_Q')^2 = \frac{4}{15} \left(\frac{3}{4}Q_1^2 + Q_2^2 + Q_3^2 + Q_4^2 + Q_5^2 \right) \tag{28'}.$$

It follows that $(\Delta T_Q)^2 = 5a_{r.m.s.}^2$. The disagreement among the various Q_{tot} reported in literature is often due to the different definitions adopted.

The observational situation can be summarised as follows:

Maps of the dipole anisotropy have been produced in the radio region with an angular resolution of 5-10 degrees at wavelengths ranging from 3.3 mm to 9 mm: as an example, we reported in Figure 12 the map obtained by the soviet experiment Relikt I, operating at 37 GHz. The dipole behaviour is evident, but there is also a significant residual contamination from the Galactic synchrotron emission.

The most sensitive maps are those (still in progress) provided by COBE at 31.5, 53.0, and 90 GHz (9.6, 5.7, 3.3 mm). In Table I we report the preliminary results of COBE, together with those listed by other authors.

To appreciate the difficulty in observing the dipole and quadrupole kinematic anisotropy in the far infrared we plot in Figure 13 the isophotes of the dipole and quadrupole patterns over an 100 μm IRAS map. Clearly the anisotropies will be strongly contaminated, unless the dust emission is rapidly decreasing at longer wavelengths. The first positive detection of the dipole anisotropy in the millimetric region was that of the Italian Group in 1978 (Fabbri et al. 1980). This positive result was due to the choice of a narrow sky modulation (6 degrees in the sky), that allowed the authors to pick up the dipole gradient observing the cleanest regions of the sky, despite the fact that they employed very wide-band bolometers, sensitive up to 400 microns of wavelength. In Figure 14 we have reported the results of this balloon-borne experiment where the sky was explored with a beamwidth of 5 degrees and a beam separation of 6 degrees. It is evident from the figure that the dipole behaviour is strongly contaminated. For this reason the authors concluded that the observed deviations from the dipole pattern (quadrupole-like structures) have presumably to be attributed to residual dust emission.

To improve this search the Italian Group arranged a multiband photometer: the Galactic contribution is subtracted from the FIR data by observing the same sky regions at shorter wavelengths (as described in section 3). In Figure 15 we report the signals obtained in a subsequent flight (de Bernardis et al. 1991b) after applying the correlation technique: the dipole behaviour is now evident.

Table 1
DIPOLE AND QUADRUPOLE ANISOTROPY PARAMETERS

$$T(\alpha, \delta) = T_0 + D_x \cos \delta \cos \alpha + D_y \cos \delta \sin \alpha + D_z \sin \delta$$
$$+ Q_1(3\sin^2\delta - 1)/2 + Q_2 \sin 2\delta \cos \alpha + Q_3 \sin 2\delta \sin \alpha + Q_4 \cos^2 \delta \cos 2\alpha + Q_5 \cos^2 \delta \sin 2\alpha$$

Coefficient	Amplitude (mK)			
	DMR[1]	Berkley[2]	Princeton[3]	Relict[4]
T_0(mK)	—			—
D_x(mK)	-3.24 ± 0.20^a	-3.37 ± 0.17	-3.07 ± 0.17	-3.08 ± 0.10
D_y(mK)	0.67 ± 0.13^a	0.63 ± 0.09	0.67 ± 0.09	0.57 ± 0.05
D_z(mK)	-0.40 ± 0.13^a	-0.49 ± 0.09	-0.45 ± 0.09	-0.44 ± 0.05
Q_1(mK)	-0.03 ± 0.07^b	0.21 ± 0.09	0.15 ± 0.08	-0.02 ± 0.03
Q_2(mK)	0.02 ± 0.07^b	0.27 ± 0.10	0.15 ± 0.11	0.04 ± 0.03
Q_3(mK)	-0.03 ± 0.07^b	0.16 ± 0.11	0.13 ± 0.07	-0.02 ± 0.03
Q_4(mK)	-0.05 ± 0.07^b	-0.10 ± 0.09	-0.06 ± 0.11	0.05 ± 0.02
Q_5(mK)	0.02 ± 0.07^b	0.05 ± 0.08	-0.01 ± 0.07	-0.15 ± 0.04
D^c(mK)	3.3 ± 0.2^a	3.44 ± 0.21	3.18 ± 0.21	3.16 ± 0.12
α(h)	11.2 ± 0.2^a	11.2 ± 0.2	11.2 ± 0.2	11.3 ± 0.2
δ(o)	-7 ± 2^a	-6 ± 1.5	-8 ± 2	-8 ± 2.5
Q^d_{rms}(mK)	$< 0.07^b$	< 0.2	< 0.2	< 0.08

364

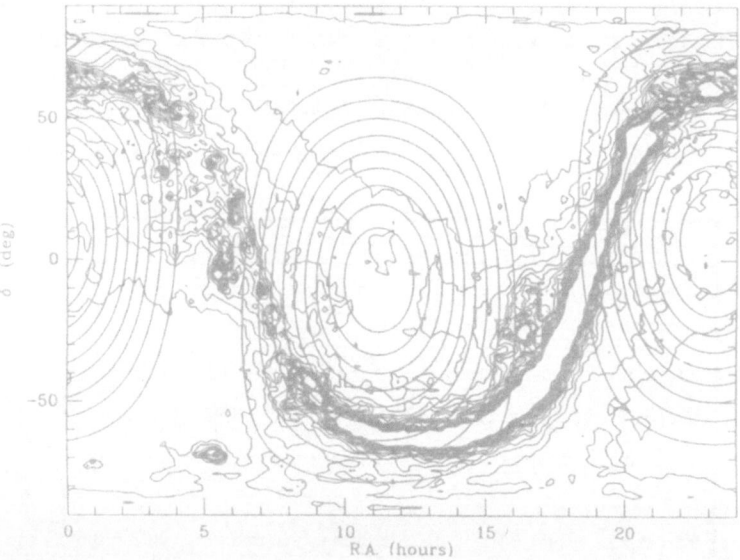

Figure 13: Isophotes of the IRAS 100 μm diffuse emission and isophotes of the CBR Dipole anisotropy.

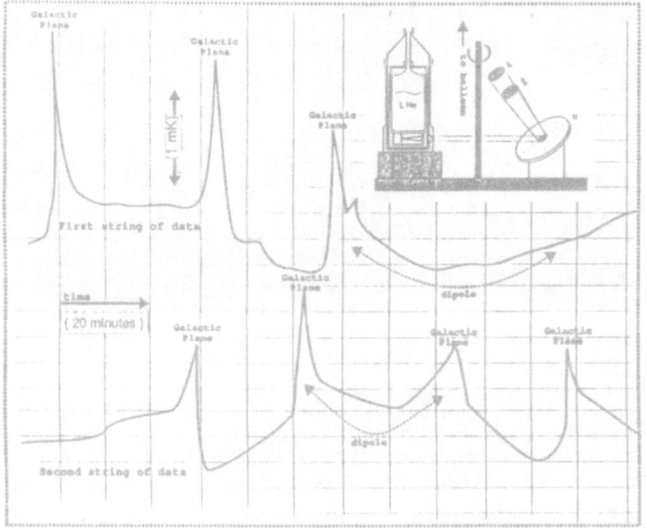

Figure 14: Sub-mm large scale anisotropy detected during flight of the italian group in 1978: the demodulated signal as plotted vs time on the chart recorder has been reproduced for two strings of data (GP = galactic plane); in the box a schematic diagram of the isotropometer is shown ($\alpha = 6°$, $\sigma = 2.2°$).

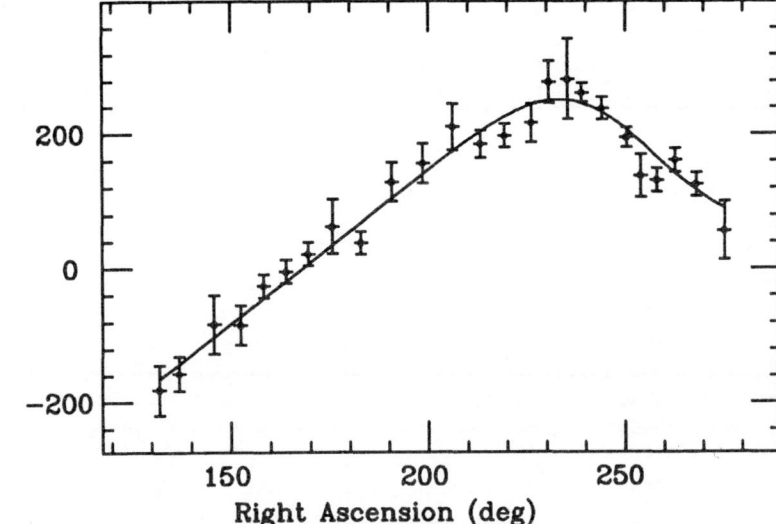

Figure 15: Large scale anisotropy of the CBR measured by a two channel isotropometer with $\alpha = 6°$, $\sigma = 2.2°$. The signals from the long wavelength channel have been cleaned for Galactic dust contamination using the signals of a short wavelength channel, as described in eq. 9 and subsequents in the text. The dipole anisotropy is evident, in good agreement with the dipole estimate obtained using the values of T_x, T_y, T_z measured by COBE (continuous line).

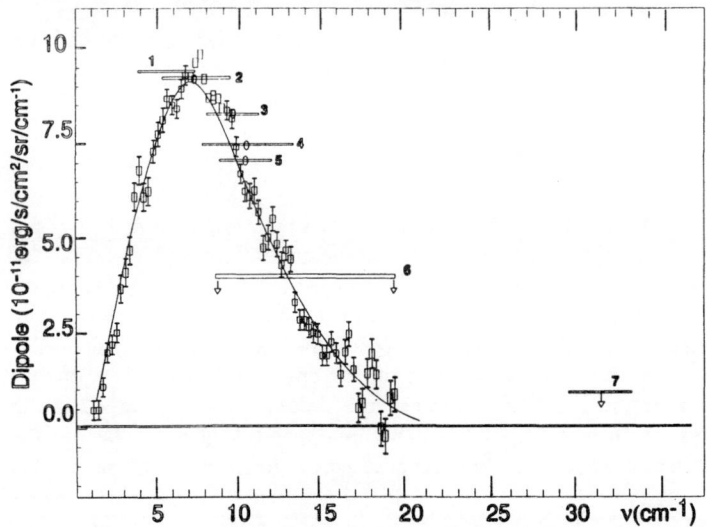

Figure 16: Dipole anisotropy of a 2.735 K Blackbody (continuous line). The experimental data have been obtained by the following CBR anisotropy experiments: COBE-FIRAS (squares); MIT (1), Bernstein et al. 1989 (2), Rome group 1977-1986 (3-7).

The situation has been greatly improved by FIRAS-COBE, where for the first time a spectrum of the Dipole Anisotropy has been obtained. In Figure 16 we have represented both the data from COBE and those relative to previous experiments.

Since the Earth motion around the Sun and the motion of the Sun in the Galaxy are known, it is not difficult to compute the motion of the Galaxy in the restframe of the LSS. The computation is a simple vector difference based on the amplitude and direction of the dipole anisotropy as measured by an Earth observer. It turns out that our galaxy is moving at the considerable speed of about 600 Km/s approximately towards the Hydra-Centaurus supercluster.

This result revitalized the interest of optical astronomers in the search for peculiar velocities, i.e. the reconstruction of the cosmic velocity field. The optical measurement of the peculiar velocities has always been affected by many systematic errors: we still do not know if our galaxy is approaching to or receding from the Virgo Cluster, the nearby huge concentration of galaxies that is supposed to influence gravitationally our Local Cluster. Despite this uncertainty, the peculiar velocity field in the Local Group and outside it up to and beyond the Virgo Cluster lies in the range from 100 to 300 Km/s , so that one is forced to conclude that the entire system, containing one thousand galaxies, is moving with respect to the last scattering surface (LSS). Therefore, while the motion of our galaxy versus LSS is a very local effect with limited cosmological relevance, the fact that our galaxy is almost at rest with respect to a large sample of other galaxies implies that this last too is moving at high speed and the dimensions of the sample are large enough to assume cosmological relevance. At present knowledge we can say that the peculiar space velocity vectors of Virgo, A1060 (Hydra I), A1367, A1656(Coma) and A2199 plus that of Local Group point generally in the same direction towards $\alpha = 10^h.9 \pm 0^h.3$, $\delta = -15° \pm 5°$ (close to the direction of the motion of our galaxy toward the LSS), with a velocity of 696 ± 96 Km/s . The cosmological relevance of these facts has been widely discussed by several authors in the present School.

4.4. THE DIPOLE ANISOTROPY AS A TOOL FOR SPECTRAL ANALYSIS

In the following we analyze the possibilities offered by the Dipole Anisotropy as an observational tool. From eq. [20] it is evident that the dipole anisotropy is sensitive to the spectral index α of the radiation, i.e. to the first derivative of the spectrum. Tiny distortions from a pure Planckian curve could result in significant changes in the Dipole Amplitude, depending on the shape of the deviation. This possibility has been experimentally exploited for the first time soon after the claim for significant spectral distortions of the CBR, raised by the Japanese group in the millimetric region. Unfortunately the deviation in this specific case tends to smooth down the planckian shape, producing a decrease in the Dipole Amplitude. This is almost completely compensated by the increase in the absolute Brightness, so that the final value of ΔI is only slightly increased. The search for this effect did not provide definite results, due to the intrinsic uncertainty in the absolute value of the dipole amplitude.

As an application of the potentiality of this technique we can investigate if the spectral distortions found in recent COBRA experiment (Gush et al. 1990, see Figure 17) are real or just due to instrumental noise, as suggested by the authors. In Figure 18 we have plotted the residuals after subtraction of a 2.735 K blackbody dipole anisotropy from the dipole data estimated using the COBRA experiment data. We also plot for comparison the same residuals in the case of the COBE-FIRAS spectrum. No evidence for significant correlation exists and we may conclude that no significant deviations of the CBR spectrum have been detected by COBRA. This conclusion only slightly improves the upper limits on CBR deviations.

Another interesting possibility offered by the Dipole Anisotropy is that of the search for emission lines at cosmological distances.

Several cosmological processes are expected to produce emission lines. Before and during the recombination phase, electronic transitions occur between H and He electronic levels: this produces spectral features in the spectrum of the Cosmic Background Radiation (CBR) (Lyubarsky and Sunyaev (1983); Zeldovich et al. (1968)). After the recombination, in the protostructures formation phase, many molecules can be formed by gas-phase reactions (Lepp and Shull 1984): they have a major role in cooling and triggering the collapse of primordial gas clouds. Due to its abundance, H_2 is particularly important (Palla et al. 1983, Shapiro and Kang 1987). Its rotational lines are excited by $H - H_2$ collisions and represent the principal cooling mechanism in the collapsing clouds. Their presence must be considered in any reasonable collapse model. The most important H_2 lines have rest wavelengths of 5.6, 6.1, 6.96, 8.02, 9.69, 12.27, 17.05 and 28.22 μm . If protoGalactic collapse took place at $5 < z < 70$, and H_2 formation took place at $z \simeq 900$ (Shchekinov and Entel 1983) these lines should now be observable in the far infrared and millimetric region. A number of other candidate primordial lines has been suggested by several authors (see e.g. Dubrovich 1977). We will consider the following species: $H, C^+, O^{++}, O, N^{++}, LiH, H_2$.

Since the strength of the cosmological lines is in general very small, their emission is hidden in the immediate background, which is dominated by Galactic dust emission.

It has been recently shown (de Bernardis et al. 1990b), that the dipole anisotropy can be used to discriminate between cosmological lines and local emission. In fact, the motion of the earth with respect to the distant matter produces a Doppler shift depending on the angle θ between the line of sight and the Earth velocity direction (i.e. the direction of the maximum of the CBR Dipole). On the other hand local (Galactic) lines have amplitudes which do not depend on the angle θ: as a result, a best fit procedure can be used to separate the cosmological signal from the local effects.

Due to the relatively narrow spectra of the line features, the dipole anisotropy of the cosmological lines could have an amplitude comparable to the absolute line emission and a peculiar spectral shape. The details of the spectral shape allow the study of the evolution processes which must have occurred during the epoch of structure formation.

The spectrum of the background observed today $i(\nu_o)$ at the observation frequency ν_o can be written

$$i(\nu_o) = \frac{c}{4\pi H_o} \int_0^{+\infty} \frac{L(\nu, z)E(z)dz}{(1+z)^5(1+\Omega z)^{1/2}} \qquad [29]$$

where $L(\nu, z)$ is the intrinsic luminosity density at the epoch z ; $E(z)$ is a weighting function which takes into account the evolution of the sources. The other constants have the usual meaning.

The large scale anisotropy due to the Doppler effect can be obtained from the spectrum using eq. [20]. If the spectral profile is quite narrow (and for $\beta \sim 1.3\cdot10^{-3} << 1$) significant dipole and quadrupole anisotropies are expected. In Figure 19 we show the spectra of

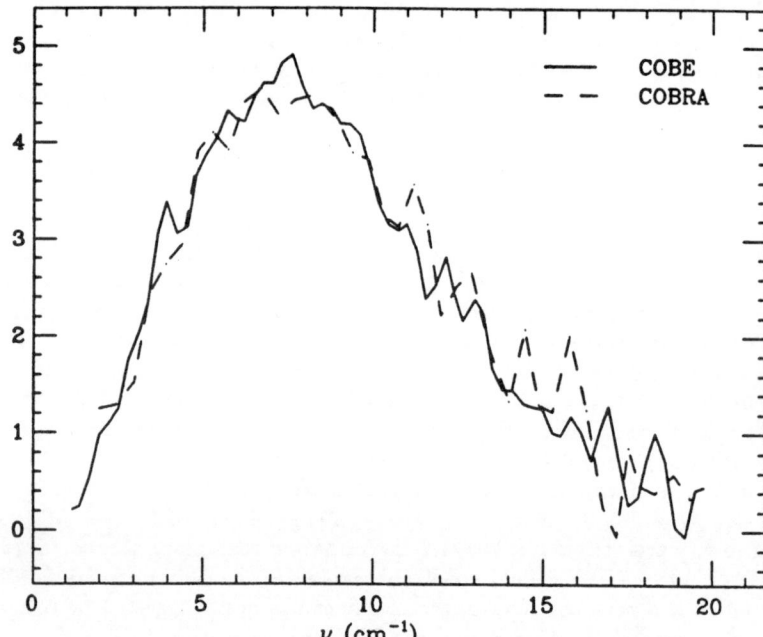

Figure 17: Dipole anisotropy spectrum as measured by COBE-FIRAS (continuous line) and estimated using the data of COBRA and equation [20] (dashed line).

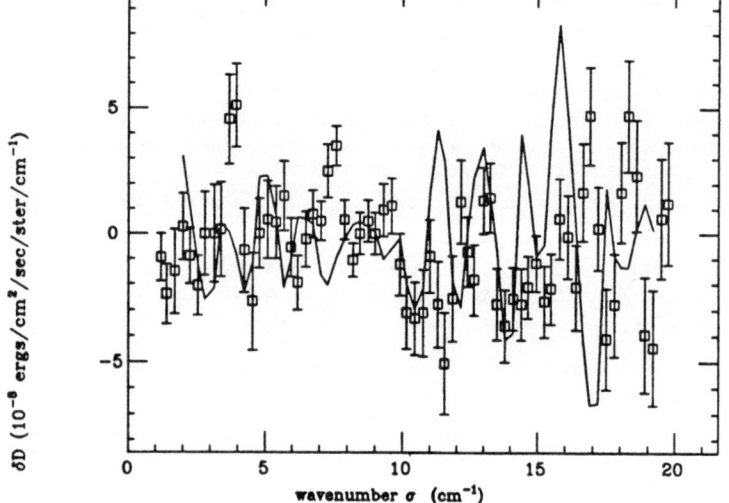

Figure 18: Residuals of the data of fig. 17 after subtraction of the Dipole anisotropy expected from a 2.735 K blackbody. Void squares refer to COBE-FIRAS while the continuous line refers to COBRA.

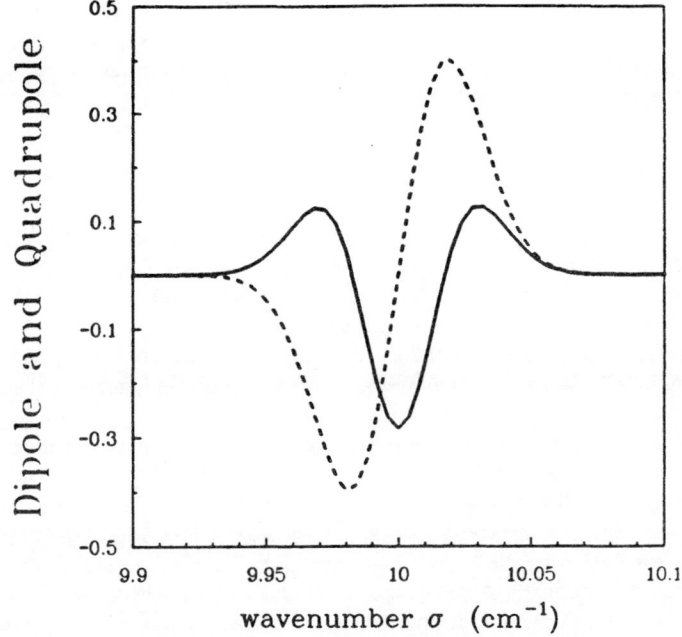

Figure 19: Dipole (dashed line) and quadrupole (continuous line) anisotropy data due to the Doppler effect on a gaussian line with center wavenumber $\sigma_o = 10 cm^{-1}$ and velocity dispersion 500 Km/s. The Dipole and Quadrupole are plotted in units of the maximum absolute brightness of the line.

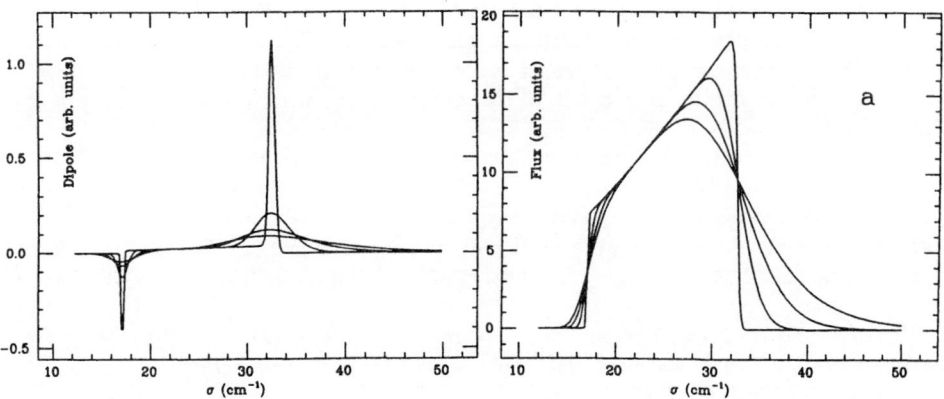

Figure 20: Absolute flux and Dipole anisotropy produced by H_2 with formation at $z \sim 20$ and destruction at $z \sim 10$, with Δz ranging between 0.1 and 2 and velocity dispersion ~ 300 Km/s.

both these terms in the case of a gaussian-like line profile $i(\nu)$. However, reasonable cosmological lines spectra are expected to be smoother, due to the wide redshift interval of the cosmological emission; in this case the spectrum of the dipole anisotropy can be computed using the first order expansion

$$D(\nu_o) = i(\nu_o)\beta(3 - \alpha) = \beta\left[3i(\nu_o) - \nu_o\frac{di(\nu_o)}{d\nu_o}\right] \qquad [30].$$

In Figure 20 we show the spectrum and the dipole anisotropy produced by H_2 with formation at $z \sim 20$ and destruction at $z \sim 10$, with Δz ranging between 0.1 and 2 and velocity dispersion $\sim 300Km/s$. It is interesting to note that the intensity spectrum becomes smoother and smoother as the formation and destruction rates decrease. The cut-on and cut-off in the absolute emission produce respectively a negative and positive line in the dipole spectrum. This is a very peculiar signature which allows discrimination between the cosmological line dipole and the spurious dipole produced by local components. The line is very sharp if the H_2 formation time is very short, and very fast hydrogen formation is expected (Shapiro and Kang 1987).

In order to search for the dipole anisotropy of cosmological lines, a far infrared spectrometer is needed. Spectra observed in regions with very low local emission must be compared, with a resolution of the order of $\sim 1\%$, in the spectral range $100 \sim 1000$ μm. Due to the presence of strong IR Atmospheric lines this experiment should be performed from a satellite platform. The measurement should be carried out with large beams. In this way line emission from interstellar molecules, which is associated to small size sources, would be diluted, while the signal from cosmological lines is integrated, greatly improving the possibility of a detection.

In order to avoid dynamic saturation due to the continuum spectrum of interstellar dust, selected couples of sky regions should be compared, where interstellar dust emission is minimum (see de Bernardis et al. 1990b).

The FIRAS experiment on the COBE satellite (Mather 1986) is a far infrared interferometer. In this instrument the radiation coming from one sky direction is compared to a cold blackbody reference. The expected calibration stability and reliability are so good ($\sim 3/10000$) that it should be possible to search for dipole anisotropies from lines in the FIRAS data.

As an example of this procedure we have already shown in Figure 16 the difference between two COBE-FIRAS spectra taken in two directions $180°$ apart: the dipole corresponding to the difference between two blackbodies (due to the Doppler shift of the CBR photons) is evident, but residual statistically significant structures are also evident. These sharp features can be fitted using a combination of local and cosmological lines (see Figure 21).

However, this is only an example of the search for cosmological lines. What should be done using the full FIRAS data-set is a dipole best fit for each frequency bin, using all the sky regions where the ratio between the dipole and the local emission is maximum. This procedure efficiently averages to zero the local emission, while enhancing the cosmological lines dipolar contributions (see de Bernardis et al. 1990b).

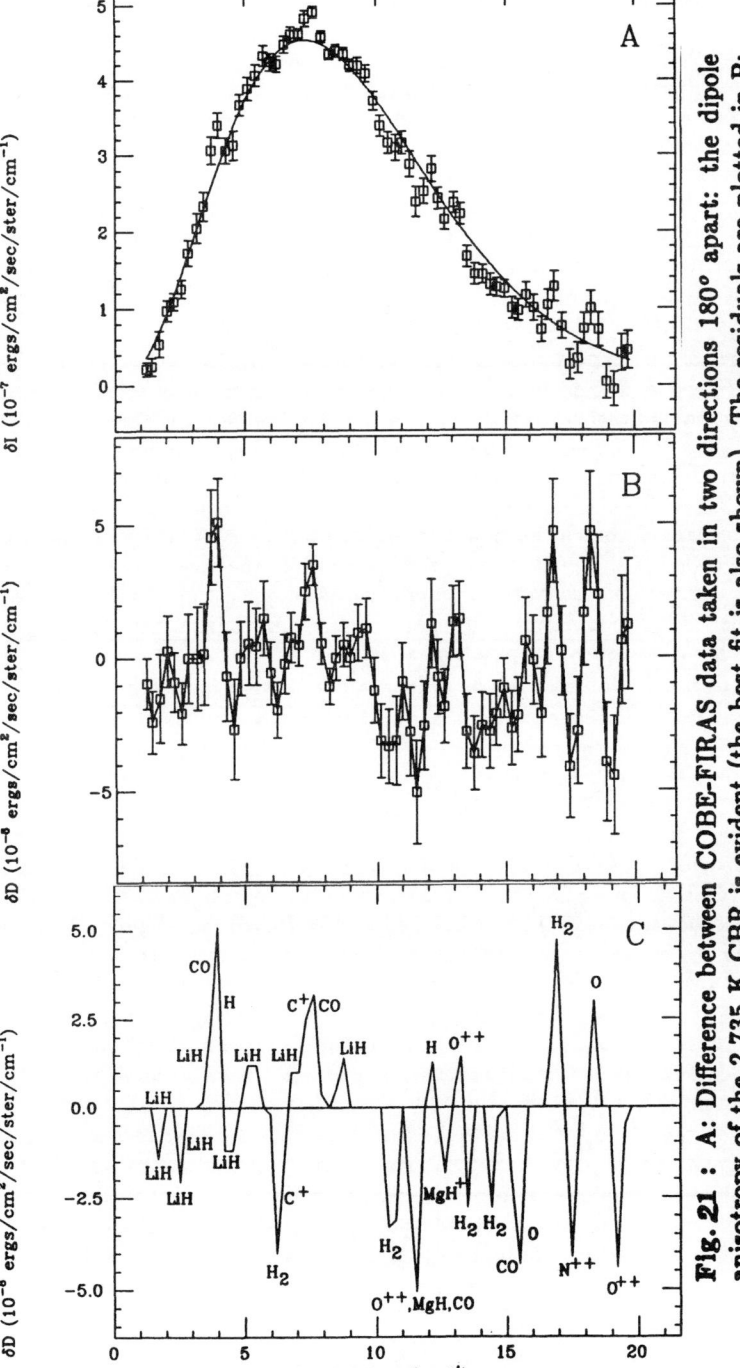

Fig. 21 : A: Difference between COBE-FIRAS data taken in two directions 180° apart: the dipole anisotropy of the 2.735 K CBR is evident (the best fit is also shown). The residuals are plotted in B: several peaks exceed significantly the detector noise. In C we plot an emission spectrum obtained using lines with the following parameters: (species, rest wavenumbers; formation z, destruction z.) CO, 3.85, 7.69, 11.53, 15.38; 0, 0. MgH, 11.46; 0, 0. MgH⁺, 12.59; 0, 0. C⁺, 63.4; 9, 7.6. O⁺⁺, 113.1, 193.1; 9, 7.6. O, 156.5; 9, 7.6. N⁺⁺, 174.5; 9, 7.6. LiH, 14.81, 29.6, 44.35, 59.03, 73.63; 16, 7.6. H₂, 1779, 5463, 9357, 11835, 12655; 900, 100. H Lyman-α and Balmer lines (Lyubarsky 1983).

5. The mm and submm Sky at Intermediate and Small Angular Scales ($\lesssim 2°$)

COBE-DMR has completed three low resolution maps of the millimetric sky while COBE-FIRAS has completed a full-sky spectral survey in the range between 1 cm and 100 μm. These maps are a standard of reference for all the future searches, due to their completeness and precise calibration. The angular resolution of these maps is 3.5°. The investigation of smaller angular scales is being pursued using non-satellite methods, mainly by means of balloon experiments in the mm and sub-mm region. In this section we will discuss which is the rationale for this search and which is the best experimental strategy; in particular we will concentrate our attention to the search for CBR anisotropies and we will investigate how to maximize the observable while minimizing the spurious noise sources.

Several physical processes are expected to produce mm and submm diffuse radiation at high Galactic latitudes. Since we do not have detailed high resolution maps of these emissions, but only a few statistical indications of their distribution, we will describe the diffuse celestial radiation Brightness $I(\vec{\gamma})$ (where $\vec{\gamma}$ is the observation direction) in terms of its smeared correlation function: $C(\theta, \tilde{\sigma}) = \langle \Delta I(\vec{\gamma}) \Delta I(\vec{\gamma}') \rangle$. Here $\vec{\gamma} \cdot \vec{\gamma}' = \cos\theta$, and $\tilde{\sigma}^2$ is the variance of the gaussian which best fits the beam profile of the experiment ($\tilde{\sigma} = 0.425°$ FWHM). We will also use the normalized correlation function $R(\theta, \tilde{\sigma}) = C(\theta, \tilde{\sigma})/C(0, \tilde{\sigma})$. R describes the correlation properties of the fluctuating Brightness, while the variance of Brightness fluctuations $(C(0, \tilde{\sigma}))$ gives an idea of their amplitude. The anisotropy level depends on the differencing order of the modulation. For example, the r.m.s. signal detected in a single difference experiment with beam separation α is $\langle \Delta I^2 \rangle^{1/2} = \sqrt{2[C(0, \tilde{\sigma}) - C(\alpha, \tilde{\sigma})]} = \sqrt{C(0, \tilde{\sigma})} \cdot \sqrt{2[1 - R(\alpha, \tilde{\sigma})]}$. In the case of a double difference experiment $\langle \Delta I^2 \rangle^{1/2} = \sqrt{1.5C(0, \tilde{\sigma}) - 2C(\alpha, \tilde{\sigma}) + 0.5C(2\alpha, \tilde{\sigma})} = \sqrt{C(0, \tilde{\sigma})} \cdot \sqrt{1.5 - 2R(\alpha, \tilde{\sigma}) + 0.5R(2\alpha, \tilde{\sigma})}$. A dual approach is based on the use of the power spectrum $P(q)$ of the temperature perturbations on the celestial sphere (q is a 2-D angular wavenumber): the contribution to the total mean square Brightness fluctuation from the wavenumbers in a logarithmic wavenumber interval around q is $q^2 P(q)$. The signal detected by an anisotropy experiment is a weighted sum of these contributions, the weighting function being, in the case of a single difference experiment, $F(q, \alpha, \tilde{\sigma}) = 2exp(-q^2\tilde{\sigma}^2)[1 - J_o(q\alpha)]$, while in the case of double difference experiments $F(q, \alpha, \tilde{\sigma}) = exp(-q^2\tilde{\sigma}^2)[1.5 - 2J_o(q\alpha) + 0.5J_o(2q\alpha)]$. In Figure 22 we plot the filter functions F of several current and planned anisotropy experiments. It is worth noting that the maximum of the Filter function is placed at angular scales similar to the beamsize ($q_{max} \sim 1/\tilde{\sigma}$), while the maximum value of the Filter function increases as the ratio $\alpha/\tilde{\sigma}$ increases: for this reason experiments with relatively large modulation amplitude will be in general more sensitive to Brightness anisotropies. In the caption of Figure 22 we list the relevant parameters of these experiments. Theoretical models give the correlation function (or the power spectrum) of the Brightness pattern in the sky; by means of the above described formalism it is possible to optimize the experimental parameters (field of view, modulation, sky scan strategy) in order to maximize the sensitivity to the target of the experiment, while minimizing the effects of other 'spurious' sources. We will describe in detail in the following five sources of mm and submm diffuse radiation at high Galactic Latitudes.

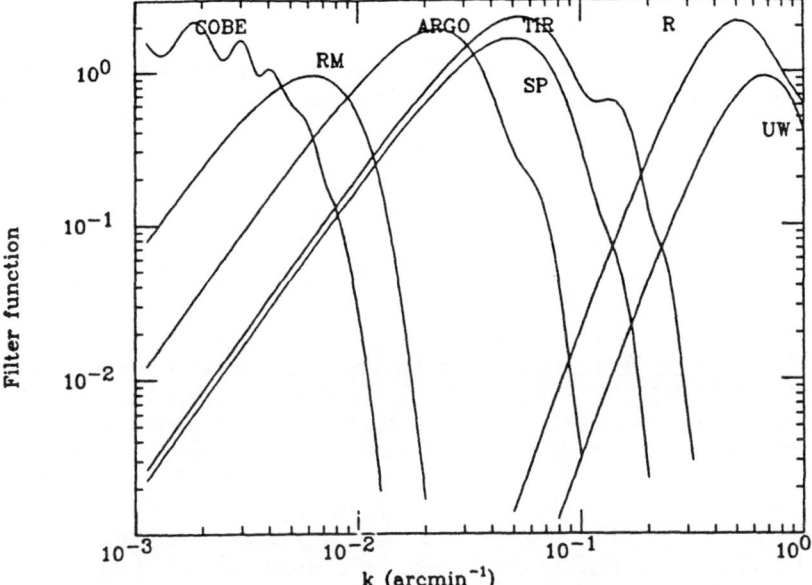

Figure 22: Filter functions of several CBR anisotropy experiments: the parameters are listed as follows: Label (Experiment), α, σ, single (2) or double difference (3) modulation: COBE (DMR), 5400, 210, 2: RM (de Bernardis et al. 1991b), 360, 130, 2: ARGO (de Bernardis et al. 1990), 140, 25, 2: TIR (Melchiorri et al. 1990), 65, 8, 2: SP (Meinhold and Lubin 1991), 60, 13, 2: R (Readhead et al. 1990), 7, 0.7, 3: UW (Uson and Wilkinson 1986).

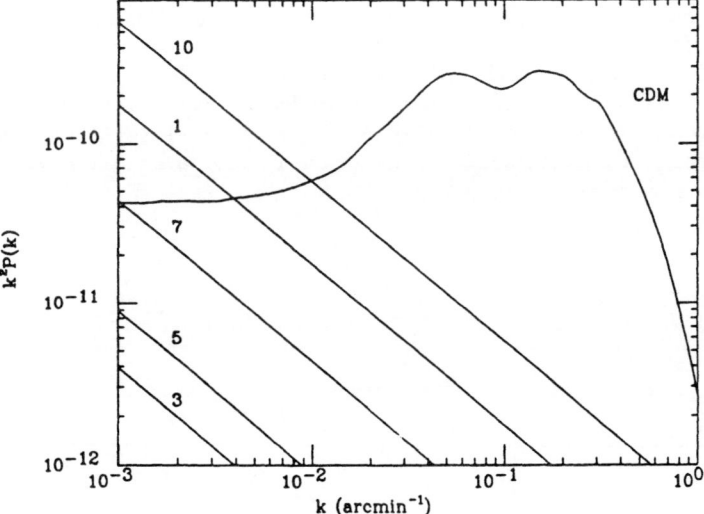

Figure 23: 2-D Power spectra of the CBR anisotropies (CDM model) and of the Galactic diffuse emission (labeled with the observation wavenumber in cm^{-1}). The best observation wavelength and angular scales can be obtained from this plot (in this case, angular scales around 10 arcmin, and wavelengths around 2-3 mm give the maximum ratio between CBR and Galactic anisotropies).

5.1. CBR ANISOTROPIES

The remarkable agreement between the CBR spectrum and a pure black-body excludes the presence of hot gas interacting with CBR photons after the recombination: small and intermediate scale anisotropies related to density perturbations on the last scattering surface are not smeared out and could be measured by sensitive enough experiments.

Small and intermediate angular scale experiments are sensitive to the most interesting part of the power spectrum of primordial density perturbations. In fact, the angular scale θ at recombination includes a sphere with a mass $M(\theta) = 1.4 \cdot 10^{12} [\theta(')]^3 (\Omega^2 h)^{-1} M_\odot$, so that observations carried out at angular scales between several arcminutes and $3°$ provide informations about the mass fluctuations between $10^{13} M_\odot$ and $10^{17} M_\odot$, which are of greatest relevance to galaxy formation (see the lecture of M. Turner in these proceedings).

Upper limits on fractional fluctuations in the thermodynamic temperature of the CBR are currently of the order of few parts in 10^{-5}, at intermediate and small angular scales (Meinhold and Lubin 1991, Readhead et al. 1989; see Vittorio and Muciaccia 1991 for a comparative analysis). A true detection should be of primary interest, because it should effectively test the theories of Galaxy formation, e.g. CDM with or without the cosmological constant; baryonic dark matter with isocurvature fluctuations; explosive Galaxy formation.

In general, the comparison between small and large scale anisotropies (mainly the quadrupole), will give informations on the 'transfer function' of the density perturbations, and, in turn, on the kind of dark matter governing the dynamics of the Universe (see e.g. the lecture of V. Lukash).

The physical processes producing CBR anisotropies from density perturbations depend on the angular scale: the adiabatic fluctuations are dominant at scales $\lesssim 30'$; the Doppler shift produced by moving clouds of electrons at the last scattering surface is important at scales $\sim 1°$; the Sachs-Wolfe effect (gravitational redshift of photons climbing the potential well of density perturbations) is dominant at large angular scales. Based on these processes, the Brightness of CBR emerging from the decoupling era has been computed numerically (Vittorio and Muciaccia 1991). The resulting angular power spectrum of CBR anisotropies in the CDM model is plotted in Figure 23: at angular scales smaller than $3'$ CBR anisotropies are strongly smeared out due to the finite thickness of the last scattering surface; the peak at around 10 arcmin is due to the Doppler effect, while the behaviour $P(q) \propto q^{-2}$ at large scales is characteristic of the Sachs and Wolfe effect.

In the case of a single difference modulation pattern the *observable* is *not* $\Delta T/T$: instead, it is the *r.m.s.* fluctuation in CBR Brightness ΔI, which is related to $\Delta T/T$ by

$$\Delta I = \frac{\Delta T}{T} \sqrt{2[1 - R(\alpha, \tilde{\sigma})]} A\Omega \int_{band} B(\sigma, T_{CBR}) \frac{xe^x}{e^x - 1} d\sigma \qquad [31]$$

or, in terms of power spectrum and filter functions,

$$\Delta I = \frac{\Delta T}{T} \sqrt{\frac{\int q^2 P(q) F(q, \alpha, \tilde{\sigma}) d\ln q}{\int q^2 P(q) d\ln q}} A\Omega \int_{band} B(\sigma, T_{CBR}) \frac{xe^x}{e^x - 1} d\sigma \qquad [32].$$

From this formula it is evident that in order to achieve high sensitivity one should maximize the throughput, the bandwidth and the ratio between modulation amplitude and beam-size. Moreover, the beam-size should be selected in order to match the peak of the CBR anisotropies power spectrum, which, in the CDM model, can be accomplished by using $\tilde{\sigma} = 10' \sim 1°$.

The expected r.m.s. fluctuations of the sky Brightness due to CBR anisotropies are listed in Table II, col. 1. By taking advantage of the wide band and multimode operation of bolometers, it is possible to build up large throughput telescopes able to detect anisotropies at a level $\sim 10^{-5}$ in few seconds of integration.

The observation of CBR anisotropies at small angular scales is usually performed observing a limited number of sky regions at high Galactic latitudes. It has been shown that for a given total integration time and a given detector noise, the best compromise between number of observed fields and integration time for each field is obtained by observing about 10 independent fields (Radford 1985, Readhead et al. 1989). Although statistically rigorous, this result has been obtained without taking into account the presence of foreground emission, which becomes more and more important as the sensitivity of the anisotropy experiments increases. In general, a large number of observed fields is essential for understanding if and which kind of foreground emission is present in the data.

5.2. ANISOTROPY INDUCED BY EXTRAGALACTIC DUST

InterGalactic gas clouds have been detected and there is some weak evidence of nearby interGalactic dust clouds (Wszolek et al. 1989) which is supported by calculation of dust ejection caused by radiation pressure in the galaxies (Greenberg et al 1987). Due to the low temperature of these dust clouds, this hypothesis could be tested effectively only by using a sub-mm survey experiment with an angular resolution $\lesssim 1°$. Moreover, primeval dust could have been produced by the first astrophysical objects evolved soon after the recombination. A distortion of the CBR spectrum should be produced by energy release at high redshifts: preGalactic and protoGalactic sources emit optical and UV radiation which could be absorbed by primordial dust and re-emitted at lower frequencies. This process results in excess emission in the submillimeter region of the CBR spectrum.

Models for dust emission have been developed by several authors (Adams et al 1990, Bond et al. 1991) according to the assumptions of interstellar-like grains (graphite, silicate and amorphous carbon) and power-law FIR dust emissivity: the total amount of emission, the peak-wavelength and the details of spectral behaviour depend mainly on input energy, dust density and optical depth, while the evolution of the sources has a negligible influence.

The recent results of the COBRA experiment (Gush et al. 1991) and of COBE-FIRAS (Mather et al 1990) strongly constrain these models: scenarios with relatively high energy release ($u_{s,T} \geq 0.04 u_{cbr}$) and relatively high dust density ($\rho_d \geq 10^{-5}\rho_{cr}$) are currently ruled out; the expected sensitivity at the end of the COBE mission ($\simeq 10^{-3} I_{pk,cbr}$) allows the testing of some other models, while short wavelength anisotropy experiments could have comparable or higher chances of testing all the reasonable models. As an example we list in Table II, col.2, the sky Brightness anisotropy expected from model M13 of Bond et al. 1991.

5.3. S-Z EFFECT FROM UNRESOLVED CLUSTERS OF GALAXIES

The S-Z effect (Sunyaev and Zeldovich, 1972) is the inverse Compton scattering of Cosmic Background Radiation photons travelling in the hot plasma present in Clusters of Galaxies.

Table II

RMS brightness anisotropy at mm and submm wavelengths from several different sources, at an angular scale of 12'. The units for the brightness are $W/cm^2/sr/cm^{-1}$. CBR = Cosmic Background Radiation (CDM standard model); SZ = Sunyaev-Zeldovich effect in unresolved clusters of Galaxies; ED = Extragalactic dust (Bond model M13); UG = unresolved evolving galaxies; G = cirrus dust and radio continuum emission.

wavenumber (cm^{-1})	CBR	SZ	ED	UG	G
5	3.2×10^{-16}	1.1×10^{-16}	1.3×10^{-15}	2.0×10^{-17}	1.0×10^{-17}
7	3.8×10^{-16}	-	1.0×10^{-15}	3.5×10^{-17}	3.0×10^{-17}
10	2.9×10^{-16}	1.3×10^{-16}	6.0×10^{-16}	1.0×10^{-16}	1.1×10^{-16}
15	1.0×10^{-16}	1.2×10^{-16}	2.5×10^{-16}	2.5×10^{-16}	2.5×10^{-16}
20	2.0×10^{-17}	4.2×10^{-17}	8.5×10^{-16}	3.3×10^{-16}	6.6×10^{-16}

Table III

Anisotropy of cosmic dust backgrounds: Bond et al. (1991) predictions for the ARGO 1989 configuration. The Brightness fluctuations are listed in $(W/cm^2/sr)$.

clustering	M1	M2	M3	M4	M5	M6	M7
Poisson	$2.6 \cdot 10^{-14}$	$6.1 \cdot 10^{-15}$	$2.6 \cdot 10^{-14}$	$7.2 \cdot 10^{-15}$	$5.0 \cdot 10^{-15}$	$6.7 \cdot 10^{-15}$	$2.8 \cdot 10^{-14}$
Fractal	$1.3 \cdot 10^{-13}$	$2.3 \cdot 10^{-14}$	$1.3 \cdot 10^{-13}$	$2.6 \cdot 10^{-14}$	$2.0 \cdot 10^{-14}$	$2.5 \cdot 10^{-14}$	$1.3 \cdot 10^{-13}$
CDM	$1.1 \cdot 10^{-14}$	$2.0 \cdot 10^{-15}$	$1.0 \cdot 10^{-14}$	$2.3 \cdot 10^{-15}$	$4.7 \cdot 10^{-15}$	$5.4 \cdot 10^{-15}$	$9.4 \cdot 10^{-15}$
clustering	M8	M9	M10	M11	M12	M13	M14
Poisson	$1.5 \cdot 10^{-14}$	$2.3 \cdot 10^{-17}$	$5.1 \cdot 10^{-17}$	$2.8 \cdot 10^{-16}$	$6.4 \cdot 10^{-14}$	$2.1 \cdot 10^{-16}$	$4.5 \cdot 10^{-16}$
Fractal	$2.7 \cdot 10^{-14}$	$2.0 \cdot 10^{-16}$	$2.2 \cdot 10^{-16}$	$2.2 \cdot 10^{-15}$	$1.1 \cdot 10^{-13}$	$7.5 \cdot 10^{-16}$	$2.8 \cdot 10^{-15}$
CDM	$4.4 \cdot 10^{-15}$	$4.8 \cdot 10^{-17}$	$4.7 \cdot 10^{-17}$	$1.8 \cdot 10^{-16}$	$2.7 \cdot 10^{-14}$	$1.1 \cdot 10^{-16}$	$2.3 \cdot 10^{-16}$

The *observable* Brightness change in CBR with respect to its average Brightness is given by

$$\Delta I = 2y \int_{band} B(\sigma, T_{CBR}) \frac{xe^x}{(e^x - 1)} \left[\frac{x/2}{tanh(x/2)} - 2 \right] d\sigma \quad ; \quad \sigma = 1/\lambda \quad ; \quad x = \frac{hc\sigma}{kT_{CBR}} \quad [33]$$

with

$$y = \int_{l.o.s} \frac{kT_e}{mc^2} n\sigma_{Th} dl \quad . \quad [34]$$

where *l.o.s.* stands for line of sight and σ_{Th} is the Thomson cross section. This effect has been observed in the radio region in a few Clusters (Birkinshaw 1990). Due to their short wavelength band, submm space experiments should be very sensitive to the positive part of the S-Z effect when the configuration has a beam-size comparable to the dimensions of clusters. Using larger beam-sizes will dilute the S-Z signal from the single cluster, but will increase the likelihood of having many clusters in the line of sight. In fact, the S-Z Brightness is independent of Cluster redshift (as long as the parameters of the plasma are constant *i.e.* evolution is absent). This fact suggests that anisotropy from unresolved Clusters of Galaxies can be quite important in the mm region, due to the S-Z effect. This has been estimated in the past (Rephaeli 1981), and re-estimated by using more recent data on the distribution of rich Clusters and on the parameters of the gas (Cole and Kaiser 1988). The effect is of the order of $\Delta T/T_{r.m.s.} \sim 10^{-5}$ at angular scales of the order of $1' \div 10'$ for CDM models; larger effects have been computed for different models, with a typical scale of the fluctuations of the order of 10' (Sunyaev 1990). Since the spectrum of the S-Z effect has a broad positive peak (between 8 and 24 cm^{-1}), one should observe it using a broad band sub-mm photometric channel and a mm channel (below 8 cm^{-1}) to detect the negative effect. In this way, the *observable* Brightness fluctuation $\Delta I_{r.m.s.}$ in the submm channel should be of the order of 10^{-15} W cm^{-2} sr^{-1} (see Table II, col. 3), well above detector noise after a few seconds of integration. The measurement of this anisotropy could allow the study of the evolution of hot gas in the clusters of galaxies in its earliest stages.

5.4. ANISOTROPY FROM UNRESOLVED GALAXIES

IRAS observations have shown that galaxies have a very high luminosity in the far infrared, with a local luminosity density of the order of one third of the optical luminosity density (Soifer et al., 1986, Saunders et al. 1990). Moreover, the sub-mm and far IR are negligibly absorbed in the Universe: these two facts strongly support the hypothesis of the presence of a far infrared background due to unresolved Galaxies. This idea is not new, and has been proposed several times in the past (Tinsley 1973, Stecker et al. 1977, Bond et al. 1986, 1991). Detailed models have recently been developed by Beichman and Helou (1991) and by Wang (1991), taking into account evolutionary effects in galaxies with significant dust emission. A strong density evolution ($\propto (1 + z)^4$), with cut off at $z_{max} \sim 5$ is ruled out by the COBE-FIRAS spectral results, while more modestly evolving models cannot be distinguished from Galactic dust emission or y-distortions of the CBR spectrum (Wright 1991; see also these proceedings). For these reasons it seems quite unlikely that absolute photometry experiments will be able to detect this far IR diffuse background. On the other hand, *anisotropy* caused by unresolved galaxies should be detectable above the detector noise and Galactic disturbance. Franceschini et al. (1989) computed the level of anisotropy ΔI expected from a random distribution of galaxies; Wang (1991) took into account the correlation properties of the distribution of galaxies and computed the r.m.s. fluctuation expected in a single-beam experiment. The expected r.m.s. level of the anisotropy produced in Wang's models is listed in Table II, col.4. Due to the different shape of correlation function R and of the spectrum, it should be possible to disentangle effectively this sky anisotropy from the others: in this way, it should be possible to have a general look at the early phases of Galaxy formation.

5.5. INFRARED CIRRUS

The "infrared cirrus" is observed in the IRAS 60 and 100 μm maps as filaments and clouds across almost the entire sky: balloon and satellite experiments are the only way of mapping these faint textures of the interstellar medium. While emission from transiently heated grains is present at wavelengths shorter than 100 μm, submm experiments will mainly map the more massive components of the cirrus clouds, i.e. 'normal' or 'fractal' grains with a temperature lower than 20 K. At high Galactic latitudes cirrus emission can represent an important obstacle in the observation of extraGalactic backgrounds, also in the 'clean' sky regions defined in Masi et al. (1991). In these regions, the r.m.s. fluctuations of IRAS 100 μm data are $\lesssim 0.2$ MJy sr^{-1} at 100 μm. At larger wavelengths, radio continuum emission from electrons becomes relevant. A certain level of correlation between dust emission and radio continuum has been found; we will assume that in order to have an order of magnitude estimate of the r.m.s. Brightness fluctuations, IRAS 100μm data map both the components. Under this hypothesis, in the 'clean' regions, these emission fluctuations simulate CBR anisotropies at a level of $\sim 4 \cdot 10^{-6}$ at 10cm^{-1}, $\sim 2 \cdot 10^{-6}$ at 5cm^{-1}, $\sim 1 \cdot 10^{-6}$ at 3cm^{-1}, $\sim 4 \cdot 10^{-6}$ at 1cm^{-1} (see also Table II, col. 5). The minimum emission is found between 2 and 7 cm^{-1}. Multi-frequency experiments can be used to go below these limits if the spectrum of the extraGalactic background which has to be observed is significantly different from the Galactic dust emission spectrum (de Bernardis et al. 1991a). The normalized correlation function of Cirrus can be computed from the power spectrum $P(q) \propto q^{-3}$ given by Gautier et al. (1991): $R(\theta)$ is quite different from that of extraGalactic sources: this will help in disentangling the different contributions in sub-mm maps obtained at high Galactic latitudes. In Figure 23 we show a comparison between the bidimensional power spectrum of Galactic high latitude emission and the power spectrum of CBR anisotropies in the standard CDM model. It is evident that at angular scales larger than 2° and at frequencies outside the $2 - 7cm^{-1}$ spectral window, Galactic emission is dominant, and multifrequency experiments are needed to extract the true CBR anisotropies. On the other hand, at angular scales smaller than 2°, Galactic contamination becomes negligible (in the 'clean' sky regions), while CBR anisotropies produce stronger signals. This fact strongly supports the study of angular scales between 2° and 10', and leaded us to start a research program devoted to the search for CBR anisotropies at angular scales smaller than 2°, in the mm and submm range.

6. The Argo and TIR Programs

We have shown in the previous paragraph that a large ratio $\alpha/\bar{\sigma}$ is convenient for effective surveys of CBR anisotropies and diffuse mm backgrounds. On the other hand, a reduction in $\tilde{\sigma}$ should reduce the system throughput, thus reducing the sensitivity to diffuse emission (as CBR anisotropies are). It is evident from this argument that the size of the telescope should be as large as possible in order to have both a large throughput and a reasonable angular resolution: diffraction limited performances are not useful in this case, while broad band bolometric detectors are preferred. In 1987 we started the development of a stratospheric balloon telescope. We flew a prototype (ARGO) with a 1.2 primary telescope and we are currently completing the 2.6 m system (TIR). The optimization of the TIR and ARGO detectors and telescopes was performed following the previous fundamental lines: maximum throughput and intermediate angular resolution; high quality off axis optical performances; mm and sub-mm operation wavelenghts; sensitivity to faint, diffuse emission gradients.

6.1. THE DETECTORS

Two operating strategies are possible for bolometric detectors: the first one is the minimization of both the radiative background and the sensor temperature, in order to maximize the responsivity; the second one is to operate at relatively high temperature ($\sim 0.3K$) but with a large throughput in order to maximize the signal accepting a non negligible background. Since we want to study diffuse radiation, we have decided to follow the second strategy, because in our case the signal enhancement due to a large throughput fully compensates the responsivity reduction due to the increased background. For the ARGO and TIR programs we have developed a multiband photometer with $A\Omega = 0.62 cm^2 sr$ and four wavelength bands ranging from 300 μm to 2 mm. Each spectral band observes the same field of view through a high efficiency network of beam splitters and filters (see de Bernardis et al. 1990c for details). The main problem in these large throughput photometers is the large amount of background radiation entering the input window: due to the large throughput, in fact, the size of the optical beam is very large and difficult to split, and care must be taken in efficient filtering of the radiation and cooling of the blocking filters. In any case these problems can be solved efficiently: large throughput bolometers have been used succesfully in the MIT balloon photometer (Page et al. 1990, Meyer et al. 1991) and in COBE-FIRAS (Mather et al. 1990).

6.2. THE TELESCOPES

We have developed two large throughput Cassegrain telescopes (De Petris et al. 1989). Both the systems operate with the same photometer. The telescope troughput is larger than the detector throughput, because a guard ring is obtained on the primary mirror, in order to improve the sidelobes rejection of the system and to avoid any offset from thermal gradients in the mirror background. Moreover, a high precision wobbling secondary mirror allows sky modulation (de Bernardis et al. 1989). It is based on two electrodynamic vibrators working in push-pull and mounted using a torque cancelling design. The secondary mirror is oversized: the edge of the secondary is at the 10% rejection level of the photometer beam-shape. In spite of the small primary $f_\#$ (0.5), the high magnification factor, and the high throughput, we obtain a large chopping angle in the sky (30' for the 2.6 m and 140' for the 1.2 m), which is essential for good modulation efficiency. The primary mirrors were obtained from a single aluminum fusion, with a $5mm$ skin and a stiffening back structure. The surface is polished up to a surface roughness of about 5 μm. The weight is slightly below 100 Kg for the 2.6 m and 35 Kg for the 1.2 m. The main problem in the search for anisotropies of the diffuse radiation is the variable offset due to thermal gradients on the primary mirror. For this reason we have protected the telescope from earth, moon and sun radiation by means of an active screen, which is based on an aluminum skeleton supporting a 1 mm alluminum skin, eight circular freon heat pipes, a 10 cm thick glass wool layer, and aluminized mylar covering the outer surface.

6.3. THE POINTING SYSTEM

The first pointing goal is an accurancy of few *arcmin* r.m.s.; in this case a two axis magnetometer is sensitive enough for azimutal pointing, which is accomplished by rotating the entire payload. Rotation of the gondola is obtained by means of an inertial wheel, while the gondola is insulated from the random rotation of the cruise balloon by means of an active mechanical Pivot. This is strategically placed at the interconnection point between the balloon and the payload and houses the necessary suspension bearings. A certain leakage torque is transmitted through the friction of the bearings. This is reduced and controlled simply by rotating continuously the Pivot by means of a torque motor. Zenithal pointing is obtained by means of a simple endless screw system. The system includes a CCD camera for the absolute pointing reconstruction. The second step in pointing accuracy will include a camera for star tracking, generating two error signals which will be used as feedback signals for both the azimuth and elevation loops. A rate integrating gyro will be used as a high frequency sensor, capable of detecting small scalar errors with respect to an ideal trim. In this way we will improve the pointing accuracy up to about 10" r.m.s..

6.4. FLIGHT TESTS and RESULTS

A first implementation of this system was flown in the ARGO 1989 experiment (de Bernardis et al. 1990d). During the flight several tests were performed: a modulator stability better than 1/1000 was achieved, in spite of the large ($> 100K$) ambient temperature change. The typical azimuth stability was about $5arcmin$ r.m.s., with a typical reorientation time of $1min$ for 90 degrees. The primary mirror ($120cm$) was isothermal within 0.01 K.

A high Galactic latitude region was observed using a broad band bolometer and a low pass filter with cut-on at ~ 350 μm; the long wavelength cut-off was around 3 mm, due to the dimensions of the bolometer cavity. The interest of these data is mainly in the relatively short wavelength of operation.

After removal of a cirrus dust component (well correlated with the dust emission mapped by IRAS, see de Bernardis et al. 1990e), we obtained a data set including 87 independent fields of view. The residual Brightness fluctuations are slightly larger than detector and Atmospheric noise, and can include a contribution due to either dust emission (a cold component not well correlated with the dust mapped by IRAS) or to other sources of extraGalactic diffuse radiation. For this reason we can use the residuals in order to obtain upper limits for the extraGalactic Brightness fluctuations at angular scales between 5 and 500 arcmin.

We used the Likelihood Ratio test in order to set upper limits to the level of the true sky fluctuations, by means of several correlation functions, and we found a 95% upper limit

$\Delta I < 6 \cdot 10^{-14}$ $W/cm^2/sr$ for our observable (which is the Brightness r.m.s. fluctuations in the band $350 \sim 3000\mu m$).

This is about a factor 4 larger than the anisotropies expected in our beam and in our band from unresolved IR Galaxies (the models of Franceschini et al. 1989 and Wang 1991 give similar results).

When converted in $\Delta T/T$, the upper limit for CBR anisotropies with gaussian correlation function is $\Delta T/T \sim 3 \cdot 10^{-4}$ for coherence angles around 45' and $\Delta T/T \sim 6 \cdot 10^{-5}$ for monochromatic waves with wavelength around 100'.

More interesting is the comparison with the anisotropy expected in the Cosmic Dust scenarios of Bond et al. (1991). The level of Brightness anisotropy expected in our band and in our beam is listed in Table III for 14 different scenarios and for 3 different clustering models (Poisson, Fractal, CDM). Models M1, M3, M5 and M6 have been excluded by the COBE-FIRAS observations of the CBR spectrum; if the spectrum fits a 2.75 K blackbody within 1% of the peak, only the models M8, M12, M13 and M14 will survive. In terms of critical density our anisotropy limit can be rewritten $\Delta\Omega_R < 1.4 \cdot 10^{-8}h^2$. If we mantain the parameters assumed in the Bond et al. paper, our upper limit rules out the models M1, M3, M7, M12 in the case of fractal clustering, and M12 also in the case of Poisson clustering. On the other hand, model M12 with Poisson clustering can be accepted if the density of sources is $n_{g*} > 0.02(h^{-1}Mpc)^{-3}$; models M1, M3, M7, M12 with fractal clustering can be accepted if the clustering length is $r_o < 3.1h^{-1}Mpc$. Similar results have been found by means of very small scale anisotropy searches at mm wavelengths.

This example shows very clearly that, with improved instrumentation, it will be possible to investigate many scenarios of structure formation by means of anisotropy experiments in the sub-mm region: this seems to be one of the most promising tools in the field of the Cosmic Backgrounds after COBE.

References

Adams F.C., Freese K., Lenon J. and McDowell J.C., 1991, *Ap.J.*, **367**, 400.

Baldecchi A., Carli B., Mencaraglia F., Bonetti A., Carlotti M., 1984, *J. Geophys. Res.*, **89**, 689.

Baldecchi A., Carli B., Mencaraglia F., Barbis A., Bonetti A., Carlotti M., 1988, *J. Geophys. Res.*, **93D**, 5303.

Beichman C.A., Helou G., 1991, *Ap.J.Lett.*, **370**, L1.

Birkinshaw M., 1990, in *The Cosmic Microwave Background: 25 years later*, N. Mandolesi and N. Vittorio eds., Astrophysics and Space Science Library **164**, Kluwer, Dordrecht, 77.

Bond J.R., Carr B.J., Hogan C.J., 1986, *Ap.J.*, **306**, 428.

Bond J.R., Carr B.J., Hogan C.J., 1991, *Ap.J.*, **367**, 420.

Boulanger F.,Pérault M., 1988, *Ap.J.*, **330**,964.

Boscaleri A., 1990, *Mem.S.A.It.*, **61**, 257.

Centurioni S., Pedichini F., Natali G., 1990, in preparation.

Cole S., Kaiser N., 1988, *MNRAS*, **233**, 637.

Dame M.T., Ungerechts H., Cohen R.S., de Geus E.J., Grenier I.A., May J., Murphy D.C., Nyman L.A., Thaddeus P., 1987, *Ap.J.*, **322**, 706.

de Bernardis P., Masi, S., Melchiorri F., Moreno, G., Vannoni, R., 1988, *Ap.J.*, **326**, 941.

de Bernardis P., Masi S., Perciballi M., Romeo G., 1989, *Infrared Phys.*, **29**, 1005.

de Bernardis P., Epifani M., Guarini G., Melchiorri B, Melchiorri F., 1990a, *Ap.J.*, **353**, 145.

de Bernardis P., Masi S., Melchiorri F., Melchiorri B., 1990b, *Ap.J.*, **357**, 8.

de Bernardis P., De Petris M., Gervasi M., Masi S., Cardoni G., De Ninno A., Scaramuzzi F., 1990c, in *From Ground Based to Space-Borne Sub-mm Astronomy*, ESA SP-314, 345.

de Bernardis P., Amicone L., Calisse P., De Luca A., De Petris M., Epifani M., Gervasi M., Guarini G., Masi S., Melchiorri F., Perciballi M., Natale V., Boscaleri A., Valmori G., Centurioni S., Natali G., Pedichini F., 1990d, in *From Ground Based to Space-Borne Sub-mm Astronomy*, ESA SP-314, 335.

de Bernardis P., Amicone L., De Luca A., De Petris M., Epifani M., Gervasi M., Guarini G., Masi S., Melchiorri F., Natale V., Boscaleri A., Natali G., Pedichini F., 1990e, *Ap.J. Lett.*, **360**, L31.

de Bernardis P., Masi S., Vittorio N., 1991a, *Galactic and CMB quadrupoles of the mm sky*, *Ap.J*, in press

de Bernardis P., Masi S., Melchiorri F., Melchiorri B., 1991b, in *Experimental tests of Inflation*, NATO Advanced Research Conference, Durham.

De Petris M., Gervasi M., Liberati F., 1989, *Appl. Opt.* , **28**, 1785.

Dubrovich V.K, 1977, *Sov. Astron. Lett.*, **3**, 128.

Fabbri R., Guidi I.,Melchiorri F.,Natale V., 1980, *Phys. Rev. Lett.*, **44**, 1563.

Fixsen D.G., Cheng E.S., Wilkinson D.T., 1983, *PRL*, **50**, 620.

Gautier T.N. III, Boulanger F., Pérault M. and Puget J.L., 1990, *Astron. J.*, in press.

Greenberg J.M., Ferrini F., Barsella B., Aiello S., 1987, *Nature*, **327**, 214.

Gush H.P., Halpern M., Wishnow E.H., 1990, *Phys. Rev. Lett.*, in press

Hasenohrl F.,1904, *Ann. Phys.* **15**, 344.

Hasenohrl F.,1905, *Ann. Phys.* **16**, 589.

Hauser M.G. et al., 1990, *in After the First Three Minutes*, *COBE-preprint*.

Lepp S., Shull J.M., 1984, *Ap.J.*, **280**, 465.

Lyubarsky Y.E., Sunyaev R.A., 1983, *Astron. and Astrophys.*, **123**, 171.

Lubin P.M., Meinhold P.R., Chingcuanco A.O., 1990, in *The Cosmic Microwave Background: 25 Years Later*, Mandolesi and Vittorio eds., Kluwer, 115.

Lubin P.M., Villela T.N., Epstein G., Smoot G.F., 1985, *Ap.J. Lett.*, **298**, L1.

Masi S., de Bernardis P., De Petris M., Epifani M., Gervasi M., Guarini G., 1991, *Ap.J. Lett.*, **366**, L51.

Mather J.C., 1986, *Capabilities of the Cosmic Background Explorer*, in 13th Texas Symposium on Relativistic Astrophysics, Chicago, Schramm ed.

Mather, J.C., Cheng E.S., Eplee R.E., Isaacman R.B., Meyer S.S., Shafer R.A., Weiss R.A., Wright E.L., Bennett C.L., Boggess N.W., Dwek E., Gulkis S., Hauser M.G., Janssen M., Kelsall T., Lubin P.M., Moseley S.H., Murdock T.L., Silverberg R.F., Smoot G.F., Wilkinson D.T., 1990, *Ap.J. Lett.*, **354**, L37.

Mathis J.S., Whiffen G., 1989, *Ap.J.*, **341**, 808.

Meyer S.S., Cheng E.S., Page L.A., 1991, *A measurement of the Large-scale Cosmic Microwave Background Anisotropy at 1.8 mm wavelength, Ap.J. Lett., submitted.*

Meinhold P., Lubin P., 1991, *Ap.J.Lett.*, **370**, L11.

Melchiorri F., Boscaleri A., Cardoni G., de Bernardis P., De Ninno A., De Petris M., Epifani M., Gervasi M., Guarini G., Mancini D., Masi S., Melchiorri B., Natale V., Natali G., Pedichini F., Scaramuzzi F., 1990, in *COSPAR Symposium* #8, The Hague; Wesselius, Barthel, Smoot eds.

Mosengeil von K., 1930, *Ann. Phys.*, **22**,377.

Owens D.K., Muhelner D., Weiss R., 1979, *Ap.J.*, **231**, 702.

Palla F., Salpeter E.E., Stahler S.W., 1983, *Ap.J.*, **271**, 632.

Page A.L., Cheng E.S., Meyer S.S., 1990, *Ap.J. Lett.*, **355**, L1.

Peebles P.J.E., Wilkinson D.T., 1968, *Phys. Rev*, **174**, 2168.

Radford S., 1985, Ph.D. Thesis, University of Seattle.

Readhead A.C.S., Lawrence C.R., Myers S.T., Sargent W.L.W., Hardebeck H.E., Moffet A.T., 1989, *Ap.J.*, **346**, 556.

Reich P., Reich W., 1986, *Astron. Astrophys Suppl. Ser.*, **63**, 205.

Reich P., Reich W., 1988, *Astron. Astrophys. Suppl. Ser.*, **74** , 7.

Reynolds R.J., 1989, *Ap.J. Lett.*, **339**, L29-L33.

Rephaeli Y., 1981, *Ap.J.*, **245**, 351.

Sachs R.K. and A. M. Wolfe, 1967, *Ap.J.* **147**, 73.

Saunders W., Rowan Robinson M., Lawrence A., Efstathiou G., Kaiser N., Ellis R.S., Frenk C.S., 1990, *MNRAS*, **242**, 318.

Shchekinov Yu.A., Entel M.B., 1983, *Sov. Astron.*, **27**, 622.

Sciama D.W., 1967, *Phys. Rev. Lett.*, **18**, 1065.

Shapiro P.R., Kang H., 1987, *Ap.J.*, **318**, 32.

Smoot, G.F., Gorenstein M.V., Muller, R., 1977, *Phys Rev. Lett.*, **39**, 898.

Smoot G., Bennett C.L., Kogut A., Aymon J., Backus C., De Amici G., Galuk K., Jackson P.D., Keegstra P., Rokke L., Tenorio L., Torres S., Gulkis S., Hauser M.G., Janssen M.A., Mather, J.C., Weiss R., Wilkinson D.T., Wright E.L., Boggess N.W., Cheng E.S., Kelsall T., Lubin T., Meyer S., Moseley S.H., Murdock T.L., Shafer R.A., Silverberg R.F., 1991, *Ap.J. Lett.*, **371**, L1.

Soifer B.T., Houck J.R., Neugebauer G., 1986, *Ann. Rev. Astron. Astrophys.*, **25**, 187.

Stecker F.W., Puget J.L., Fazio G.G., 1977, *Ap.J. Lett.*, **214**, L51.

Strukov I.A., Skulachev D.P., 1988, *Astrophysics and Space Physics Review*, **6**, 147.

Sunyaev R.A., Zeldovich Y.B., 1972, *Comm. Astrophys. Sp. Phys.*, **4**, 173.

Sunyaev R.A., 1990, in *COSPAR Symposium* #8, The Hague; Wesselius, Barthel, Smoot eds.

Temi P., de Bernardis P., Masi S., Moreno G., Salama A., 1988, *Ap.J.*, **337**, 528.

Tinsley B.M., 1973, *Astron. Astrophys.*, **24**, 89.

384

Vittorio N., Muciaccia P.F., 1991, *Observational bounds on the Cosmic Microwave Background Anisotropy at Small and Intermediate Angular Scale*, preprint.

Wang B., 1991, *Integrated Far Infrared Background from Galaxies*, preprint.

Wijnbergen J.J., Léna P., Celnikier L.M., 1978, *Infrared Phys.*, **18**, 157.

Wright E.L., 1987, *Ap.J.*, **320**, 818.

Wright E.L., Mather, J.C., Bennet C.L., Cheng E.S., Shafer R.A., Fixsen D.G., Eplee R.E., Isaacman R.B., Read S.M., Boggess N.W., Gulkis S., Hauser M.G., Janssen M., Kelsall T., Lubin P.M., Meyer S.S., Moseley S.H., Murdock T.L., Silverberg R.F., Smoot G.F., Weiss R.A., Wilkinson D.T., 1991, *Preliminary spectral observations of the Galaxy with a 7° beam by the COBE*, *Ap.J.*, *submitted*

Wright E.L., 1991b, 'COBE' in *Texas-ESO-CERN Symposium*.

Wszolek B., Rudnicki K., Masi S., de Bernardis P., Salvi A., 1989, *Astrophysics and Space Science*, **152**, 29.

Zeldovich Ya.B., Kurt V.G., Sunyaev R.A., 1969, *Sov. Phys. JETP*, **28**, 146.

UPPER LIMITS ON THE CMB VARIATION FROM 1300μm OBSERVATIONS

Rolf Chini
Max-Planck-Institut für Radioastronomie
Auf dem Hügel 69
D-5300 Bonn 1
Federal Republic of Germany

ABSTRACT. As a by-product of extremely sensitive 1300μm observations of radio-quiet quasars at the IRAM 30m MRT, an estimate of the small scale isotropy of the Cosmic Microwave Background (CMB) has been achieved. From a sample of 25 quasars only 5 objects could be detected above the 3σ level. As the observed 1300μm flux density of four detections is consistent with an extrapolation of the non-thermal radio spectrum these data have been omitted from the present investigation. Treating the remaining data, i.e. 20 null results and 1 detection, like measurements of empty fields and using them as probes for the small scale CMB isotropy we derive a value of δT/T ≤ 2.6E-4 for a scale of 30" at a 95% significance.

1. Introduction

Measurements of the isotropy of the CMB are performed at many wavelengths and at many spatial scales. So far there is no evidence for background fluctuations but even the upper limits obtained so far are valuable for constraining various models about the early times of our universe. The experiment described in the following was not intended to contribute to this field of research. However, when the null results concerning radio-quiet quasars were in the literature many colleagues encouraged us to use the data and test their implications on the isotropy of the CMB. The importance of the present observations with unprecedented sensitivity lies in the combination of the short wavelength of 1300μm and the small scale of 30". This short contribution should be treated as a description of the system and its capabilities rather than a serious approach to a CMB experiment. Future observations in the submm range, carried out with even higher sensitivity as the present ones on carefully selected fields in the sky might contribute significantly to the question of isotropy.

M. Signore and C. Dupraz (et al.), The Infrared and Submillimetre Sky after COBE, 385–390.
© 1992 *Kluwer Academic Publishers.*

2. Observations

2.1 Technical details

The observations discussed in the following have been carried out during November 1987 and April 1988 at the IRAM 30m telescope (MRT) on Pico Veleta. We used the MPIfR bolometer (Kreysa, 1990) at a wavelength of 1300µm. The detector consists of a ^{3}He-cooled Ge-bolometer operating at 0.27K. In order to achieve highest sensitivity for point sources the system is designed to work in the diffraction limit; such a configuration yields the highest spatial resolution – 11" at the MRT – and keeps the background on the bolometer minimal. A fundamental sensitivity limit is given by the photon noise of the thermal background radiation. The passband of the system was defined by a resonant mesh filter (low pass) and a short piece of cylindrical waveguide was served as a high pass filter. This combination yields a bandwidth of 60GHz (HPW) perfectly matched to the atmospheric window at 1300µm.

Ground-based submm observations require signal modulation in order to cancel atmospheric emission fluctuations. Undoubtedly, chopping with the subreflector is the best way of doing this. Due to technical problems with the wobbling secondary at the time of our observing runs we had to use a focal plane chopper operating at 8.5Hz and giving two beams with a separation of 30" on the sky. In the Nasmyth focus of the MRT, this configuration has the disadvantage that the chopping direction on the sky will change with elevation. Normally, when chopping with the secondary, the chopping direction is chosen to be fixed in azimuth to eliminate possible sky gradients. Nevertheless, it could be achieved that after, a careful adaption of the system to the telescope, no measurable imbalances remained.

The MRT has a Gaussian beam shape of 11" HPBW at 1300µm although its surface is imperfect at these short wavelengths. The imperfect reflector surface results in scattering sidelobes and in reduced aperture and beam efficiencies. The latter one was determined to be about 25%, i.e. only this fraction of the incoming flux from an astronomical source is reflected into the receiver. Nevertheless, scattering sidelobes are usually not very prominent, because the bolometer is operated in a beamswitched mode and the small chopper throw of about three times the beamwidth is an efficient high pass spatial filter.

The sensitivity of the MPIfR bolometer system in the absence of any atmospheric distortion was determined to be 20mJy in 1s of integration time at a S/N of 1. Most of the time, however, this sensitivity is degraded by skynoise and/or refraction. Sky noise is observed on clear sky as a low frequency imbalance of the detector signal caused by cells of uncondensed water vapor drifting across the two beams. This picture is supported by the observation that sky noise increases linearly with the beam separation and the beamwidth, e.g. when using oversized feedhorns or smaller telescopes. Such a behavior is expected when the two beams separate rather early on their path through the atmosphere. Refraction effects can only be observed when tracking a source; they are equivalent to the "seeing" in optical astronomy and result in a slow irregular wandering of the source relative to the position of the beam.

This decreases the signal-to-noise ratio even for bright sources as the planets severely. The average sensitivity achieved during the course of the quasar observations was 70mJy in 1s with a S/N of 1.

The observations of the quasars were done by performing ON-ON measurements at the position of the source. After a pointing calibration the object was centered in one of the two beams. After an integration time of 10sec the telescope was moved by 30" in order to center the object into the second beam and to integrate for another 10s. This procedure, called one cycle, was repeated ten times giving in total 200s of pure integration time. Then the observation was stopped and a mean value with a statistical error was calculated from the 10 ON-ON pairs. In the following, we treat such a procedure as "one independent measurement" of the object. Depending on the performance of the telescope, a new pointing calibration was done by means of a nearby bright quasar and/or a new measurement was started.

The atmospheric transmission has to be determined via "skydips", a procedure that measures the thermal emission of the atmosphere at different elevations and enables us to calculate the atmospheric opacity. This procedure must be done rather frequently in order to trace any fluctuations of the τ-values and thus interrupts the measurements of the program sources regularly. Typically, τ ranges between 0.1 and 0.2 during good observing conditions in the winter season on Pico Veleta; higher values are normally accompanied by clouds which prevent 1300μm observations. During other seasons τ can rise to about 0.5 without significant cloud formation. In order to calibrate the incoming flux of arbitrary units to an absolute scale, ON-ON observations of the planets Uranus and Neptune were made, adopting brightness temperatures of 101 and 104K, respectively.

2.2 The observed fields

As mentioned earlier the present experiment was never dedicated to probe the isotropy of the CMB. We rather intended to elucidate the origin of the submm/FIR radiation in radio-quiet quasars as described in the lecture "Submm Emission from Galaxies and Quasars". For that purpose we selected all radio-quiet quasars from the list of Neugebauer et al. (1986) which are accessible from the northern hemisphere and which had been detected at least in three IRAS bands. It was our aim to obtain meaningful upper limits for this entire sample of 25 quasars rather than to detect a few individual objects with significant S/N. Therefore, we repeated the measurements for an objects until we reached a statistical error of approximately 1mJy (1σ). This value was a compromise between the telescope time available for that project and the implications that could be derived from an upper limit of that order.

At the end of the two observing runs we had collected between 20 and 100 independent measurements for each quasar. The individual observations were reduced to the absolute flux density scale and then summed together weighted by their statistical errors. In this way all 25 quasars could be observed and 5 of them were detected above the 3σ level; the upper limits and the positive results are given in Table 1 of Chini et al. (1989).

3. Results

Concerning the implications for the radio-quiet quasars, the described experiment was a great success because the widely debated question about the physical origin of the submm/FIR emission could now be answered. From the steep spectral index between 100 and 1300μm we concluded that i) The FIR emission must be thermal radiation from dust. ii) The expected thermal emission at 1300μm – extrapolated from the dust temperature at 60 and 100μm – must be of the order 0.64 ± 0.44mJy, which explains our failure to detect most of the sources. iii) The positively observed 1300μm emission in five of the quasars must be due to non-thermal Synchrotron radiation. The latter assumption could be corroborated by radio data because for four of the quasars, our observed 1300μm value was consistent with an extrapolation of the radio spectrum. Unfortunately, there are no radio data for the fifth quasar PG1411+44 that we had detected with a flux density of 3.9 ± 1.0mJy so that the nature of this value remains uncertain. Nevertheless, these results finally led to the idea to treat the 21 measurements as if they were performed on empty fields of the sky and to ask for the implications on the isotropy of the CMB. Fig. 1, taken from Kreysa and Chini (1988, 1989), summarizes the final results for the individual quasars. S_{field} denotes the flux density obtained by averaging all independent measurements weighted with their statistical errors.

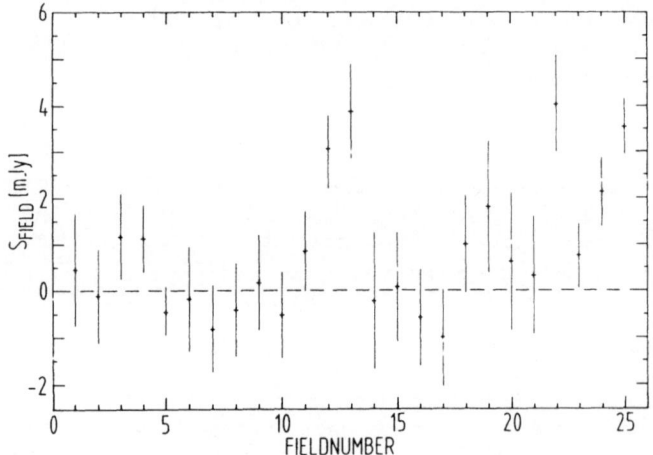

Fig.1: Final 1300μm flux densities for the 25 fields observed

Apart from the five detections already mentioned, Fig. 1 shows no statistically significant signal in any of the remaining 20 fields. In order to evaluate the errors in Fig. 1 we follow the procedure described by Uson and Wilkinson (1984) and investigate the scatter in the measurements. Fig. 2 contains the result for the 21 fields and shows that the data are consistent with normally distributed errors.

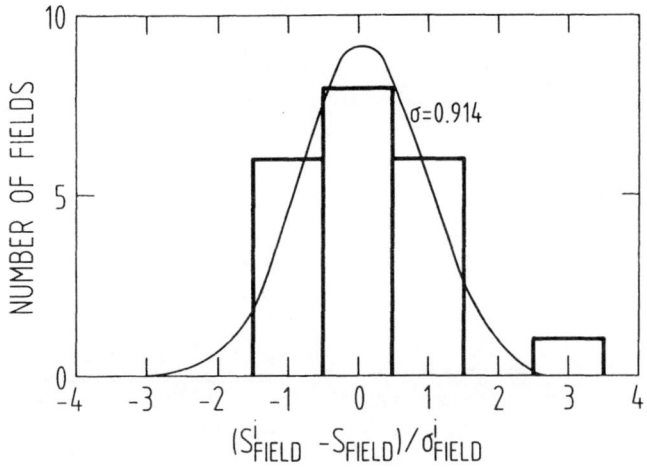

Fig.2: Scatter of results for 21 fields

The upper bound for sky roughness is derived with the statistics given by Boynton and Partridge (1973). This method asks whether there is evidence for a true statistical sky signal, σ_{sky}, which has increased the scatter among the final values S_{field} in Fig. 1 by more than what is expected from the statistical errors also shown in Fig. 1. We derive an upper limit for the fluctuation of the sky of 0.94mJy at a 95% confidence level. As demonstrated by Uson and Wilkinson (1984) ON—ON measurements test the statistical sky fluctuations with an efficiency of $(1.5)^{1/2}$ at the scale of the beam separation; this correction reduces the sky fluctuations to $\delta S_{rms} = 0.77$mJy. Finally, we have to convert the flux density scale to a temperature scale. For a blackbody of 2.75K, observed at 230GHz (1300μm), the relation is

$$\delta T/T = 0.245 \; \delta S/S \tag{1}$$

The total flux S of the CMB contained in our 11" beam is 0.733Jy. This yields our final result

$$\delta T/T = 2.6 \; 10^{-4} \tag{2}$$

at a 95% significance and for a scale of 30".

4. Conclusions

This lecture should demonstrate that ground based submm observations are also a promising tool to investigate the small scale isotropy of the CMB. Although the present experiment was not dedicated to have cosmological relevance we were astonished about the resonance among CMB experts. For the future we plan two kinds of cosmological experiments:

i) Sensitive isotropy experiments with a careful selection of the fields. Due to several technical changes we could improve the sensitivity of our bolometer system. A major progress could be achieved by moving the system to another location in the receiver cabin of the MRT. It turned out that distortions which were previously thought to be due to skynoise originated from vibrations of two mirrors. The new beam path is absolutely free from any microphonics. Likewise, we have used the wobbling secondary very successfully during our last observing runs. Altogether, it seems that we have gained a factor of two in sensitivity which make a new attempt rather promising. A seven channel bolometer array has been tested some weeks ago and will be used for the new isotropy experiments. It is hoped, that the larger number of channels can be used to cancel out skynoise more efficiently. In addition, this configuration allows to measure the CMB isotropy simultaneously on different scales. Finally, one may think to perform the experiment at shorter wavelengths either at the MRT or in the future at the 10m SMT.

ii) Measurements of the Sunyaev-Zeldovich effect with a two-channel bolometer. This configuration will allow the simultaneous measurement of two wavelengths above and below the crossover point at 1270μm.

References

Chini, R., Kreysa, E., Biermann, P.L.: 1989,
    Astron. Astrophys. **219**, 87
Kreysa, E.: 1990, in "From Ground-Based to Space-Borne Submm Astronomy", Proc. 29th Liège International Astrophysical Colloqium
    July 1990, Eds. N. Longdon, B. Kaldeich, ESA SP-314, p. 265
Kreysa, E., Chini, R.: 1988, in Proc. 3rd ESO-CERN Symp. on "Astronomy, Cosmology and Fundamental Particles", Bologna
Kreysa, E., Chini, R.: 1989, in Proc. on "Particle Astrophysics Workshop", Berkeley, Dec. 1988
Neugebauer, G., Miley, G.K., Soifer, B.T., Clegg, P.E.: 1986,
    Ap.J. **308**, 815

MAPPING THE SKY WITH THE COBE DIFFERENTIAL MICROWAVE RADIOMETERS

M. A. JANSSEN
S. GULKIS
Mail Stop 169-506
Jet Propulsion Laboratory
4800 Oak Grove Drive
Pasadena, California 91109
USA

ABSTRACT. The Differential Microwave Radiometers (DMR) instrument on COBE is designed to determine the anisotropy of the Cosmic Microwave Background by providing all-sky maps of the diffuse sky brightness at microwave frequencies. The principal intent of this lecture is to show how these maps are generated from differential measurements.

1. Introduction

The measurement objective of the Differential Microwave Radiometer instrument on the Cosmic Background Explorer satellite is to provide precise maps of the microwave brightness of the sky on large angular scales, to be used for the purpose of deducing the anisotropy of the Cosmic Microwave Background (CMB). The instrument has been described in detail by Smoot et al. (1990), and consists of six differential microwave radiometers with two independent radiometers at each of three frequencies: 31.5, 53, and 90 GHz (wavelengths 9.5, 5.7, and 3.3 mm). These frequencies span a window in which the CMB dominates foreground galactic emission by a factor of more than 1000. The overall idea of the multiple frequency approach is to identify and account for the small but important galactic components by their spectral signature, thereby enabling us to derive maps of the CMB by itself. These maps will allow us to infer the distribution of matter and energy in the early universe as discussed in other contributions to this proceedings.

In Section 2 we outline the general design of the DMR experiment. The presentation is at a tutorial level and the following questions are addressed: 1) How does one describe the microwave brightness of the sky? 2) What is the relationship of the large-scale brightness distribution from all sources to that due to the CMB itself? 3) How do radiometers work and what are their limitations in mapping the sky? 4) How are sky brightness maps generated from differential measurements on the sky and what are the expected limitations of such maps?

In Section 3 we give more detail on how the actual DMR data are turned into maps. An early version of the mathematics of this inversion was presented by Torres et al. (1989) and is expanded upon here. This aspect of the DMR analysis is important not only for the generation of maps but in the understanding of the systematic errors which may ultimately limit their cosmological significance.

M. Signore and C. Dupraz (et al.), The Infrared and Submillimetre Sky after COBE, 391–408.
© 1992 *Kluwer Academic Publishers.*

2. Fundamental Considerations in the DMR Experiment

2.1 RADIATION

Radiant energy propagating through space is conveniently described by the specific intensity (usually shortened to just "intensity"), I_ν, which is the instantaneous radiant power (energy per unit time) crossing a unit area normal to the direction of propagation, per unit frequency interval at a specified frequency, in a given direction per unit solid angle (Chandrasekhar,1960; Krause, 1966). The intensity is illustrated in Figure 1. The radiant energy emitted by a surface is called the brightness, and is the flow of energy across a unit area per unit frequency from a source viewed in an element of solid angle $d\Omega$. Note that brightness and intensity have the same units, and the equivalence of the two in free space is illustrated in Figure 2. For example, the brightness of a blackbody emitter viewed in a particular direction from a point in free space is also the intensity I_ν at that point along the same line of sight.

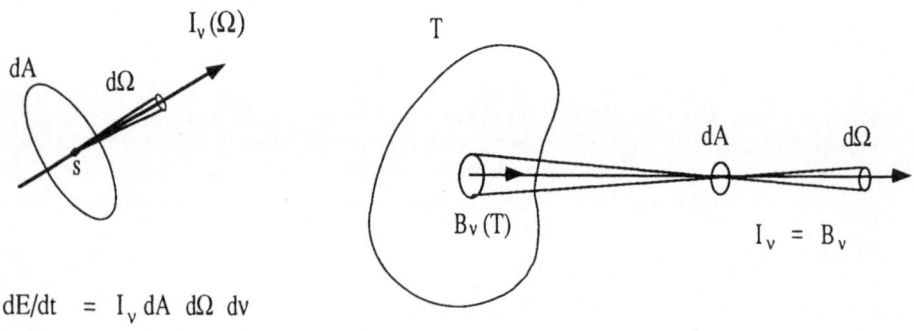

Figure 1   Figure 2

The brightness of a blackbody is of particular importance to us. This is given by the Planck function

$$B_\nu(T) = \frac{2h\nu^3}{c^2} \frac{1}{e^{h\nu/kT} - 1} \tag{1}$$

where the various symbols have their usual meanings.

Other units are sometimes used to describe brightness. For example, the flow of energy per unit wavelength is given by

$$B_\lambda = B_\nu |d\nu/d\lambda| = B_\nu c/\lambda^2. \tag{2}$$

Another scaling may be made to give power per unit of photon energy $h\nu$ instead of per unit frequency,

$$B_E = B_\nu d\nu/d(h\nu) = B_\nu/h. \tag{3}$$

The Planck law for these units may be easily found using Equation 1. An important variant in the units used for radiant intensity involves the Rayleigh-Jeans limit $h\nu \ll kT$, which applies almost always in the microwave region (caution - see below). As $h\nu/kT$ goes to zero Equation 1 approaches the limit

$$B_\nu(T) = \frac{2\nu^2 kT}{c^2} = \frac{2kT}{\lambda^2} \qquad (4)$$

The linear dependence of brightness on temperature leads to a scaling of the intensity

$$T_b(\nu) = \frac{\lambda^2}{2k} I_\nu. \qquad (5)$$

where T_b is called the brightness temperature. One must be careful with the use of this term because there are two definitions which are used in the literature:

1. The brightness temperature is strictly a scaling defined by Equation 5, and is sometimes called the equivalent Rayleigh-Jeans temperature.

2. The brightness temperature is the temperature of an equivalent blackbody which gives the observed intensity, and is sometimes identified as the equivalent thermodynamic temperature.

These two definitions are equivalent in the Rayleigh-Jeans limit, but this is not a good approximation when $h\nu/kT \leq \sim 1$, which is the case at the DMR frequencies. We will use the former definition here, noting that we are always talking about intensity when we mention brightness temperature, and not thermodynamic temperature.

2.2 THE MICROWAVE SKY

Figure 3 shows the expected microwave background spectrum with its various major components. The cosmic component assumes a Planckian spectrum at 2.735 K (Mather et al., 1990) The fall-off of the curve at higher frequencies illustrates the departure from the Rayleigh-Jeans law and demonstrates that we are indeed plotting on an intensity scale. Emission due to galactic synchrotron radiation from energetic electrons in interstellar magnetic fields, and due to bremsstrahlung from thermal electrons in ionized HII regions, dominates the background at very low frequencies. At high frequencies the dominant source is cold interstellar dust, the absorption (and corresponding emission) cross section of which increases with frequency. The plotted curves show the resulting brightnesses near the galactic plane, and are less near the galactic poles.

The decrease of the galactic electron emission with frequency along with the increasing dust emission combine to create a broad window throughout the microwave region where the cosmic background dominates. The sum of the galactic terms is a minimum with respect to the cosmic background at a frequency of about 50 GHz. The vertical lines show the frequencies of the DMR radiometers, and illustrate the intention of the experiment design to bracket this minimum. A major step in the generation of cosmic background maps is to identify the galactic contributions from their frequency dependences in these maps and remove them. Radioastronomical maps of synchrotron and HII emission at lower frequencies and dust emission maps from FIRAS at high frequencies will be important elements in this process.

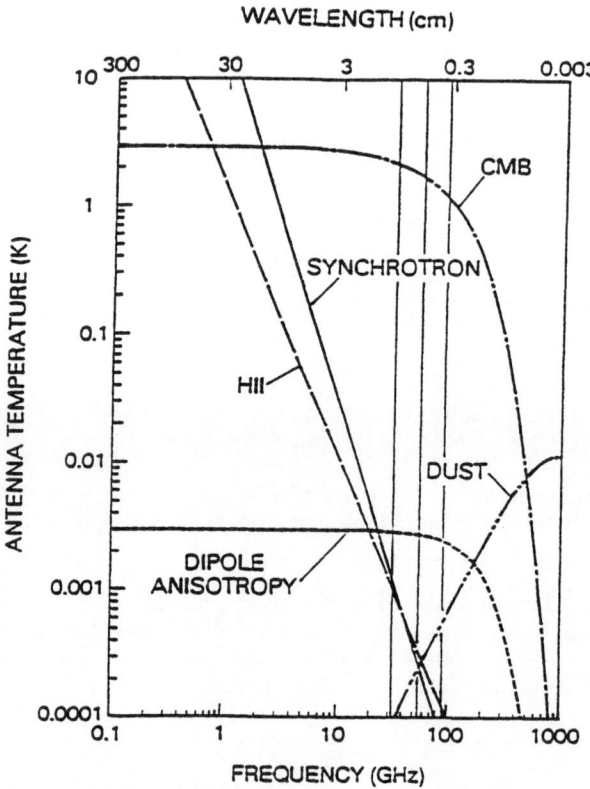

Figure 3

As far as we know at this point in our analysis of the COBE data, the cosmic component of the microwave background is a perfect blackbody — with one major exception: the isotropy of the radiation can only be maintained in a single rest frame, and it would be unlikely that this is the same as ours. It can be shown in general that a Doppler-shifted blackbody spectrum remains a blackbody spectrum, but at a different temperature (Bracewell and Conklin, 1968). At a given frequency, the relativistically correct adjustment to the temperature is (Peebles and Wilkinson, 1968)

$$T(\theta) = T_o \frac{(1 - \beta^2)^{1/2}}{(1 - \beta \cos\theta)} \tag{6}$$

where $\beta = v/c$ and θ is the angle, measured in the observer frame, between the velocity v of the observer with respect to the blackbody rest frame and the direction in which the intensity is measured. Expanding in orders of β, we have

$$T(\theta) = T_o [1 + \beta \cos\theta + \beta^2(\cos2\theta/2) + O(\beta^3)] \quad . \tag{7}$$

The first-order term in β, $\beta \cos\theta$, results in the so-called Doppler dipole anisotropy, and its magnitude for a velocity of a few hundred km/sec (our best current value is 365 ± 18 km/sec; Smoot et al., 1991a) is indicated in Figure 3. This is small (10^{-3} of T_{CMB}), but very large in terms of the sensitivity of the DMR radiometers. Another component of interest (not shown on the figure) is the modulation on this dipole due to the motion of the Earth around the sun. Since this velocity is ~30 km/sec, the modulation is about 10% of the dipole amplitude and is an important effect to consider in our measurement. Finally, the β^2 term in Equation 7 leads to a quadrupole anisotropy which is about 10^{-6} of T_{CMB}. This will be near the limit of measurement for the DMR experiment.

2.3 MICROWAVE RADIOMETRY

A radiometer is by definition a device used to measure radiant intensity. Its main elements include an antenna to collect power over a restricted range of solid angle, a receiver to convert this power into a linearly related voltage, and circuitry to average this voltage and produce digital output. The power collected by the antenna is described as the antenna temperature

$$T_a = \frac{\int_{4\pi} G(\Omega)\, T_b(\Omega)\, d\Omega}{\int_{4\pi} G(\Omega)\, d\Omega} \tag{8}$$

where G is the relative antenna gain in the direction of the solid angle element Ω, and $kT_a B$ is the power in physical units actually entering the receiver in the bandwith interval B. Note that $T_a = T_b$ when the latter is uniform throughout the antenna beam.

We will not attempt to describe the internal workings of the receiver in any detail here. Many microwave receivers (including the DMR receivers) use a heterodyne technique in which in all the power is collected in two rectangular bandwidths B which are symmetric around a "local oscillator" frequency v_o, and translated to a lower frequency (an "intermediate frequency"). This power may then be readily amplified and converted into a voltage (e.g., see Krause, 1960). In this process some power is generated within the receiver itself and amplified along with the signal power, so that the mean output voltage can be represented by

$$V = G_v (T_a + T_r) = G_v T_s \tag{9}$$

where T_r represents this added noise power, and G_v is a typically very linear calibration constant. This constant is typically determined by observing blackbody targets at known temperatures. The total (calibrated) noise power is given by the sum $T_s = T_a + T_r$ and is called the system temperature.

The radiometer output voltage is intrinsically noisy! The reason for this is that the thermal power, including that generated within the receiver itself, is noise power. One way to describe this power is that in any interval of time τ and for a bandwidth B, the output voltage may be Fourier decomposed into $B\tau$ independent frequencies of random complex amplitude. The sum of squares of these amplitudes is proportional to the total power in the band and is uncertain because of the inherently random nature of each amplitude. The relative uncertainty can be written as inversely proportional to the square root of the number of samples (see Evans and McLeish, 1977, for a full derivation), or

$$\frac{\Delta P}{P} = \frac{1}{\sqrt{n}} = \frac{1}{\sqrt{B\tau}} . \tag{10}$$

Since $\Delta P/P = \Delta T_{noise}/T_s$, this gives the radiometer noise formula

$$\Delta T_{noise} = \frac{T_s}{\sqrt{B\tau}} \tag{11}$$

where ΔT_{noise} is the rms uncertainty in any measurement of the calibrated noise power averaged over the bandwidth B and time interval τ.

2.4 MEASURING THE MICROWAVE SKY

Let us assume that we have an ideal radiometer which we will use to map the sky. The rules are simple: we are only concerned with the intrinsic noise given by Equation 11 (our ideal radiometer has a bandwidth B and a system temperature T_s), we can choose our beamwidth, and we have a time t in which to complete our map. If we choose an antenna which gives us a beam solid angle Ω_b, then our map will have $N = 4\pi/\Omega_b$ pixels, and a map with uniform noise will be obtained if we observe each pixel for a time $\tau = t/N$. The uncertainty of each pixel temperature will then be

$$\Delta T_{pix} = \frac{T_s}{\sqrt{Bt/N}} . \tag{12}$$

Equation 12 is plotted as the solid line in Figure 4 for radiometer parameters corresponding to the DMR 53 GHz channel A radiometer, which is our best, and a two year mission time. As expected, the map noise is inversely proportional to the resolution at small angles, deviating slightly at large angles where the spherical geometry becomes important. One obvious point illustrated in this plot is that small angular scales are not particularly well studied with all sky mapping, and the cost of providing large apertures necessary to provide high resolution would not be a good bargain. Small-scale anisotropies may be studied over limited regions from the ground because atmospheric fluctuations are less restricing on these scales. At any resolution given by our choice of antenna, however, lower resolution maps may be obtained without penalty.

The DMR beams are about 7 degrees wide and provide the sensitivity shown in the figure. The sensitivity on larger angular scales is estimated by the curve and can be conceptually achieved by spatial averaging. For example, the plot indicates a sensitivity of about 1.5 μK for 180 degree resolution, which corresponds (loosely) to the dipole anisotropy. Generally, any spherical harmonic amplitude may be determined with this same approximate sensitivity.

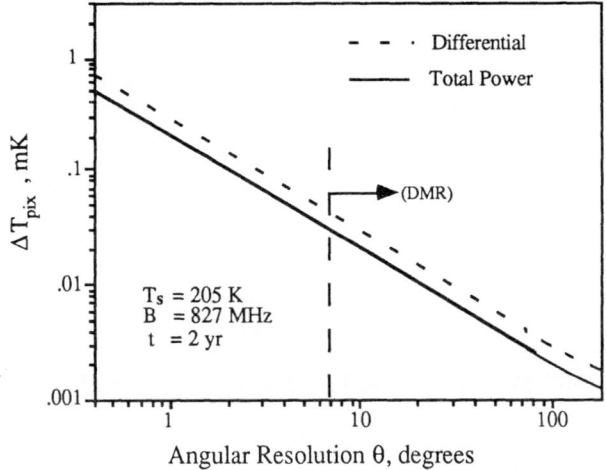

Figure 4: Map pixel noise with ideal total power and differential radiometers

2.5 A ROBUST APPROACH USING DIFFERENTIAL RADIOMETERS

The above gives us an idea how ideal our radiometer should be to enable such a mapping. In short, we expect to measure large-scale variations which are $\sim 10^{-6}$ times T_{CMB}, and more to the point, $\sim 10^{-8}$ times the output noise of the radiometer T_r. This means that our ideal radiometer must be stable to this level during the entire mapping! It turns out that this is about six orders of magnitude too optimistic for practical radiometers. The generic solution for this class of small signal measurement is to chop the signal against a reference signal, so that gain variations in the radiometer act only on the difference signal and not the total power. Microwave radiometers which are designed to do this are called "Dicke-switched" radiometers after their inventor R. Dicke of Princeton, and usually incorporate an internal temperature-controlled absorptive load as the reference. The measurement uncertainty for such a radiometer is

$$\Delta T_{noise} = \frac{2T_s}{\sqrt{B\tau}} . \tag{13}$$

where the factor of two relative to the total power radiometer case described by Equation 11 results from the facts that the actual observation time on the source is reduced by half, and that the difference between two noisy signals is being measured, both of which lead to a root two increase in the noise.

The DMR solution is to use the sky itself as the "reference". Each radiometer has two identical horn antennas which are pointed 60° apart on the sky, and all measurements it makes are of the differences between brightnesses of regions of the sky which are separated by 60°. Before we turn to the question of how this leads to sky maps, we must point out that the measurement problem is still not solved. Differential measurements are still not ideal: small loss differences in the two signal paths along with more complicated asymmetries in the circuit electronics lead to an effective radiometric "offset", which is typically on the order of 1 K. This offset is still susceptible to radiometric gain variations so that we really win back only two or three orders of magnitude of the eight with which we have to contend. However, this offset may be cancelled by a further symmetrization of the measurement in which the spacecraft is spun on an axis which is aligned with the bisector of the angle between the antenna beam axes. Consequently, if the spacecraft rotates sufficiently rapidly, then any differential measurement D_{ij}, where i and j index the pixels on a sky map, may be paired with the measurement D_{ji} taken after the spacecraft rotates 1/2 turn to eliminate the offset. Specifically, we have

$$D_{ij} = T_i - T_j + O \tag{14}$$

where T_k is the brightness temperature of the kth pixel and O is an offset which can be considered constant on the time scale of a spacecraft rotation period, whereupon

$$S_{ij} = (D_{ij} - D_{ji})/2 = T_i - T_j \tag{15}$$

yields the differential signal S_{ij} without the offset.

2.6 OBTAINING MAPS FROM DIFFERENTIAL MEASUREMENTS

There are still many sources of error which we must deal with; meanwhile, we have at least established a plausible measurement approach. The next question to be addressed is how differential measurements can be used to generate maps of the sky.

We will set up here a simplified version of the DMR experiment which is easy to analyze and illustrates many important features of the mapping process. First, we imagine that our "double-switching" scheme works successfully and we have a set of differential measurements S_{ij} which contain only random noise. Secondly, we will ignore details about geometry and orbits, and assume that our measurements are somehow uniformly distributed around the sky. Hence we can say that we have a set of data

$$S_{ij} = T_i - T_j + \sigma \tag{16}$$

where each pixel i is associated with a subset $\{j\}_i$ of the total pixel set, and σ is the measurement uncertainty for each and all difference measurements S_{ij}. If the horn separation angle is 60°, for example, then a given pixel i will be associated with differential measurements on a ring of pixels 60° from i, which we denote as the subset $\{j\}_i$, unique to i (see Figure 5). We may obtain σ from the following argument: the error on D_{ij} is given by Equation 13 for $\tau = t/NM$, where each subset $\{j\}_i$ is assumed to contain M elements, and τ is the time spent measuring this differential temperature. The error on S_{ij} is reduced by a further factor of root two by the averaging of Equation 15, hence

$$\sigma = \frac{\sqrt{2}\, T_s}{\sqrt{Bt/NM}}. \tag{17}$$

Now, we may form the chi-squared sum

$$\chi^2 = \frac{1}{2\sigma^2} \sum_{i=1}^{N} \sum_{\{j\}_i}^{M} (T_i - T_j - S_{ij})^2 \tag{18}$$

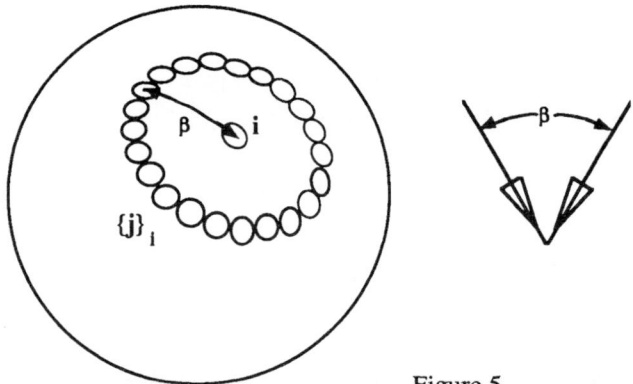

Figure 5

where the factor of 1/2 in front is required to account for the fact that each (i, j) pair appears twice in the summation.

The map may now be determined as the set of T_k which minimizes χ^2, which in turn is the solution to the set of equations formed by differentiating Equation 18 with respect to the T_k and setting the result to zero:

$$T_k - \frac{1}{M} \sum_{(j)_k}^{M} T_j = \frac{1}{M} \sum_{(j)_k}^{M} S_{kj} . \tag{19}$$

These are called the normal equations, and form a set of N linear equations to be solved for the T_k's, which we can think of as the map vector **t**. They can be expressed in matrix form as

$$M \cdot t = D \tag{20}$$

or graphically,

$$
\begin{pmatrix}
1\ 0 \dots 0 \dots -1/M \ \dots 0 \dots \dots 0 \dots \dots 0 \dots 0 \\
0\ 1 \dots 0 \dots . \ 0 \dots 0 \dots -1/M \dots 0 \dots 0 \dots 0 \\
0\ 0\ 1 \dots 0 \dots 0 \dots -1/M \dots 0 \dots \dots 0 \dots \dots 0 \\
\cdot \\
\cdot \\
\cdot \\
\cdot \\
\cdot \\
\cdot \\
0 \dots \dots -1/M \dots 0 \dots \dots 0 \dots \dots 0 \dots 1\ 0\ 0 \\
0 \dots 0 \dots \dots 0 \dots 0 \dots \dots 0 \dots -1/M . \ 0\ 1\ 0 \\
0 \dots \dots 0 \dots 0 \dots -1/M \dots \dots 0 \dots \dots 0\ 0\ 1
\end{pmatrix}
\begin{pmatrix}
T_1 \\
T_2 \\
T_3 \\
\cdot \\
\cdot \\
\cdot \\
\cdot \\
\cdot \\
\cdot \\
T_{N-2} \\
T_{N-1} \\
T_N
\end{pmatrix}
=
\begin{pmatrix}
D_1 \\
D_2 \\
D_3 \\
\cdot \\
\cdot \\
\cdot \\
\cdot \\
\cdot \\
\cdot \\
D_{N-2} \\
D_{N-1} \\
D_N
\end{pmatrix}
$$

where the D_k's represent the sums on the right-hand side of (19) and each row of the N x N matrix **M** contains only M nonzero elements of value -1/M. **M** is therefore sparse, and large (N ≈ 1000 for 7° resolution, while N = 6144 is actually used in the current DMR map as discussed in Section 3). The solution

$$\mathbf{t} = \mathbf{M}^{-1} \cdot \mathbf{D} \tag{21}$$

is conceptually straightforward, but its implementation requires some tricks because of the size and nature of $\mathbf{M}$. Iterative techniques are typically used for such sparse matrices (e.g., see Franklin, 1968): an initial guess T_k^0 is used with the direct equations (20) to compute an improved guess T_k^1, and so forth until some convergence criterion is met. For example, the Gauss-Seidel method has proven useful for generating the successive estimates T_k^i (see Section 3.4).

The error matrix is the inverse of the curvature matrix:

$$\frac{1}{2} \frac{\partial \chi^2}{\partial T_l \partial T_k} . \tag{22}$$

This has elements M/σ^2 for $l=k$, $-1/\sigma^2$ for l in $\{j\}_k$, and zero otherwise. This is a difficult matrix to invert or interpret, but we may make some general observations. First, since $M \ll N$, the diagonal elements will be approximately equal to σ^2/M, while the off-diagonal elements will be of order $1/M^2$ (1 in $\{j\}_k$) or $1/M^3$ (otherwise). If M is sufficiently large that we can ignore these off-diagonal terms, then the individual pixel noise is readily found to be

$$\Delta T_{pix} = (\sigma^2/M)^{1/2} = \frac{\sqrt{2} T_s}{\sqrt{Bt/N}} , \tag{23}$$

which is a factor of root two worse than for the total power radiometer case in Equation 12. The root two comes from the fact that the measurements are of differences, but there is not a second root two as in the Dicke noise formula of Equation 13 because the sky is measured all the time. Equation 23 is shown as the dashed curve in Figure 4 for the choice of parameters indicated.

2.7 THE CHOICE OF HORN OPENING ANGLE

The horn opening angle β is a tradeoff between Earth signal pickup in the sidelobes for large horn opening angle, and the loss of sensitivity to large-scale background structure for a small opening angle. This problem was studied early in the definition of the mission (e.g., Janssen, 1978), with the conclusion that $\beta = 60°$ is a broad optimum and neither effect is sharply sensitive to this choice.

For the dipole anisotropy, any set of differential measurements will suffer the same noise, but the signal will decrease when the angle between the beams becomes small. The ratio of signal to noise thus decreases for small angles. Detailed calculations have shown that the sensitivity to the dipole is reduced by a factor of about 1.4 at $\beta = 60°$ relative to what one would expect from uniform, uncorrelated noise consistent with the approximation given in Equation 23. Further, the sensitivity to all spherical harmonics does not reach a smooth minimum with varying β, but oscillates approximately $\pm 20\%$ around the same "uniform noise" case after a sufficiently large value of β is reached. This dependence arises from the off-diagonal elements in the error matrix which we ignored above, and illustrates that they aren't entirely negligible.

On the other hand, the further the beam axes lie from the spacecraft spin axis, the greater will be the pickup of Earth emission diffracted around the COBE shield. This is a difficult calculation to carry out because of the complicated geometry of the shield edge with respect to the horns; however, estimates indicate that the induced signal is not a sharp function of β.

3. A Technique for the Inversion of DMR Data

3.1 THE DMR DATA SET

In the previous section we demonstrated that differential measurements could be used to produce maps of the relative sky brightness for an idealized case. We will formalize that process here for nonuniformly sampled data from the real DMRs flying in the COBE orbit. We showed that intrinsic radiometer noise sets a fundamental limitation on the certainty of such maps, and also considered the radiometric offset as it led to the choice of the differential approach to the measurements. We will now proceed to show how maps may be obtained in the case of real data, and also how an important class of systematic errors may be included in the analysis.

In general the time-dependent, uncalibrated output of a COBE differential radiometer can be described as the sum of several terms:

$$S(t) = G(t) \left\{ \Delta T(t) + \sum_{p=1}^{m} W_p(t) + \sum_{q=1}^{n} U_q(t) + O(t) + N(t) \right\} \tag{24}$$

where:

ΔT = The instantaneous difference in antenna temperature between the two horns due to the "ideal", unknown brightness distribution of the sky. We define this ideal brightness distribution as that due only to sources outside the solar system as it would appear to an ideal radiometer from a reference frame fixed with respect to the center of mass of the solar system. We take this distribution to be the objective of our present analysis and leave its subsequent cosmological interpretation to other work (for example see Section 2.2 above, and Smoot , 1991b).

W = A set of well known corrections to this differential temperature due to varying observer motions or local sources. Doppler effects caused by the motions of the Earth and spacecraft are accounted for here, as may be some weak but effectively predictable signals caused by solar system objects such as Jupiter, and the Moon when it is sufficiently far from the beam axis.

U = The set of all other effects, generally unknown, which influence the measured differential signal. We categorize these as systematic errors and consider their determination as an important aspect of the mapping problem.

O = offset voltage (e.g., see Section 2.5).

N = "instantaneous" noise voltage. This term only has strict meaning when we consider finite averaging times below.

G = the radiometer gain in appropriate units. If the output is in volts, for example, then the gain might have units of volts/Kelvin. The terms appearing in brackets in Equation 24 are written as calibrated quantities for convenience so that the gain factors out; however, this doesn't restrict their generality.

The continuous signal S(t) is measured as a set of uniform averages, or "integrations", over contiguous fixed time intervals of 0.5 sec. Each integration is converted to digital units by a digitizer. The beam smearing in 0.5 sec results in a small loss of spatial resolution as the beams scan the sky, which we won't consider further here. The digitization interval is set to be small compared to the noise in one integration, so that there is no significant added noise because of the digitization. The resulting discrete data set can be described by making the approriate modifications to Equation 24:

$$S_i = G_d(t_i) \left\{ \Delta T_i + \sum_{p=1}^{m} W_p(t_i) + \sum_{q=1}^{n} U_q(t_i) + O(t_i) + N_i \right\}$$ (25)

where the gain G_d now also accounts for the digital conversion, t_i is the time at the midpoint of the ith integration, and i ranges from 1 to N_{obs}. N_{obs} is 2 x the number of seconds since launch, which can be a very large number (e.g., $\sim 10^8$ in a two-year mission).

3.2 THE LINEARIZED EQUATIONS

Three assumptions must be made to reduce Equation 25 to a set of linear equations to be solved for the sky brightness distribution and the systematic errors. These are:

1. We consider the gain $G_d(t_i)$ and the offset $O(t_i)$ as known. First, the experiment design includes redundant methods for determining the gain and tracking its variations (Smoot et al., 1990, Bennett et al., 1991). Also, the offset O is slowly varying and can be determined by averaging over the time series long enough to reduce both the noise and the real signal (which is modulated by the spacecraft rotation) to negligible proportions.

2. We assume that each systematic error U_q may be written in the form

$$U_q(t_i) = E_q f_q(t_i)$$ (26)

where $f_q(t_i)$ is a known functional time dependence and E_q is an unknown amplitude. This presumes that we are clever enough to predict the functional form of the systematic errors as they depend on time, although we might be uncertain about their amplitudes. For example, the major systematic error which we have identified is the susceptibility of the radiometer output to an externally applied magnetic field, and can be expressed in this form (see discussion in Section 3.4).

These assumptions lead to the form

$$D_i = \Delta T_i + \sum_{q=1}^{n} E_q f_q(t_i) + N_i$$ (27)

where we define the corrected and calibrated differential measurement as

$$D_i \equiv \frac{S_i}{G_d(t_i)} - O_i - \sum_{p=1}^{m} W_p(t_i) .$$ (28)

3. Finally, we must pixelize the map. The quadrilateralized-cube pixelization scheme described by Wright (1991) is used with a choice of $N_{pix} = 6 \times 32 \times 32 = 6144$, which

yields a mean pixel size of approximately 2.6° square. As with the finite integration time, this also tends to slightly degrade the resolution of the resulting map. The desired skymap thus becomes the discrete set of 6144 unknown pixel brightness temperatures T_k, which we use to represent each ideal map differential

$$\Delta T_i = T_{l(i)} - T_{m(i)} \tag{29}$$

The index $l(i)$ denotes the pixel pierced by the positive horn of the differential pair at the time t_i, while $m(i)$ denotes the pixel pierced simultaneously by the negative horn. With this substitution for ΔT_i in Equation 27, the latter becomes N_{obs} linear equations in the unknowns T_k and E_q:

$$D_i = T_{l(i)} - T_{m(i)} + \sum_{q=1}^{n} E_q f_q(t_i) + N_i \tag{30}$$

3.3 A MATRIX SOLUTION

Equation 30 is the general basis for a least-squares solution. Formally, we wish to minimize the chi-square sum

$$\chi^2 = \sum_{i=1}^{N_{obs}} \frac{\left\{ D_i - \left[T_{l(i)} - T_{m(i)} + \sum_{q=1}^{n} E_q f_q(t_i) \right] \right\}^2}{\sigma_i^2}. \tag{31}$$

for the map T_k and error coefficients E_q, where σ_i is our best estimate of the noise term N_i. If we neglect the noise term in Equation 30, it may be expressed in matrix form as

or

$$V \cdot t = d \tag{32}$$

where V is a rectangular matrix of dimension $N_{pix}+n$ by N_{obs}. The ith row of V contains 1 and -1 at the map pixel indices $l(i)$ and $m(i)$ respectively, and $f_q(t_i)$, which we write as f_{iq}, at the indices $N_{pix} + q$. t is the generalized map vector whose elements are the desired map pixel brightnesses for $j \leq N_{pix}$ and the systematic error coefficients E_q for $j = N_{pix} + q$. d is the data vector of length N_{obs}. This is a highly overdetermined set of linear equations which can be solved in a least squares sense by appealing to the chi-square sum (Equation 31) for σ_i = constant, which can be shown to lead to the normal equations (e.g., see Franklin, 1968, p. 50 ff.)

$$M \cdot t = V^t \cdot V \cdot t = V^t \cdot d \tag{33}$$

with the solution

$$t = (M^{-1} \cdot V^t) \cdot d \tag{34}$$

where V^t denotes the transpose of the matrix V, and the "cross-products", or "moment" matrix $M = V^t \cdot V$ is a square matrix of dimension $N_{pix}+n$. More generally, the least squares solution may be found for the case of arbitrary noise σ_i by direct differentiation of Equation 31, leading to the equivalent of Equation 33 for arbitrary σ_i :

$$\sum_{j=1}^{N_{pix}+n} \left(\sum_{i=1}^{N_{obs}} \frac{v_{ik} v_{ij}}{\sigma_i^2} \right) t_j = \sum_{i=1}^{N_{obs}} \frac{v_{ik} D_i}{\sigma_i^2} \tag{35}$$

where the v_{ij}'s are the elements of the matrix V; i.e., 1 and -1 for the positive and negative horn pixels, zero for the rest of the skymap pixels, and the f_{iq} for $k > 6144$.

The skymap portion of the matrix M (where both j and $k \leq 6144$) accounts for the vast majority of it elements. For this region it is clear from Equation 35 and the definition of the v's that the sum over j of the product $v_{ik} v_{ij}$ is nonzero only when pixels j and k can be connected by a 60° angle. Also from this equation one sees that the diagonal elements of M count the number of times that the respective pixels are observed, while each $j \neq k$ element is minus the number of times the difference between pixels j and k was measured. Out of the total number of elements $\sim 4 \times 10^7$, the nonzero elements number only about 10^6 (~2% of the total) with the present pixelization, so the matrix is sparse. To make the connection with the simple case we studied in Section 2.6, we note that the system of equations expressed in 33 becomes identical to that in Equation 20 if we uniformly sample the pixels and ignore systematic errors.

Finally, the matrix is singular because any constant skymap (with zero E_q's) satisfies $M \cdot t = 0$. This aspect of M is fundamental but simply expresses the fact that the differential measurement approach is inherently incapable of determining the absolute level of the sky emission; e.g., if a constant value is added to each T in the χ^2 sum it remains unchanged. Formally one may approximate Equations 33 as

$$(M + \varepsilon I) \cdot t = V^t \cdot d \tag{36}$$

where I is the identity matrix and ε is a small factor which removes the singularity of M. We have found the solution to converge for ε as small as 0.001, whereas the diagonal elements of M are typically $\sim 10^4$. This is essentially equivalent to adding one more measurement to each pixel which says that its temperature should be zero.

It is impractical to consider a standard matrix inversion approach because of the large size of M. There are a variety of iterative approaches which have been developed, however, in which an initial guess t^0 in a system of the form of Equation 33 is used to generate a better estimate t^1, which is used to generate an even better estimate t^2, and so on until the solution (possibly) converges. A commonly used approach for generating the successive estimates is the Gauss-Seidel method: Each linear equation k is solved for t_k using, for k = 1, the previous guess t^i alone, and for successive k's the new elements of t^{i+1} as they become available. Convergence criteria are difficult to establish, however. We have relied on exercises using our inversion algorithms on simulated data sets for which we know the answers to ensure the stability of our operational solutions, and have found no hint of unstable tendencies.

Computationally speaking, the recovery of maps from the DMR data set is not exceptionally demanding in spite of the huge (N_{obs}) number of equations implicitly involved. This is because the normal equations (Equations 33 or 35) are conveniently collapsed over the time index i; further, because of the sparseness of M, only about 10^6 elements must be stored for each DMR channel. Hence the elements of M and the vector $V^i \cdot d$ may be obtained as running accumulations with time, and may be used at any time as input to the Gauss-Seidel algorithm to produce maps.

3.4 SYSTEMATIC ERRORS

The quantity $M^{-1} \cdot V^i$ in Equation 34 may be thought of as a matrix which projects the N_{obs}-space vector of data d onto the generalized map vector t, which exists in a much smaller subspace of $N_{pix}+n$ dimensions. This view brings into play concepts which are useful to understand the relationships between systematic errors and the skymap. To illustrate, let us imagine that we haven't allowed for any systematic errors and t represents only the skymap. In the event that d in fact contains a systematic error vector s (e.g., $d = d_0 + s$ where d_0 is purely due to the sky), we must consider whether the projection of s has 1) all, 2) some, or 3) no components which lie in the subspace of t. If the latter, then we may proceed with ignorant bliss because there will be no consequence in the map. If the first case applies, however, then s would be indistinguishable from something that belongs in our skymap t.

A simple example would be a demon who adds a small, constant voltage V_{demon} to our signal everytime the positive horn crosses pixel x, and minus the same voltage when the negative horn crosses the same pixel. If we knew everything about the demon except for the magnitude of the added voltage, we would be helpless because the equations would be linearly dependent and there would be no unique solution for either the pixel temperature or the voltage increment. However, if the demon were known to work a standard 8-hour shift, then the added signal could be readily determined by its predictable modulation. It would have a skymap component if we didn't allow for it — pixel x would appear to have a temperature increment of $V_{demon}/3G_v$ — but its inclusion in t-space as a signal of unknown magnitude but known time dependence would permit its determination independently of the skymap. Hence the signal from the union demon would fall into the middle of the above categories.

The magnetic susceptibility, or the tendency of the radiometer output to respond to the strength and orientation of an external magnetic field, produces an important systematic error which also falls into this category, having components with varying degrees of orthogonality to the skymap. The susceptibility is largely due to the Dicke switch which alternates the input to the receiver from one horn to the other. This is a ferrite device which is generically susceptible to an external magnetic field such as that due to the Earth. The magnetic susceptibility error signal can be written

$$S_{mag} = \mathbf{B} \cdot \boldsymbol{\chi} \qquad (37)$$

where the Earth field vector **B** can be accurately computed in the frame of the Dicke switch as a function of time, and the linear susceptibility χ expressed with unknown coefficients which are three respective E_q's in the generalized map vector **t**.

The error signal due to the magnetic field component parallel to the spacecraft spin axis (X-axis) is very nearly orthogonal to the skymap because it is not modulated by the spacecraft spin. Although a strong signal may result when two given pixels are observed, the same strong signal is still seen after half a spacecraft rotation when the pixels are reversed and a supposed sky signal should also reverse. The r-axis (radial, or orthogonal to the plane of the horn opening angle) and t-axis (tangential, or parallel to this plane) magnetic susceptibilities are modulated by the spacecraft spin and their effects on the skymap have been studied in simulations. For the magnitudes of the susceptibilities which were measured prior to launch (Smoot et al., 1991a, 1991c), these lead to important but different systematic effects on the skymaps. For example, the r-axis susceptibility produces a signal consistent with a systematic East-West gradient around a circle of constant declination, while the t-axis susceptibility produces a North-South gradient. The former also tends to be orthogonal to the skymap because of the impossibility of having a brightness distribution which consistently increases all the way around a circle, although the sizable second-order geometric effects in the relation of the COBE orbit to the Earth's magnetic field lead to a significant quadrupole term. The first-order effect of the t-axis susceptibility is to produce a dipole, however, so that one can say that this error has a significant projection onto the skymap. Nevertheless, these susceptibility terms are significantly modulated by the 24-hour rotation period of the Earth due to the offset between the rotational and magnetic field axes of the Earth. Figure 6 shows that this modulation is large and provides a signal which is clearly orthogonal to the map.

It is clear that certain types of error are not directly handled by this formalism. The magnetic susceptibility is a second-order effect in the Dicke switch inasmuch as the ferrite is magnetically saturated and its internal field is thousands of times stronger than the modulated external field. It is conceivable that it is subject to as yet unknown time-dependent effects such as vacuum outgassing or radiation dose that lead to a variation of the respective susceptibility coefficients with time. Nevertheless, this case may be studied by obtaining solutions for **t** over intervals and looking at the time dependence of the coefficients.

At present other errors we have identified don't fall as neatly into this scheme as does the magnetic susceptibility case. The signal due to Earth emission diffracted around the shield is an important error and should be a separable function which depends on the nadir angle of the spacecraft spin vector and the rotation angle of the spacecraft; however, the model function is hard to calculate in a believable way. Another case is that of the Moon, which produces a signal so large that the data simply must be discarded when it is within 15° - 20° of the main beam, and may be corrected outside of this range by knowledge of the Moon brightness and the antenna patterns (we have not implemented this correction yet and currently discard all data taken within 25° of the Moon). Errors in this process will be due to uncertainties in our knowledge of the antenna patterns and the Moon brightness, which can't be modeled easily. Other possibilities for errors include radio frequency interference (RFI), or any other kind of signal with unpredictable time dependencies. Nevertheless, the fact that such signals are correlated with the position of the spacecraft with respect to the Earth or Moon would allow estimating functions to be used and limits to be set on potential map errors. An alternate procedure which we currently use is to produce maps of the sky in geocentric or other relevant coordinates in which the error will be fixed, so that the signal may be built up against the noise and its behavior better characterized.

Finally, errors which have no correlation with any quantitative aspect of the experiment will not be detectable in any case. On the other hand, such errors are likely to have little impact on the maps other than to add a small increment of noise.

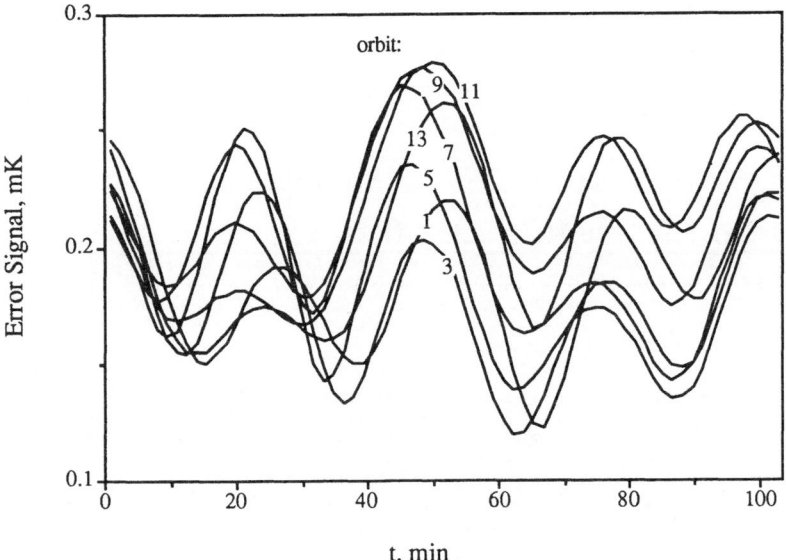

Figure 6: Error signal envelope due to a t-axis magnetic susceptibility of 1 mK/Gauss, as a function of time for several orbits in a 24-hr period. Every other orbit is plotted. The error signal is modulated by the 0.8 rpm spacecraft rotation and only the absolute envelope is shown. Because the orbit precesses at the rate of only 1° per day, any real sky signal must repeat from one orbit to the next, whereas the 24-hr modulation on this error signal is nearly a factor of two.

ACKNOWLEDGEMENTS

The mapping approach reported here represents the efforts of many individuals over the years, including the members of the Science Working Group and the many professionals associated with the COBE Project, with particular mention of the DMR Principal Investigator George Smoot. Ned Wright introduced the idea of including systematic errors in the map solution and contributed significantly to the development of the mathematics. Sergio Torres implemented the early computer algorithms for the map solution, which were later taken over and skillfully refined by Phil Keegstra. Appreciation is expressed to Al Kogut and Luis Tenorio for their efforts in providing materials for this lecture, and to Charley Backus for a lesson in linear algebra. Special thanks are due to Charley Lineweaver for his careful reading of the manuscript and his many helpful comments. The preparation of this paper represents one phase of the research carried out at the Jet Propulsion Laboratory, California Institute of Technology, under contract NAS7-100 to the National Aeronautics and Space Administration. The COBE program is supported by the Astrophysics Division of NASA's Office of Space Science and Applications.

REFERENCES:

Bennett, C.L., et al., 1991. In preparation.

Bracewell, R.N. and E. K. Conklin, 1968. *Nature* **219**, 1343-1344.

Chandrasekhar, S., 1960. *Radiative Transfer*, Dover Publications, Inc., New York .

Evans G.and C. W. McLeish, 1977. *RF Radiometer Handbook*, Artech House, Inc., Dedham, Massachusetts.

Franklin, J. N,.1968. *Matrix Theory*, Prentice-Hall, New Jersey.

Janssen, M.A., 1978. COBE Memorandum #4007.

Krause, J.D., 1960. *Radio Astronomy*, McGraw-Hill Book Company, New York .

Mather, J.C. et al., 1990. *Ap. J.* **354**, L37-L40.

Peebles, P.J.E. and D. T. Wilkinson, 1968. *Phys. Rev.* **174**, 2168.

Smoot , G.F., et al., 1990. *Ap. J.* **360**, 685-695.

Smoot, G.F. et al., 1991a. *Ap. J.* **371**, L1-L5.

Smoot, G. F., 1991b. Lecture I, this proceedings.

Smoot, G. F., 1991c. In *After the First Three Minutes*, AIP, eds. Holt, Bennett, &Trimble.

Torres, S., et al., 1989. In *Data Analysis in Astronomy III*, ed. V. Di Gesu et al. Plenum, New York, 319.

Wright, E. L., 1991. This proceedings.

FUNDAMENTAL AND PRACTICAL LIMITS TO THE SENSITIVITY OF SUBMILLIMETER ASTRONOMICAL OBSERVATIONS

J.M. LAMARRE
Institut d'Astrophysique Spatiale
B.P. 10
91371 Verrières le Buisson cedex
France

ABSTRACT. Unwanted radiation emited by sources in the foreground and immediate environment of the detectors is much more important than what astronomers want to observe. The success or the failure of submillimeter observations depends mainly on the techniques used to remove this parasitic radiation and its fluctuations such as photon noise. We analyse with some detail the formation of photon noise in multimode instruments and show that the usual simplified equations may give significantly wrong results in practical cases. Other sources of fluctuations, such as the atmospheric noise, and the way used to partly remove them are also described.

Keywords : Submillimeter Waves, Astronomy, Photon Noise

1. Submillimeter Sources of Noise

1.1. 25 YEARS TO ACHIEVE A GOOD SUBMM MEASUREMENT OF THE COSMIC BACKGROUND

In 1965 Amo A. Penzias and Robert W. Wilson [1] measured the cosmic background radiation (CBR) at a wavelength of 7 cm. This was the unwanted result of a search for noise reduction on the large antenna of Holmdel (New Jersey). Ten years later, in 1975, Woody et al. [2] published the first sub-millimetric spectrum of the remnant radiation. The wavelength range of the observation (0.4 mm to 4 mm) covered the region of the maximum spectral brightness of the CBR and showed without any ambiguity the expected spectrum of a ≈ 3 K Blackbody. This was obtained with a sophisticated cryogenic instrument aboard a stratospheric balloon, developed especially for this purpose and using all the technological and conceptual refinements available in the sub-millimeter wavelength range. Nevertheless, the obtained spectrum showed noticeable deviations from a perfect blackbody. The history of published measurements showing excesses or depletions in the 3K Blackbody spectrum has been exposed by B. Carr during his lecture [3]. We know it now, all of them were due to measurement errors.

The nearly perfect spectrum of the satellite COBE comes 25 years after the discovery of Penzias and Wilson. This long delay significantly shows how difficult it has been to obtain results of high quality in the submillimeter range.

M. Signore and C. Dupraz (et al.), The Infrared and Submillimetre Sky after COBE, 409–421.
© 1992 *Kluwer Academic Publishers.*

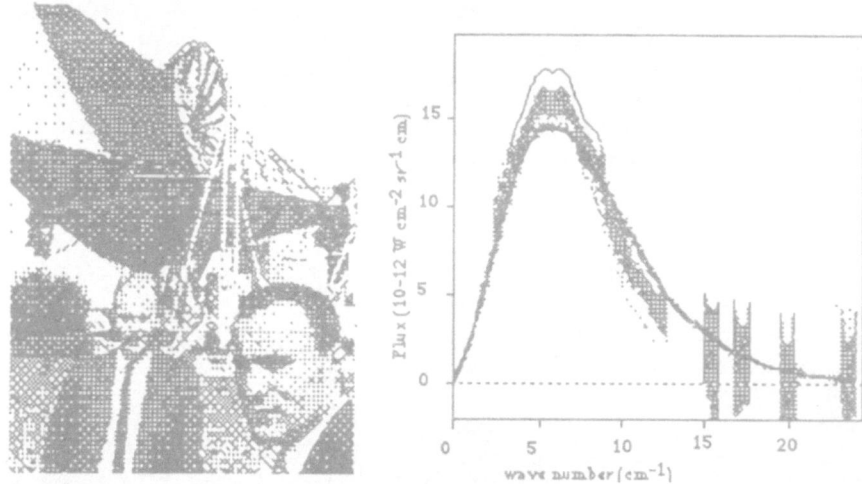

Figure 1 : Left) The discovery of the remnant radiation at 3 K has been made in 1965 by Penzias and Wilson at a wavelength of 7 cm. They did not expect to find what was considered for a moment as an inexplicable residual noise. Thanks to discussions with astyrophysiscians of Princeton University, they could make the link with the theory of universe expansion.
Right) Ten years later, in 1975, Woody et al. published this sub-millimeter spectrum of the cosmic background. It was obtained by a balloon borne cryogenic instrument specialy designed for this purpose.

1.2. SOURCES OF UNWANTED RADIATION

It is true that the sub-millimeter range is somewhat late with respect to its neighbour wavelength ranges, the Infrared and Radio domains and that it corresponds to a technological gap in the electromagnetic spectrum, region where it is difficult, for example, to build very sensitive detectors. Nevertheless, the main problems encountered by observers of the CBR is the large amount of unwanted radiation coming from objects situated in the foreground with respect to the cosmologic sources. Some of these sources are extensively described in other papers of this edition (cf. contributions of B. Carr [3], F. Boulanger [4] and others).

It is convenient to separate these sources of unwanted background in three families :

- Astronomical sources : Extragalactic (emission of galaxies), galactic (dust in the galaxy, even at high galactic latitudes), solar system (zodiacal scattered and thermally emitted radiations)

- The atmosphere is a huge source for all ground based submillimeter observations. Its emission is an important fraction ($\geq 0,1$) of that of a blackbody at ≈ 270 K and therefore larger than any astronomical source by orders of magnitude. Even at stratospheric balloon altitudes, the atmospheric emission is still a source strong enough to jeopardize the shortwave part of the spectrum. The first measurement of the CBR with a precision better than 10% (COBE) had to be done out of the atmosphere.

- Instrumental sources are unavoidable, even for cryogenically cooled satellite experiments. The temperatures that can be reached in a detection system are :

Ground based telescope : 200 K (winter at south pole) to 300 K

Passively cooled "warm" telescope in space : 100-200 K.

Cryogenically cooled Telescope in space : 4-10 K
Cryogenically cooled instrument : 2-4 K

The radiation from blackbodies at these temperatures are still strong at $\lambda = 1$ mm. It is thus not possible to reduce down to negligeable values the sub-mm emission from the instrument by cooling it, as it is usually done for infrared experiments (IRAS, ISO).

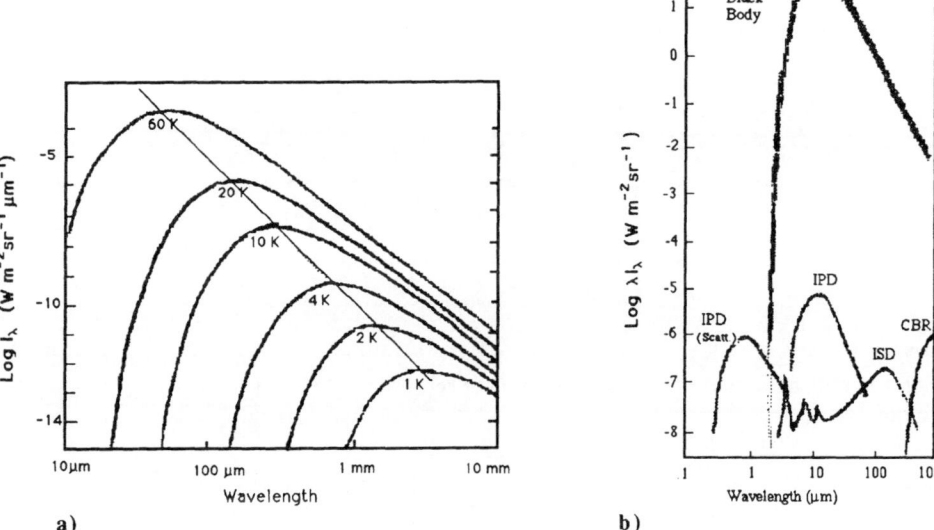

Figure 2 : a) Even for very low temperature instruments (2 to 4 K), the maximum thermal emission of the detector neighbourhood will have its maximum in the submillimeter range. (b) The atmospheric emission is an important fraction of that of a blackbody at three hundred Kelvin. It is by far the main foreground source for ground based submillimeter observations. The brightness of interplanetary dust (IPD), interstellar dust (ISD) and the cosmic background radiation (CBR) are much smaller.

1.3. FLUCTUATIONS OF THE UNWANTED FLUX

The measurement of small sources cancelled by a very large foreground source is a common challenge for groundbased, radio and infrared astronomical observations. Special techniques have been developed to remove the continous part of this emission. The most used is the sky chopping which consists of making all measurements as the difference of brightness between two regions of the sky observed alternately with a frequency of several Hertz. Then only fluctuations of this radiation are detected. These fluctuations have two main sources : one is photon noise and the other one changes in the physical conditions of the source of the radiation, especially temperature changes. In section 2, the basic equations of thermal radiation and its fluctuation will be derived. In section 3, other sources of noise will be analyzed, with a special interest in atmospheric noise.

2. Blackbody Radiation and Fundamental Fluctuations

2.1. BASIC QUANTITIES

Let us consider an element of area dA emitting in a solid angle $d\Omega$ at an angle Θ of its normal a radiative power dW_ν per element of frequency $d\nu$. The brightness, or specific intensity of this source is given by :

$$I_\nu = dW_\nu / \cos\Theta \, dA \, d\nu \, d\Omega$$

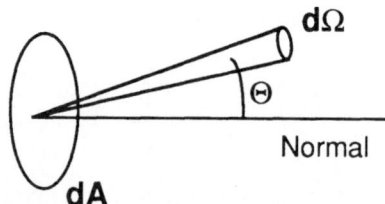

Figure 3 : definitions

The important property of the brightness is that this quantity is kept constant when the radiation is propagating through a non-scattering non-absorbing medium. It is the basic quantity in photometry and radiative transfer (Chandrasekar, "Radiative Transfer"). The brightness of a blackbody is given by the Planck function.

$$I_\nu = 2 \, h\nu \, (\nu^2/c^2) \, (e^{h\nu/kT}-1)^{-1} \qquad (J \, s^{-1} \, Hz^{-1} \, m^{-2} \, sr^{-1})$$

I_ν is the energy by unit of time, by unit of frequency, by unit of surface and by unit of solid angle radiated by a blackbody.

2 is the number of polarizations

$h\nu$ is the energy of a photon of frequency ν

$(\nu^2/c^2) = 1/\lambda^2$ is the number of space modes by unit of beam throughput. This can be quickly verified for example by computing the diffraction pattern for a beam of aperture number N. The Full Width Half Maximum diameter of the diffraction pattern is equal to $1.22 \, \lambda N$. The beam throughput occupied by one radiation mode is then :

$$A\Omega = (\pi/4)^2 \, (1/N)^2 \, (1.22 \, \lambda N)^2 \qquad \approx \lambda^2$$

$(e^{h\nu/kT}-1)^{-1}$ is the Bose-Einstein occupancy factor which gives the mean number of bosons by cell in the phase space, i.e. for photons by mode of radiation.

This gives a simple view of the Planck function : energy = number of photons per mode x energy of photons x number of modes. This picture will be usefull to derive the fluctuations of the blackbody radiation.

2.2. PHOTON NOISE

2.2.1. Usual expression of photon noise. The expression for photon noise was first derived only for the blackbody radiation [5,6]. Later [7] the use of the partition theorem allowed the extension of the theory to sources with any spectrum. For bosons, the variance $<\delta N^2>$ of the number N of particles occupying g cells of the space of phases [8] is given by :

$$<\delta N^2> = <N> + <N^2>/g \qquad (1)$$

The domain of validity of this formula has been extended [9,10,11,12] by showing that it was also applicable to the number of photons detected with a quantum efficiency less than one. On the other side, one must stress the point, as A. Kastler did when speaking of lasers, that this formula is valid only if the total number of particles is distributed among a large number g of cells in the space of phases, i.e. if the radiation contains a large number of modes.

Let us define a measurement process as follows : a photon detector of area A_d counts photons during a time T in a band Δv around the frequency v coming from a source seen with a beam étendue Ω from the detector. The beam throughput of this process is equal to

$$G = \int A_d \cos\Theta \, d\Omega \approx A_d \, \Omega \qquad (2)$$

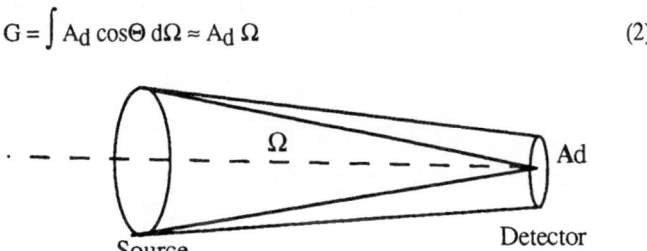

Source   Detector

Figure 4 : geometry of the measurement process

The total number of cells in the phase space defined by this process is :

$$g = G(v/c)^2 \, \Delta vT \qquad (3)$$

The quantities $G(v/c)^2$ and ΔvT are dimensionless. They are respectively the number of space modes (see §II.1) and the number of time modes. After [13] this expression is valid only if both terms are much larger than one and if the source and the detector have no high space frequency content.

Let us suppose now that these conditions are met, i.e. the beam is produced by an incoherent source and is uniformly distributed over a beam throughput large with respect to the étendue of coherence. This excludes from our argument all instruments having an angular resolution not far from the diffraction limit, then nearly all the instruments used in astronomy. Let Q_v be the mean optical spectral power effectively detected. $Q_v = \eta W_v$ where η is the quantum efficiency of the detector.

$$<N> = Q_v \, \Delta vT / hv \qquad (4)$$

The fluctuation of the detected energy is :

$$<dW^2> = (h\nu^2) <N> (1+ <N>/g)$$

$$<dW^2> = (Q_\nu \, h\nu + Q_\nu^2 \, c^2 / G\nu^2) \, \Delta\nu \, T \tag{5}$$

The fluctuations of the electrical signal of the detector are such that the electrical Noise Equivalent Power is equal to :

$$NEP_{ph}^2 \, B = <dW^2> / T^2 \tag{6}$$

where B is the equivalent bandwith of the measuring process and is equal to 1/2T for a perfect integrator of duration T [14]. By integration over the optical spectrum we obtain :

$$NEP_{ph}^2 = 2 \int Q_\nu \, h\nu \, d\nu + 2 \int Q_\nu^2 \, (c^2 / G\nu^2) \, d\nu \tag{7}$$

For unpolarized radiation, the number of modes has to be multiplied by two and it comes

$$NEP_{ph}^2 = 2 \int Q_\nu \, h\nu \, d\nu + 2 \int Q_\nu^2 \, (c^2 / m \, G\nu^2) \, d\nu \tag{8}$$

where m is the number of polarizations (m = 1 or 2). This formula is usually found in the litterature with m = 2 and with a developed expression of Q_ν for a blackbody :

$$Q_\nu = 2 \, A_d \, \Omega \, h\nu \, (\nu^2/c^2) \, (e^{h\nu/kT}-1)^{-1} \tag{9}$$

A few papers [9,10,15] dedicated to bolometric detection treat the case of grey bodies, of cold filters and of non perfect photon detectors. These cases can be described by the simpler equation (8) of this paper. Q_ν is then to be computed taking in account the emissivity of the source $\varepsilon<1$, the quantum efficiency of the detector $\eta <1$ and the transmission of the cryogenically cooled filters $\tau < 1$.

$$Q_\nu = 2 \, \eta \, \varepsilon \, \tau \, A_d \, \Omega \, h\nu \, (\nu^2/c^2) \, (e^{h\nu/kT}-1)^{-1} \tag{10}$$

In equation (8) NEP_{ph}^2 is an electrical NEP, that is the incident power that would give a signal equal to the noise amplitude spectral density if the quantum efficiency of the detector were equal to 1. To obtain a "true" (or "optical") NEP, one should divide NEP_{ph} by the quantum efficiency η of the detector.

The first term in the right hand side of (8) corresponds to <N> in equation (1). This is the shot noise produced by a Poisson process, were the detected photons are not correlated. The shot noise is the largest term in equation (8) if hv/kT >> 1, i.e. for λT << 20 mm K, which is the case for all current thermal sources in the visible part of the spectrum. In these conditions, the other term in (8) is small. It is called "excess noise" by the opticians.

The second term dominates at radio wavelengths. It is directly linked to the term $<N> / g$ of equation (1) and then to the boson factor of the radiation. This term is proportionnal to the square of Q_v. This means that we cannot consider the different sources contributing to Q_v as statistically independant and simply add the variances of the sources of noise. The case of the addition of different uncorrelated physical sources producing correlated noise is very frequent in astronomy, where the emission of the observed source mixes with the thermal emission from the sky and to that of the telescope and the instrument itself. The underlying physical process is the interference between photons in the detection system. Electrical fields add instead of power to produce fluctuations. This picture has been extensively used by R. Hanbury Brown and R.Q. Twiss to compute photon noise in the case where the number of space modes g is small.

2.2.2. The semi-classical Theory. The quantity (c^2 / Gv^2) is the inverse of the number of modes occupied by the beam in our measuring process, or in terms of optical sciences, the space coherence of the beam. If the source angular dimensions decrease towards zero, this expression increases towards infinity, which is obviously wrong since the minimum number of modes of a radiation is one. The actual maximum value of its inverse should be one. We are then looking for an expression of the space coherence of the beam that would be correct even for small values of G. The photon noise would be given by :

$$NEP_{ph}^2 = 2 \int Q_v \, hv \, dv + 2 \int Q_v^2 \, \Delta(v) \, dv \tag{11}$$

where $\Delta(v)$ is less than one and will be called a partial coherence factor.

Most of the theory necessary to establish the expression of $\Delta(v)$ has been made by R. Hanbury Brown and R.Q. Twiss in their famous work on intensity interferometry. The basic principle used here is to establish what is the correlation coefficient between interferences all over the area of the detector. The detail of the work is a little bit complex and will be found in references [16,17,18 and 13]. The final result can be expressed in terms of the specific intensity $I_v(x,y,\xi,\eta,v)$ between different points of the source $S(\xi,\eta)$ and the detectors $P(x,y)$.

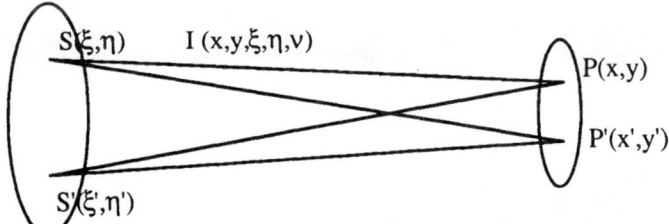

Figure 5 : definition of coordinates on the source and the detector

$$\Delta(v) = Q_v^{-2} \iiiint\iiiint [I_v(x,y,\xi,\eta,v) \, I_v(x,y,\xi',\eta',v) \, I_v(x',y',\xi,\eta,v) \, I_v(x',y',\xi',\eta',v)]^{1/2}$$

$$xR^{-4} \cos\{(2p/cR) \, [(x-x')(\xi-\xi')+(y-y')(\eta-\eta')]\} d\xi d\xi' d\eta d\eta' dx dx' dy dy' \tag{12}$$

where R is the distance between the source and the detector. The photon noise is then given by :

$$NEP_{ph}^2 = 2 \int Q_v \, h\nu \, dv + (1+P^2) \int Q_v^2 \, \Delta(v) \, dv \tag{13}$$

where P is the degree of polarization. The term $(1+P^2)$ is applicable either in the case of fully polarized light (P=1) either in the case of single mode radiation for any type of polarization.

Starting from these equations, one can :

- show the symetrical behaviour of the source and the detector
- reduce the order of the integral to 4, using bidimensionnal Fourier Transform
- find the relation between $\Delta(v)$ and the coherence factor defined by Zernike [19]
- show that for large and uniform sources equation (13) reduces to equation (8)
- show that for small source and detector $\Delta(v) \approx 1$.
- find a scaling parameter μ to make all practical computations easier.

We obtain a set of formulae and definitions [13] that make possible (even if uneasy) the computation of the photon noise for cases that astronomers do meet when designing instruments.

2.2.3. Application to practical cases. It is now possible to see how much wrong we are when we use the expression (8) instead of the exact solution. Let us plot on the same graph $\Delta(v)$ and the ratio betwee $\Delta(v)$ and its approximation (c^2 / Gv^2) versus the scaling factor μ. This factor μ is equal to 0.3 for a circular source when the diameter of the detector is equal to the Full Width Half Maximum of the diffraction pattern. For other circular detectors, μ is proportionnal to the detector diameter.

For a uniform source with circular limits, the use of the approximation (8) always yields an overestimation of the interference part of the photon noise.

a) Case of a circular uniform source

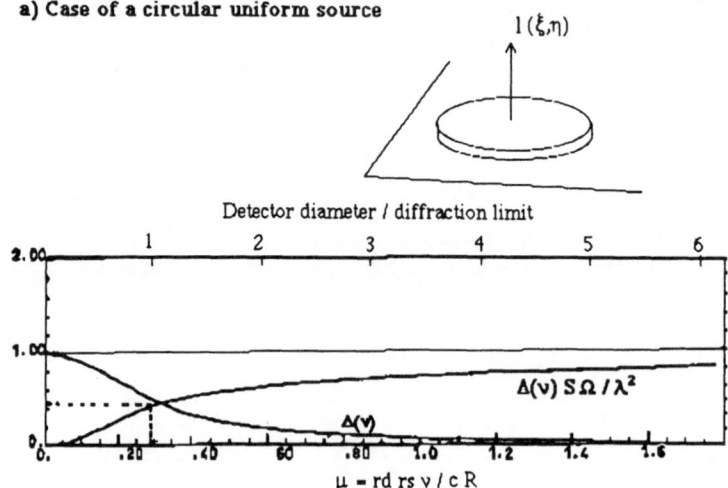

Figure 6

b) case of a Cassegrain telescope

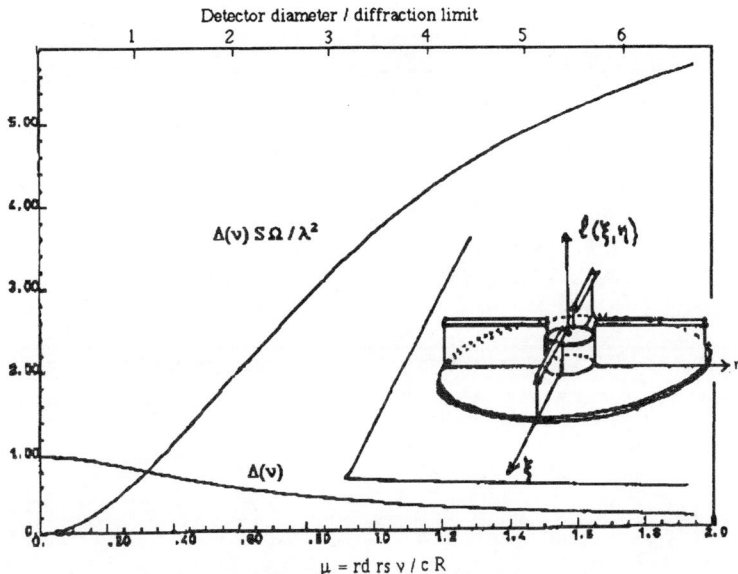

Figure 6 : variation of the partial coherence factor with the diameter of the field of view in two different cases. (a) Uniform source illuminating an uniform detector. (b) Emission from a Cassegrain telescope.

The situation is very different for localized sources isolated or embeded in a large uniform source. We have taken an example frequent in astronomy to illustrate this point. The supports of the secondary mirror in a Cassegrain telescope are represented by linear distributions along a diameter of the instrumental pupil. Then the formula (8) under-erstimates the photon noise for field of view larger than ≈ 1.6 times the FWHM of the diffraction pattern. Let us consider now a complete picture of the emission of a ground based telescope, including the four legs of the "spider" and the emission from the central hole in the primary mirror (figure 6). We can see that the approximation is very optimistic, especially for large throughput experiments.

3. Atmospheric Noise

3.1. SKY CHOPPING, BEAM SWITCHING AND OTHER METHODS FOR FOREGROUND SUBSTRACTION

As shown in figure 2, the brightness of the foreground sources is much larger than that of the astronomical sources. It is thus necessary to remove this unwanted radiation at some stage in the measurement process. The basic principles for that are always differential measurements.

To remove the emission from the instrument itself, one will make the difference between the source to be measured and a reference one that may be a known instrumental reference, such as a blackbody, positionned whenever wanted in front of the instrument. This solution has been used for example in the case of the FIRAS experiment in COBE. The reference source is often an other region of the sky. Then the measured quantity is the brightness difference between different regions of the sky. This solution was chosen for the DMR experiment in COBE (cf. the paper by Jansen in this issue). The continuous componenet of the emission is then unknown and a complete image of the sky is recovered only by a complex data reduction process.

For ground based observations, it is usual to remove the emission coming from the atmosphere and from the instrument by periodic observations of a reference field of view at the same elevation that the observed source but not too far from it, to keep as constant as possible the atmospheric emission. This practice is called sky chopping if the frequency of the change of field of view is several hertz and beam switching for lower frequencies. The need for frequencies larger than a few Hertz comes often from the noise spectrum of the detection system that may show a 1/f behaviour at low frequencies (see fig 7).

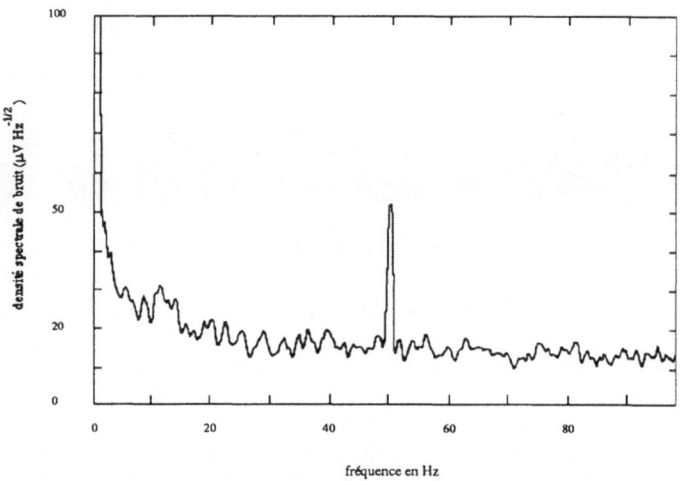

fréquence en Hz

Figure 7: amplitude spectral density of the noise of a bolometer cooled at 0.3 K. The 1/f noise is significant for frequencies less than 20 Hz.

The alternate observation of a reference and of the source is not enough. Residual parasitic signals are often observed. They may come from temperature gradients in the instrument or in the telescope, from changes in the response of the instrument during long observations periods, or from any type of unsymetrical quantity in the measurement process. It is then necessary to remove them by a double weighing operation, the part of the source and the reference beeing periodicaly interchanged. It is striking that all the experiments in COBE have two systems of modulation working at different frequencies and with complementary principles.

When sky chopping is used, it is usual to alternately point the telescope in the two directions that will put the source in the two chopped fields of view. This method, called the ON-OFF or the ON-ON observation is described by figures 8 and 9.

These modulation systems cannot remove all the sources of unwanted signal. There may be some residual source of signal, coming for example from the second derivative of the sky emission, that can either introduce additionnal noise, either look like significant signal and be interpreted as such. We consider that only the detection of empty sky with fluctuations much lower than the detected signal can give a proof that a source has actually been detected.

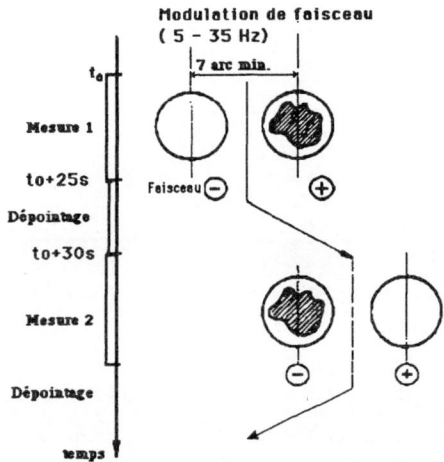

Figure 8 : Measurement made by alternate pointing and sky chopping (ON-OFF)

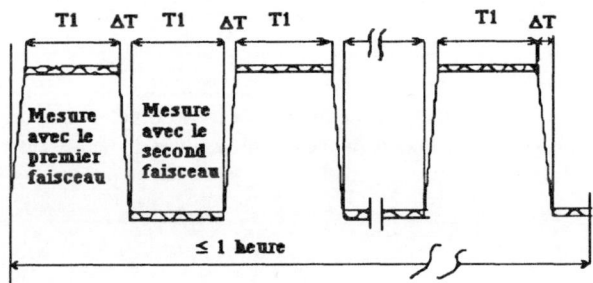

Figure 9 : Sequence of ON-OFFs

3.2. ATMOSPHERIC NOISE

The fluctuations with time and line of sight of the brightness temperature of the sky is very important with respect to the astronomical sources, even at the best observatories in the world. The comparison of two fields of view distant only by small angles with a high enough frequency will reduce it by an important amount. The atmospheric noise is now a limitation to the sensitivity of ground based submillimeter instruments in many cases. The development of new methods able to reduce this noise is necessary to achieve the ultimate sensitivity allowed by the best detectors.

420

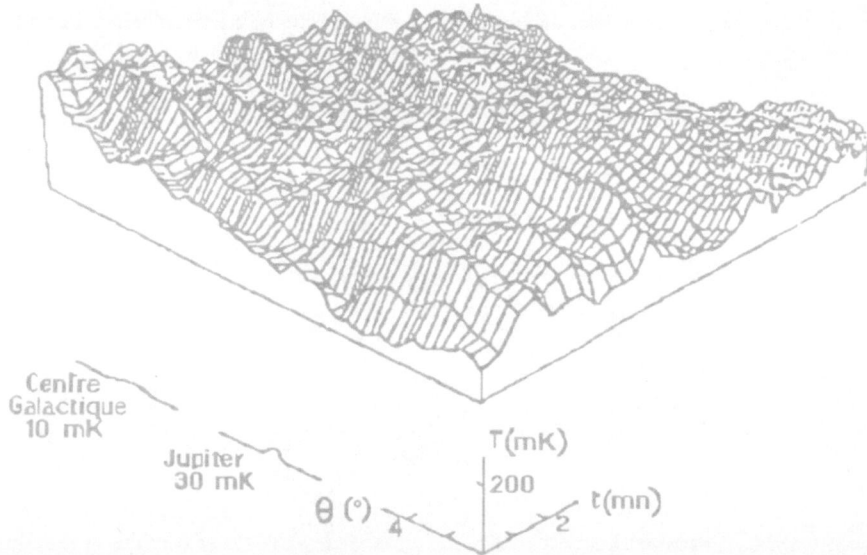

Figure 10 : measurement of the fluctuations of the atmospheric brightness temperature of the sky at Mauna Kea (Hawaii) with time and with the observed direction. (after F. Pajot, Thesis)

4. Conclusion

Unwanted radiation emited by sources in the foreground and immediate environment of the detectors is much more important than what astronomers want to observe. The success or the failure of submillimeter observations depend mainly on the techniques used to remove this radiation and its fluctuation.

References

1. A.A. Penzias and R.W. Wilson (1965) *Ap. J.* 142, 419.
2. D.P. Woody, J.C. Mather, N.S. Nishioka and P.L. Richards (1975) *phys. Rev. Let.,* 34, 16, 1036.
3. B. Carr, in the same volume
4. F. Boulanger, in the same volume
5. A. Einstein (1909) *Phys. z.,* 10, 185.
6. A. Einstein (1909) *Phys. z.,* 10, 817.
7. R. Furth, "uber Strahlungsschankungen nach der Licht Quantumstatistik" (1928) *Z. f. Phys.,* 50, 310.
8. A. Kastler (1964) "Le caractère de "Bosons" des photons et les fluctuations d'un faisceau lumineux", *Compte rendu de la Troisième Conférence Internationale d'Electronique quantique,* P. Grivet et N. Blœmberger éd., p. 4, Dunod, Paris.

9. J.C. Mather (1982) "Bolometer Noise : Nonequilibrium Theory", *Appl. Opt.*, 21, 6, 1125.

10. R.W. Boyd (1982) "Photon Bunching and the Photon-noise-limited Performance of Infrared Detectors", *Infrared Phys.*, 22, 157.

11. R.H. Kingston (1978) *Detection of Optical and Infrared Radiation*, Springer, New York.

12. K.M. Van Vliet (1967) "Noise Limitation in Solid State Photodetectors", *Appl. Opt.*, 6, 1145.

13. J.M. Lamarre (1986) "Photon Noise in Photometric Instruments at Far-infrared and submillimeter Wavelengths", *Appl. Opt.*, 25, 870.

14. R. Bracewell (1965) *The Fourier Transform and its Applications*, Mc Graw Hill, New York.

15. V.D. Gromov (1983) "The Quantum Limit of Sensivity of Radiation Detectors with Nonisothermal Background" (in russian)., *Publications of the Academy of Science of USSR*, IKI, pr. 784.

16. R. Hanbury Brown and R.Q. Twiss (1956) "A Test of a New Type of Stellar Interferometer on Sirius", *Nature*, 178, 1046.

17. R. Hanbury Brown and R.Q. Twiss (1957) "Interferometry of the Intensity Fluctuations in Light : I. Basic Theory : the Correlation Between Photons in Coherent Beams of Radiation", *Proc. Roy. Soc.*, 242 A, 300.

18. R. Hanbury Brown and R.Q. Twiss (1958) "Interferometry of the Intensity Fluctuations in Light : II. An Experimental Test of the Theory of Partially Coherent Light", *Proc. Roy. Soc.*, 243 A, 291.

19. F. Zernike (1938) "The Concept of Degree of Coherence and its Applications to Optical Problems", *Physica* , 5, 785.

**Observations at (sub)millimetre wavelengths:
effect of atmosphere and telescope sidelobes.**

M. Guélin
*Institut de RadioAstronomie Millimétrique,
300 rue de la piscine,
F-38406 St Martin d'Hères,
France*

1. Introduction

The subject of millimetre wave observations is a vast one and I don't have the ambition to cover it, even crudely. Books exist on the radio atronomy fundamentals[1-4], including interferometric techniques[5], as well as descriptions of newly built millimeter-wave instruments[6,7]. The reader is referred to them: most of what they tell is applicable to millimetre-wave lengths. For a quick overview of the subject, the reader is referred to a lecture given by D. Downes three years ago[8].

Here, I will focus on two topics of special importance for millimetre-wave measurements requiring accuracy (e.g. line intensity ratios, background inhomogeneities,...). The first deals with atmospheric absorption and related problems; the second with telescope sidelobes.

2. The terrestrial atmosphere at millimetre wavelengths

A plane wave propagating along the z direction in the atmosphere can be represented by its electric field vector $\mathbf{E}$:

$$\mathbf{E}(z,t) = \mathbf{E}_m e^{-j2\pi\nu(t-n_c\frac{z}{c})},$$

where $n_c = n + jn_i$ is the complex refraction index.

The real part, the true refraction index, $n = c/v_p$ (where v_p is the phase velocity of the wave), is often expressed in terms of the refractivity,

$$N = 10^6(n-1) \simeq 0.223\rho_a + 1760\rho_w/T.[1]$$

The right hand sum, which was empirically derived, is known as the Smith-Weintraub equation. It separates the contribution of the dry air component (first term, where ρ_a is the air density, expressed in gm^{-3}) from that of water vapor (second term, where ρ_w is the density of water vapor).

M. Signore and C. Dupraz (et al.), The Infrared and Submillimetre Sky after COBE, 423–443.

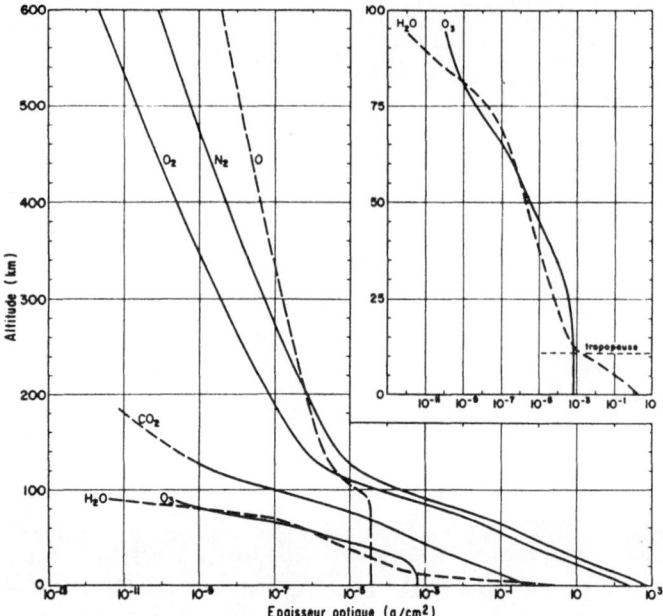

Figure 1: Zenith column densities of the main atmosphere components in function of the altitude–after Craig (1965)

The imaginary part of the refraction index, n_j causes an exponential attenuation of the wave amplitude, and is related to the power absorption coefficient κ_ν by:

$$n_j = \frac{c\kappa_\nu}{4\pi\nu}.$$

At millimetre wavelengths, by clear weather, atmospheric absorption is dominated by rotational and fine structure transitions of molecules in their ground electronic and vibrational state. The strongest transitions are electric dipole transitions from polar molecules, such as water. Intrinsically weaker (typically by a factor of 10^{2-3}), but of considerable practical importance, are the magnetic dipole transitions from radicals.

Of the major molecular constituants of the atmosphere (see Fig. 1), only water vapor and ozone, owing to their bent structure, have a non-zero electric dipole moment. Molecular nitrogen, an homonuclear species, and CO_2, a linear symmetric species, have no permanent electric or magnetic dipole moment in their lowest energy states. These latter, as is the case for 99.9% of gaseous molecules, are singlet states, with electrons arranged two-by-two with opposite spins.

Molecular oxygen, although homonuclear, hence with zero electric dipole moment, has a triplet electronic ground state, with two electrons paired with parallel spins. The resulting electronic spin couples efficiently with the magnetic fields caused by the end-over-end rotation of the molecule, yielding a "large" magnetic dipole moment, $\mu^{mag} = 10^{-20}$ emu. The magnetic dipole transitions of O_2 have intrinsic strengths $\sim 10^{2-3}$ times weaker than the water transitions. O_2, however, is 10^{2-3} times more abundant than H_2O, so that the atmospheric lines of the two species have comparable intensities.

Ozone is formed from the reaction of O_2 with atomic oxygen, the latter resulting from the photodissociation of O_2 in the upper atmosphere. Ozone is inexistent below 10 km; its opacity becomes comparable to that of H_2O above 15 km (see Fig. 1), where its relative abundance may reach 10^{-5}. At short millimeter wavelengths, it causes narrow absorption lines (mostly above 230 GHz).

The other trace atmospheric components CH_4, SO_2, CO, N_2O, NO, etc.. have very low abundances and, even polar, play no significant role, their maximum opacities being typically more than 10 times smaller than those of ozone.

2.1. HYDROSTATIC EQUILIBRIUM

At hydrostatic equilibrium,

$$dp/dh = -g\rho,$$

$$p = \frac{\rho R_a T'}{m_a} [2]$$

where ρ is the density at an altitude h, p is the pressure, $T' \simeq T$ the air "virtual" temperature, and $R_a \simeq R$ the gas constant. $m_a \simeq 29$ is the average molecular weight, and g the local gravitational field.

In the "standard US atmosphere" model, T, the temperature of the air, is given between h=0 and 11 km (the troposphere) by:

$$T = T_o - 6.5(h - h_o)$$

(T in Kelvin, h in km), and is constant from 11 to 20 km (the tropopause).

For relatively small $h - h_o$, $T \simeq T_{ave} = (T(h) + T(h_o))/2$, we find Laplace's hydrostatic formula:

$$\rho \simeq e^{\frac{-gm_a h}{RT_{ave}}} = \rho_o e^{-h/h_o},$$

where ρ_o is the density at sea level and $h_o = RT/m_a g = 8.4(T/288)$ km, the scale height.

The gas column density (expressed in g.cm^{-2}) along the vertical above a point at sea level is:

$$M_o = N_o m_a = \int \rho dh = \rho_o h_o [3]$$

and that above a point at an altitude h:

$$M = M_o e^{-h/h_o}.$$

2.2. WATER VAPOR

The scale height of water, h_w, which results from a fast evaporation/condensation process, is small ($\simeq$ 2 km) compared to the equilibrium scale height $h_o = 8.4$ km. At h= 2.9 km, the altitude of the Pico Veleta observatory, where is located the IRAM 30 m millimeter-wave telescope (Fig. 2), the water vapor column density M_w (or w, "amount of precipitable

Figure 2: The IRAM 30 m-diameter millimetre-wave telescope, located near Granada (Spain) at an altitude of 2.9 km.

water", when expressed in $g.cm^{-2}$, or cm of water) is reduced by a factor of 4, with respect to sea level. At $h = 4$ km, the altitude of Mauna Kea observatory, where are located the JCMT and CSO sub-millimetre telescopes, this factor reaches 6–7. Note, however, that because the latitude of Mauna Kea (Hawaii) is smaller than that of Pico Veleta (southern Spain), the sea-level amount of precipitable water is ~ 1.6 times larger, so that the altitude advantage is almost cancelled.

The value of w on a site can be estimated directly by measuring the air pressure p_{tot} (in mbar), temperature T (in kelvin) and relative humidity RH (in percent), and using the formulas [2] and [3] adapted for water (the air pressure p is replaced by the partial water vapor pressure e):

$$w[mm] = \rho_w[gm^{-3}]h_w[km][4]$$

$$\rho_w[gm^{-3}] = \frac{em_w}{RT} = 216.5e[mbar]/T[K].$$

e is related to $e_s(T)$, the water vapour pressure in saturated air, by:

$$e = RH \cdot e_s/100.$$

e_s is a rather complex function of T, for which an approximate analytical expression, as well as tabulated values, can be found in the litterature (see e.g. ref.[9], p.120). To the zero order, it can be expressed by:

$$e_s[mbar] \simeq 6(T/273)^{18}.$$

At a frequency ν close to a rotational transition $\nu_{lu} = (E_u - E_l)/h$, the optical depth is:

$$\tau_\nu = \int \kappa_\nu(T, P)dh.$$

κ_{lu}, the absorption coefficient of the transition lu is given by the standard asymmetric top formula (see e.g. ref. [10], p.102):

$$\kappa_{lu} = \frac{8\pi^2 h^{1.5}}{3c(kT)^{2.5}}\left(\frac{\rho_w}{m_w}\right)g_I\sqrt{ABC}\,\mu^2 S_{lu}T^{-2.5}\left(1 - \frac{h\nu}{kT}\right)e^{E_l/kT}\nu_{lu}^2\Phi(\nu - \nu_o), [5]$$

Replacing A, B, C, the rotational constants (here in Hz), and $\mu = 1.85$ debye $= 1.85\ 10^{-18}$ esu.cm, the electric dipole moment, by their values for H_2O, and setting $(1 - \frac{h\nu}{kT}) = 1$,

$$\kappa_{lu}^w[cm^{-1}] = 5.7 10^{-24}(\rho_w/m_w)g_I S_{lu}(T/273)^{-2.5}e^{E_l/kT}\nu_{lu}^2\Phi(\nu - \nu_o),$$

S_{lu} is the transition intrinsic strength, $g_I = 3/2$ or $1/2$ is the nuclear relative statistical weight of the ortho and para levels (see below), and $\Phi(\nu - \nu_o)$ the line profile. ν_{lu} is now in GHz.

The line profile is given to a good approximation by the well known Van Vleck and Weisskopf collisional profile (here, multiplied by π)[10,p.342]:

$$\Phi(\nu - \nu_o) = \frac{\Delta\nu}{(\nu - \nu_o)^2 + (\Delta\nu)^2} + \frac{\Delta\nu}{(\nu + \nu_o)^2 + (\Delta\nu)^2}, [6]$$

$$\Delta\nu = \frac{1}{2\pi\tau},$$

where $\Delta\nu$ is the line width and τ the mean time between molecular collisions. At the centre of the line, the second term of [6] becomes negligible, and

$$\Phi(\nu - \nu_o) = 1/\Delta\nu = 2\pi\tau \sim \frac{2\pi}{\frac{\rho_a}{m_a}v\sigma^2}, [7]$$

$v \sim \sqrt{T}$ is the molecular velocity and σ the collisional cross section.

Note that the density in [5] is ρ_w, whereas that in [7] is ρ_a: the absorption coefficient at the centre of the line is proportional to ρ_w/ρ_a. It is independant of the total air pressure, as long as this ratio (hence RH) stays constant. This is not the case, of course, away from the centre, since $\Delta\nu \sim \rho$: as the density drops, the lines become narrower and narrower. In the far wings of the line, the second term in [6] and the contribution from the wings of other lines cannot anymore be neglected. In fact, a better fit to the water emision in the far wings is reached if [6] is replaced by another collisional line shape, called the "kinetic profile"[2,p.155].

The rotational energy level diagramme of vapor is shown on Fig. 3. Each level is denoted, as usual for asymmetric top molecules, by three numbers $J_{K_{-1},K_{+1}}$. J, which is a "good" quantum number, represents the total angular momentum of the molecule; by analogy with symmetric tops, K_{-1} and K_{+1} stand for the rotational angular momenta around the

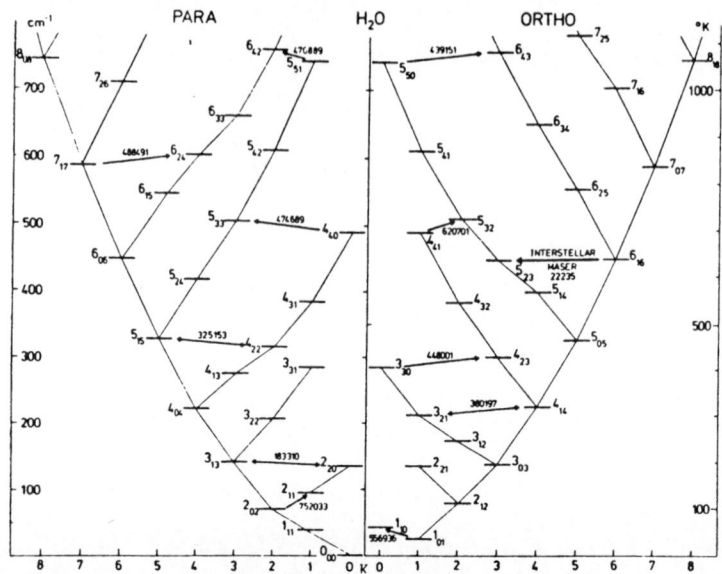

Figure 3: The rotational energy level diagramme of water vapor

axis of least and greatest inertia. Allowed radiative transitions obey the selection rules $\Delta J = \pm 1, \Delta K = \pm 1, 3$, with K_{-1}, K_{+1} : $odd, odd \leftrightarrow even, even$ or $o, e \leftrightarrow e, o$. The levels with K_{-1} and K_{+1} of the same parity are called $para$ levels, those of opposite parity, $ortho$ levels. Transitions between ortho and para levels are forbidden; due to the presence of two symmetrical hydrogen nuclei, the ortho levels have a nuclear statistical weight 3 times larger than the para levels of same J (see e.g. TS, $ibid$).

The opacity and width of the main $H_2^{16}O$ lines are large: $\tau_o = 60$ and $\Delta\nu \simeq 20\text{GHz}$, for the fundamental line in "normal" conditions of p and T, and for $\rho_w = 1\text{g.m}^{-3}$ and $w_w = 1$ mm (dry weather). These lines dominate most of the millimetre and submillimetre atmospheric attenuation; deviations from theoretical lineshapes in the far wings (typically 1/10th of intensity) are accounted for by an "empirical continuum". The rare isotopomer $H_2^{18}O$, a few hundred times less abundant than $H_2^{16}O$, makes a negligible contribution.

2.3. MOLECULAR OXYGEN

The spin of 1 makes of the ground electronic state of O_2 a triplet state ($^3\Sigma$). **N**, the rotational angular momentum couples with **S**, the electronic spin, to give **J** the total angular momentum: **N** +**S** = **J**. The **N**·**S** interaction (and the electronic angular momentum–electronic spin interaction **L**·**S**) split each rotational level of rotational quantum number $N \geq 1$ into three sublevels with total quantum numbers

$$J = N + 1, J = N \text{ and } J = N - 1,$$

the $J = N + 1$ and $J = N - 1$ sublevels lying below the $J = N$ sublevel by approximately $119(N+1)/(2N+3)\text{GHz}$ and $119/(2N-1)$ GHz, respectively[10,p.182]. Note that the two identical ^{16}O nuclei have spins equal to zero and obey the Bose-Einstein statistics; there are only odd N rotational levels in such a molecule.

The magnetic dipole transitions obey the rules $\Delta N = 0, \pm 2$ and $\Delta J = 0, \pm 1$. Transitions within the fine structure sublevels of a rotational level (i.e. $\Delta N = 0$) are thus allowed. The first such transition is the $(J, N) = 1, 1 \leftarrow 0, 1$ transition, which has a frequency of 118.75 GHz. The second, the $1, 1 \leftarrow 2, 1$ transition, has a frequency of 56.26 GHz. It is surrounded by a forest of other fine structure transitions with frequencies ranging from 53 GHz to 66 GHz [11]. The first "true" rotational transition, the $N = 3 \leftarrow 1$ transitions, have frequencies above 368 GHz (368.5, 424.8, and 487.3 GHz).

The rare isotopomer $^{18}O^{16}O$ is not homonuclear, hence has odd N levels and a non-zero electric dipole moment. This latter, however is vanishingly small $(10^{-5}D)$. $^{18}O^{16}O$, moreover, has a very low abundance (few hundred times smaller than the main isotopomer), so that its magnetic dipole transitions (even the $\Delta N = 2$, which have stronger intrinsic strengths), can be neglected.

The line opacity and absorption coefficients of $^{16}O_2$ are given by relations similar to [5] and [6].

2.4. OZONE

As noted above, ozone is mostly concentrated between 11 and 40 km altitude; it shows large seasonal and, mostly, latitude variation. Because of its high altitude location, its lines are narrow: at 25 km, ρ_a, hence $\Delta \nu$, is reduced by a factor of 20 with respect to see level; moreover, the dipole moment of ozone ($\mu = 0.53$ D), 3.5 times smaller than that of H_2O, further reduces the ozone linewidths.

Because of their small widths and despite the small ozone abundance, ozone lines have significant peak opacities, especially above 230 GHz ($\tau_o = 0.2 - 0.3$).

2.5. THE ATMOSPHERIC MILLIMETRE WAVE ABSORPTION SPECTRUM

The water vapor, oxygen and ozone opacities at zenith, calculated for sea level, using the standard US atmosphere model are shown on Fig. 4. One may note the importance of the water line wings above 150 GHz, compared to those of O_2 (a consequence of the absence of electric dipole moment in the latter molecule) and O_3.

Calculations of zenith atmosphere opacity at 2.5 and 2.9 km, the altitude of the IRAM sites, have been made by J. Cernicharo[12]. A computer programme repeating these calculations has been installed on-line on the IRAM telescopes of Pico Veleta and Plateau de Bure; it is activated at each calibration or skydip and allows to interpret the observed sky emissivity in terms of water and oxygen contributions and of signal and image sideband opacities. (During skydips, the antenna is pointed successively at different elevations and the emission of the sky measured at each step; the sky emission variation is fitted by an exponential function of the air mass, and the atmosphere opacity and average temperature readily derived). Note that the opacities derived from sky emissivity observations do not always agree with those calculated from the measurement of p, T, and RH on the site, as water vapor is not at hydrostatic equilibrium.

Some of the results for the band 20 –350 GHz are shown on Fig. 5. One recognizes from left to right, the (blended) forest of fine structure transitions from O_2, near 60 GHz, the $1, 1 \leftarrow 0, 1$ fine structure line of O_2 at 118.75 GHz, the third lowest lines of para water (still 200 K above the ground level), at 183.31 GHz, and the fourth ortho water line

430

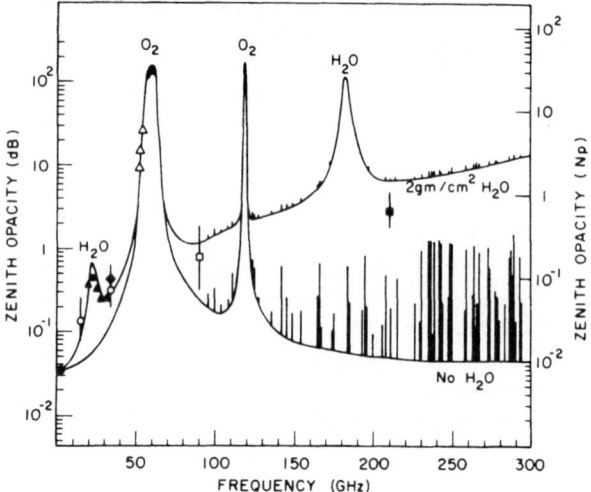

Figure 4: Zenith opacity of oxygen and ozone, with and without water vapor. Dots and triangle are measured opacities– from [2]

(420 K above ortho ground level), at 325.15 GHz. The fundamental line of ortho water $(1_{10} \leftarrow 1_{01})$, at 556.94 GHz is visible of Fig. 6. The water and oxygen lines delineate the 4 atmospheric "windows" of the millimetre spectrum (called the 3 mm, 2 mm, 1.3 mm and 0.8 mm windows). Water is seen to dominate completely atmospheric absorption above 150 GHz.

Less than one millimeter of precipitable water vapor corresponds to exceptionally good winter weather conditions on the sites of Pico Veleta and Plateau de Bure. Such conditions seldom happen even on Mauna Kea. Two to three millimeter of water are standard by clear winter nights at these observatories; six to ten millimeter of water are typical of clear summer nights.

The typical zenith atmosphere opacities, in the dips of the 1.3 mm and 0.8 mm windows (e.g. at the frequencies of the $J = 2 - 1$ and $1 - 0$ rotational transitions of CO, 230.54 and 345.80 GHz), are respectively 0.15–0.2 and 0.5–0.7 in winter. The astronomical signals at these frequencies are attenuated by factors of respectively $\simeq 1.2$ and 2 at zenith, 1.3 and 2.8 at 45 degree elevation, and 1.7 and 6 at 20 degree elevation. Larger attenuations are the rule in summer and in winter by less favorable conditions.

The $J = 1 - 0$ line of CO, at 115.27 GHz, is close to the 118.75 GHz oxygen line. Although this latter is relatively narrow, it raises by $\simeq 0.3$ the atmosphere opacity (which is 0.35–0.4). The atmosphere attenuation is then intermediate between those at 230 and 345 GHz (by dry weather, however, it is more stable than the latter, since the water contribution is small). The measurement of accurate CO line intensity ratios (even not considering the problems linked to differences in beam size and receiver sideband gain ratios) requires therefore good weather, a high source elevation, and a careful monitoring of the atmosphere.

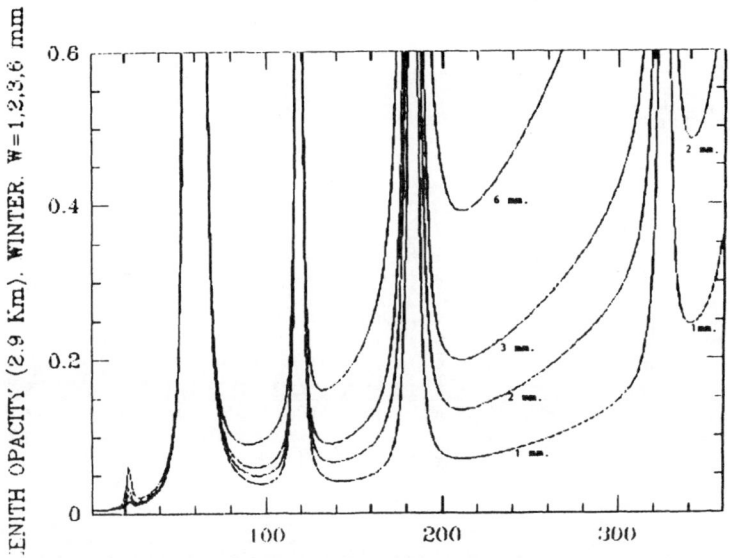

Figure 5: Zenith opacity of the standard winter atmosphere at an altitude of 2.9 km for 1, 2, 3 and 6 mm of precipitable water, in function of frequency (GHz) – from [12]

A catalogue of lines intensities in several standard astronomical sources, measured with the IRAM 30 m telescope has been published[13]. The lines intensities were calibrated by the "chopper-wheel" method, following the above recipes. The reader is referred to this report for details.

2.6. THE SUBMILLIMETRE BAND

Fig. 6 shows an extension of Fig 5. to 1 THz; this time, one has plotted the atmospheric transparency, $1 - e^{-\tau}$, rather than the optical depth τ. (The calculations actually were made for a column density of 1 mm of H_2O and for $h=3.5$ km –the altitude of Mt. Graham (Arizona) where is being erected the 10 m MPIfR/University of Arizona submillimeter telescope)[14]. One sees clearly on this figure, between the lowest lines of H_2O, the four submillimetre atmospheric windows, at 0.75, 0.65, 0.45 and 0.35 mm wavelength.

Fig. 7 shows the atmospheric transmission between 100 and 470 GHz, *measured* at Pico Veleta (altitude 2.9 km) in winter 1987. The observations were made by scanning the absorption spectrum of Saturn with a Fourier Transform spectrometer. The agreement with the calculations for 1.5 mm of H_2O (0.8 mm a zenith) is quite good[14].

A 4-month survey of sky emissivity at Magdalena Mountains, another site above 3 km in the SW USA where should be located the N.R.A.O. Millimeter Array, has shown that conditions of ≤ 1 mm of H_2O are fulfilled during less than 20% of time in winter. This statistics, which is probably typical of sites with altitude between 3 and 4 km, shows that submillimetre observations above 420 GHz (in particular observations of the $J = 4 - 3$ and higher J lines of CO, as well as of the fine structure lines of CI and CII) will be difficult from there.

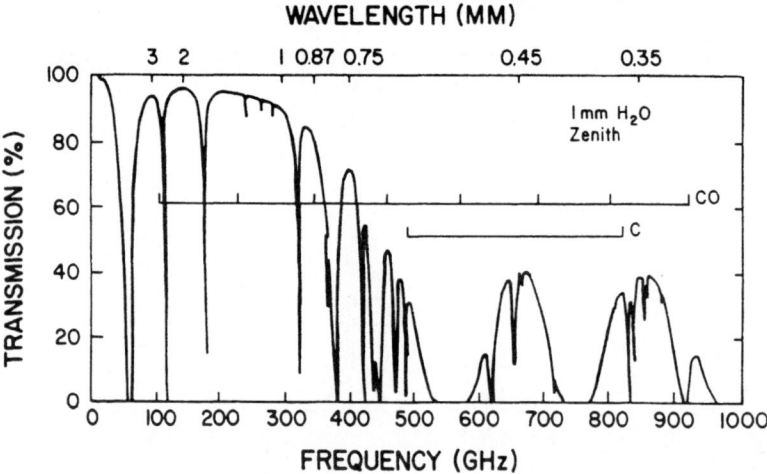

Figure 6: Zenith atmospheric transmission from an altitude of 3.3 km for $w = 1\text{mm}$ – from [14]

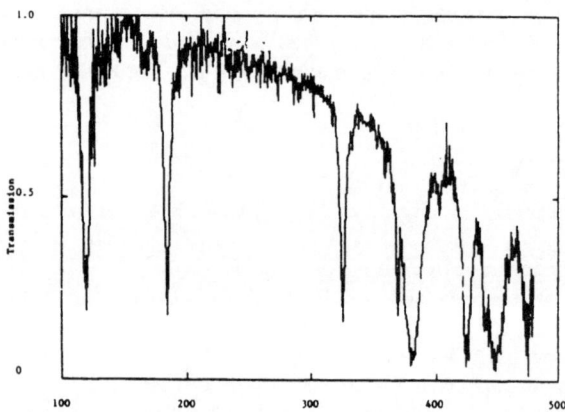

Figure 7: Atmospheric transmission during observations of Saturn, at 28° elev., with the IRAM 30 m telescope– x scale is frequency in GHz from ref. [14]

The best ground submillimetre sites are probably in the high plateaux of Antartica. Sky-emissivity measurements, made at 700 μm in the winter 1987 (Stark et al. 1989) at the NSF South-Pole Basis, show atmosphere optical depths as low as 0.11–0.16. There are plans to build a 2 m-diameter submillimetre telescope on that site.

2.7. RADIO ASTRONOMICAL REFRACTION AND SEEING

A wave travelling through wet air is delayed with respect to the same wave in free space by (see [1]):

$$\Delta t = \int (\frac{1}{v_p} - \frac{1}{c})dz = 1/c \int (n-1)dz = \frac{10^{-6}}{c} \int N(z)dz.$$

The delay due to the water vapor is, from relation [1]:

$$\Delta t_w = \frac{10^{-6}}{c} \int 1760 \frac{\rho_w}{T} dz,$$

or, for a ∼constant temperature,

$$\Delta t_w = \frac{10^{-6}}{c} \frac{1760 \rho_w}{T}$$

(ρ_w in g.m^{-3}).

As noted above, water vapor is not at hydrostatic equilibrium, even on scales of tens of meters. 20% variations of ρ_w between bubbles of sizes 50–100 m are not rare. Such moisture variations cause delay variations and a local bending of the incident wavefront. For a baseline of length D the bending angle is:

$$\delta\theta[rad.] = \frac{c\delta \Delta t_w}{D}.$$

If the telescope diameter is approximately equal to the bubble size, the image can move by an angle

$$\delta\theta[arcsec.] = \frac{360}{T} \delta \rho_w.$$

For a bubble with an extra moisture $\delta\rho_w = 3$ g.m^{-3}, the image can move by as much as 4", an effect often observed at Pico Veleta and Mauna Kea (see Fig. 8 and ref.[15]). The timescale for such motions is D/v (where v is the wind speed) or 3–10 seconds. Amplitudes larger than 20" have sometimes been observed (e.g. in summer, when the inversion layer reaches the telescope). They can lead to substantial distortions of the primary beam for a telescope like the 30 m telescope (whose full width at half-power is $\simeq 12''$ at 230 GHz).

This millimetre-wave seeing (also called "anomalous refraction"[15,16]) has been noticed during the very first 3 mm observations with 30 m telescope in September 1984: we noticed that it was not possible to calibrate the telescope gain on Venus at sunset, the source, despite its size, getting out of the 24"-wide beam every 15 sec. or so!

434

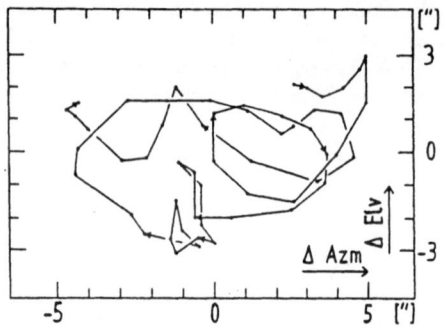

Figure 8: Image motion observed at 3 mm with the IRAM 30 m telescope. –from [15].

3. The far field sidelobes of a single-dish telescope; the error beam

3.1. THE RADIATION PATTERN OF A PERFECT REFLECTOR

The theory of diffraction[17] tells us that the electric field produced at large distances (or in the far field) by an aperture is nothing else but the 2-dimensional Fourier Transform (FT) of the electric field distribution through this aperture:

$$F(l, m) = \int \int E(\frac{u}{\lambda}, \frac{v}{\lambda}) e^{-j(2\pi/\lambda)(lu+mv)} d(\frac{u}{\lambda}) d(\frac{v}{\lambda}),$$

where u, v are the space coordinates in the aperture plane and l, m the angular coordinates of the far field. $F(l, m)$ is referred to as the angular spectrum produced by the aperture distribution. $| F(l, m) |^2$ is the angular power spectrum; it is equal to the auto-correlation function of the aperture distribution $E(u, v)$.

Similarly, the "grading" of a reflector (i.e. the current intensity and phase distribution over its aperture plane) and the electric voltage in the far field (field or directivity pattern) are related by a 2-d Fourier Transform (more simply expressable as a Hankel transform, if the grading has an axial symmetry). The angular power pattern (or beam pattern) is the square of the field pattern and is equal to the autocorrelation function of the grading.

For a uniformly illuminated slot aperture of width L, the grading is the rectangle function
$g(u, v) = 1$ for $-L/2 \leq u \leq +L/2$
$g(u) = 0$ elsewhere. The field pattern at a wavelength λ is

$$F(l, m) \sim \frac{sin(\pi Ll/\lambda)}{\pi Ll/\lambda}.[8]$$

Its square, the power pattern, is an Airy function:

$$F(l, m) \sim \frac{sin^2(\pi Ll/\lambda)}{(\pi Ll/\lambda)^2}.[9]$$

Maximum for $l = 0$, it is reaches zero at intervals, $l = \pm 1/L, 2/L, 3/L...$, and has regularly spaced secondary maxima: $m = \pm 3/2L, 5/2L,...$ Those, are called the first, second, etc.

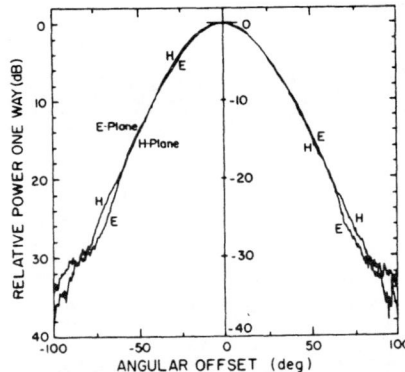

Figure 9: The beam pattern of a corrugated feed horn – from [2].

secondary lobes. The portion of the power pattern comprised between the first positive and negative zeros $(-1/L \leq l \leq +1/L)$ is called the "main beam".

A square aperture with uniform grading has a field pattern of the form:

$$F(l,m) \sim \frac{sin(\pi Ll)}{\pi Ll} \cdot \frac{sin(\pi Lm)}{\pi Lm}.$$

and a power pattern which is the product of two Airy functions [9].

Finally, the field pattern of an uniformly graded circular aperture is a Bessel function of order 1:

$$F(\Theta) = \pi D^2/2 \frac{J_1(\Theta)}{\Theta}$$

where $\Theta = \frac{\pi D}{\lambda}\sqrt{l^2 + m^2}$, is the dimensionless angle coordinate.

The shape of the power pattern along a meridian plane is not very different from that of [9], showing regularly spaced nulls and secondary maxima. Its main lobe, down to 1/5 th of its power is well approximated by a gaussian.

In practice, the reflector grading of a radio astronomical antenna is not uniform, since it results from the illumination of the reflector by a horn. This latter, which has a circular or rectangular aperture, produces a field pattern and a power pattern of the types described above, hence approximately gaussian. (see e.g. Fig. 9). The grading is maximum near the reflector centre and tapers off as e^{-r^2}, until it reaches the edges of the reflector.

The FT of a gaussian being a gaussian, the field and power patterns produced by an under-illuminated reflector (i.e. with a very low illumination at its edges) are also very close to gaussians. They decrease smoothly to zero without noticeable secondary lobes.

In a multi-purpose antenna, the horn will not under-illuminate the reflector. The grading will not be close to zero at the edges of the reflector, but of course will vanish outside the reflector's surface. Intuitively, the far field power pattern will be somewhere between a gaussian and an Airy function and will show some mild secondary lobes. The intensity of these latter depend on the value of the illumination at the reflector edges, referred to as the "taper" in relative power units.

More precisely, the grading along a meridian plane can be seen as the product of a gaussian (the horn field pattern) by a rectangle (the reflector). According to the convolution theorem[17,p.108], its FT, the antenna directivity pattern, is the convolution of a gaussian (the FT of a gaussian) by the Airy function [9] (the FT of the rectangle function). The narrower is the horn field pattern, the broader is its FT: one sees that increasing "taper" (i.e. decreasing the illumination) at the edges of the reflector smoothes the main beam and the sidelobes by a broader function and reduces their intensity.

Four (not independant) parameters characterize an antenna from the astronomical point of view: G_o, the on-axis gain (i.e. the efficiency for detecting point sources), η_b, the beam efficiency (the efficiency for detecting weak extended sources), $HPBW$, the main beam half-power width (which defines the resolving power) and η_s, the fraction of the energy radiated in the near sidelobes.

The on-axis gain is defined as $G_o = \frac{4\pi A_e}{\lambda^2}$, where A_e, the "effective area" is the physical area of a perfect reflector with the same sensitivity to point sources. It decreases when the main beam gets broader due to reflector under-illumination. Narrowing the main beam can be done by decreasing the taper, but this increases the sidelobes and, mostly, the spillover of the horn outside the reflector's edge. Getting the highest gain, or the highest "aperture efficiency" ($\eta_a = \frac{2A_e}{\pi D^2}$), means thus a trade off between under- and over-illumination of the reflector.

For telescopes with large F/D ratios (see below) illumination tapers of 8 to 14 decibels (or 1/6 to 1/16 of the power at the reflector centre) provide the best gain along the telescope axis but first sidelobes between 1 and 2 percent.

The beam efficiency, defined as the fraction of the power radiated in the main beam

$$\eta_b = \frac{\int_{mainbeam} \mid F(l,m) \mid^2 dldm}{\int_{4\pi} \mid F(l,m) \mid^2 dldm},$$

does not depend on the $HPBW$, but mostly on the horn spillover and on the level of the sidelobes. It is increased by under-illuminating the reflector. For general purpose telescopes, designed to observe point sources, as well as extended sources with a high angular resolution, one tries to maximize both G_o and η_b: at taper of 14 dB is probably the best compromise.

3.2. THE DUAL-REFLECTOR ANTENNA – CASSEGRAIN TELESCOPE– EFFECT OF OBSTRUCTION

The main reflectors of large radiotelescope are paraboloids. In a paraboloid, the distance from the focus to the edges of the reflector is larger than that to the refelector vertex (i.e. than the focal distance, F). This results in an extra attenuation (called "space attenuation") at the edges of the reflector, hence in a decrease of G_o and A_e. This decrease cannot be compensated by a reduction of the illumination taper, as this would mostly increase the spillover.

Maximizing G_o thus requires increasing F/D. There are other advantages in increasing the F/D ratio, such as a less critical positioning of the receiver horn phase center and the possibility to use focal plane arrays, and this solution is generally adopted.

For practical reasons, large F/D ratios are achieved with the help of an hyperbolic (Cassegrain) or elliptical (Gregorian) subreflector. The latter, following ray optics, is placed so that one of its foci coincides with the paraboloid focus and the other with the receiver

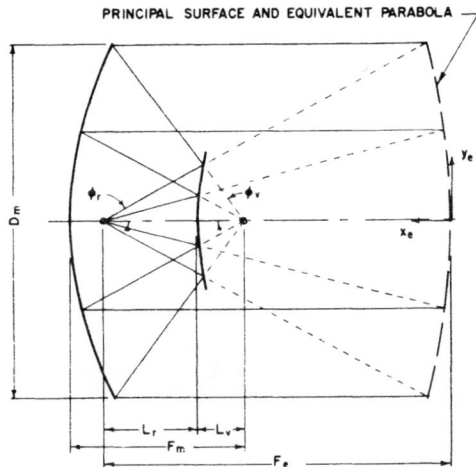

Figure 10: Geometry of the equivalent paraboloid – from [18].

horn center of phase. The Cassegrain arrangement, more compact, is prefered in radio telescopes.

The image of the main reflector, observed from the horn through the hyperboloidal subreflector, has moved to a distance $F_e = F\frac{e+1}{e-1}$, where e is the ellipticity of the hyperboloid. F_e is called the "equivalent focal length" and $m = F_e/F = \frac{e+1}{e-1}$, the "magnification". Many of the properties of the dual-reflector antenna (to the exception of subreflector blockage, however) can be transposed from those of the single paraboloid reflector by replacing F by F_e and using the "equivalent parabola" concept[18] (see Fig. 10).

In many radio telescopes (see e.g. Fig. 2), the second focus of the subrefelector is located behind the paraboloid vertex, in an enclosed, temperature regulated room. For the IRAM 30 m telescope, $F/D = 0.35$ and $m = 27.8$. The large equivalent F_e/D (9.7) minimizes, as discussed, the space losses and make relatively uncritical the positioning of the receivers (as long as their horn pattern are centred on the subreflector) – a notable advantage, since three receivers are used simultaneously on this telescope. Note, however, that the positioning of the subreflector is critical: a displacement of as little as 1 mm is enough at 230 GHz to decrease G_o by 10%; the 230 GHz receiver horn would have to be moved by m^2 mm along its axis to yield the same result.

The Cassegrain telescope has two main drawbacks: i) the blockage of the reflector by the subreflector and its supporting legs (usally, a quadrupod) and ii), for large F_e/D ratios, the large size of the (diffraction) beam waist in the focal plane (less of a problem, however, at millimetre wavelengths).

The effect of reflector blockage is illustrated on Fig. 11. Owing to the linearity of the FT, the field pattern of the blocked reflector is (to a good approximation at millimetre wavelengths) the superposition of the pattern of the unblocked reflector, minus the pattern of the subreflector and quadrupod (not shown on Fig. 11). The subreflector pattern has about the same shape than that of the main reflector (i.e. than [8]), but is much wider (by

438

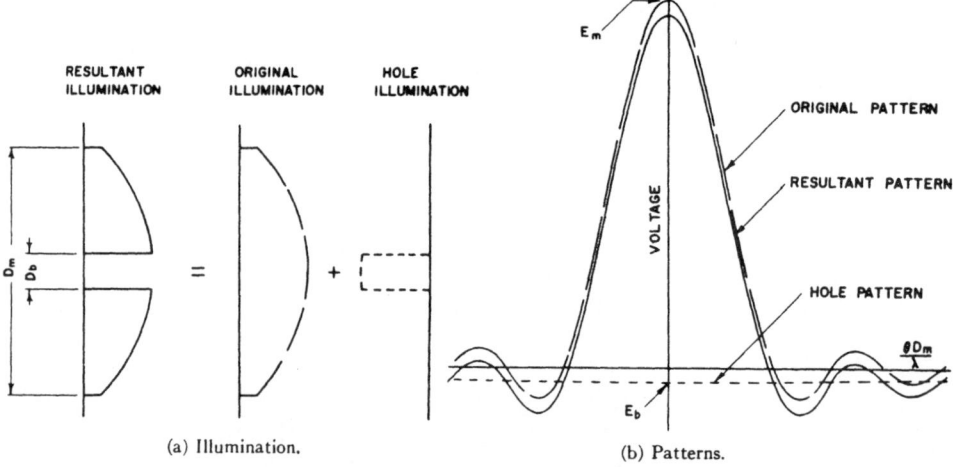

(a) Illumination.

(b) Patterns.

Figure 11: Effect of aperture blocking by the subreflector – from [18].

the ratio of the reflector sizes, e.g. $d/D = 15$).

For the beam (power) pattern, which is the square of the field pattern, the effect of the blockage is to reduce the intensity of the main beam and even sidelobes, while raising the intensity of the odd sidelobes. Mostly, moreover, it scatters power into a wider beam.

Note that for an uniformly illuminated reflector, the decrease in amplitude of the field pattern, due to blockage, is proportional to the reduction in unblocked area, i.e. to $(1 - (d/D)^2)$, so that the gain loss of the main beam (hence the decrease in A_e) is proportional to $(1 - (d/D)^2)^2 \simeq (1 - 2(d/D)^2)$. For a reflector illuminated with an edge taper of 14 dB, this gain loss is $\simeq (1 - 2a(d/D)^2)$, where $a \simeq 2$.

The effect of the quadrupod legs blockage can be treated in a similar way. The leg shadow, viewed from the horn, appears as two dark, perpendicular lanes crossing the entire reflector. Their field pattern has the shape of a cross and must be substracted from the reflector pattern. Once squared, the cross pattern adds a non-axisymmetrical contribution to the antenna beam pattern.

3.3. SURFACE ERRORS– THE ERROR BEAM

So far, we have assumed that the reflector had exactly the shape of a paraboloid. Gravity, wind and temperature gradients make the reflector change its shape with elevation and time. Resulting deviations from a paraboloid have however large coherent lengths (e.g. are all forward in the upper half of the reflector and backward in the lower half). They lead to astigmatism and loss of gain, but, as long as they remain small, not necessarily to a decrease in beam efficiency. They can be minimized by an homologous design of the supporting backstructure: this latter, in the 30 m telescope, has been calculated to keep a shape close to paraboloidal (with changing F, however) when the telescope points at different elevations. Moving the subreflector, so that it stays aligned with the reflector axis, then allows to greatly reduce the gain losses. To minimize temperature gradients due to sunshine, etc..., the reflector backstructure is enclosed by isolating panels and its temperature regulated by cold/hot air circulation[6]. Residual astigmatism makes occasionally that the 30 m telescope

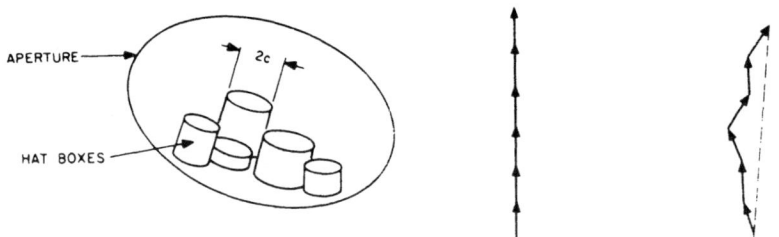

Figure 12: "Hatbox" model of the reflector surface errors. Each element could be a mispositioned panel –from [19].

HPBW is different in the horizontal and vertical planes.

The roughness of the pannels which compose the reflector surface (there are 420 panels of average size 1 m× 2 m) and the positioning error of these panels have more subtle effects.

To get a first idea of these effects, let us suppose that the reflector pannels are all displaced by $+\epsilon$ or $-\epsilon$ from the ideal paraboloid[3,p.49]. The phase error caused by each pannel on the nearby element of the reflected wave will be $\delta\phi = \pm 4\pi\epsilon/\lambda$. If half of the elements are displaced in one way, the other half in the opposite way, the distant field pattern will be $F'(l,m) \sim F(l,m)\cos\delta\phi$, where $F(l,m)$ is the perfect reflector field pattern (according to the "shift theorem"[17], the FT of two symetrically shifted impulse symbols, II(x) is a cosinusoid). Squaring this expression, we find that $\mid F'(0,0) \mid^2$, hence the on-axis gain and the effective area, are reduced with respect to the perfect reflector by a factor

$$\cos^2\delta\phi \simeq 1 - \delta\phi^2. \quad [10]$$

A more sophisticated analysis[19] assumes a gaussian positioning error distribution for the surface elements (or "hat boxes"). These latter are supposed to have a size (or correlation length) 2c, small with respect to the reflector size, but large compared to λ (see Fig. 12).

The reflected electric field will not anymore be represented by a series of vectors with constant amplitude and phase (as for a perfect reflector) or as the sum of two vectors of same amplitude and opposite phase shifts (as in the case just discussed), but as sum of constant amplitude vectors with different phase errors. If those can be described by a a gaussian distribution of standard deviation $\delta\phi = 4\pi\epsilon/lamba$, the gain loss can be expressed as

$$\frac{G'}{G} = \frac{A'_e}{A_e} \simeq \frac{\eta'_b}{\eta_b} \simeq e^{-\delta\phi^2} + o(c, D, \lambda)[11]$$

where $o(c, D, \lambda)$ is a correction term which accounts for the non-zero length of the correlation length c.

Relation [11], known as Ruze's formula[19], gives for small c a very simple expression of the gain (or aperture) and beam efficiency losses (note that for small $\delta\phi$, it reduces to relation [11]). The loss in efficiency is plotted on Fig. 13 for various values of $\delta\phi$ and c. One can see that for c $\gg \lambda$, the efficiency decreases by 1/4, for a surface error $\epsilon = \lambda/20$. It is customary to say that the shortest wavelength at which a telescope can operate is $\lambda = 16\epsilon$.

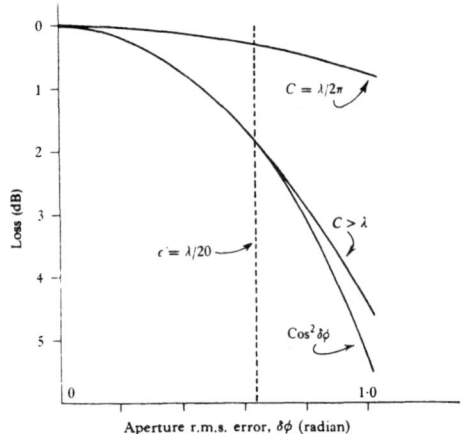

Figure 13: Loss in aperture and beam efficiencies due to reflector surface errors, according to relation [11].

The energy lost in the main beam due to small-size surface errors is scattered into an "error beam" of very roughly gaussian shape and of width $\simeq \lambda/2c$.

Fig. 14 shows the beam pattern of the IRAM 30 m telescope[20] at 0.86 mm (350 GHz), i.e. at a wavelength $\simeq$ 14 times larger than the r.m.s. surface error measured by holographic techniques. A large fraction (60%) of the power at these short wavelengths is scattered outside the main beam over a region of size 110-120" (the map is noisy, however). The size of the error beam suggests a correlation length c of 1.5 m, equal to the average size of the pannels. The error beam is thus readily explained in terms of panel positioning errors.

3.4. THE COMPLETE BEAM PATTERN

Fig. 15 shows a scan across the Moon, made at 230 GHz with the same telescope, and the derivative of this scan. The Moon, in the first approximation can be considered as an uniform temperature body with a straight, sharp edge. The power measured across the Moon edge is thus a finite one-dimensional integral of the power in the beam

$$\int_{-\infty}^{l} \int_{\infty}^{infty} \mid F(l,m) \mid^2 dldm$$

and its derivative is closely related to the beam. One sees clearly on Fig. 15 several beam components: the sharp main beam ($HPBW \simeq 12''$), which contains at the two third of the power at this wavelength, the first sidelobes, the "error beam" ($HPBW \simeq 150''$) caused by errors in panel positioning *and* reflector blockage by the subreflector, and lower, broader "beam" of size $> 300''$, probably caused by the roughness of the pannels and blockage by the quadrupod legs. At a larger scale, one would detect the horn spillover above the reflectors' rim.

A more accurate measurement of the "error beam" integrated power has been made for the 30 m telescope by observing the galaxy M51 in the 2–1 line of CO [22] . The molecular

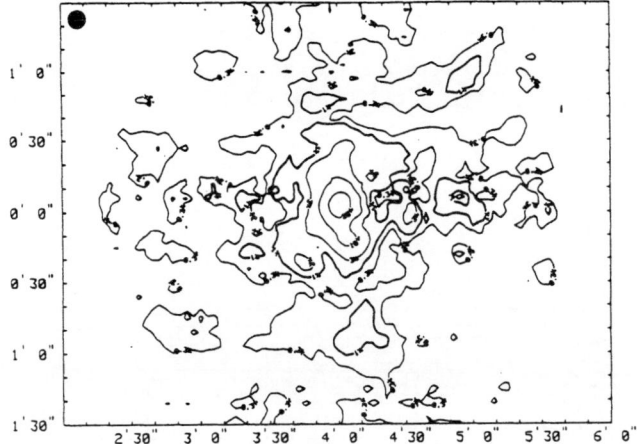

Figure 14: The 30 m telescope beam pattern at 350 GHz, observed with Mars at 53° elevation; the black circle represents the HPW of the main beam– from [20].

442

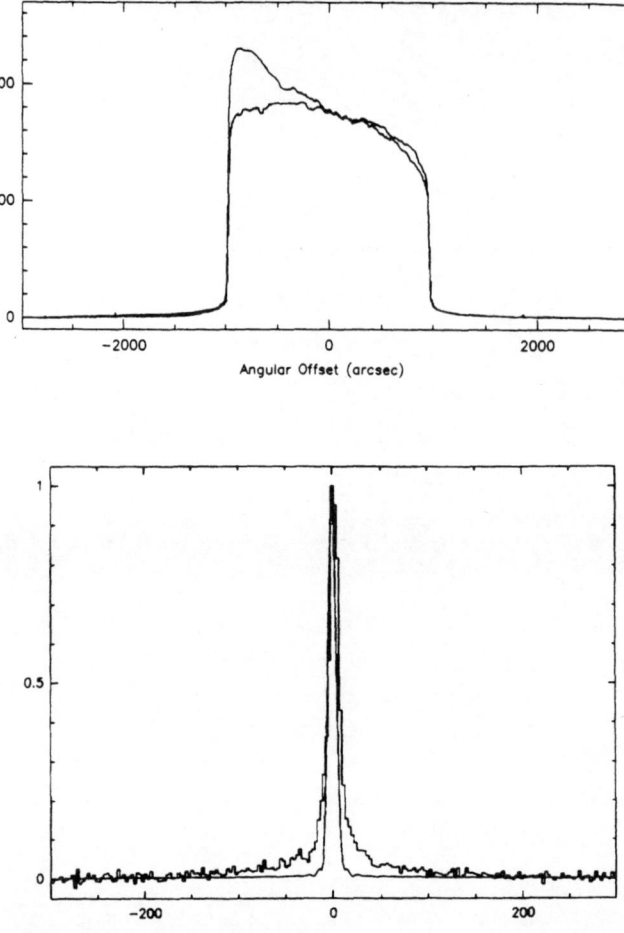

Figure 15: Power recorded at 230 GHz during a scan across the Moon, made with the IRAM 30 m telescope, and its derivative –from [21].

gas in M51 is in a region of size 3'x3' and is mostly concentrated into narrow spiral arms. Arms and interarm regions are resolved at 230 GHz by the 12"-wide main beam of the 30 m telescope.

Little gas (and CO emission) is present between the arms, and it has a narrow velocity spread (e.g. 10 kms^{-1}). The velocity spread over the entire galaxy, due to rotation, is >130 kms^{-1}. By observing an interarm region near the galaxy minor axis, one observes a line profile which is the superposition of the narrow emission of the interarm gas, observed by the main beam, and of the broad emission of the entire galaxy, observed by the error beam. By comparing the intensity of the broad component with that one would observe with a main beam of HPW 150", one derives an error beam-to-main beam power ratio of $\approx 1/4$".

Clearly, the different lobes of a telescope and their variation with frequency and elevation must be considered carefully when measuring line intensity ratios or the flux of weak sources in confused areas.

References

[1] Kraus, J.D. *Radio Astronomy*, Cygnus-Quasar Books 1982.

[2] Meeks,M.L.,*Astrophysics, part B: Radio telescopes*, Academic Press 1976.

[3] Christiansen, W.N., Hogbom, J.A. *Radiotelescopes* Cambridge University Press, 1969.

[4] Rohlfs, K. *Tools of Radio Astronomy* Springer 1986.

[5] Thompson, A.R., Moran, J.M. and Swenson, Jr. *Interferometry and Synthesis in Radio Astronomy*, Wiley-Interscience 1986

[6] Baars, J.W.M., Hooghoudt, B.G., Mezger, P.G. and de Jonge, M.J.. 1987, Astron. Astrophys.

[7] Guilloteau et al. 1991, *The Plateau de Bure Interferometer* Astronomy. Astrophys. in press.

[8] D. Downes *Introductory Courses in Galaxies' Evolution and Observational Astronomy*, eds. Appenzeller et al., Springer 1989.

Allen, C.W. *Astrophysical Quantities* The Athlone Press 1973.,p.120.

[10] Townes, C.H. and Schawlow, A.L. 1975, Dover. ??llen, C.W., *Astrophysical Quantities*, University of London press, 1981.

[11] Lovas and Tiemann 1974 –Phys. Chem. Ref. Data **3**, 609.

[12] Cernicharo, J. *ATM: a Program to Compute Theoretical Atmospheric Opacities*, IRAM Internal Report n° 52, 1985.

[13] Mauersberger, R., Guélin, M., Martin-Pintado, M., Thum, C., Cernicharo, J., Hein, H. and Navarro, S. *Astron. Astrophys. Suppl.* **79**, 217.

[14] Martin, R.N. et al. *A submillimeter facility* MPIfR, 1988. report.

[15] Downes, D. and Altenhoff, W.J. *Proceedings IAU/URSI Symposium on "Radio Astronomical Seeing"* ed. J.E. Baldwin and W. Shouuan 1989

[16] Altenhoff, W.J., Baars, J. W.M., Downes, D. and Wink, J.E. 1984, *Astron. Astrophys.***184**, 381.

[17] Bracewell, R.N. *The Fourier Transform and its Applications* McGraw-Hill 1978.

[18] Hannan, P.W., *Reflector Antennas*, ed A.W. Love, p. 136, IEEE press 1978.

[19] Ruze, J. *Reflector Antennas*, ed A.W. Love, p. 300, IEEE press 1978.

[20] Kreysa, E., et al. 1991, MPIfR memo 72.

[21] Garcia-Burillo, S., Guélin, M. and Cernicharo, J. 1992, *Astron. Astrophys., submitted* – see also Guélin, M., Hein, H. and Liechti, S. 1990, *Spectral line calibrations on the 30m telescope* IRAM 30m tel. tech. note 2.

EARLY MEASUREMENTS OF THE CMB
(COSMIC MICROWAVE BACKGROUND)

PIERRE ENCRENAZ
Laboratoire de Radioastronomie, Ecole normale supérieure
24, rue Lhomond
F–75231 Paris Cedex 05, France
and DEMIRM, Observatoire de Meudon
F–92195 Meudon Principal Cedex, France

ABSTRACT : Half a century ago, the temperature of the CMB was measured by optical means by Mc Kellar. The significance of these results had to wait twenty-five years before it was recognised by the direct measurements of Penzias and Wilson.

In order to determine the temperature of the CMB, you may try to measure directly the power received from the sky at the focal plane of a radio-telescope, or to measure the effect that this radiation has on the population of rotational levels of molecules sensitive to these photons.

As early as 1937, Swings and Rosenfeld [1], and Adams [2] did detect extremely narrow lines in absorption in front of stars. The CN, CH, and CH^+ radicals were responsible for these features. The strengths of the lines observed at 3874.0, 3874.6, and 3875.8 Å, permit the determination of the temperature of the radiation field populating the rotational levels of the CN radical (Fig. 1). A value of 2.3 K was deduced by Mc Kellar [3] and this value was published by G. Herzberg in his famous book "Spectra of Diatomic Molecules" [4] as early as 1950 (Fig. 2). It is unfortunate that the physical meaning of this measurement was not recognised by him. A complete review of this work is in [5].

At the same time, predictions of the value of the temperature T_c of the CMB was made by Gamow [6] and Alpher *et al.* [7] on theoretical grounds – but again the connection to Herzberg's value was not established. Later on, Novikov and Doroshnikov, then Peebles, independently rediscovered the theory of this background.

As radioastronomy gained momentum in the late forties, different measurements were made using the relatively poor receivers available at that time. Very elaborate techniques were developed to get rid of the receiver noise. In particular E. Leroux, in his 1955 thesis, did measure a value of $T_{sky} = 3\pm2$ K (Fig. 3) – see [8], [9] for a description of his clever method. Measurements were also made by Shaanov [10], Findlay [11], Shakeshaft [11], and De Grasse *et al.* [12]. The conflicting values obtained did not give a unique temperature for the CMB, and no connection to cosmology was established.

Using the horn-reflector antenna of Bell Telephone Laboratories (BTL) developed for the Telstar program, E. Ohm had access to the finest available equipment (travelling wave masers at 4.08 GHz) to precisely measure the atmospheric attenuation. He found a value of $T_c = 3$ K, but increased the error bar to include O.K. in the result [13]. A.B. Crawford, then director of the radio-astronomy group at BTL, gave as a first task to Arno Penzias – who had just graduated from Columbia – and to Robert Wilson – who had just graduated

M. Signore and C. Dupraz (et al.), The Infrared and Submillimetre Sky after COBE, 445–448.

446

from Cal Tech –, the more precise measurement of this excess noise. After years of very delicate analysis – subtracting all possible contributions from the atmosphere, side-lobes, losses in the antenna and waveguides, and mismatch to the rotary joint –, they came to the conclusion that a signal remained whose origin could not be accounted for. B. Burke, then professor at MIT, who was visiting BTL, told them that the Princeton group under R. Dicke was building a dedicated experiment to measure T_c. The connection was established and after a thorough analysis of the equipment at BTL, Dicke concluded that Penzias and Wilson had indeed measured the temperature of the background radiation. They decided to write two consecutive papers in the Astrophysical Journal [14, 15] (Fig. 4).

This result made the front page of the New York Times and of the world press, and gave a major push to Observational Cosmology.

REFERENCES

[1]   Swings, P., Rosenfeld, L.: 1937, *Astrophys. J.* **86**, 483
[2]   Adams, W.S.: 1938–1939, *Ann. Rep. Dir. Mount Wilson*
[3]   Mc Kellar, A.: *Publ. Dominion Astrophys. Obs. Victoria B.C.* **7**, 251
[4]   Herzberg, G.: 1950, *Spectra of Diatomic Molecules*, 2nd edition, D. Van Nostrand, Inc., p. 495
[5]   Thaddeus, P.: 1972, *Ann. Rev. Astron. Astrophys.* **10**, 305
[6]   Gamow, G.: 1948, *Phys. Rev.* **74**, 505
[7]   Alpher, R.A., Bethe, H.A., Gamow, G.: 1948, *Phys. Rev.* **73**, 803
[8]   Denisse, J.-F., Lequeux, J., Leroux, E.: 1957, *C.R.A.S.* **244**, 3030, and Leroux, E.: 1955, unpublished thesis (Université de Paris)
[9]   Le Floch, A., Bretenaker, F.: 1991, *C.R.A.S.*, and 1991, *Nature* **352**, 198
[10] Lukash, V.N.: 1991, private communication
[11] Lasenby, A.: 1991, private communication
[12] De Grasse, R.W., Hogg, D.C., Ohm, E.A., Scovil, H.E.D.: 1959, *Proc. Nat. Electronics Conf.* **15**, 370
[13] Ohm, E.A.: 1961, *B.S.T.J.* **40**, 1065
[14] Dicke, R.H., Peebles, P.J.E., Roll, P.G., Wilkinson, D.T.: 1965, *Astrophys. J.* **142**, 414
[15] Penzias, A.A., Wilson, R.W.: 1965, *Astrophys. J.* **142**, 419

447

Thus far this method has only been applied to the sun, first by Birge (806). Various investigators have obtained rather discordant results for the temperature of the reversing layer but the three most recent determinations [Adams (735), Bitzer (815), Hunaerts (1081)] based on different molecules give within 200° the same result of 4500° K. This agrees fairly well with the excitation temperature obtained from atomic lines [see Wright (1561)].

As was first pointed out by Birge (806), in an atmosphere in which the temperature varies the higher rotational lines will indicate a higher temperature than the lower ones, corresponding to different layers of the atmosphere. In this way, using the CH band 4315 Å, Richardson (588) found a temperature of 6080° K for the lowest part of the reversing layer of the sun and 4430° K for the uppermost part. This corresponds to a temperature gradient of about 13°/km.

Interstellar space. In high dispersion spectra of distant stars a few exceedingly sharp lines occur which have been shown to be due to absorption in interstellar space [see the reviews by Beals (792) and Adams (737a)]. While some of these lines were readily identified as due to Ca II, Ca I, Na I, K I, Fe I, and Ti II (where for the last two only the lowest multiplet component of the ground state gives observable lines) several others remained unidentified for some time until it was realized, first by Swings and Rosenfeld (1454) and McKellar (1205) (1206) that they are due to *interstellar molecules* in their lowest rotational levels. Thus the lines 4300.32 Å and 3874.61 Å found by Dunham and Adams (918)(736) agree exactly with the lines $R_2(1)$ of the $^2\Delta—^2\Pi$ (0–0) band of CH and $R(0)$ of the $^2\Sigma—^2\Sigma$ (0–0) band of CN, respectively. Further lines of CH [$Q_{2i}(1)$, $Q_2(1)$, $R_2(1)$] of the $^2\Sigma—^2\Pi$ (0–0) band) and of CN [$R(1)$ and $P(1)$ of the $^2\Sigma—^2\Sigma$ (0–0) band] predicted by McKellar (1205) were subsequently found by Adams (737).' Three of the remaining lines (4232.58, 3957.74, and 3745.33 Å) suggested a v'-progression of a diatomic molecule not known at the time [see McKellar (1206)]. It was later established by laboratory investigations of Douglas and Herzberg (902) to be the CH⁺ molecule. At present only two sharp interstellar lines at 3934.29 and 3579.04 Å remain unidentified. The first of these may be due to NaH [see McKellar (1206)].

The observation that in interstellar space only the very lowest rotational levels of CH, CH⁺, and CN are populated is readily explained by the depopulation of the higher levels by emission of the far infrared rotation spectrum (see p. 4 :) and by the lack of excitation to these levels by collisions or radiation. The intensity of the rotation spectrum of CN is much smaller than that of CH or CH⁺ on account of the smaller dipole moment as well as the smaller frequency [due to the factor v^4 in (I, 48)]. That is why lines from the second lowest level ($K = 1$) have been observed for CN. From the intensity ratio of the lines with $K = 0$ and $K = 1$ a rotational temperature of 2.3° K follows, which has of course only a very restricted meaning.

It is interesting to observe that no lines due to C¹³H or C¹³N have been found. From the intensity of the observed C¹³H and C¹²N lines Wilson (1544a)

---

* This paper contains several beautiful spectrograms showing the interstellar lines.

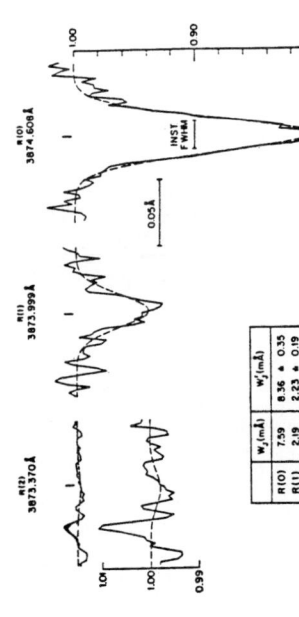

318 THADDEUS

R(2) 3873.370Å R(1) 3873.999Å R(0) 3874.608Å

0.05Å

INST FWHM

	W_λ(mÅ)	W'_λ(mÅ)
R(0)	7.59	8.56 ± 0.35
R(1)	2.19	2.23 ± 0.19
R(2)	0.077	0.077 ± 0.055

FIGURE 7. The interstellar CN lines in ζ Oph observed with a high-resolution Fabry-Perot spectrometer (Hegyi, Traub & Carleton 1972). The region of R(2) was scanned for many nights, that of R(0) and R(1) for much shorter times.

Figure 1. The interstellar CN lines observed towards ζ Oph [5].

Figure 2. Reproduction of page 496 of [4].

(23) $\quad \Delta T_a (-3) := 218 - 0,77 \ T_c$

et on obtient, sur la courbe 26,

(24) $\quad \Delta T_a = \dfrac{1}{k} (86 \times 4,97) = 215°K$

En résumé, on a trois équations donnant T_c :

$v_o = 0 \qquad 137 = 138 - 0,485 \ T_c \longrightarrow T_c = 2°K$

(25) $\quad v_o = 5 \qquad 50 = 51,3 - 0,485 \ T_c \longrightarrow T_c = 2,7°K$

$v_o = -3 \qquad 215 = 218 - 0,77 \ T_c \longrightarrow T_c = 3,9°K.$

les coefficients $1/k$, ρ, ρ', ρ'' et T_c. Mais la bonne cohérence des valeurs obtenues pour T_c montre que les valeurs prises pour ces coefficients sont correctes avec une bonne approximation. Si on diminuait le coefficient $1/k$ on obtiendrait des valeurs négatives pour T_c, quelles que soient les valeurs prises pour ρ' et ρ'' qui interviennent de façon différente dans les 3 équations précédentes, le coefficient ρ' intervenant notamment de façon opposée dans les deux dernières équations. De même, une augmentation de $1/k$ de quelques pour cent donnerait des valeurs de T_c incohérentes. Enfin, un coefficient de réflexion du sol non nul donnerait $T_c < 0$.

En fait, on devrait déduire, de plusieurs équations de ce genre,

Il est difficile de déterminer l'erreur sur cette valeur de T_c, basée sur la cohérence de différentes mesures. Nous pensons que l'erreur absolue doit être de l'ordre de 2°K, en prenant :

$$\boxed{T_c = 3°K}$$

Comme nous l'avons dit, une méthode plus rigoureuse et plus précise serait de placer l'antenne dans une boîte assimilable à un corps noir. Le nombre de fréquences résonnantes dans une boîte de volume $V = 1m^3$ est $dN = \dfrac{8\pi V \nu^2}{c^3} \ d\nu$, soit 1 par Mhz.

Une telle boîte peinte à l'intérieur d'une couche absorbante serait donc un bon corps noir qui permettrait un étalonnage précis du récepteur. Les boîtes à couches antiréflechissantes sont sans intérêt car elles ont des fréquences privilégiées.

Stabilité dans le temps de la température du récepteur.

Comme le montrent les mesures faites sur la Lune, ch. VII, la température de bruit du récepteur $T_{réc}$ est stable à au moins $3. \ 10^{-2}$ par mois ; à condition de respecter les conditions posées dans la définition du gain courant ($L = 0$, $T_a = 0$) le coefficient k est stable à la même précision.

Figure 3. Reproduction of page 50 of [7] (Leroux's thesis).

Thu Sept 29 '64
Orinoco Dacha
745 - 8th Street
Boulder, Colorado

Dear Dr. Penzias,

Thank you for sending me your paper on 3°K radiation. It is very nicely written except that "early history" is not "quite complete". The theory of what is now known as "primeval fireball" was first developed by me in 1946 (Phys. Rev. 70, 572, 1946 ; 74, 505, 1948 ; Nature 162, 680, (1948). The prediction of the numerical value of the present (residual) temperature could be found in Alpher & Herman paper (Phys. Rev. 75, 1093; 1949) who estimate it as 5°K, and seen in my paper (Kong. Dansk. Ved. Sels. 27 no.10, 195 with the estimate of 7°K. Seen in my popular book "Creation of Universe" (Viking 1952) you can find (app. 42) the formula $\Delta T = 1.5 \cdot 10^{10}/t^2$ °K, and the apparent limit of 50 °K. Thus you see the world did not start with almighty Dicke. Sincerely G Gamow)

Figure 4. Fac-simile of the letter of Gamow to Penzias, recalling that he had predicted the existence of the CMB seventeen years prior to Dicke.

SUMMARY OF MEETING

B.J.CARR
Astronomy Unit
Queen Mary & Westfield College
Mile End Road
London E1 4NS

1. INTRODUCTION

One of the striking features of this meeting has been the wide range of topics covered. We have heard talks on the early universe (9 lectures), large-scale structure (3 lectures), dust in galaxies (5 lectures), infrared to submillimetre backgrounds (5 lectures), microwave anisotropies (7 lectures), and various aspects of instrumentation (5 lectures). In consequence, we have grappled with the latest developments in particle physics, astrophysics and cosmology and touched on the physics of all scales from the Planck length to the horizon length.

At first sight it might seem surprising that a meeting on the "infrared and submillimetre sky" should spawn such diversity. However, further reflection makes it clear why this should be. As indicated in Figure (1), the infrared/submillimetre band corresponds to a range of wavelengths (10^{-4} to 10^{-1} cm) which is almost exactly midway between the Planck scale (10^{-33}cm) and the horizon scale (10^{28}cm). (The midpoint is 30 microns.) It is also midway between the scales associated with dust grains (10^{-6}cm) and telescopes (10^{1}cm) and midway between the GUT scale (10^{-27}cm) and the galaxy scale (10^{22}cm). There is therefore a literal sense in which the infrared/submillimetre band plays a central role in the Universe!

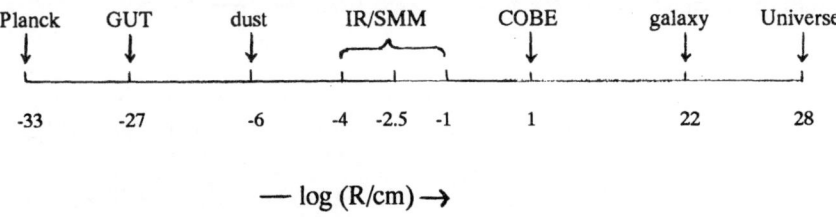

$$\longleftarrow \log (R/cm) \longrightarrow$$

Figure (1) : illustrating the central role of infrared/submillimetre radiation in the Universe

The crucial role of the COBE results, in particular, is summarized in Table (1). This provides an overview of their many implications for cosmology. Most of these implications have been explored in depth at this meeting but it may be helpful to bring them all together. Recall that FIRAS measures the CBR temperature T and puts upper limits on the deviations from a black-body spectrum over the waveband 500μ to 5000μ

449

M. Signore and C. Dupraz (et al.), The Infrared and Submillimetre Sky after COBE, 449–461.
© 1992 *Kluwer Academic Publishers.*

(as measured by the chemical potential μ, the Compton parameter y and the excess intensity ΔI at wavelength λ); DMR measures the CBR temperature anisotropies $\Delta T/T$; and DIRBE constrains the background radiation density (Ω_R in critical density units) at various wavelengths in the infrared band.

T=2.735± 0.06 K. The precise temperature is not crucial, although it does permit greater accuracy in cosmological nucleosynthesis calculations [1] and in predicting the density of relict particles. The significant point is the precision with which the black-body spectrum has been established.

$\mu < 10^{-2}$: This puts an upper limit on energy released in the redshift range $10^6 > z > 10^4$, with μ being a measure of the energy input relative to the CBR density. This places interesting constraints, for example, on superconducting strings, evaporating black holes [2], and the dissipation of initial density perturbations [3].

$y < 10^{-3}$: This places constraints on energy released in the redshift range $z < 10^4$, both before and after decoupling, the y parameter being a measure of the electron pressure integrated over redshift. The limit may already exclude the explosion [4] and mock gravity [5] scenarios for large-scale structure formation, and it constrains dark matter scenarios involving decaying elementary particles. It also restricts the thermal history and present density and temperature of any hot intergalactic medium, implying in particular that a smooth IGM could only provide a small fraction of the observed X-ray background.

$\Delta I(\lambda) < 10^{-2} I(\lambda_{peak})$: This places a direct constraint on any submillimetre photons produced in the period after decoupling, as distinct from indirect constraints (of the kind discussed above) associated with the interactions of pre-existing CBR photons. One important source of photons in this band is dust .

$(\Delta T/T)_{dipole} = 10^{-3}$: The dipole component, measured by both DMR and FIRAS, is usually interpreted as being due to our peculiar motion with respect to the cosmological rest frame. Attempts to explain this motion in terms of the gravitational influence of nearby mass aggregations provide an important probe of large-scale structure. The dipole result also feeds into the evidence for large-scale streaming motions.

$(\Delta T/T)_{quad} < 3 \times 10^{-5}$: The quadrupole limit constrains any source of global anisotropy in the Universe. In particular, it restricts the amount of shear and rotation, as well as the amplitude of gravitational waves with wavelength comparable to the current horizon size.

$(\Delta T/T)_{\theta > 7^o} < 5 \times 10^{-5}$: The small-scale anisotropy limits constrain, for example, the form of the density fluctuations at decoupling, various features of non-linear structure at the present epoch, and cosmic strings. They also constrain models for spectral distortions produced by dust.

Ω_R limits : The main cosmological importance of the background radiation limits is that they constrain scenarios for primeval galaxies, the evolution of infrared and normal galaxies, and scenarios involving pregalactic radiation which has been reprocessed by dust.

TABLE (1): Some cosmological implications of the COBE results

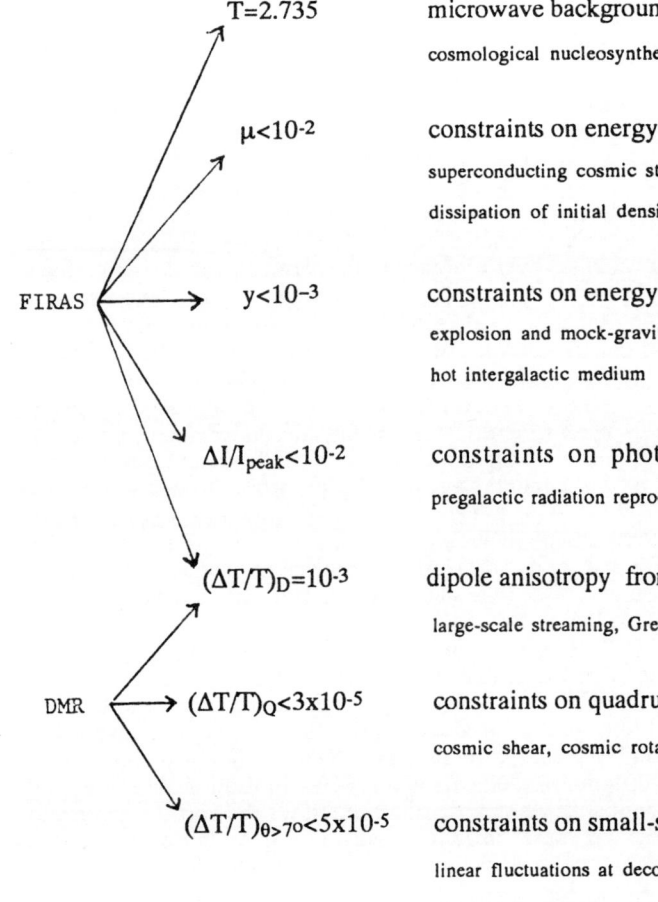

T=2.735   microwave background temperature

cosmological nucleosynthesis, relict particle density

$\mu<10^{-2}$ constraints on energy release for $10^6<z<10^4$

superconducting cosmic strings, evaporating black holes,

dissipation of initial density fluctuations

$y<10^{-3}$ constraints on energy release for $z<10^4$

explosion and mock-gravity scenarios, decaying particles,

hot intergalactic medium

$\Delta I/I_{peak}<10^{-2}$ constraints on photon production for $z<10^3$

pregalactic radiation reprocessed by dust, IRAS galaxies

$(\Delta T/T)_D=10^{-3}$ dipole anisotropy from peculiar motion

large-scale streaming, Great Attractor

$(\Delta T/T)_Q<3\times10^{-5}$ constraints on quadrupole anisotropy

cosmic shear, cosmic rotation, gravitational waves

$(\Delta T/T)_{\theta>7^o}<5\times10^{-5}$ constraints on small-scale anisotropy

linear fluctuations at decoupling, non-linear fluctuations

at present epoch, large-scale structure, strings, dust

FIRAS

DMR

DIRBE   → Ω_R limits   constraints on near-IR and far-IR backgrounds

primeval galaxies, IRAS galaxies, pregalactic radiation

Besides its importance for cosmology, COBE also has important implications for understanding the Galaxy. FIRAS can probe the nature of the interstellar and interplanetary dust grains (in particular, it can look for submillimetre cooling lines); DMR can study synchroton emission and look for very cool dust; DIRBE can study molecular clouds and seek the 3.3μ PAH feature.

2. THE EARLY UNIVERSE

This subject attracted nine lectures, larger than any other topic. Most of the discussion was in the context of the standard hot Big Bang theory and the lectures covered most of the key epochs between the Planck time and galaxy formation.

The Planck Era and Grand Unification. Veneziano discussed the connection between fundamental strings, gravity and the constants of nature. Quantum string theories clearly have many advantages over ordinary field theories - for example, their action need contain no fundamental constants - and he mentioned some of their interesting implications for the Planck era. He also discussed the connection between quantum and classical strings and raised the interesting question of whether inflation produces cosmic strings or vice-versa. However, I was somewhat perplexed by the suggestion that the expansion of the Universe may be an illusion produced by string shrinkage! Sanchez brought us back to earth by stressing the importance of relating fundamental strings (which only exist in higher dimensions) to the real 4-dimensional world. She discussed whether fundamental strings can produce cosmic strings by compactification, although this idea seems to be incompatible with various astronomical observations. One of the most important recent developments in the particle/astrophysics arena is the implication of LEP that the Grand Unified Theories need supersymmetry in order that all the coupling constants cross at the same energy. The unification energy then needs to be about 10^{16}GeV, a result whose implications is discussed below.

Inflation. The attractions of the inflationary scenario were summarized with characteristic zeal and humour by Turner: it solves the flatness problem, the horizon problem, the monopole problem and it provides a possible source for the density fluctuations required to explain large-scale structure. However, Turner also emphasized that the simplest variants of inflation are unsatisfactory: "old" inflation faces the exit problem (i.e. the bubble formation rate is never large enough for the entire Universe to complete the phase transition) and "new" inflation faces the reheat problem (i.e. the coupling constant is so small that the universe is not hot enough for baryosynthesis to occur after inflation). The neatest solution seems to be to invoke "extended" inflation; in this the nucleation rate grows with cosmic time, so that it is small enough to provide the required amount of inflation at early times but large enough to allow inflation to end at late times. In order for the coupling constant to change, one has to introduce a dilaton field but, as Turner stressed, this is a natural feature of supergravity theories. However, it requires G to decrease with time and this may raise problems of its own.

Lukash introduced some further variants of inflation: in particular, chaotic, eternal and designer inflation. In his discussion of the designer variant he stressed the important point that inflation may in principle produce fluctuations with an arbitrary spectrum, the scale-invariant form only applying for exponential inflation in the slow roll-over approximation, and this may be relevant in explaining certain features of large-scale structure. A simple illustration of this, noted by Turner, is power-law inflation, in which the cosmic scale factor grows as t^n and the horizon-scale fluctuations go as $\lambda^{1/n}$. Lukash's memorable cartoon of eternal inflation throws light on the obsession he has shown for ball games throughout the meeting!

One interesting prediction of the inflationary scenario, stressed by both Turner and Smoot, is that there should be gravitational waves whose amplitude on entering the particle horizon depends only on the temperature of the inflationary phase transition (i.e. T_{GUT} if it is a Grand Unified phase transition). In particular, the waves entering the horizon now will induce a quadrupole anisotropy and the COBE constraint on this already implies $T_{GUT} < 10^{17}$GeV. This just satisfies the LEP result mentioned in Veneziano's talk, which highlights an intriguing connection between measurements in the macroscopic and microscopic domains. Inflation would also produce gravitational waves with wavelengths well below the horizon scale and - in the extended scenario - vacuum popping would produce an interesting background at wavelengths around 10^4cm. However, both of these backgrounds would be very hard to detect at present.

I would like to return to the question of whether inflation necessarily requires the density parameter Ω to be 1, an issue which I feel was rather glossed over. As Ellis has stressed [6], Ω must always deviate from 1 eventually unless it has always been exactly 1 (in which case inflation does not explain anything): it will tend to 0 in an open Universe and infinity in a closed Universe. Even though there is an extended period in which Ω is close to 1, it does not satisfy this condition most of the time, so it seems to me that one needs some extra condition (possibly an anthropic constraint) to explain why we live in such a period. Note, however, that one can only have small CBR anisotropies when Ω is close to 1, so there are perhaps good empirical reasons for believing $\Omega=1$. Another issue which received little attention is the question of why the cosmological constant should be so small today. Although one often assumes $\Lambda=0$ at the present epoch, there are neither convincing theoretical arguments nor compelling observational reasons for believing this. If Λ is zero, the explanation may require quantum cosmology [7].

Cosmic Strings. Bouchet reviewed the attractions of cosmic strings: if the dimensionless parameter $G\mu$ (which measures the mass per unit length of the string) is about 10^{-6} (as expected in the GUT scenario), then it has been claimed that closed strings (which form by self-intersections) can generate seeds for galaxies, while open ones can make large-scale structure via wakes. Strings can also produce gravitational waves (mainly by the shrinkage of loops) and characteristic gravitational lensing effects. However, the simulations of Bouchet and his collaborators show that earlier estimations of the importance of these effects may have been over-optimistic. Because they contain small-scale kinks, strings tend to break up into much smaller loops than was originally expected and this means that their cosmological effects are less important. In particular, the loops decay much more quickly through gravitational radiation losses and this implies that the millisecond pulsar constraint on the gravitational wave background gives a weaker limit on $G\mu$ than originally claimed (viz. $G\mu < 4 \times 10^{-5}$). This limit was discussed in detail by Sanchez. Even if cosmic strings turn out to be unimportant for large-scale structure, they could still leave an important signature via the CBR anisotropies they induce, another issue considered by Sanchez. The OVRO results currently require $G\mu < 10^{-5}$ and Bouchet's discussion of whether the non-Gaussian nature of string-induced anisotropies could be probed by using topological measures like genus was particularly interesting.

Spectral Distortions. Salati discussed how any energy input in the period before decoupling would produce distortions in the CBR spectrum, a topic also touched on by Silk. This is crucial for understanding the cosmological implications of the FIRAS results and he must be congratulated on making technical calculations seem so comprehensible. Even the derivation of the Kompaneetz equation seemed easy! For $z>10^6$ any heat input would be completely thermalized and merely boost the entropy per baryon associated with the CBR. For $10^6 > z > 10^4$, the Compton interactions of energetic electrons with the

CBR photons would produce a Bose-Einstein spectrum with a chemical potential μ; this is because there is no time for Bremmstrahlung processes to provide the extra photons required to produce a black-body spectrum except at very low frequencies. For $z<10^4$, complete relaxation to a Bose-Einstein distribution is not possible but Compton effects still remove photons from the Rayleigh-Jeans part of the spectrum and redeposit them in the Wien part, so one gets a y distortion. Sanchez gave an interesting application of these results when she showed how the FIRAS limits on y and μ constrain scenarios involving superconducting cosmic strings.

Relict Freeze-out. Salati and Raffelt showed why one would expect elementary particle relicts to be left over from the first second of the Big Bang and discussed which of these would be good candidates for the dark matter required by inflation. It is well-known that light neutrinos can make up the critical density only if their mass m_v is around 100eV, while heavy neutrinos would need a mass of about 2GeV. However, the LEP results imply that a heavy neutrino must be of the Dirac type (with the particle and antiparticle being different) with a mass $m_v>46$GeV. Unfortunately, even this possibility is now excluded by the germanium experiments, so it seems that no heavy neutrino could provide the dark mass. Fortunately, the light neutrino is currently enjoying a comeback. This is because the MSW effect can explain the solar neutrino problem if the muon neutrino has a mass of order 10^{-2}eV; the tau neutrino mass would then be of order 100eV, as required to give $\Omega_v=1$.

If the MSW effect goes away, one will need some more exotic particle to provide the dark mass. Many possible candidates arise in the supersymmetry theories (eg. the photino) and, in this case, one might hope to detect the ones in the halo via terrestial experiments (eg. using ionization and cryogenic detectors), as discussed by Turner and Raffelt. However, it will be several years before one could expect such experiments to yield positive results. The axion remains a viable candidate and this could be detected by converting halo axions into microwave photons in a magnetic cavity. Pilot experiments have already been completed and proposed experiments (if approved) will either detect or exclude an axion mass in the expected range of $0.6-16\mu V$ within the next 5 years. Decaying elementary particles are unattractive in my view, because they require fine-tuning of the decay time, but the possible detection of a 17keV neutrino is obviously an interesting development since, as discussed by Salati, such a particle would have to decay in order to avoid having too large a density.

Reheating. Blanchard emphasized that one expects an intergalactic medium with a density parameter of at least 0.02 in the standard CDM picture and he discussed its likely thermal history. The Gunn-Peterson test implies that it would need to be reheated by a redshift of at least 3 and the question arises of how this heating occurred. This is a puzzle because, as Blanchard emphasized, ordinary stars and quasars would not suffice. Barcons discussed the constraints on the thermal history imposed by the COBE limit on y and radio limits on free-free emision. He stressed the point that a hot IGM cannot produce the X-ray background if it is smooth. A clumped IGM might do so but the X-ray anisotropy limits would then require it to be clumped on such a small scale that the clumps would be identified with galaxies anyway! Another puzzle stressed by Blanchard is that one would expect too much gas to have cooled by the present epoch in the simplest hierarchical scenario. Indeed there should not even be enough gas left over to make the visible material seen in galaxies. One way around this is to invoke reheating by a first generation of massive stars, since this would incresase the cooling time for the surviving gas. However, I would also like to stress the possibility that most of the gas may indeed have cooled and thereby gone into a population of low mass brown dwarfs.

3. LARGE-SCALE STRUCTURE

Three lectures were devoted to large-scale structure and the associated dark matter problem. Most attention focussed on the "Cold Dark Matter" scenario, which has come under considerable flack recently as a result of new observations of galaxy clustering and large-scale streaming.

Dark Matter. Understanding the origin of large-scale structure depends crucially on the nature of the dark matter required in galactic halos and clusters of galaxies. If one believes in inflation, there must also be dark matter distributed throughout intergalactic space. Raffelt discussed the possible candidates. Cosmological nucleosynthesis constraints suggest that baryonic dark matter could only have $\Omega_B \sim 0.1$ and this means that the intergalactic dark matter would certainly need to be non-baryonic. If it consisted of hot particles (like light neutrinos), galaxies would form too late, so cold particles (like axions or photinos) are now favoured. The $\Omega_B \sim 0.1$ condition would still suffice to provide the dark matter in galactic halos and, as Blanchard emphasized, there must be some dark baryons since the density of visible material is only $\Omega_V \sim 0.01$. Therefore one probably needs both baryonic and non-baryonic dark matter. The only plausible baryonic candidates would be black holes or brown dwarfs. Raffelt stressed that a powerful signature of   brown dwarfs would be their gravitational lensing effects. However, even if such objects do inhabit the halo, one should note that elementary particle relicts could still dominate the density there (indeed this would be expected if the relicts were cold) and, in this case, the lensing rates indicated by Raffelt would be over-optimistic. The best signature of black holes would be the background radiation generated by their precursor VMOs and Carr discussed the prospects of detecting this background with COBE.

The CDM Scenario. The standard CDM scenario assumes, firstly, that the dark matter is cold and, secondly, that the primordial density fluctuations have the scale-invariant Harrison-Zeldovich form. Despite its initial attractions, this picture has encountered severe difficulties, as discussed by Juszkiewicz. The basic problem is that the form of the galaxy correlation function $\xi_G(r)$ and application of the cosmic virial theorem on scales below 10Mpc require a bias parameter b=2.5, while streaming motions on scales above 10Mpc require b<1.5. This has been known for several years but the problem has been compounded by recent studies of the velocity correlation function $\xi_V(r)$, the APM angular correlation function for galaxies w(θ), and the QDOT redshift survey, all of which indicate more large-scale power than is compatible with CDM above 10Mpc. The evidence for large-scale streaming itself was reviewed by Gorski, with Silk reporting results of Mathewson et al. [8] which indicate coherent motion on scales even larger than the Great Attractor. However, CDM enthusiasts do not give up easily and possible solutions, as discussed by Lukash, are designer inflation or a mixture of hot and cold dark matter. It is also conceivable that some form of astrophysical modulation could provide a large-scale feature in the galaxy correlation function. A non-zero Λ, originally proposed in order to solve the age problem and to eliminate the need for bias or exotic particles, would also help. However, the observational evidence for non-zero Λ is ambiguous. At one stage it seemed to be required to fit the B counts of distant galaxies but the K counts now seem to favour $\Lambda=0$ and $\Omega=1$, in which case the B objects cannot be identified with the precursors of galaxies [9].

4. DUST IN GALAXIES

A topic which is more directly relevant to infrared/submillimetre observations, and to the results of COBE in particular, is dust. Two lectures focussed on the dust in the Milky Way and two more on the dust in other galaxies.

Interstellar Dust. One of the most memorable pictures shown in the school must surely have been the DIRBE map of the Galactic Plane, showing both the thin disk of dust emission and the Galactic bulge. Boulanger's talk emphasized several interesting features of this dust. At high Galactic latitude there is a good correlation between the 100μ intensity and the HI column density at $0.5°$ resolution, with the ratio showing a clear "cosecb" distribution at $|b|>50°$. This indicates that the gas at high latitude is mostly HI, with the dust and gas being well mixed on 100pc scales. However, pulsar dispersion measurements indicate that there is also a diffuse distribution of HII with a vertical scale-height of about 1.5kpc. One important point is the existence of windows with little infrared emission at high latitude and this is useful for constraining the extragalactic background. I was impressed that a simple three-grain model of the interstellar dust (with PAHs below 15A, large grains above 50A, and small grains in between) with a temperature of about 30K fits the data so well. But I was then confused that Wright's much longer fractal grains seemed to fit the data just as well! Although most of these results were known from IRAS, this is clearly a field in which COBE will have a tremendous impact. Particularly exciting is the ability of FIRAS to pick up submillimetre lines (as emphasized by Wright, it has already made the first ever observation of NII) and the possibility that DIRBE will detect PAH features and molecular clouds. Some of us were bothered over whether spatial structure in the Galactic dust distibution could contaminate the galaxy angular correlation function $w(\theta)$ and it seems to me that this issue is still not completely resolved.

Zodiacal Dust. Boulanger and Wright also told us about the interplanetary dust. It now seems that one can separate zodiacal and interstellar dust at the 0.5% level by using their different spatial and spectral distribution, although no model fits to better than 10%. The zodiacal dust grains are somewhat larger and hotter, with sizes between 0.01μ and $10^3\mu$ and a temperature of 200-300 K. The crucial advantage of DIRBE over IRAS is its extensive spectral coverage and, as emphasized by Wright, the fact that one can use chopping to obtain a signal which is unaffected by AC-DC non-linearity. (The problem here is that the IRAS point source calibration cannot be applied to diffuse sources). However, modelling of the DIRBE data has hardly began and most attention was therefore paid to the IRAS results. In particular, we learnt how IRAS has detected zodiacal bands formed from asteroid collisions and a number of comet trails. Boulanger also raised the question of whether COBE will be able to detect Kuiper's belt.

Dust in Other Galaxies. Among the highlights of the meeting must surely have been the exciting results of Chini on dust in other galaxies. Submillimetre observations in various atmospheric windows provide a unique way of discriminating between thermal and non-thermal emission. Chini's observations of spirals show a clear turn-over above 100μ, thus providing evidence for thermal emission from dust with a temperature around 20K. What is surprising is the amount of gas involved, the mass in cold gas corresponding to 90% of the total gas mass. The infrared luminosity for these spirals was found to be well-correlated with the gas mass, with a light-to-gas-mass ratio $L/M_{gas}=5$; this is the sort of value expected for normal star formation. Chini has now extended this work to starburst galaxies and quasars. He finds that starburst galaxies typically have T=30K and $L/M_{gas}=100$; this indicates that gas is turning into stars with very high efficiency and that their IMF is shifted towards larger masses. His survey of the quasars in the IRAS catalogue shows that both radio-quiet and radio-loud ones give evidence for a far-infrared thermal bump, although - in the latter case - it is occasionally hidden by synchroton emission. The dust is confined within the central kiloparsec and has T=35K and L/M_{gas}=1000, the sort of value associated with a circumnuclear disk heated by an active nucleus. Thus spirals, starburst galaxies and quasars all show evidence for dust but with the different physical conditions being indicated by the different L/M_{gas} values.

5. INFRARED TO SUBMILLIMETRE BACKGROUNDS

Besides the CBR itself, one could expect various extragalactic sources of background radiation in the infrared/submillimetre band and these would appear as distortions of the CBR spectrum at long wavelengths. Five lectures focussed on the detection and theoretical origins of such backgrounds.

Observational Searches. Carr emphasized the difficulty of detecting an extragalactic background in the infrared/submillimetre bands, mainly because of the problem of subtracting the foregrounds from interstellar and interplanetary dust. Despite various false alarms in the past, there is currently no definite detection apart from the CBR itself and all one can claim is upper limits. Although there have been several exciting developments recently, with the improved COBE limits in the far-IR and the Noda et al. results [10] in the near-IR, it must be recalled that one's ability to find an extragalactic background is still limited by the precision with which one can model foregrounds. Melchiorri stressed how one can use the Doppler effect associated with our peculiar motion to detect or at least limit the density of any extragalactic background and this is a particularly powerful method if one is seeking cosmological lines. Barcon's talk raised the issue of what one means by a "background" : any background must ultimately be generated by discrete sources - even if they are individuals atoms - so it is really an issue of resolution. In distinguishing between the diffuse component of a background and the integrated emission of discrete sources, it is obviously important to have a good statistical description of the source population and the fluctuations due to source clustering, so Barcon's review of this topic was useful.

Galaxies. As Carr and Wright emphasized, the near-IR background from ordinary galaxies is already close to the upper limits permitted by Noda et al. The far-IR background from IRAS galaxies is much less certain since this depends on how the density and luminosity of IRAS sources evolve as one goes to higher redshifts. Even with no evolution, one would expect the IRAS background to be well above the DIRBE sensitivity but one will probably need SIRTF to separate this from the dust emission of our own galaxy. For as Wright stressed, the dust in our own and other galaxies is likely to be very similar, so one cannot distinguish between them except via spatial distribution.

Pregalactic Sources. The era between decoupling and galaxy formation could provide a wide variety of sources of background radiation (eg. VMOs, brown dwarfs, decaying particles or black hole accretion). As discussed by Carr, this would either appear in the near-infrared or, if it was reprocessd by cosmic dust, in the far-infrared/submillimetre bands. These scenarios are speculative but some of the pregalactic sources are inevitable, so there must be a background at some level. COBE is already constraining these scenarios and it may eventually eliminate some of them altogether. This is particularly relevant to the scenario in which galactic halos consist of the black hole remnants of VMOs. Wright raised the question of whether the entire CBR could be generated by pregalactic stars and whiskers, as advocated anew in a recent paper by Arp et al. [11], but he concluded that this is very unlikely.

Dust Anisotropies. If there is background due to cosmological dust, an important prediction is that the anisotropies associated with it could dominate the primary CBR anisotropies at small angular scales. As discussed by Carr, the anisotropies are associated with the dust clumpiness and temperature variations and they could be distinguished from CBR effects by their wavelength dependence. Relevant upper limits on $\Delta T/T$ were mentioned by Chini (IRAM gives 2.6×10^{-4} at 30" and 1300μ), Lasenby (JCMT gives 1.4×10^{-3} at 40" and 800μ) and de Bernardis (ARGO gives 4×10^{-5} at 100' and 380μ) and these all imply interesting constraints on galaxy formation scenarios.

6. CBR ANISOTROPIES

Obviously the most significant background in the submillimetre band is the CBR itself and seven lectures were devoted to discussing the temperature anisotropies associated with this. As with the CBR spectral distortions, one currently only has upper limits on the anisotropies (except for the dipole term), so most of the lectures discussed the theoretical implications of these limits.

Dipole Effect. The only positive detection of an anisotropy in the CBR is the dipole term with $\Delta T = 3.3 \text{mK}$, as discussed by Janssen and others. This is usually attributed to the peculiar motion of the Earth relative to the cosmic rest frame, although Turner raised the possibility that it could arise because the Universe is "tilted" (in the sense that the CBR rest frame moves with respect to the rest frame of the cosmic expansion.) Although the dipole effect was first measured more than 15 years ago, as described in the nice review by Melchiorri, it has been confirmed by DMR with greater precision than before. I was particularly impressed by one of Smoot's pictures, showing the accuracy with which one can now detect the 10% dipole modulation due to the Earth's motion around the Sun. It is amusing to speculate that, were the Earth completely cloud-covered (thereby thwarting ground-based observations), we might only now have discovered the Copernican Principle! Melchiorri explained why a dipole effect leads to a kinematic quadrupole term. It was surprising to learn that people of the stature of Pauli and Planck miscalculated this but gratifying to realize that - after Melchiorri's elucidating discussion - nobody at this school is likely to repeat their mistake!

Quadrupole Effect. Smoot showed that the COBE quadrupole limit is now $(\Delta T/T)_Q < 3 \times 10^{-5}$ and this severely constrains any global deviations from isotropy. In particular, it puts upper limits on the present rotation rate ω and shear σ of the Universe: $\omega/H < 10^{-7}$ and $\sigma/H < 10^{-7}$ where H is the Hubble parameter. It also constrains the amplitude of gravitational waves on the present horizon scale h_H to be less than 3×10^{-5}; since $h_H = (T_{GUT}/m_{Pl})^2$ in the inflationary scenario, this implies that the GUT phase transition temperature must be below 10^{17}GeV (cf. the LEP result discussed earlier). Gorski discussed how the quadrupole limit can be used to constrain large-scale peculiar motions. POTENT suggests that the bulk flows should be 388 ± 67 km/s at R=4000 km/s and this implies that the quantity a_2 - which is less than $(\Delta T/T)_Q$ by a factor $\sqrt{5}/4\pi$ - must exceed 2.5×10^{-6} in any model, whatever the form of the fluctuations.

Small-Scale Anisotropy. Martinez-Gonzalez and others discussed the observational constraints on the small-scale anisotropies from a variety of terrestial, space and balloon experiments. Although there is still no definite detection of anisotropies on small angular scales, they must exist at some level and several talks emphasized the implications of this. Silk summarized the various theoretical contributions to $\Delta T/T$. If the density fluctuations at decoupling have amplitude $\Delta \equiv \delta\rho/\rho$ on a scale $\lambda \equiv L/ct$ relative to the horizon size, then there are five distinct contributions: a contribution of $\Delta/3$ from adiabatic fluctuations (important for $\theta > 5'$), a contribution of $\Delta\lambda$ from velocity effects (important for $\theta > 10'$), a Sachs-Wolfe contribution of $\Delta\lambda^2$ (dominant for $\theta > 2^{\circ}$), a 2nd order contribution of $\Delta^2\lambda$ which pertains if reionization occurs (important for $\theta > 1'$), and a 2nd order "integrated gravitational effect" (IGE) of $\Delta^2\lambda^3$ which arises from non-linear structures at the present epoch. The last effect has been rather neglected in the literature and was discussed in detail by Martinez-Gonzalez. He showed, for example, that the IGE is 4×10^{-6} for the Great Attractor, 4×10^{-7} for Bootes and 10^{-5} for the Great Wall. There was some confusion over whether the IGE contribution should be included as part of the Sachs-Wolfe effect. This is mainly a semantic issue but it seems to me that it should be. Juszkiewicz also considered the importance of reionization and the implication of having the plasma in compact protogalactic clouds rather than a smooth IGM.

Several people discussed the implications of the current small-scale anisotropy limits for cosmological models. Baryon-dominated and baryon-isocurvature models are already excluded and Silk and Lasenby showed how interesting constraints can be placed on the density fluctuation spectrum in more general scenarios. Another theoretical issue which received attention was the question of whether the fluctuations are expected to be Gaussian or non-Gaussian and Martinez-Gonzalez discussed various topological measures (such as pips, spots and genus) which might be used to test this.

Observational Prospects. The race to make the first positive detection of a small-scale CBR anisotropy is clearly hotting up but it is not clear whether satellites (represented by Smoot), balloons (represented by Melchiorri and de Bernardis) or terrestial telescopes (represented by Lasenby) will achieve this first. All three groups face the problem of Galactic synchroton emission below 30GHz since variations in this could easily masquerade as a cosmic anisotropy. This has already led to false alarms, as illustrated by Lasenby's tale of the Tenerife experiment: the new 408 MHz and 1420 MHz maps show that the signal originally reported at 10.4 GHz was undoubtedly Galactic in origin and this emphasizes the importance of multifrequency coverage. Lasenby also stressed the importance of using likelihood methods in obtaining upper limits from the data. Balloon experiments have the additional problem of atmospheric emission and Melchiorri discussed the large-scale anisotropies associated with atmospheric noise. He showed that balloons give very competitive limits on the degree scale, although the throw angle must be less than 20° in order to stay in the clean regions at Galactic latitude $|b| > 40^{\circ}$ and one must select the waveband carefully. de Bernardis discussed the technical problems in looking for small-scale anisotropies with balloons and showed how these could be tackled using correlation techniques, telescope wobbling, high-frequency dust observations and monitoring of the atmosphere. In particular, he mentioned the results of the 1989 ARGO experiment.

What one really needs to know, of course, is the ultimate sensitivity of these sorts of experiments. The current Tenerife limit is $\Delta T/T < 3 \times 10^{-5}$ on the 5° scale but the 15 GHz survey may bring this down to 1×10^{-5} within a year and Lasenby emphasized that both ground-based interferometry and space experiments could eventually reach 10^{-6}. For example, the CAT experiment - which uses dense pack interferometers at 15GHz with simultaneous measurements at 11, 13 and 15 GHz to remove the atmosphere - could reach this level on scales between 6' and 2°, while RELICT2 could reach it on scales above 10°. The standard model (in which large-scale structure arises from gravitational instability) would have to be rejected altogether if fluctuations were not found at the 10^{-6}

level. It seems rather frustrating that the minimum anisotropies predicted by theory and the optimum sensitivity permitted by technology should be so close!

Zeldovich-Sunyaev Effect . One of the most difficult observations in cosmology must surely be the detection of the Zeldovich-Sunyaev distortion arising from the hot gas in clusters. As discussed by Lasenby, a definite deficit in the Rayleigh-Jeans temperature has now been found for 3 out of 20 clusters studied. In some of these cases, one can use the temperature decrement together with the X-ray data to get the value of the Hubble parameter. A2218 gives $H_0=24\pm10$, which is rather low, but A665 gives $H_0=(40$ to $50)\pm12$. There are several sources of error in these determinations (eg. non-sphericity and substructure) and one really needs detailed 2-dimensional maps. The use of the new 5km Ryle telescope promises to be very useful in this context. X-ray and Zeldovich-Sunyaev measurements also allow one to infer the radial velocity of a cluster and eventually this may allow one to map the velocity field of clusters.

7. INSTRUMENTATION

Five lectures focussed on instrumentation issues. As a theoretician, I am hardly qualified to comment on these contributions and it is difficult for me to sift out the most important developments but I will do my best.

COBE. The star of the show was obviously COBE and we were fortunate in having three members of the COBE team to tell us about it. Smoot gave us a general introduction to the three instruments, while Janssen focussed on the DMR in particular. He taught us about the advantages of heterodyne radiometers and explained why one needs differential radiometers and spacecraft spin in order to attain the stability of 10^{-8} required for proper maps. His discussion of the sources of systematic noise (where contributions even at the μK level were considered) vividly conveyed the marvellous technology associated with the instrument. Wright focussed on data analysis and taught us about COBE cubes and chopping.

Sources of noise . Lamarre discussed the various sources of noise (atmospheric, instrumental and detector) which plague observations in the infrared to submillimetre bands. He stressed that - even with cryogenic cooling - instrument noise is unavoidable for submillimetre telescopes and he explained the importance of sky chopping and beam switching. His discussion of photon noise was particularly interesting and the 300 equations of Hanbury Brown and Twiss made me realize that theoretical cosmology does not have a monopoly on difficult mathematics. Guelin focussed in more detail on the problems posed by atmosphere for submillimetre telescopes and he discussed the importance of siting them at high altitude. Finally, Lamarre also mentioned some of the exciting prospects that await us with future experiments like ISO, SIRTF, LDR, FIRST, and PRONAOS.

ACKNOWLEDGMENTS

Finally it is my pleasant duty to thank various people who have made this school so enjoyable and successful. Firstly, I must thank the secretaries (Daniele Choupin and Brigitte Rousset) and the dining-room staff (Michele and her colleagues) who looked after all of our pressing day-to-day needs. Secondly, and most importantly, I must thank the organizers Monique Signore and Christophe Dupraz, whose great efforts and organizational skills - during both the school itself and the months preceding it - have made it run so smoothly.

I first met Christophe when he collected me at the railway station. He subsequently made nearly a hundred other trips to the station, all in an effort to ensure someone's rapid arrival or departure. Indeed I do not think I ever walked down to the town without his passing me at least twice in both directions! The fact that Christophe seems to have been in a constant state of motion prompts me to formulate what may be termed the "Dupraz Uncertainty Principle": this states that if you know where he is, you cannot know where he is going, and that if you know where he is going, you cannot know where he is! (My apologies to another well-known astronomer to whom this Principle has been applied.)

Monique, of course, has displayed a very different style: always calm and sedate but firmly in control and never afraid to impose her authority when necessary (as when speakers tried to overrun their allotted time). As the chairman, Monique was the only person - apart from me - whose duties required that she sit through every talk and - unlike me - she always arrived on time! For this and her many other fine qualities she cannot be praised too highly. It is the combination of Christophe's frantic dynamism and Monique's calm control which have made this meeting so successful and I know that everybody will join me in thanking them.

REFERENCES*

1. Olive,K.A.,Schramm,D.N.,Steigman,G. & Walker,T.,1990,Phys.Lett.B.236, 454.
2. MacGibbon,J., & Carr,B.J., 1991, Ap.J., 371, 447.
3. Barrow,J.D. & Coles,P., 1991, Mon.Not.R.Astr.Soc., 248, 52.
4. Levin, J.J., Freese,K. & Spergel, D.N., 1991, Ap.J., in press.
5. Hogan,C.J., 1989, Ap.J., 340,1.
6. Ellis, G.F.R. & Madsen, M.S., 1991, Class.Quant.Grav., 8, 667.
7. Weinberg, S., 1989, Rev.Mod.Phys., 61, 1.
8. Mathewson,D.S., Ford,V.L. & Buchhorn,M., 1991, Ap.J.Lett., in press.
9. Gardner,J., 1991, in *The Early Observable Universe From Diffuse Backgrounds*, ed. B.Rocca-Volmerange (Editions Frontiers).
10. Noda, M. et al., 1991, Nagoya University Preprint.
11. Arp, H.C., Burbidge, G., Hoyle, F., Narlikar, J.V. & Wickramasinghe, N.C., 1990, Nature, 346, 807.

*References are included here only if they do not appear elsewhere in this volume

AUTHOR INDEX

SUBJECT INDEX